# Justice and food security in a changing climate

# Justice and food security in a changing climate

EurSafe 2021
Fribourg, Switzerland
24-26 June 2021

edited by:
Hanna Schübel
Ivo Wallimann-Helmer

Wageningen Academic
Publishers

**EAN: 9789086863624**
**e-EAN: 9789086869152**
**ISBN: 978-90-8686-362-4**
**e-ISBN: 978-90-8686-915-2**
**DOI:10.3920/978-90-8686-915-2**

**First published, 2021**

# Acknowledgements

Both the planning of EurSafe2021 and the origin of this book were profoundly influenced by the Covid-19 pandemic, and we are extremely pleased to see that so many contributions still arrived. We would like to thank all authors for sharing their work and insights and all reviewers for their much-appreciated expertise on the vast range of topics. We are grateful for the support and encouragement received from the EurSafe board and Svenja Springer, in preparing this conference despite the challenging circumstances. Our special thanks go to Lucie Benoit, Tanja Carillo, Patrick Hohlwegler, Melisa Kaymaz, and Amélie Pochon, who all contributed substantially to preparing and realizing this EurSafe conference and its proceedings. We also want to mention the financial support from the Swiss National Science Foundation and the University of Fribourg, without which this conference would not have been possible.

## Reviewers

We would like to thank our reviewers for their work on both the abstracts and the short papers, on which they provided valuable insights and helped to improve the contributions to this book.

S. Aerts
F.H. Arion
V. Beusmann
B. Bovenkerk
D.E. Dumitras
C.D. Dürnberger
T. Carrillo
L. Escajedo San-Epifanio
M. Gjerris
H.G. Grimm
K. Hagen
M. Huth
A. Inza-Bartolomé
A. Jelenkovic
J.I.M. Jitea
M. Kaiser
N.C. Karafyllis
T. Kortetmäki
M. Magalhães Sant'ana
F.L.B. Meijboom
S.M. Meisch
K. Millar
S. Monsó
I.C. Muresan
L. Neto
A.-C. Olsson
F. Pirscher
C.B. Pocol
E. Rebato
H. Röcklinsberg
P. Sandin
H. Schübel
B. Skorupinski
S. Springer
P.B. Thompson
P.M.R. Vaz-Pires
D.A.N. Vodnar
A. Wallenbeck
I. Wallimann-Helmer
K. Weich
A.M. Willemsen

# Table of contents

## Section 3.
## Adapting agriculture to sustain food security

**Section 4.**
**Animal ethics, veterinary ethics, and food security**

## Section 5.
## Methodology and further challenges to environmental ethics



## Section 6.
## Covid-19: new directions for ethics and food security?

## Agriculture and ethical imperatives in a changing climate

These last twelve months have been like no other, yet like a metronome the timely delivery of this volume of papers represents the proceedings of the 16th Congress of the European Society for Agricultural and Food Ethics (EurSafe), which is taking place between 23-26 June 2021 as originally planned over two years ago. This book is a testament to the hard work of not just the individual contributors in this volume, but in particular it represents the impressive work of the team at the University of Fribourg, Switzerland, our virtual hosts for this 2021 EurSafe Congress, who have remained committed and passionate about the delivery of this event.

The theme of the Congress is 'Justice and Food Security in a Changing Climate', which is a very timely theme for 2021. The EurSafe community explored aspects of climate change and the ethical implications for food and agriculture at the 10th EurSafe Congress held in Tübingen, Germany in 2012 (Conference focus: Climate change and sustainable development: ethical perspectives on land use and food production). However, this year our hosts have encouraged the community to extend our focus and explicitly consider the dimensions of justice, as a particularly important area of the discussion in terms of current practices and the future of agriculture at a time of climate crisis.

Many of the Congress participants were looking forward to the delights of 'vegan' fondue and Vully wines. Other participants were looking forward to a brisk morning walk to the Congress venue inhaling the freshness of the Fribourg region air. Most importantly, all participants were looking forward to the stimulation of the discussion and debate that is always present and much enjoyed at a EurSafe Congress. Fortunately, this activity of intellectual debate and an important component of the research eco-system will still be possible at the Congress, albeit online. Personally, I was looking forward to thanking our Swiss organisers in person, not just for the programme, selection of keynote speakers and the adaption of the Congress to an online platform, but for their enthusiasm and drive that has ensured the EurSafe traditions have continued even during these challenging times.

This volume includes contributions that cover some of the recurring topics presented at EurSafe Congresses, spanning the ethics of animal production, veterinary practice and use of gene editing and genetic modification technologies as well as covering specific ethically relevant issues that relate to agriculture and food production systems. As such, the papers presented in this volume represent an important contribution to current debates in the field of agricultural and food ethics. The Congress theme has not been represented as widely as we may have hoped in terms of the number of papers which address the theme directly. However going forward this theme should challenge our community to more widely engage in debates that cut across the field of environmental ethics to this community of research and practice working on issues related to food and agriculture. Considering the climate challenges we face as a global community, this theme of 'climate, agriculture and justice' may be seen by many of us as an ethical imperative for our work. Alongside the important theme of the Congress, when considering the notion of justice, I note a further aspect of the Congress programme this year. As I write it is International Women's Day 2021 this week, and that of course affects the lens through which I view the Congress programme. As such when reading the backgrounds of our exciting keynote speakers, I note a much greater inclusion of women (4:1) and wider range of career stages than we observed in the EurSafe Congress programmes 15-20 years ago. As a Society, we have endeavoured to build an inclusive Society that celebrates work based on the merit of the contribution. Going forward I hope our Society continues to endeavour to find ways to encourage and include a plurality of contributions and voices.

There will undoubtedly be challenging times going forward, more than ever we need structured and systemic approaches to help us map, analyse and respond to the ethical issues that are and will be raised by the complex choices that we face when considering our agriculture and food futures. Hence as a reader of this book, I hope you find this edited volume stimulating and relevant for your work. I would like to thank our Fribourg e-hosts and EurSafe Congress organisers, in particular Professor Ivo Wallimann-Helmer and Hanna Schübel. The intellectual drive, commitment and warm-hearted willingness to deliver the Congress during a time of a global pandemic has been remarkable, inspiring and is greatly appreciated by the EurSafe Community. Looking forward, I hope you are able to join the members of EurSafe at a future Congress and let us all hope that the future Congress organisers, for example for September 2022, will have the opportunity to choose whether they welcome delegates in person or in a EurSafe virtual world.

*Professor Kate Millar*
President of the European Society of Agricultural and Food Ethics and
Director, Centre for Applied Bioethics, Schools of Biosciences and Veterinary Medicine and Science, University of Nottingham, UK

# Introduction: Justice and food security in a changing climate

*I. Wallimann-Helmer and H. Schübel*

Climate change is leading to more frequent intense weather events such as floods and droughts and will very likely impact agricultural food production, particularly in such tropical regions as Sub-Saharan Africa (SSA) and other drylands (IPCC, 2014; Richardson *et al.,* 2018; UNCCD, 2017). Because these regions already suffer the highest food insecurity, this change will very likely aggravate the current global injustice in food security (D'Odorico *et al.,* 2019; FAO, IFAD, UNICEF, WFP and WHO, 2018; Gilson and Kenehan, 2019). Even though the world community has agreed in the Sustainable Development Goals to end hunger and malnutrition in all of its forms by 2030 (United Nations, 2015), the number of chronically undernourished people increased from 775 million in 2014 to 821 million in 2017 (FAO, IFAD, UNICEF, WFP and WHO, 2018). Ongoing climate change is likely to exacerbate this situation.

These observations explain why we wanted the EurSafe 2021 conference in Fribourg to focus on the pressing questions of justice and food security in a changing climate. Our aims were to attract all experts in the field and to foster new research and projects. However, these pressing questions seem not to be researched either as much or as explicitly as one might expect. This becomes clear when considering the contributions to the conference. Relatively few papers deal directly and explicitly with the questions we consider to be key in the context of food security and climate change, but many contributions to the conference deal with aspects relevant to tackling the ethical challenges concerning food security in a changing climate.

In what follows, we first introduce what we consider to be the main aspects to be researched in the fields of food security, justice, and climate change before we link together the chapters and papers of this volume that directly and explicitly deal with ethics, food security, and climate change. This allows us to further clarify how the justice and ethics challenges identified are affected both by animal ethics, veterinary ethics, and their methods and by the consequences and challenges of the current pandemic. The first section introduces what we consider to be the main challenges in the nexus between climate change, justice, and food security, and the second and third sections introduce the more general themes of the chapters of this book.

## Food security, climate change and the relevance of justice

Climate action concerns at least three areas of policy making: mitigation, geoengineering, and adaptation. Mitigation is the most prominent area and concerns reducing the production of new emissions by changing lifestyles, transportation, and energy consumption more generally. Geoengineering has often been discussed as a futuristic techno-fix to the challenge of climate change by technological intervention into the climate system. However, a recent report by the IPCC shows that the target of the Paris Agreement cannot be realized without such technology. Finally, climate adaptation concerns necessary measures to minimize the risks of negative impacts of already changing climatic conditions. Measures in all these three areas of climate action can undermine food security. However, climate ethics has seldomly researched these fields of ethical conflict.

### *Mitigation, global markets, and world hunger*

To reach the target set in the Paris Agreement in 2015 'to limit global warming to well below 2 °C' by 2100 (UNFCCC, 2015), the economic system and energy production in both developed and developing countries must change radically (Rockström *et al.,* 2017). Recent studies have shown that mitigation action is likely to increase food insecurity (Hasegawa *et al.,* 2018). Many mitigation efforts require agricultural land, which increases prices for land and food (Hasegawa *et al.,* 2015; Humpenöder *et al.,* 2018; Stevanović *et al.,* 2017). The global poor face the disastrous consequence that this aggravates global hunger even more. What needs to be analysed here are the ethical implications of trade-offs between mitigation and the threats to food security from mitigation efforts. This research field has barely been analysed in empirical science and has yet not directly investigated by normative theory at all.

#### *Geoengineering agricultural production, and land*

Current efforts to mitigate climate change are likely to result in a global average mean temperature of about 3 °C above pre-industrial levels by 2100 (UNEP, 2018). Furthermore, two thirds of all models showing that reaching the Paris target is possible assume that the climate will be manipulated by geoengineering technologies such as negative emission technologies (NETs) (Anderson and Peters, 2016). The development and implementation of NETs present particular challenges to food security (Dooley and Kartha, 2018; Minx *et al.,* 2018; Kortetmäki and Oksanen, 2016; Obersteiner *et al.,* 2018). NETs are very likely to increase food insecurity because many of them demand vast areas of agricultural land (Boysen *et al.,* 2017; Kreidenweis *et al.,* 2016; Smith *et al.,* 2016). This may lead to conflicts between NETs and agricultural food production, which will involve many ethical concerns that currently remain unresearched. First attempts can be found in this volume.

#### *Sociotechnical adaptation of agriculture to sustain food security*

Agricultural systems need to adapt to changing climatic conditions if they are to sustain food security (FAO, IFAD, UNICEF, WFP and WHO, 2018; Hasegawa *et al.,* 2018; Kline *et al.,* 2017). The increased competition for agricultural land from both mitigation action and NETs further complicates such adaptation. Thus, the resilience of communities depends not only on changing climatic conditions but also on such factors as global trade and social stability (Aldrich and Meyer, 2014). Strategies for adapting agricultural production to improve the resilience of communities are of key importance, particularly in tropical and dryland regions (FAO, IFAD, UNICEF, WFP and WHO, 2018). The challenges of implementing adaptation strategies in agriculture have not concerned ethics much so far. However, this volume contains many contributions that highlight the ethical implications of climate adaptation for agriculture and food production.

## Climate change justice and food security

Overall, one of the main challenges when dealing with food security and justice in a changing climate is the interrelatedness of topics. This is nicely exemplified by our keynote papers, which show the interrelatedness between food security and the various areas of climate action (Section 1). This discussion continues with contributions dealing with both climate mitigation and geoengineering and the challenges these kinds of measures pose for food production (Section 2). Finally, a vast number of challenges to food systems and food security arise either because the systems must be adapted to changing climatic conditions or because their adaptive transformation potentially risks undermining the functioning of traditional food systems (Section 3).

#### *Section 1. The climate change challenge to food security*

The contributions of keynote papers provide an insight into geoengineering, justice in climate adaptation, responsibility, and food system adaptation. Teea Kortetmäki shows the links and conflicts between climate engineering and food security and argues that justice reflections are crucial in climate engineering deliberations. Considering the unequal contributions to climate change, Cristian Timmermann argues for the need for fair prices to implement adaptation technologies as a demand of justice. Angela Kallhoff addresses the challenge of a climate emergency and argues that we need to adopt forward-looking approaches to climate cooperation and address the future in new ways. Finally, Guntra Aistara argues for the importance of networks when adapting our food systems to climate change and the importance of building on the knowledge systems of small farmers.

#### *Section 2. Climate mitigation, geoengineering, and food security*

To avoid dangerous climate change, the economic system and energy production most countries must change radically. Recent studies have shown that mitigation action is likely to increase food insecurity. Many mitigation efforts require agricultural land, which increases prices for land and thus for food. Approaches to relaxing this tension include the expansion of arable land in the Arctic and possibilities of technologies to maintain the production of traditional goods. However, climate change may require new approach to food security because many states are already feeling climate pressures on their food systems. These challenges are made more difficult negative emission technologies probably need to be implemented at large scale to meet the Paris Agreement. Whether the negative impacts of these technologies on food security are morally more severe than those potentially resulting from mitigation measures needs discussion.

#### *Section 3. Adapting agriculture to sustain food security*

Changing climatic conditions require that many food systems adapt the traditional patters of livelihood, food production, and water use. Climate adaptation raises important issues about value in food production and about cultural identity. These questions accompany many challenges to food security because changing values and identity often means transforming traditional food systems. This is why the right of communities to decide on their own how and what food to produce, termed food sovereignty, is especially important in transforming food systems to face negative climate impacts. Not involving the people most directly concerned risks transforming food systems in ways that are unacceptable to those people. However, because food systems tend to be fragile, their transformation without destruction is not a task easy to fulfil.

### The relevance of animal ethics, veterinary ethics, and Covid-19 to food security

Not all the papers deal directly with food security during climate change. However, they all bear some relevance to this nexus in one way or the other. Papers on animal ethics, veterinary ethics, and animal welfare and food production all provide perspectives on food security because animals are involved in food production systems in many diverse ways, and the conditions that affect their health may change if climate alters (Section 4). This is also the case for the various contributions on ethics as a methodology and further challenges in changing environmental conditions because climate change is a phenomenon so radical that we might think our ethical approach needs to change in a similarly radical way (Section 5). Finally, we include several contributions that analyse the implications of the Covid-19 pandemic for food production and veterinary ethics (Section 6). These contributions together indicate that a great deal of crucial research is already underway to tackle all the interconnected challenges that accompany climate change and its impact on food security.

#### *Section 4. Animal ethics, veterinary ethics, and food security*

Animals will be affected by climate change in various ways, including new and more widespread infectious diseases, changes in the systems of production, and the application of novel animal biotechnologies.

These are challenges that concern various aspects of animal ethics, veterinary ethics, and animal well-being in food production. Contributions from the perspective of animal ethics provide both concrete case studies on ethical challenges in food production and theoretical contributions on the status and rights of animals, such as Kantian animal ethics. Contributions in veterinary ethics investigate key questions of veterinary practice that might be challenged due to transformations of food systems in response to climate change. This links those investigations on animal well-being in food production with its bearing on food security.

#### *Section 5. Methodology and further challenges to environmental ethics*

Discussions on food security and animal well-being also entail methodological considerations, questions of teaching, vegan food systems, and fundamental questions about the rights of nonhuman entities and resource exploitation. Although methodological considerations are more detached from challenges to food security in a changing climate, these considerations become essential because climate change tends to corrupt our moral deliberation (Gardiner *et al.,* 2010). In this context, new approaches to ethical challenges and adjustments in education are needed. Because not only food systems but also our whole relation to nonhuman nature might require a fundamental shift from meat consumption, a reconsideration of the status of nature and what this implies for the design of food systems might also become relevant.

#### *Section 6. Covid-19: new directions for ethics and food security?*

The ongoing Covid-19 pandemic poses new questions for ethics and food security. Several contributions address the challenges of the emergence of the SARS-CoV-2 as a zoonosis and its implications for the global food system and food justice. Pandemics are spread by wildlife hunting and trading that extend human civilisation into wildlife habitats, but they are also facilitated through intensive farming practices. This raises many issues about the ethical implications of food production and the food traditions that currently ensure food security. By contrast, the current pandemic has shown the fragility of food systems that can lead to hunger and malnutrition if no appropriate safety nets exist. Undoubtedly, these developments pose questions to be addressed by further research on the interrelatedness between food security, justice, and solidarity.

Overall, we hope to have been able to show that the issue of food security implies a vast number of challenging and conflicting ethical questions that in our view have not been adequately researched to date. We would be pleased if the contributions in this volume that engage more directly with the nexus between food security, justice, and climate change inspire future innovative research and lead to new and important debates in applied ethics and climate justice. We also believe that many of the contributions that engage more indirectly with food security and climate change provide materials that are important for meeting the challenges at this nexus. This may motivate rapid engagement with future ethical investigations.

## References


Aldrich, Daniel P.; Meyer, Michelle A. (2014): Social Capital and Community Resilience. In *American Behavioral Scientist* 59 (2), pp. 254-269. https://doi.org/10.1177/0002764214550299

Anderson, Kevin; Peters, Glen (2016): The trouble with negative emissions. In *Science (New York, N.Y.)* 354 (6309), pp. 182-183. https://doi.org/10.1126/science.aah4567

Boysen, Lena R.; Lucht, Wolfgang; Gerten, Dieter; Heck, Vera; Lenton, Timothy M.; Schellnhuber, Hans Joachim (2017): The limits to global-warming mitigation by terrestrial carbon removal. In *Earth's Future* 5 (5), pp. 463-474. https://doi.org/10.1002/2016EF000469

D'Odorico, Paolo; Carr, Joel A.; Davis, Kyle F.; Dell'Angelo, Jampel; Seekell, David A. (2019): Food Inequality, Injustice, and Rights. In *Bioscience* 69 (3), pp. 180-190. https://doi.org/10.1093/biosci/biz002

Dooley, Kate; Kartha, Sivan (2018): Land-based negative emissions. Risks for climate mitigation and impacts on sustainable development. In *Int Environ Agreements* 18 (1), pp. 79-98. https://doi.org/10.1007/s10784-017-9382-9

FAO, IFAD, UNICEF, WFP and WHO (2018): The state of food security and nutrition in the world. Building climate resilience for food security and nutrition. FAO. Rome.

Gardiner, Stephen M.; Caney, Simon; Jamieson, Dale; Shue, Henry (eds) (2010): Climate Ethics. Essential Readings. ebrary, Inc. Oxford, New York: Oxford University Press.

Gilson, Erinn.; Kenehan, Sarah (eds) (2019): Food, Environment, and Climate Change. Justice at the Intersections. Lanham: Rowman & Littlefield International.

Hasegawa, Tomoko; Fujimori, Shinichiro; Havlík, Petr; Valin, Hugo; Bodirsky, Benjamin Leon; Doelman, Jonathan C. *et al.* (2018): Risk of increased food insecurity under stringent global climate change mitigation policy. In *Nature Clim Change* 8 (8), pp. 699-703. https://doi.org/10.1038/s41558-018-0230-x

Hasegawa, Tomoko; Fujimori, Shinichiro; Shin, Yonghee; Tanaka, Akemi; Takahashi, Kiyoshi; Masui, Toshihiko (2015): Consequence of climate mitigation on the risk of hunger. In *Environmental science & technology* 49 (12), pp. 7245-7253. https://doi.org/10.1021/es5051748

Humpenöder, Florian; Popp, Alexander; Bodirsky, Benjamin Leon; Weindl, Isabelle; Biewald, Anne; Lotze-Campen, Hermann *et al.* (2018): Large-scale bioenergy production. How to resolve sustainability trade-offs? In *Environ. Res. Lett.* 13 (2). https://doi.org/10.1088/1748-9326/aa9e3b

IPCC (2014): Climate Change 2014. Synthesis Report. Contribution of Working Groups I, II and III to the Fifth Assessment Report of the Intergovernmental Panel on Climate Change [Core Writing Team, R.K. Pachauri and L.A. Meyer (eds)]. IPCC, Geneva, Switzerland, 151 pp.

Kline, Keith L.; Msangi, Siwa; Dale, Virginia H.; Woods, Jeremy; Souza, Glaucia M.; Osseweijer, Patricia *et al.* (2017): Reconciling food security and bioenergy. Priorities for action. In *GCB Bioenergy* 9 (3), pp. 557-576. https://doi.org/10.1111/gcbb.12366

Kortetmäki, Teea; Oksanen, Markku (2016): Food Systems and Climate Engineering. A Plate Full of Risks or Promises? In: C.J. Preston (ed.) Climate Justice and Geoengineering. Ethics and Policy in the Atmospheric Anthropocene. Lanham: Rowman & Littlefield International, pp. 121-135.

Kreidenweis, Ulrich; Humpenöder, Florian; Stevanović, Miodrag; Bodirsky, Benjamin Leon; Kriegler, Elmar; Lotze-Campen, Hermann; Popp, Alexander (2016): Afforestation to mitigate climate change. Impacts on food prices under consideration of albedo effects. In *Environ. Res. Lett.* 11 (8). https://doi.org/10.1088/1748-9326/11/8/085001

Minx, Jan C.; Lamb, William F.; Callaghan, Max W.; Fuss, Sabine; Hilaire, Jérôme; Creutzig, Felix *et al.* (2018): Negative emissions – part 1: research landscape and synthesis. In *Environ. Res. Lett.* 13 (6), p. 63001. https://doi.org/10.1088/1748-9326/aabf9b

Obersteiner, Michael; Bednar, Johannes; Wagner, Fabian; Gasser, Thomas; Ciais, Philippe; Forsell, Nicklas *et al.* (2018): How to spend a dwindling greenhouse gas budget. In *Nature Clim Change* 8 (1), pp. 7-10. https://doi.org/10.1038/s41558-017-0045-1

Richardson, Katy J.; Lewis, Kirsty H.; Krishnamurthy, P. Krishna; Kent, Chris; Wiltshire, Andrew J.; Hanlon, Helen M. (2018): Food security outcomes under a changing climate. Impacts of mitigation and adaptation on vulnerability to food insecurity. In *Climatic Change* 147 (1-2), pp. 327-341. https://doi.org/10.1007/s10584-018-2137-y

Rockström, Johan; Gaffney, Owen; Rogelj, Joeri; Meinshausen, Malte; Nakicenovic, Nebojsa; Schellnhuber, Hans Joachim (2017): A roadmap for rapid decarbonization. In *Science (New York, N.Y.)* 355 (6331), pp. 1269-1271. https://doi.org/10.1126/science.aah3443

Smith, Pete; Davis, Steven J.; Creutzig, Felix; Fuss, Sabine; Minx, Jan; Gabrielle, Benoit *et al.* (2016): Biophysical and economic limits to negative $CO_2$ emissions. In *Nature Clim Change* 6 (1), pp. 42-50. https://doi.org/10.1038/NCLIMATE2870

Stevanović, Miodrag; Popp, Alexander; Bodirsky, Benjamin Leon; Humpenöder, Florian; Müller, Christoph; Weindl, Isabelle *et al.* (2017): Mitigation Strategies for Greenhouse Gas Emissions from Agriculture and Land-Use Change. Consequences for Food Prices. In *Environmental science & technology* 51 (1), pp. 365-374. https://doi.org/10.1021/acs.est.6b04291

UNCCD (2017): The Global Land Outlook 2017, first edition. United Nations Convention to Combat Desertification. Bonn, Germany.
UNEP (2018): Emissions Gap Report 2018. United Nations Environment Programme. Nairobi.
UNFCCC (2015): Paris Agreement. United Nations Framework Convention on Climate Change. Paris.
United Nations (2015): Transforming Our World: The 2030 Agenda for Sustainable Development.

# Section 1.
# Keynote papers:
# the climate change challenge to food security

# 1. Food justice, food security, and climate engineering

*T. Kortetmäki*
*Department of Social Sciences and Philosophy, University of Jyväskylä, P.O. Box 35, 40014 Jyväskylä, Finland; teea.kortetmaki@jyu.fi*

## Abstract

This chapter provides the contextual outline and introduction to the keynote talk on climate engineering, food security, and justice. Instead of summarising the talk as a whole, the chapter focuses on introducing the background research that is essential for constructing arguments with relation to the topic. The first section introduces climate engineering, followed by a description of different methods to it, and the existing scientific findings on the connections between climate engineering and food security. The summation articulates the main starting points for reasoning about climate engineering and justice from the viewpoint of food security.

**Keywords:** climate change, solar radiation management, carbon dioxide removal

## Food security, justice, and climate change

Food security refers to the ability of all people to meet their dietary needs and food preferences for an active and healthy human life at all times. Food security has four pillars: availability, access, utilisation, and stability. Food security is a necessary (though not sufficient) condition for a just society. Thus, food security is crucial in any approach to justice regardless of the definition of justice. It is a central concern to food and environmental justice. Framing food security merely as a matter of sufficient supply or distribution is inadequate. Hunger and famine, the gravest forms of food insecurity, are often entitlement rather than food availability failures (Sen, 1982). Another case in point is the presence of food insecurity even in the wealthy world with plenty of food (Riches and Silvasti, 2014). Food (in)security is, then, broadly linked to justice and related with other inequities.

Climate change mitigation and adaptation are crucial for global food security. Climate change impairs food security by impacting crop productivity and failures, food prices, pests and diseases, supply chain risks, and the decreased nutritional quality of crops (Mbow *et al.,* 2019). Because of these impacts, states act unjustly if they do not try to manage their contribution to climate change. Justice requires states to address climate change and its harmful impacts (among other reasons) to protect future food security. However, climate actions do not always promote food justice (Kortetmäki, 2019b) or food security (Kortetmäki, 2019a). This is most evident for solutions that create land use conflicts or raise energy and food prices. Justice and food security require that climate action must protect / not impair / not prevent food security (different approaches to justice endorse different strengths in formulating the criterion in this respect). On the other hand, counterfactual reasoning suggests that almost any effective climate action is 'more just' from the viewpoint of food security than inaction, which will significantly impair food security and has already done so (Mbow *et al.,* 2019).

## The rise of climate engineering discourse

In the present reality, a greater problem than the relationship between mitigation actions and food security is the insufficiency of action. Communities are far beyond achieving emission reductions required to avoid dangerous climate change (that also poses serious risks to food security). This has given rise to a discourse that first seemed science fiction but has gained increasing foothold in scientific


EurSafe 2021 – DOI 10.3920/978-90-8686-915-2_1, © Wageningen Academic Publishers 2021

arenas: climate engineering, the manipulation or modification of the atmosphere and processes therein, as distinct from mitigation and adaptation.

The main argument for climate engineering is that the humanity is running out of time to prevent the dangerous impacts of climate change by mitigation and adaptation measures alone (Preston, 2016). Authors who suggest climate engineering as a potential solution rarely label it as desirable; rather, it is the least bad realistic option. Some suggest that concern for the developing countries requires researching SRM as an option to alleviate near-term climate damages to global South (Horton and Keith, 2016: 89; for criticism on this view, Hourdequin, 2018). It is possible to provide an additional, optimistic argument for climate engineering from the viewpoint of food justice: it could relieve the urgency and pressure for low-carbon transition in food systems and support the preservation of food values and traditions that conflict with decarbonisation demands (Kortetmäki, 2019b; Kortetmäki and Oksanen, 2016).

Climate engineering calls for considerations of climate justice due to its global impacts and related uncertainties; intergenerational questions are also pertinent to it (Preston, 2016: viii-ix). Because climate engineering is usually posed as the least bad option or as a necessary measure to avoid catastrophic outcomes even if mitigation is done, climate engineering is a question of non-ideal justice: it focuses on the real world situation and feasible alternatives, rather than looking what a perfectly just society/world would look like (Svoboda, 2017). Non-ideal justice also often requires 'comparative' approaches: climate engineering should not be looked as an isolated phenomenon but contrasted with contrafactual reasoning and other available options (Preston, 2016).

## What is climate engineering?

Climate engineering is the intentional management of atmospheric chemical composition or solar radiation impacts in order to alleviate the harmful impacts of anthropogenic climate change. Climate engineering is a highly technological endeavour with constantly developing ideas; hence, this introduction does not aim to provide a state-of-the-art description of it but paint the big picture needed to reason about food security and justice with relation to it. Climate engineering is an umbrella term that comprises two technically very distinct (Preston, 2016) major approaches: carbon dioxide removal and solar radiation management. While both are categorised as distinct from mitigation and adaptation due to their large-scale impacts, most carbon dioxide removal (CDR) techniques actually lie on the continuum of mitigation activities (although the emphasis shifts to taking previous emissions 'back' rather than just preventing future emissions).

CDR techniques reduce the atmospheric carbon dioxide levels the rise of which is among the core problems of climate change. CDR methods fall into two categories. 'Terrestrial carbon sequestration' includes afforestation and deforestation, bioenergy with carbon capture and storage (BECCS), biochar, enhanced weathering, and direct carbon dioxide capture by for example artificial trees. 'Marine carbon sequestration' includes ocean fertilisation, biomass storage, and enhanced weathering (McCormack *et al.*, 2016). Many CDR techniques are also discussed without the label of climate engineering because they are practiced or considered for other reasons: for example, biochar may improve soil properties and food production in some conditions though mechanisms for this are not well known (Wang and Wang, 2019; IPCC, 2018). These methods become climate engineering when applied on atmospherically significant scales. The IPCC 1.5 °C special report (2018) labels CDR as net emission technologies (not engineering), which highlights the difference between CDR and SRM. CDR is viewed as essential for limiting warming to 1.5 °C.

Solar radiation management (SRM) techniques aim to reduce the solar radiation energy on Earth surface by reflecting radiation back into space. This would reduce the direct and indirect harmful impacts of radiation energy increase ('the greenhouse effect'). SRM techniques include 'atmospheric radiation management' with sunshades, aerosol injections, and enhanced marine cloud albedo, and 'radiation management on Earth's surface' with albedo enhancement (McCormack *et al.,* 2016). Stratospheric aerosol injection, which could be relatively cheap and effective in restricting radiation impacts, has received most attention among these. The IPCC special report (2018) mentions SRM but identifies it as having low feasibility. This, however, is labelled as a 'low agreement' statement, highlighting how controversial SRM is.

## Climate engineering and food security

Some climate engineering techniques would be deployed in the agricultural sector and impact directly food system activities. 'Agricultural geoengineering' means climate engineering that utilises techniques related to agriculture (Kortetmäki and Oksanen, 2016). Many terrestrial CDR techniques are such: it is a matter of scale whether they are climate engineering or not. Some small-scale albedo enhancement can also be done at the farm level (albedo crops; in this case it is not climate engineering due to the scale). Notably, the effectiveness of agricultural geoengineering, even in large scale applications, is a fraction of global solar radiation management solutions (or emission mitigation potential in food systems): they are additional climate action tools rather than alternative pathways to the low-carbon future.

### *Agricultural geoengineering and food security*

Agricultural geoengineering methods for CDR are generally safe according to current knowledge. They comprise a 'natural' continuum from soil management methods that aim to increase soil carbon sequestration in farming. Methods improving soil properties may also promote food security by increasing crop productivity and food availability (Kortetmäki and Oksanen, 2016: 132). Biochar is suggested promising in this respect (Wang and Wang, 2019). Some other CDR methods, such as BECCS, require high-biomass crops: therefore, CDR does not always go hand in hand with increased food production but would require either the appropriation of some of the present agricultural land for CDR or land clearing, which would reduce the total effectiveness of CDR. It is in principle possible to target CDR to low-producing pastures to minimise conflicts with food production. However, the availability of such land in appropriate locations and the feasibility of policies to ensure CDR deployment only therein are matters of low confidence (IPCC, 2018, Ch. 5). In sum, some forms of CDR could benefit agriculture and food security if applied properly. Yet, agricultural CDR has not been tested in large scale and its overall benefits depend on other land use and soil management practices. The risk of the displacement of food crops is significant and must be addressed to make it a feasible method. Other concerns include price impacts. For example, biochar application is still expensive: if it becomes the privilege of wealthier regions, its benefits for food security are not enjoyed by deprived regions with soil degradation who would benefit much more from biochar. This would only increase the existing disparities in production capacities and resilience. The risk of food price increases would also threaten access to food; prices depend on complex dynamics but some CDR scenarios estimate doubling food prices. Land grabbing is also a relevant consideration. Problems relate to the difficulty of creating reliable global markets and regulation systems for CDR (IPCC, 2018: 494-497). Risks, if realised, hit the already vulnerable communities more likely.

Albedo enhancement is another option. Albedo determines how much a surface reflects back solar energy. Snow has albedo close to 1.0; vegetation cover albedo is typically around 0.2. Crop albedo could be enhanced by 0.02-0.08, which could have locally significant benefits (Davies-Barnard, 2014): 0.04 albedo increase could have a cooling effect of 1 °C locally and seasonally in North America and Europe (yet only 0.11 °C globally). Regional cooling could maintain productivity during the hottest seasons for heat-sensitive crops like wheat (Davies-Barnard, 2014). Nutritional contents of some heat-sensitive crops also decline by warming: cooling could support the utilisation pillar of food security in important crops. Albedo enhancement could thereby improve food security with relation to availability, utilisation, and supply stability. Relatively low risks suggest the method could be desirable. Yet, justice considerations call attention to food system activities around the creation and use of new varieties, including the availability of varieties, intellectual property rights, and possible GM related considerations (Kortetmäki and Oksanen, 2016: 124-125).

### *Global-scale solar radiation management*

Solar radiation management is the hot potato of the climate engineering discussion. It represents the highest hopes and fears. The benefits of the best case scenario of SRM deployment involve high efficiency, low price, and rapid effects. The last point has been employed to argue that SRM should be researched as a method to alleviate the short-term climatic harms in global South (Horton and Keith, 2016); this argument involves the concerns of the short-term impacts of climate change on food security in global South. Large-scale technological solutions like SRM, however, require consideration from the food systems viewpoint for both the estimated (direct and indirect) impacts and the type and magnitude of related risks (Kortetmäki and Oksanen, 2016).

Many climate change related risks to food security relate to the temperature increase (Mbow *et al.,* 2019; other factors matter, too). Reducing warming by SRM could benefit food security in terms of availability, utilisation, and stability pillars. However, the indirect SRM impacts on insolation (direct sunlight decrease) and precipitation are of concern for food production. Studies on SRM and agriculture are scarce because required modelling is in its infancy. Recent research has studied the effects of 'natural SRM analogues', volcanic eruptions, on maize, rice, soy, and wheat production (Proctor *et al.,* 2018). Results were mixed: impacts were both positive, by temperature decrease, and negative, by insolation effects. The confidence interval was very wide. Authors conclude that 'SSAs [stratospheric sulphur aerosols] SSAs used in SRM may not be able to substantially lessen the risks that climate change poses to global agricultural yields and food security' (Proctor *et al.,* 2018: 482).

There is a concern that SRM could significantly decrease precipitation. Studies have found precipitation decrease both globally (Laakso *et al.,* 2017) and regionally, for example, in Southern Sahel and Western Africa by up to 10% (Da-Allada *et al.,* 2020). The Sahel region has witnessed frequent droughts causing crop failures, food shortages, and famines in nearly 30 countries and causing at worst regional death rates of 3% / month (Epule *et al.,* 2014). This raises a worry: would SRM impair food security in perilously food insecure, very vulnerable regions? Then again, climate change is also a driver of Sahel droughts and not acting upon it sufficiently may have equally catastrophic impacts. In sum, the present knowledge suggests that SRM unlikely improves food security despite temperature decrease impacts. Its risks must be better understood and manageable to justify SRM from the food security viewpoint. Moreover, SRM would not help ocean acidification (by the $CO_2$ level increase) that is a major driver for fish population depletion and thereby threatens food security in coastal regions.

A key concern is the presence and extent of anticipated and unanticipated risks of SRM methods, and the risks and costs of terminating global-scale SRM techniques in case something goes wrong (Kortetmäki and Oksanen, 2016). These risks are not just a matter of research: seeing the actual, long-term impacts of SRM with certainty would require its full-scale application and long-term observations, which cannot do. Climate engineering fights other risks with new risks. Climate engineering discussions must weigh the acceptability of different types of risks from the viewpoint of food security as a fundamental right; including vulnerable groups in discussions is crucial (Preston and Carr, 2019; Hourdequin, 2018).

## Conclusion

- Food security is a fundamental right and crucial for contemplating justice and climate engineering. Reasoning should be comparative vis-à-vis other scenarios for action.
- Distinguishing carbon dioxide removal and solar radiation management is crucial.
- Agricultural geoengineering by CDR may improve food availability, utilisation, and stability. Higher-albedo crops may improve regional food security. While reduced climatic risks to food security in such ways also benefit all communities, justice requires ensuring the access of less advantaged communities to those methods.
- Global SRM is effective, rapid, and very risky. It is estimated not to alleviate the climate change induced risks to agriculture. SRM is estimated to reduce precipitation regionally, also in regions with frequent drought-prone food scarcity. Deploying a method that would harm the already most harmed appears very unjust unless the alternatives are even worse.
- Justice reflections need to address the politics of risk. Deliberating them requires the promotion of procedural and recognitive justice in climate engineering deliberation.
- The more mitigation actions lag, the weightier are justifications to consider climate engineering due to the impacts of climate change on justice and food security.

## References

Da-Allada, C.Y., Baloïtcha, E., Alamou, E.A., Awo, F.M., Bonou, F., Pomalegni, Y., Biao, E.I., Obada, E., Zandagba, J.E., Tilmes, S. and Irvine, P.J. (2020). Changes in West African Summer Monsoon Precipitation Under Stratospheric Aerosol Geoengineering. Earth's Future 8(7).

Davies-Barnard, T. (2014). Cooling the Earth with crops. In: Hester, R. and Harrison, R. (eds) Geoengineering of the Climate System. Royal Society of Chemistry, Cambridge, UK, pp. 105-130.

Epule, E., Peng, C., Lepage, L. and Chen, Z. (2014). The causes, effects and challenges of Sahelian droughts: A critical review. Regional Environmental Change 14: 145-156.

Hourdequin M. Climate Change, Climate Engineering, and the 'Global Poor': What Does Justice Require?. Ethics, Policy & Environment. 2018 Sep 2;21(3):270-88.

IPCC Intergovernmental Panel on Climate Change (2018). Global Warming of 1.5°C. An IPCC Special Report [Masson-Delmotte, V., P. Zhai, H.-O. Pörtner, D. Roberts, J. Skea, P.R. Shukla, A. Pirani, W. Moufouma-Okia, C. Péan, R. Pidcock, S. Connors, J.B.R. Matthews, Y. Chen, X. Zhou, M.I. Gomis, E. Lonnoy, T. Maycock, M. Tignor, and T. Waterfield (eds)]. Available at: https://www.ipcc.ch/sr15/

Kortetmäki, T. and Oksanen, M. (2016). Food Systems and Climate Engineering: A Plate Full of Risks or Promises? In Preston, C. (ed.) Climate Justice and Geoengineering: Ethics and Policy in the Atmospheric Anthropocene. Rowman & Littlefield, New York, United States, pp. 121-135.

Kortetmäki, T. (2019a). Nobody's Fault? Structural Injustice, Food, and Climate Change. In Food, environment, and climate change, ed. E. Gilson, S. Kenehan, 47-61. Rowman & Littlefield, New York, United States, pp. 47-61.

Kortetmäki, T. (2019b). Tensions between food justice and climate change mitigation. In: Vinnari, E. and Vinnari, M. (eds) Sustainable governance and management of food systems: ethical perspectives. Wageningen Academic Publishers, Wageningen, the Netherlands, pp. 53-58.

Mbow, C., C. Rosenzweig, L.G. Barioni, T.G. Benton, M. Herrero, M. Krishnapillai, ... Y. Xu, 2019: Food Security. In: Shukla, P.R., Skea, J., Calvo Buendia E., Masson-Delmotte, V., Pörtner, H.-O. ... Malley (eds) Climate Change and Land: an IPCC special report on climate change, desertification, land degradation, sustainable land management, food security, and greenhouse gas fluxes in terrestrial ecosystems. Available at: https://www.ipcc.ch/srccl/chapter/chapter-5/

Laakso, A., Korhonen, H., Romakkaniemi, S. and Kokkola, H. (2017). Radiative and climate effects of stratospheric sulfur geoengineering using seasonally varying injection areas.(Report). Atmospheric Chemistry and Physics 17: 6957.

McCormack, C. G., Born, W., Irvine, P. J., Achterberg, E. P., Amano, T., Ardron, J., ... and Kissling, W. D. (2016). Key impacts of climate engineering on biodiversity and ecosystems, with priorities for future research. Journal of Integrative Environmental Sciences 13: 103-128.

Preston, C., and Carr, W. (2019). Recognitional Justice, Climate Engineering, and the Care Approach. Ethics, Policy & Environment 21: 308-323.

Proctor, J., Hsiang, S., Burney, J., Burke, M., and Schlenker, W. (2018). Estimating global agricultural effects of geoengineering using volcanic eruptions. Nature 560: 480-483.

Riches, G. & Silvasti, T. (eds). (2014). *First world hunger revisited: Food charity or the right to food?* Palgrave Macmillan, Basingstoke, UK, 248 pp.

Sen, A. (1982). Poverty and famines: an essay on entitlement and deprivation. Oxford university press, Oxford, UK, 270 pp.

Svoboda, T. (2017). The ethics of climate engineering: solar radiation management and non-ideal justice. Routledge, New York, United States, 186 pp.

Wang, J., and Wang, S. (2019). Preparation, modification and environmental application of biochar: a review. Journal of Cleaner Production 227: 1002-1022.

# 2. Adapting agriculture to a changing climate: a social justice perspective

*C. Timmermann*
*Institute of the History, Philosophy and Ethics of Medicine, Ulm University, Germany; cristian.timmermann@gmail.com*

## Abstract

We are already past the point where climate change mitigation alone does not suffice and major efforts need to be undertaken to adapt agriculture to climate change. As this situation was both foreseeable and avoidable, it is urgent to see that particularly people who have historically contributed the least to climate change do not end up assuming most of the costs. Climate change will have the worst effects on agriculture in the tropical region in the form of droughts, extreme heat waves and massive storms. The historical unequal contributions to climate change and its unjustly distributed consequences morally oblige us to distribute global adaptation costs fairly. Yet to have a full understanding of the social implications of adapting to climate change we need to look beyond making adaptation technologies accessible and available. Using a social justice framework, I defend fair prices of adaptation technologies as a demand of justice in exchange, an allocation of research attention proportional to urgent global needs as distributive justice, inclusive technology development and governance as contributive justice and addressing the interests of future generations as intergenerational justice.

**Keywords:** climate change, farm management, sustainable stewardship, responsible innovation

## Introduction

We are past the point where mitigation efforts alone will suffice to confront climate change – major efforts to adapt agriculture to the new climatic conditions are unavoidable. Yet climate change adaptation does not have to, nor should, be devoid of efforts to mitigate climate change (Nightingale *et al.*, 2020). Depending on the agricultural pathway we choose to follow, food production systems can be a major capturer of atmospheric carbon, both above and below ground (Koohafkan *et al.*, 2012). To influence the decision on which food future we ought to follow, an important step is to identify the different ethical reasons to justify such a transition. Such an approach needs to acknowledge that agriculture is not only about increasing food supply. Working the land has a deep impact on the environment and involves multiple issues for social justice (Noll, 2018; Thompson, 2017). It is also a form of cultural expression, a way of life and a form to relate to nature (Berry, 1977, 2015).

## Inequalities and climate change

In the media we can find particularly striking images of rising sea levels, wildfires due to long droughts, record breaking hurricanes and tropical cyclones, and humans and animals suffering from extreme heat waves. These images coming from all over the world make clear that climate change is a global threat. Yet climate change is far from being a great equalizer – rather the opposite is true. Particularly in agriculture we can observe that climate change increases inequalities at multiple levels:

- *Global inequalities.* The tropical regions of the world, which are also the areas with the highest rate of hunger and malnutrition, need to make the strongest efforts to adapt agriculture to higher temperatures, erratic precipitation and stronger storms (McMichael, 2017). Adaptation involves massive investments in irrigation and the adoption of new crop varieties. This is also the region with


EurSafe 2021 – DOI 10.3920/978-90-8686-915-2_2, 

historically the lowest contribution to climate change. In addition, arable land becomes increasingly scarce for the local population, due to direct effects of climate change, such as desertification and salinisation, and indirect effects, such as land grabbing to cultivate crops for biofuels and land use changes destined to off-set emissions from high-income countries and conservation efforts (Corson and MacDonald, 2012).
- *Local inequalities.* As case studies in the Caribbean show, lower precipitation induces poorer farmers with no irrigation systems to concentrate their cultivation in a shorter period of the year and to avoid water-intensive crops (Popke *et al.*, 2016). As a consequence, poorer farmers harvest similar crops at the same time, leading to a short-term oversupply that drastically reduces their market value and to food waste. Richer farmers who usually count with irrigation systems can avoid cultivating crops during times when these will end up in oversupply and can risk planting water-intensive crops (idem.). As a result, richer farmers will obtain higher premiums for their crops, and will thereafter likely be able to buy land and labour from distressed farmers, exacerbating inequalities.
- *Gender inequalities.* Women play the central role in *in situ* agrobiodiversity conservation and development (Glazebrook *et al.*, 2020). Seed exchange networks empower women. When agricultural development mandates impose commercial seed varieties, especially women lose important cultural, social and knowledge exchange networks, and become more dependent on external factors. Hunger is also a major threat for subsistence farmers and here women and girls are particularly vulnerable (Shiva, 2009).

Countries and communities who have historically contributed the least to climate change should not end up assuming most of the costs and should have more options than buying at high prices from countries with a high emission history the technologies they need to adapt to the new climatic conditions (Biddle, 2016). Both the unequal historical contribution to greenhouse gasses emissions and the unevenly distributed consequences of climate change morally oblige us to distribute the global costs of confronting climate change fairly. However, revising agricultural practices to respond to climate change involves much more than distributing costs.

## A social justice framework

Innovation is generally seen as the answer to maintain food production capabilities. To provide a different lens to identify how innovation affects society I have developed a framework built on six dimensions of social justice (Timmermann, 2020). This framework includes four forms of justice:
- *Justice in transaction.* Exchanges need to be free from deceit, avoid harmful relations of dependency, not lead to harm and be based on fair prices.
- *Distributive justice.* There needs to be a fair distribution of research attention that corresponds to the urgency of the social and environmental problems.
- *Contributive justice.* Participation is intrinsically valuable and a prerequisite for good science and governance. Capacity-building is needed to facilitate meaningful participation. Innovation systems need to be inclusive and systematically incorporate local knowledge.
- *Intergenerational justice.* The interests of future generations need to be considered, particularly in relation to efficient resource use, maintaining biodiversity and avoiding relationships of unilateral dependency.

And two processes to establish justice before and after the innovation processes and their diffusion:
- *Procedural justice.* The innovation processes need to respect codes of good scientific practice and principles of access and benefit sharing. There needs to be a democratization of research agendas.
- *Restorative justice.* Reconciliatory processes need to take place to do justice to those intentionally or accidentally wronged or disproportionally disadvantaged by innovation processes and diffusion.

## Technological futures

To visualize the potential of this framework I introduce three technological futures. These categories are highly simplified for illustration purposes:

- *Conventional approach*: One option is to continue with business as usual, relying on high external inputs and progressive adapting agriculture as yields fall and harvests fail. Climate change is seen as a reality, but agriculture is not given any special role in mitigation. Increasing yields is seen as a goal as it also allows to give back land to nature and thus store carbon and harbour biodiversity (Springmann *et al.*, 2018).
- *Agroecology*: Climate change can be an opportunity to transform the food system into a more sustainable and just system (Koohafkan *et al.*, 2012). By using farmers' traditional ecological knowledge together with principles identified in agroecological studies, farms can be designed to benefit from ecological synergies to produce food. Ideally farmland should be both a nature reservoir and a food system. This allows it to have richer soils and maintain larger biomass capturing additional carbon.
- *Emergency synthesis*: Recognizing climate change as an emergency dictates the use of whatever means achieves the fastest reduction of emissions. The rationale is that climate change will cause such disastrous outcomes that we need to sink emissions without much regard to other near-term social justice issues. A strong position would prohibit certain types of land conversion, or even dictate rewilding, when areas have a strong potential for providing ecosystem services that capture carbon (Davidson, 2017).

## Justice considerations

A preliminary analysis would suggest the following considerations. From a 'justice in transaction' perspective, innovation needs to be acquired on a voluntary basis and be free from deceit (cf. Walton, 2014). Prices need to be fair for both innovators and users. The acquisition of externally produced inputs would have to come to some advantage to the users to compensate for any disadvantages of not being able to use previously internally produced inputs. This of course does not exclude that a well-informed farmer may also voluntarily purchase a climate-friendly innovation to reduce emissions. As climate-friendly technologies produce a series of public goods, subventions benefiting users and innovators are compatible or even demanded by principles of justice, particularly for sectors that already have done their share in reducing emissions.

'Distributive justice' would need to consider global research needs. Strategies that are conducive to reducing the plight of the worst-off and at the same time contribute to climate change mitigation should receive the largest share in research attention. This offers a strong argument to rectify the weak funding agroecological studies receive (Tittonell, 2014). Yet considering the state of emergency, one could argue for some exceptions from such rule for innovations with a high potential in mitigating climate change, even when such technologies would give already well-off farmers an additional advantage.

Participation plays a fundamental role for 'contributive justice' and offers strong grounds to condemn innovation systems that are insufficiently inclusive. There are also epistemic grounds on why inclusiveness should be aimed at: local knowledge can help make innovation more sustainable and fill critical knowledge gaps (Ziegler and Ott, 2011). To benefit from inclusivity, biases and prejudices need to be confronted to create an environment that is receptive to new ideas without regarding where they come from. During an emergency we might be tempted to sacrifice inclusive technology development. Yet, particularly in relation to climate change, many innovations have failed to achieve their expected results and have even been sabotaged for not having fully considered important local factors (Bremer, 2017).

In terms of 'intergenerational justice', there is a strong argument in favour of addressing environmental sustainability by taking emergency measures to mitigate climate change. Future generations should be able to secure their basic needs. Yet intergenerational justice is not only worried about future food availability. We should avoid placing future generations in a position of unilateral dependency by completely outsourcing from farms the production of key inputs and thereby imperilling food sovereignty. Moreover, as we have been bequeathed agrobiodiversity as cultural heritage (Sievers-Glotzbach *et al.*, 2020), it is not up to one generation alone to decide to forfeit such millenary legacy to make up for decades of climate inaction. Similarly, being a farmer is also a way of relating to the land and the environment, which allows people to exercise certain values, such as care, empathy and reciprocity, and thereby maintain intrinsically valuable forms of relationships (cf. Jax *et al.*, 2018). Food futures that do not allow such relationships to the land come with major (agri)cultural losses.

To achieve these demands of justice, two additional processes need to be followed. As a matter of 'procedural justice', we should ask ourselves: how is the problem phrased? Is this phrasing supported by all stakeholders or is it a result of highly unequal power relations? For instance, the term climate change adaptation places responsibilities to individuals and communities instead of targeting causes (Ribot, 2011). Some farming systems can be given a lesser role depending on how their contribution is framed. Phrasing the question as 'how can biotechnology best contribute' and 'how much can organic agriculture contribute' already gives a farming system a higher role than the other (Tittonell, 2014).

To address issues of 'restorative justice', we need to identify those who disproportionately suffer from the chosen innovation strategies. A reconciliatory process should aim at recognizing their burden and improving their situation. Such processes are particularly important when dealing with an emergency situation that obliges to make major sacrifices. Even in emergencies technology needs to be critically assessed. The prioritization of one technological pathway needs to follow well-established criteria and be publicly justified. Climate change is not caused by a generation alone, but needs to be fixed by a single generation. This allows us to pass on some of the costs involved in preventing the catastrophic outcomes of continuous inaction.

## Conclusion

Up to what level we can justify emergency measures to confront climate change will ultimately depend on how seriously we take climate change as a threat. Erratic climatic conditions jeopardize food production and as food is a basic need, we could justify major short-term injustices in the innovation system to secure future food availability and thus liveability. Radical change that at the same time makes food systems fairer is still possible. Yet further delays in making the necessary arrangements for a major food transition will oblige us to make greater and greater sacrifices in justice in innovation. Even when making decisions in a state of emergency, we still have room to do our best to address as many dimensions of justice as possible.

## References


Berry, W., 1977/2015. The unsettling of America: Culture & agriculture. Counterpoint Press, Berkeley.

Biddle, J.B., 2016. Intellectual Property Rights and Global Climate Change: Toward Resolving an Apparent Dilemma. Ethics, Policy & Environment 19: 301-319.

Bremer, S., 2017. Have we given up too much? On yielding climate representation to experts. Futures 91: 72-75.

Corson, C. and MacDonald, K.I., 2012. Enclosing the global commons: the convention on biological diversity and green grabbing. Journal of Peasant Studies 39: 263-283.

Davidson, M.D., 2017. Equity and the conservation of global ecosystem services. Sustainability 9: 339.

Glazebrook, T., Noll, S. and Opoku, E., 2020. Gender matters: Climate change, gender bias, and women's farming in the global South and North. Agriculture 10: 267.

Jax, K., Calestani, M., Chan, K. M., Eser, U., Keune, H., Muraca, B., O'Brian, L., Potthast, T., Voget-Kleschin, L., and Wittmer, H., 2018. Caring for nature matters: a relational approach for understanding nature's contributions to human well-being. Current Opinion in Environmental Sustainability 35: 22-29.

Koohafkan, P., Altieri, M.A. and Gimenez, E.H., 2012. Green agriculture: foundations for biodiverse, resilient and productive agricultural systems. International Journal of Agricultural Sustainability 10: 61-75.

McMichael, A., 2017. Climate Change and the Health of Nations: Famines, Fevers, and the Fate of Populations. Oxford University Press, Oxford.

Nightingale, A.J., Eriksen, S., Taylor, M., Forsyth, T., Pelling, M., Newsham, A., Boyd, E., Brown, K., Harvey, B. and Jones, L., 2020. Beyond Technical Fixes: climate solutions and the great derangement. Climate and Development 12: 343-352.

Noll, S., 2018. Balancing Food Security and Ecological Resilience in the Age of the Anthropocene. In: E.C. Gilson and S. Kenehan (eds), Food, Environment, and Climate Change: Justice at the Intersections. Rowman & Littlefield, Lanham, pp. 179-192.

Popke, J., Curtis, S. and Gamble, D.W., 2016. A social justice framing of climate change discourse and policy: Adaptation, resilience and vulnerability in a Jamaican agricultural landscape. Geoforum 73: 70-80.

Ribot, J.C., 2011. Vulnerability before adaptation: Toward transformative climate action. Global Environmental Change 21: 1160-1162.

Shiva, V., 2009. Women and the gendered politics of food. Philosophical Topics 37: 17-32.

Sievers-Glotzbach, S., Tschersich, J., Gmeiner, N., Kliem, L. and Ficiciyan, A., 2020. Diverse Seeds–Shared Practices: Conceptualizing Seed Commons. International Journal of the Commons 14.

Springmann, M., Clark, M., Mason-D'Croz, D., Wiebe, K., Bodirsky, B.L., Lassaletta, L., de Vries, W., Vermeulen, S.J., Herrero, M. and Carlson, K.M., 2018. Options for keeping the food system within environmental limits. Nature 562: 519-525.

Thompson, P.B., 2017. The spirit of the soil: Agriculture and environmental ethics. Routledge, New York.

Timmermann, C., 2020. Social justice and agricultural innovation. Springer, Cham.

Tittonell, P., 2014. Ecological intensification of agriculture – sustainable by nature. Current Opinion in Environmental Sustainability 8: 53-61.

Walton, A., 2014. Do Moral Duties Arise from Global Trade? Moral Philosophy and Politics 1: 249-268.

Ziegler, R. and Ott, K., 2011. The quality of sustainability science: a philosophical perspective. Sustainability: Science, Practice, & Policy 7: 31-44.

# 3. The climate emergency: elements of a repair of the future

*A. Kallhoff*
*University of Vienna, Department of Philosophy, NIG 306, Universitätsstr. 7, 1010 Vienna, Austria; angela.kallhoff@univie.ac.at*

## Abstract

The prospects of climate change have spurred the claim of a state of emergency. This contribution argues that even though the facts speak in favour of that interpretation, an emergency has severe implications that should be avoided. Instead of relying on that political experiment, it is much more important to rethink our capacities to act together in favour of climate goals. Section one explains the implications of an emergency scenario against the background of political ethics. Section two adds an alternative, without losing sight of the dramatic situation of climate change. Instead of reiterating the shortfalls of climate change, recent approaches of climate cooperation explain forward-looking ways to climate cooperation. Section three explains new models of climate cooperation; section four addresses methods of a 'repair of the future' that go beyond risk-assessment. In order to achieve a turn-around regarding climate stability and food security, it is obligatory not only to emphasize the human capacity to cooperation, but also to start imagining new ways to address the future.

**Keywords:** climate emergency, climate cooperation, climate goals, climate risks

## Introduction

Claiming a state of emergency is not only a very dramatic thing to do, it also has severe consequences. An emergency situation is one in which usual institutional procedures can be bracketed. Instead of following rules and institutional constraints, an emergency situation emphasizes leadership.

In political philosophy, Michael Walzer explores the conditions and effects of a state of emergency and argues an emergency ethics. Its context is the ethics of war. He explores what he terms 'supreme emergency', recalling most difficult times during in WWII when survival was at stake. 'Supreme emergency is a time for heroic decision, when nations and leaders are measured by the measures they take; but it is also a desperate time, when the measures taken are ones we would avoid if we possibly could.'(Walzer, 2004: 33) Even though supreme emergency is a special case of emergencies, I shall take Walzer's interpretation as a starting-point for discussing its application to the climate scenario. Actually, some cities have already proclaimed a climate emergency; and the rumour has spread that the national climate adviser in the US, John Kerry, also thinks about doing so for the US. There might be further cities and nation states that will follow. In our context, the most important question is not whether the proclamation of emergency is justified, but rather whether it is useful. The current development of climate change is still dramatic. And the forecasts also predict that food security is at stake.

In both regards, a thorough discussion of the implications of an emergency scenario is helpful. I shall argue that on the one hand, it is important to address the climate scenario as a dangerous situation. Unfettered climate change endangers the life on planet earth in a severe way; and this has also severe consequences for the food sector. On the other hand, it is debatable whether the emergency call really opens ways to support the repair of the future. This step has two sides. It is important to take the climate change scenario seriously. But in another respect we need much more innovative and creative methods to reassess our practices and our current institutions.


EurSafe 2021 – DOI 10.3920/978-90-8686-915-2_3, © Wageningen Academic Publishers 2021

The debate on the climate emergency is a broad debate. In the context of this contribution, I shall focus on three elements. Firstly, I will portray the implications of an emergency scenario with respect to its ethical elements. Secondly, I will then discuss its usefulness with respect to climate change. Thirdly, I will discuss further ways to start a discussion about the 'repair of the future'. Overall, one lesson from the emergency discussion is that the action we need to see now is a joint effort. The fourth section is dedicated to proposals to take this lesson seriously.

## Michael Walzer on emergency ethics

A state of emergency is a situation in which the life of a population is under severe threat. Walzer's analysis of that scenario relates to the years of 1940/41 when the Nazi victory in WWII was frightening close. Walzer interprets this situation as a supreme emergency, a phrase that Winston Churchill coined when defending his decisions. Walzer notes: 'A supreme emergency exists when our deepest values and our collective survival are in imminent danger, and that was the situation in those years.' (Walzer, 2004: 33) The question that really troubles him is the following: 'Can moral constraints have any hold upon us at such time?' (Walzer, 2004: 33) In order to understand the moral trouble that comes with emergency, we need to understand that Walzer talks about a situation in which the life of some is intentionally set at risk in order to defend the lives of many. A moral leader can accept in a very exceptional situation the deaths of some, in particular when they are soldiers who defend their homeland. But he cannot accept the death of many (Walzer, 2004: 41). And even though there might be situations in which choices are tragic and leaders will necessarily get out of a situation with 'dirty hands' (Walzer, 2004: 46) the only justification of letting down innocent lives is an imminent threat to the life of the community (Walzer, 2004: 46). Even in wartime, this situation is rare. Walzer summarizes his thoughts by highlighting the moral consequences in those moments: ''Supreme emergency' describes those rare moments when the negative value that we assign – that we can't help assigning – to the disaster that looms before us devalues morality itself and leaves us free to do whatever is militarily necessary to avoid disaster, so long as what we do doesn't produce an even worse disaster.' (Walzer, 2004: 40)

Let us pause for a second. And let us try to connect these thoughts with climate change. Actually, it is not that far-fetched to recall the negative value that we assign and cannot help assigning to the disaster that looms before us. Recent data on climate change do not only underpin the view that climate change is catastrophic; they also support the view that the life of communities, even of peoples is at stake here.[1] The most telling examples are of course examples of people losing their homeland because of rising sea levels. But there are many more facets of climate change that render the claim of an imminent threat to whole populations reasonable.

Before concluding too quickly that we might apply the concept of a 'supreme emergency', it is also important to look at the conclusions that Walzer draws. Firstly, he states that supreme emergency should be treated as the absolute exemption. Note that a situation in which moral laws and insights are overridden does not mean that emergency laws are in place. Instead, another likely scenario is chaos and also a literal emergency law. And presupposed a whole people faces the real threat of not being able to survive any longer, everything might not only happen, but under law of emergency also be allowed. This destroys any legal order, not to speak of any basic morality. Secondly, Walzer highlights that supreme emergency is a communitarian doctrine (Walzer, 2004: 45). Here, Walzer reiterates his argument from 'Just and Unjust Wars. A Moral Argument With Historical Illustrations' that the only justification for life-threatening acts of defence is the defence of a community of people who together and as single persons are endowed with human rights (Walzer, 2006: 53-58). In short, the emergency situation is not only an exemption from existing law. It is also an exemption that should be handled with care. It opens

[1] For data on climate victims and the rise in numbers, see https://www.lancetcountdown.org/2020-report/.

ways not only to an emergency ethics, but also to a situation in which the most fundamental agreements on the level of a nation state and beyond are set at risk.

## A climate emergency?

In order to discuss the option of the climate emergency thoroughly, it also needs to be noted that the position of Michael Walzer is controversial. In particular, he speaks about 'supreme emergency' and not about an emergency law more generally. Therefore it is necessary to look into the broader discussion of emergency, too.

An important trait of an emergency scenario is that the usual and normal rule of law is bracketed. Two options can be distinguished: on the one hand, authors favour an interpretation that is an *in-law view*. Even though emergency allows for specific rules, it does not transcend the general frame of law. Instead, it works against a narrow interpretation of law and supports the capacity of leaders to adjust rules to current situations. A second interpretation instead claims that an emergency situation necessarily 'suspends law'. In an emergency situation, a legitimate authority is free to design new laws and to install emergency rules, once a situation of emergency is agreed upon.[2] Here again, emergency has very deep implications for political institutions and their performances.

A third perspective on emergency is possible best suited to discuss the climate emergency. Different from a focus on moral rules and their suspension as proposed by Walzer and different from a frame that discusses the implications for the rule of law, an emergency scenario can also be explained as a situation in need of a new ethics. *An emergency situation is a scenario in which the usual norms and moral guidelines do not suffice in order to address a scenario that threatens not only the lives of persons, but also the livestock*. And this is precisely the scenario that the world community faces. Climate change threatens the integrity not only of the climate system, but also of the most important natural goods that guarantee water supply and food supply. An important insight in the discussion of an emergency scenario highlights the capacity not only to anticipate that situation, but also to start developing an ethics that helps to address the looming emergency in a constructive and forward-looking way.

However, Walzer appears to be right in one important respect: The only justification for emergency measurements is a situation in which the lives of whole communities are at stake. More importantly, even in an emergency situation, normative limits need to be respected. The reply to an emergency situation is only justified if it rests on the will to protect the rights of persons and to do so for all persons who are seriously affected by climate change.

It is difficult to outline the consequences of these insights shortly. But at a minimum, two important lessons need to be learned. In my view, it is obligatory to rethink climate action in terms of something that will not only do harm to individuals, but that threatens community life. It has to be answered by also recalling communal rights and communal duties. It is a big problem that climate action has primarily been addressed in terms of shortfalls in collective action. Instead, it is time to elaborate on climate cooperation. In addition, a forward-looking model of burden-sharing with respect to truly joint climate goals is obligatory. And it is important to stop thinking exclusively about risks and precaution and everything else that the assessment of technology and of future threats includes. Instead, forward-looking methods to stop environmental exploitation need to be combined with a new vision of procurement and of a repair of the future. Both aspects, the need to think about cooperation and the importance of forward-looking methods shall be addressed in turn.

[2] For a discussion of both options and a comprehensive discussion of emergency law, see (Kaiser, 2020).

## Climate cooperation

An emergency ethics needs to serve the people; and it should be strong in doing so, because the disaster that looms at the horizon is not only life-threatening, it will also destroy communal values and the bonds of morality that societies enjoy. The answer to climate change needs to be 'climate cooperation'. In order to achieve climate cooperation, and important step forwards is the rejection of reservations against cooperation that have been reiterated again and again. In addition, it is important to develop concepts for climate cooperation that go beyond small-scale initiatives – even though they are also important.

Part of the negative story of failures in climate cooperation are the notorious dilemmas that surround the natural commons. As long as we think of people as rational and self-interested persons and as long as we allow a design of institutions that supports this claim, the dogma of a 'tragedy of the commons' is all that can be attested (Gardiner, 2001, Hardin, 1968). But Elinor Ostrom explains that communities are able to design rules that protect valuable resources (Ostrom, 1990). In addition, social philosophy defends the strong claim that almost everything is wrong with an interpretation of the social world as constituted of monades. Instead, our social world is built on cooperation and on institutions that support cooperation (Gilbert, 1989, 1996; Pettit and Schweikard, 2006; Searle, 1995). Philosophers have developed a model of cooperation that helps to understand the presuppositions for joint action. In particular, when groups of people wish to realize a goal together and when they hold the view that the goal is worth the joint effort, they can succeed in realizing that goal (Bratman, 2014, 1993; Tuomela, 2010a,b). These insights in how groups of people succeed in realizing shared goals can be applied to climate ethics. In that view, climate cooperation is not only a reasonable, but also a successful option – presupposed we have a very clear view about the goals we want to achieve (Kallhoff, 2021).

A second element in starting to repair the future is the rejection of the stories that have been explained in some theories of economics and that all amount to one lesson: Whenever an actor really engages for an environmental agenda, his efforts will be spoiled by someone else. It is not the place here to discuss the rebound effects and the green paradox in length, but a thorough analysis not only of the story, but also of the still widely accepted dogmas of non-cooperative behaviour is necessary and helpful in order to understand the dogmas underlying that flawed view (Van der Werf, 2012). Another severe problem that forestalls the repair of the future is the theoretical reduction of cooperation to market-style exchange practices. The discussion of public goods demonstrates that shared goods are an essential element in each society; moreover, it is reasonable and possible to take care of shared goods, including shared environmental resources; this common focus on collective goods even helps to keep the social order intact (Kallhoff, 2011, 2014).

A third element in the emphasis on cooperation is – of course – a thorough exploration of the conditions for successful climate cooperation. Instead of reiterating the narrative of social dilemmas such as free-riding and passing externalities to future generations, theories on climate justice help to understand one important lesson: In order to achieve goals together, it is important to allocate proactive burdens according to principles of justice. 'Climate justice' does not only refer to the distribution of emission budgets, but it also claims support for climate victims on the one hand and a fair share in burdens for future goals on the other hand (Caney, 2014). The achievement of goals of mitigation and of adaptation is costly; and it is important to defend principles that allocate the costs and burdens not only according to historic responsibility, but also according to capacities to act, according to principles of mutual aid and also of liability where necessary (Kallhoff, 2021). As always, the concept of justice can be used in two different ways. It can be used in a purely distributive manner and support the view that the atmosphere has been overconsumed as a sink for greenhouse gases by some, in particular rich nations. In order to understand the full story, it is also important to regard climate goals as particularly urgent goals. And in that respect, burden-sharing justice also needs to include the view that those who are capable and well-equipped to engage for change should do so.

In short, a repair of the future presupposes that we get rid of some dogmas that reiterate the story of the human animal as a non-cooperative, lonely and ultimately egoistic being. Instead, we should start to recapitulate the value of civilized life as the outcome of cooperation and of the joint effort to achieve goals together. And this insight can and should be applied immediately to climate cooperation and to environmental cooperation broadly conceived.

## Two ways towards a repair of the future

In order to achieve a turn-around in the climate scenario, it is very important to train our mind not only in optimism, but also in ways to think of the future as something that we can shape together. And this is also very important when addressing food security in a changing climate.

The repair of the future can be understood in two different ways. On the one hand, one might think of a repair as 'safeguards against threats' that are indeed severe: The worldwide water shortage is growing (Arnell, 1999; Feitelson, 2017); the effects of climate change include desertification of huge areas of the world; and rising sea levels threaten many big cities and whole countries.[3] The important ways to address these future threats need to include a concerted effort to reduce greenhouse gases to a zero-emission level and to invest in adaptation. A concerted response to risks is a necessary part of a repair of the future. However, climate change has taught the lesson that humankind is very bad in addressing risks that are likely to occur in the future. The rationale is obvious: People are not very good in working on risks that might occur in the future. Yet, risks that effect community life and that relate to a shared ethics might look differently.

The other way to a repair of the future relates to capacities of imagination. It is important to develop methods and also ways of thinking that transcend the focus on negative events that are likely to occur. I think it is a positive signal that even in sciences on new technologies, true exercises in imagination start to gain ground. They comprise the exercise of 'socio-technological imaginaries', a methods developed by Jasanoff which tries to paint prospects that combine technological development with concepts of good social life (Jasanoff and Kim, 2015). Another development that opens new ways to think about the future is green utopianism. Far from just dreaming of another world, recent investigations in green utopianism help to develop scenarios that combine social needs with a deep understanding of the beauty and value of landscapes and the natural environment (Wright, 2010). Part of this new way to engage in a forward-looking imagination is also the recent dialogue on green virtues, including the virtues of gardening (Barry, 1999, Brook, 2010, Kallhoff and Schörgenhumer, 2017).

Even though the focus on risks and their prevention and control is necessary, it needs to be completed by methods of imagination of a green future that go beyond the risk-assessment and help to imagine a good future. The real threat is the continuation of dogmas of the isolated human being that seeks its self-interest and is incapable of cooperation. Climate change, instead, should be taken as a final call to cooperation and to a true effort of imagination of a healthy planet.

[3] For prospect climate damage, see https://www.ipcc.ch/sr15/chapter/spm/ For a recent assessments of regional climate damage, see https://oxfamilibrary.openrepository.com/bitstream/handle/10546/620870/mb-who-takes-the-heat-230919-en.pdf.

## References


Arnell, Nigel W. (1999). Climate change and global water resources. *Global Environmental Change, 9*, 31-49.

Barry, J. (1999). *Rethinking green politics: Nature, virtue, and progress*. SAGE.

Bratman, M. (2014). *Shared agency: A planning theory of acting together*. Oxford University Press.

Bratman, Michael E. (1993). Shared Intention. *Ethics, 104*(October), 97-113.

Brook, I. (2010). The Virtues of Gardening. In D. O'Brien (ed.) *Gardening – Philosophy for Everyone. Cultivating Wisdom* (pp. 13-25). Wiley-Blackwell.

Caney, S. (2014). Two Kinds of Climate Justice: Avoiding Harm and Sharing Burdens: Two Kinds of Climate Justice. *Journal of Political Philosophy, 22*(2), 125-149. https://doi.org/10.1111/jopp.12030

Feitelson, Eran. (2017). A hierarchy of wter needs and their implications for allocation mechanisms. In Ziegler, Rafael & Groenfeldt, David (eds), *Global Water Ethics. Towards a Global Ethics Charter* (pp. 149-166). Routledge.

Gardiner, Stephen. (2001). The Real Tragedy of the Commons. *Philosophy and Public Affairs, 30*, 387-416.

Gilbert, M. (1989). *On Social Facts*. Princeton University Press.

Gilbert, M. (1996). *Living Together. Rationality, Sociality, and Obligation*. Rowman and Littlefield.

Hardin, Garrett. (1968). The Tragedy of the Commons. *Science, 162*(3859), 1243-1248.

Jasanoff, S. & Kim, S.-H. (eds). (2015). *Dreamscapes of modernity: Sociotechnical imaginaries and the fabrication of power*. The University of Chicago Press.

Kaiser, A.-B. (2020). *Ausnahmeverfassungsrecht*. Mohr Siebeck.

Kallhoff, A. (2011). *Why democracy needs public goods*. Lexington, a Division of Rowman & Littlefield Publ.

Kallhoff, A. (2014). Why societies need public goods. *Critical Review of International Social and Political Philosophy, 17*(6), 635-651. https://doi.org/10.1080/13698230.2014.904539

Kallhoff, A. (2021). *Climate Justice and Collective Action*. Routledge.

Kallhoff, A. & Schörgenhumer, Maira. (2017). The Virtues of Gardening: A Relational Account of Environmental Virtues. *Environemntal Ethics, 39/2*, 193-210.

Ostrom, E. (1990). *Governing the Commons. The Evolution of Institutions for Collective Action*. Cambridge University Press.

Pettit, P. & Schweikard, D. (2006). Joint Actions and Group Agents. *Philosophy of the Social Sciences, 36*(1), 18-39. https://doi.org/10.1177/0048393105284169

Searle, John. (1995). *The Construction of Social Reality*. The Free Press.

Tuomela, R. (2010a). *The philosophy of social practices: A collective acceptance view*. Cambridge University Press.

Tuomela, R. (2010b). *The philosophy of sociality: The shared point of view*. Oxford Univ. Press.

van der Werf, E. (2012). Imperfect Environmental Policy and Polluting Emissions: The *Green Paradox* and Beyond. *International Review of Environmental and Resource Economics, 6*(2), 153-194. https://doi.org/10.1561/101.00000050

Walzer, M. (2004). *Arguing about war*. Yale University Press.

Walzer, M. (2006). *Just and unjust wars: A moral argument with historical illustrations* (4th ed.). Basic Books.

Wright, E. O. (2010). *Envisioning real utopias*. Verso.

# 4. Networked climate sovereignties: linking agroecology, food sovereignty, and climate justice

*G.A. Aistara*
*Central European University, Department of Environmental Sciences and Policy, Quellenstrasse 51, 1100 Vienna, Austria; aistarag@ceu.edu*

## Abstract

Agroecology, organic agriculture, and food sovereignty are often regarded as well-intentioned but unrealistic niches that are ultimately unable to transform food systems. As a result, they are discounted as unfeasible options for meeting global food demand for a growing population in the face of climate change. Yet wide-scale support for agroecology and supporting structures for food and seed sovereignty have rarely, if ever, been implemented. Rather, practitioners of agroecology, organic farming, and food sovereignty are marginalized through strict regulations and free trade agreements, as well as ongoing trends towards technological fixes, unjust subsidies for large-scale conventional commodity production, and the financialization of agriculture. In this paper I extrapolate what we might learn from how small farmers in small countries have adapted to global changes for how we may move towards more environmentally sound, socially just, and climate resilient ways of feeding the world. Drawing upon experiences working with organic farmers in Latvia and Costa Rica for over a decade, particularly surrounding issues of agrobiodiversity conservation and seed sovereignty, I argue that adaptation of our food systems to climate change must happen through nested and mutually interdependent networks across scales. This will require supportive rather than restrictive governance structures that recognize and build upon farmer knowledge systems, connections to place, and networks of diversity, while forging spaces where innovative practices and transformative values may restructure productive activities, markets, and supply chains. Lack of attention to these elements creates a set of injustices across scales that result in frustrated and fractured movements and restricted farmer, community, and state sovereignties. The current interlinked global crises will require the involvement of nested actors, knowledge systems, and networks of organic sovereignties working across scales to connect agroecology, food sovereignty, and climate justice.

**Keywords:** agroecology, food sovereignty, agrobiodiversity, regionalization, free trade

## Introduction

Food security papers often ask: 'how can we feed 9 billion people in the face of climate change?' and present increased food production as a moral imperative (Fouilleux *et al.*, 2017). This genre glosses over issues of food distribution, waste, deskilling, rural unemployment, urbanization, biodiversity, soil quality, nutrition, fossil fuel use, and climate change (Fouilleux *et al.*, 2017, Wezel *et al.*, 2020). Moreover, while food security focuses on availability and access, food sovereignty includes the rights of food producers to decide on production techniques and participate in decision making, reminding us that food security cannot be achieved without food sovereignty (Jarosz, 2014).

Agroecology, at once a scientific discipline, a set of production practices, and a social movement for transformative change, holds the potential to address many of the ills of the current system: reduce greenhouse emissions, sequester more carbon in soils, replenish soil nutrients and microorganisms, reduce irrigation needs, diversify diets, crops, and landscapes, maintain meaningful employment and livelihoods in rural areas, and many others (Altieri *et al.*, 2015, Gliessman, 2016; Wezel *et al.*, 2020). Even


EurSafe 2021 – DOI 10.3920/978-90-8686-915-2_4, © Wageningen Academic Publishers 2021

a complete conversion to organic agriculture, in spite of climate change impacts, could supply enough calories to feed 9 billion people, with parallel reductions in meat consumption and food waste (Muller *et al.*, 2017). Nevertheless, organic and agroecological approaches remain marginalized by policies and subsidies that sustain the status quo of industrial monocultures (Searchinger *et al.*, 2020). Broader institutional support for agroecological approaches is comparatively scarce and disperse.

Scholars analysing the success of agroecological approaches are understandably interested in 'scaling up' from agroecological successes (Mier *et al.*, 2018). Yet the secret to successful initiatives may be their embeddedness in local knowledge systems, socio-ecological networks, and specific political-cultural-economic contexts (Aistara, 2018). Therefore it is crucial to understand farmer and movement motivations in order to maintain and link successful initiatives across scales.

We should ask, then: 'how can we produce and distribute sufficient amounts of 'culturally appropriate' and 'healthy' foods employing 'environmentally sound' and 'socially just' methods in the face of climate change?' Deciding who, and on what scale, will define these parameters is at the heart of discussions of food sovereignty. Given inherent inequalities in historic greenhouse gas emissions, predicted climate change impacts, and power dynamics of current policy negotiations, food security and food sovereignty must also be placed in the context of climate justice (Schlosberg and Collins, 2014).

Therefore, in this paper I explore the relationship between agroecology, food sovereignty, and climate justice. I draw upon observations of small organic farmers in different parts of the world for over a decade[1] – and bring these into dialogue with the work of some of my students and other scholars on climate change adaptation, in order to consider how we might move towards more transformative food systems. I contend that achieving food security in the face of climate change can only happen through networking across scales to link existing local and regional agroecological and food sovereignty movements, resulting in networked climate sovereignties.

## Lessons from 'Organic Sovereignties'

In my book 'Organic Sovereignties' (2018), I explored how organic farmers in Latvia and Costa Rica navigated entry into free trade regions such as the European Union (EU) and the Central American Free Trade Agreement (CAFTA), respectively. I argued that these local organic movements, in contesting various elements of organic regulations and the harmonization of legislation due to regionalization, attempt to negotiate spaces of sovereignty within these transnational processes. These struggles entail five main elements: 'memories of place, networked diversities, autonomies of practice, transformative values', and 'nested environmental justice', which I explain below. I reflect here how these main tenets of organic sovereignties may help us imagine broader transformations in food systems in the face of climate change.

In thinking about sovereignty beyond the state, I have been inspired by Catelino (2008) and Erazo (2013), who define sovereignty not as autonomy, but as a negotiation of subjectivities and relationships, creating intermingled autonomy and interdependence between indigenous groups and the state. Latham (2000: 2-13) has suggested that we explore social sovereignties as 'the structures of relations that set the terms for – or are constitutive of – a domain of social existence.' I identified five such structures of relations that made up organic farmers' struggles for sovereignties, which I review here.

[1] While there are debates about the relationship between organic agriculture and agroecology, particularly in industrialized countries, in my work with organic farmers in peripheral regions, I include it as a subset of broader agroecological and food sovereignty approaches.

First, the definition of what is considered 'culturally appropriate' food must go much beyond questions of cuisine to the broader landscapes in which food is produced, and how these are related to farmer cultural and historical memories and connections to place.

*Memories of place*: Organic farmers engage in a dynamic re-evaluation of landscape and farm histories that influence their subjectivities, knowledge, and practices. Understanding the landscape as a place in which farmers' embodied memories, knowledge, and practices interact with ecological processes and socio-economic realities (Ingold, 2000; Massey, 2005) can help us understand why farmers often have such significant ties to particular farmscapes. While in Latvia this meaning was embedded in land farmers received back after de-collectivization to re-establish lost family farmsteads, in Costa Rica they were embedded in farmer seeds passed down through generations and brought along with them as they settled new lands. In both cases, this connection to places, memories, and landscapes as a foundation for reinventing the future is fundamental for understanding what motivates farmer practices.

Second, healthy food is connected to diversity – of varieties, species, landscapes, socio-ecological networks, and livelihoods that can lead to more nutritious and healthy food.

*Networked diversities*: Farmers collaborate not only with other farmers and consumers but also with other species to create networks of diversity on their farms and in their landscapes. Diversity seen as a process of multi-species networking builds upon local place-based knowledge and creates new forms of socio-economic diversity for farm livelihoods. For instance, organic farmers in Costa Rica built their seed exchange practices around ecological webs in their landscapes involving seeds, plants, pollinators, and the soil, as well as changing social networks among kin and other like-minded farmers. This influenced the genetic make-up of the exchanged seeds, the diversity of farm landscapes, the nature of farmer social networks and livelihood options, increasing access to a diversity of healthy foods.

Third, 'environmentally friendly practices' at the farm, market, and movement levels are often threatened by some of the very policies that purport to support them.

*Autonomies of practice:* Farmers' place-based socio-ecological knowledge can be disrupted by standardized policy solutions harmonized to fit international treaties. For instance, in Costa Rica, the organic movement had protected the right to save and exchange seeds in a new Law on Organic Agriculture. This was superseded by CAFTA, however, which required joining the International Union of Plant Variety Protection (UPOV), which protects breeders' intellectual property rights on seeds. The 'Farmers' exception' was so limited as to function more as a farmers' exclusion, restricting farmer seed-saving, breeding, and exchange practices. This creates hierarchies between breeders and farmers and limits farmers' autonomy of practice within their own farms. In Latvia, criteria for receiving the Agri-environmental support payments sometimes discounted the most innovate farmer practices.

*Transformative values*: At the market level, movements cultivate visions for alternative futures that transform social relationships. Farmers in both Costa Rica and Latvia attempted to take control over the supply chain, but came up against various market, administrative, and policy barriers that acted as a glass ceiling preventing further development of the sector. Standardized policy solutions and hegemonic market forces push farmers and their movements toward more conventional practices, constituting a disciplining of organic imaginaries and another form of 'conventionalization' of organic agriculture.

Finally, 'social and environmental injustices' are nested across scales and can only be addressed through interdependent movements and broader policy reforms.

*Nested justices:* Organic movements often struggled to simultaneously achieve the three pillars of environmental justice: recognition of their way of life, participation in decision-making, and distribution of environmental benefits. When these goals clashed and farmers' innovate practices were constrained, they experienced nested injustices, and protested in various ways. For example, the same seed laws were being 'harmonized' to comply with the UPOV treaty, resulting in protests, in both Latvia and Costa Rica. Farmers feel that the state's relationship to international treaties and trade regions take precedence over the state's relationship to its citizens and farmers. These policies constrain farmer practices, alternative market visions, and locally adapted state policies. These are contested as infringements on sovereignty, and creating alternatives requires building interdependence among movements.

As Fraser (2001) notes, movements must be multivalent and advocate for parity of participation, whereby recognition is based not on identity politics, but rather on guaranteeing equal social status in a way that does not preclude others from participating. Movements must work with a multi-tiered conception of sovereignty (Fraser, 2001), that reemphasizes the need to affirm interdependence with other movements in order to regionalize and globalize protest against nested injustices.

## From organic sovereignties towards climate justice

These insights about organic sovereignties can serve as a starting point for imagining transformative politics linking agroecology, food sovereignty, and climate justice. This will require: (1) developing adaptive capacities; (2) building and redefining resilience; and (3) situating resilience within nested climate justice discourses.

### *Adaptive capacities: embodied knowledge in place*

'Memories of place' and 'autonomies of practice' can serve as the foundations of embodied knowledge that is necessary to adapt to climate change. As Tschakert (2012) has noted, we need to understand climate change as embodied experiences that are deeply embedded in concrete places, cultures, and environments.

Local knowledge is not only tied to memories of place and embeddedness in local socio-ecological networks, but also embodied and situated in subjects, inflected by their gender, race, class, and other aspects of their positionality. For example, Gonda (2019) argues that climate change adaptation policies must take into account farmer emotions, subjectivities, and knowledge politics in order to facilitate emancipatory socio-environmental transformations. In Nicaragua, climate change initiatives reinforced stereotypes of women's vulnerability, and efforts to mandate women's participation added to their workloads, as they did not ensure simultaneous relief of domestic chores. And as Castillo (2021) shows in Costa Rica, households with asymmetric gendered power dynamics among members are less likely to achieve resilience.

### *Resilience through diversity*

'Networked diversities' and 'transformative values' are foundational to building a more transformative concept of resilience. While resilience, understood as the ability of a system to recover from shocks, can reinforce the status quo, Tschakert and Tuana (2013) note that we need more situated and relational understandings of resilience – and I would also say more process-based. The processes of networking diversities are crucial for building functional relationships that help foster resilience.

Farmers build resilience through working with diverse crops and varieties, which allow for adaptation to climate change. In 2019 in Central Mexico I visited a farmer, José, who is the fourth generation of farmers using agroecological methods to grow heritage varieties of corn. Responding to climate uncertainties, he plants them in three different zones and controls cross-pollination by staggering planting time and place. He maintains certain varieties, while allowing others to cross-pollinate and continue to adapt. Diversity is embodied in José's practices and networked relationships, rather than defined merely by the number of varieties he maintains. Rather than maintaining the old productionist paradigms (Fouilleux *et al.*, 2017) of support for monoculture agriculture, in the age of climate change, policies need to be grounded in facilitating diversity for resilience (Altieri *et al.*, 2015).

Nevertheless, contradictions may arise between efforts to achieve agroecology, food sovereignty, and resilience due to persistent structural inequalities (Kerr *et al.*, 2106). Holt-Jimenez *et al.* (2021) have suggested redefining agroecological resilience to take into account different scales and broader agrarian vulnerability based on social and political structures that agroecological approaches alone cannot fix.

### *Nested climate justices*

Following Massey (2005), 'places' are not merely local, but inherently global and relationally connected. Thus, climate justice, like environmental justice, must be nested across scales. Interdependence across movements and scales can be developed through farmer-to-farmer initiatives (Anderson *et al.*, 2019; Holt-Jimenez *et al.*, 2021).

A visit to a community seed bank in Guatemala in 2019 illustrated the nested and structural nature of injustices that movements seek to contest. The organization wished to conserve local seed varieties that have been threatened by monoculture agriculture, neoliberal policies, displacement during the war, discrimination of indigenous knowledge systems, and climate change. These were further compounded by efforts to implement UPOV-based seed laws, initially defeated as the Monsanto laws in 2014 (Griest, 2015) but reappearing as a new law on GMOs in 2020. While agrobiodiversity conservation in community seed banks can increase resilience to climate change, it cannot overcome larger structural barriers (Porcuna-Ferrer *et al.*, 2020). As Tsing (2015) has observed, movements must be polyphonic, reflecting various voices, and work relationally to network across scales, incorporating transformative values and conceptions of justice.

## Conclusion: networked climate sovereignties

In conclusion, in order to link agroecology, food sovereignty, and climate justice, we need supportive policies to foster a network of movements that respect different people's embodied knowledges and 'memories of place'; are grounded in the 'networked diversities' of integrated socio-ecological systems; allow for 'autonomies of practice' for different groups within overarching goals for climate change mitigation and adaptation; support 'transformative values' as visions of alternative futures; and work within a multi-tiered framework of sovereignty as 'nested justices' – that are relational and networked across scales, balancing autonomy and interdependence among actors.

## References

Aistara, G.A. (2018). Organic sovereignties: Struggles over farming in an age of free trade. Seattle, University of Washington Press.

Altieri, M.A., Nicholls, C.I., Henao, A. & Lana, M.A. (2015). Agroecology and the design of climate change-resilient farming systems. Agronomy for sustainable development, *35*(3), 869-890.

Anderson, C.R., Maughan, C. & Pimbert, M.P. (2019). Transformative agroecology learning in Europe: building consciousness, skills and collective capacity for food sovereignty. Agriculture and Human Values, 36(3), 531-547.

Castillo, E.C. (2021). Gender and the lived experiences of climate change: Vulnerabilities, adaptation capacities, and resilience of smallholder farms in Costa Rica. Vienna: Central European University, Department of Sociology and Social Anthropology.

Cattelino, J. (2008). High stakes: Florida Seminole gaming and sovereignty. Durham, NC, Duke University Press.

Erazo, J. (2013). Governing indigenous territories: Enacting sovereignty in the Ecuadorian Amazon. Durham, NC: Duke University Press.

Fouilleux, E., Bricas, N. and Alpha, A. (2017). 'Feeding 9 billion people:' Global food security debates and the productionist trap. Journal of European Public Policy, 24:11, 1658-1677.

Fraser, N. (2001). 'Social justice in the knowledge society: Redistribution, recognition, and participation.' Presented at the conference 'Gut zu Wissen,' Heinrich-Böll-Stiftung, May.

Gliessman, S. (2016) Transforming food systems with agroecology. Agroecology and Sustainable Food Systems, 40:3, 187-189,

Gonda, N. (2019). Re-politicizing the gender and climate change debate: The potential of feminist political ecology to engage with power in action in adaptation policies and projects in Nicaragua. Geoforum, 106, 87-96.

Griest, R. (2015). Guatemalan protests against Monsanto law. In Swarthmore global nonviolent action database. Available at: https://nvdatabase.swarthmore.edu/content/guatemalan-protests-against-monsanto-law-2014

Holt-Giménez, E., Shattuck, A. & Van Lammeren, I. (2021). Thresholds of resistance: agroecology, resilience and the agrarian question. The Journal of Peasant Studies, 1-19.

Ingold, T. (2000). The perception of the environment: Essays on livelihood, dwelling and skill. New York: Routledge.

Jarosz, L. (2014). Comparing food security and food sovereignty discourses. Dialogues in Human Geography, *4*(2), 168-181.

Kerr, R.B., Nyantakyi-Frimpong, H., Lupafya, E. and Dakishoni, L., 2016. Food sovereignty, agroecology and resilience: competing or complementary frames. Global Governance /Politics, Climate Justice and Agrarian/ Social Justice: Linkages and Challenges. The Hague.

Latham, R. (2000). Social sovereignty. Theory, culture & society, 17(4), 1-18.

Massey, D. (2005). For space. London: SAGE Publications.

Mier, M., Giménez Cacho T., Giraldo O,F., Aldasoro, M., Morales, H., Ferguson, B.G., Rosset, P., Khadse, A. and Campos. C. (2018). Bringing agroecology to scale: key drivers and emblematic cases, Agroecology and Sustainable Food Systems, *42*(6), 637-665.

Muller, A., Schader, C., Scialabba, N.E.H., Brüggemann, J., Isensee, A., Erb, K.H., Smith, P., Klocke, P., Leiber, F., Stolze, M. and Niggli, U. (2017). Strategies for feeding the world more sustainably with organic agriculture. Nature communications, 8(1), 1-13.

Porcuna-Ferrer, A., Fiala, V., Freyer, B., van Etten, J., Vernooy, R. & Probst, L. (2020). Do community seed banks contribute to the social-ecological resilience of communities? A case-study from Western Guatemala. International Journal of Agricultural Sustainability, 18(3), 232-249.

Searchinger, Timothy D.; Malins, Chris; Dumas, Patrice; Baldock, David; Glauber, Joe; Jayne, Thomas; Huang, Jikun; Marenya, Paswell. (2020). Revising public agricultural support to mitigate climate change. Washington, DC: World Bank.

Schlosberg, D. and Collins, L.B., 2014. From environmental to climate justice: Climate change and the discourse of environmental justice. *Wiley Interdisciplinary Reviews: Climate Change*, *5*(3), pp.359-374.

Tschakert, P.(2012). From impacts to embodied experiences: Tracing political ecology in climate change research. Geografisk Tidsskrift Danish Journal of Geography, 112(2),144-158.

Tschakert, P. & Tuana, N. (2013). Situated resilience: Reframing vulnerability and security in the context of climate change. Transformation, 3, 4.

Tsing, A. (2015). The mushroom at the end of the world: On possibility of life in capitalist ruins. Princeton and Oxford, Princeton University Press.

Wezel, A., Herren, B.G., Kerr, R.B., Barrios, E., Gonçalves, A.L.R. & Sinclair, F. (2020). Agroecological principles and elements and their implications for transitioning to sustainable food systems. A review. *Agronomy for Sustainable Development*, *40*(6), 1-13.

# Section 2.
# Climate mitigation, geoengineering, and food security

# 5. Vegetable farming, climate change, and food security in the Arctic

*D.L. Friedrich*[1,2]
[1]*University of Vienna, Department of Social and Cultural Anthropology, Vienna, Austria;* [2]*The Arctic Institute, Center for Circumpolar Security Studies, P.O. Box 21194, Washington, DC 20009, USA; doris.friedrich@thearcticinstitute.org*

## Abstract

The last years have been witness to an increasing number of projects, initiatives, and farms in the Arctic, experimenting with a variety of techniques to grow vegetables (referring here to all plants used as food) to make the most out of the Arctic's often harsh conditions. In addition to the use of more traditional greenhouses for vegetable production, newer technologies such as hydroponics and geodesic domes help Arctic farms thrive. As infrastructure is often limited in many remote Arctic regions, the transportation of fresh, healthy food can be lengthy, inadequate, and difficult, rendering the supply unreliable or at times impossible. This greatly reduces the availability of healthy foods and food security in many Arctic communities. Removing the troublesome transportation from the equation, better quality produce, improved health and well-being, and increased food security are among the benefits of regional or local vegetable production in the Arctic. Another benefit of locally growing vegetables is decreasing greenhouse gas emissions stemming from the long transport routes of imported food. An important factor that presents opportunities as well as drawbacks for food procurement and vegetable production in the Arctic is climate change due to associated warmer temperatures in some areas, but also more extreme weather events. This contribution examines the linkages between Arctic vegetable farming, climate change, and sustainability, as well as vegetable farming's repercussions on Arctic communities' food security, health, and well-being. The example of 'Arviat Goes Green' illustrates the developments and effects of these linkages.

**Keywords:** agriculture, gardening, sustainability, health

## Introduction

Despite appearing unsuitable for growing vegetables (vegetables here refer to all plants used as food) due to its climate, soil quality and other conditions, the Arctic has a history of gardening and vegetable farming that reaches back into the late 18$^{th}$ century, when Russian settlers established colonies in Alaska (Loring and Gerlach, 2010). After some cycles of growth and decline, the blooming of a number of projects in various parts of the Arctic in the last two decades indicates a renewed interest (Nobel, 2013). Many of these projects arose from the desire to improve the health and well-being of Arctic communities, to increase their food security and independence, and to establish a more sustainable system of food production, both in terms of environmental and economic sustainability (S. Tagalik, one of the directors of Arviat's community organization 'Aqqiumavvik Society', and B. Vidmar, founder of 'Polar Permaculture', personal communications). As a result, communities and projects' originators researched and experimented with ways to grow vegetables in the highly diverse conditions of various Arctic regions (see, e.g. Gardens in the Arctic n.d.). In addition to fields outside, many have come to rely on greenhouses, growboxes and more novel technical solutions such as hydroponics, and vertical farming, sometimes in combination with urban agriculture (Naumov *et al.*, 2020). Despite the increasing numbers of vegetable farms and news articles reporting on many of the initiatives, relatively little scientific literature has explored and analysed the social and cultural aspects, reasons, and repercussions


EurSafe 2021 – DOI 10.3920/978-90-8686-915-2_5, © Wageningen Academic Publishers 2021

of gardening and vegetable farming in the Arctic (Chen and Natcher, 2019; Loring and Gerlach, 2010; Naumov *et al.,* 2020; Skinner *et al.,* 2014).

After examining the development of gardening and vegetable farming in the Arctic, this contribution will establish the link to food security, the overall well-being of Arctic communities, and climate change and sustainability.

## Development of vegetable farming in the Arctic

Vegetable farming has experienced a variety of development trajectories in different Arctic countries and territories. In the Arctic regions of European countries, agriculture developed already in the pre-industrial era. By contrast, in Alaska, Canada and the Arctic regions of Russia, the expansion of the oil and gas and mining industries and the associated sharp population increase incited the development of agriculture only a few decades ago (Naumov *et al.,* 2020), with the exception of the often overlooked earlier small-scale farming of Indigenous people and settlers (Loring and Gerlach, 2010). Nowadays, the value and importance assigned to agriculture also varies greatly, which Naumov *et al.* (2020) gauge by the surface area dedicated to agriculture and its contribution to the economy. Northern Canada's provinces of Saskatchewan and Alberta have the highest share of agricultural surface of territories in the Arctic with 36.1 and 28.9% respectively.

It is important to stress that in most if not all of the Indigenous communities in the Arctic regions where gardening and agriculture developed, they disrupted the communities' way of life and, what might seem counterintuitive at first, with it their possibilities to meet their nutritional needs (Lamalice *et al.,* 2018).

## Food security and vegetable farming in the Arctic

Food insecurity can be defined as the 'inadequate availability and access to healthy food of sufficient nutritional quality' (Lamalice *et al.,* 2018: 327). In Arctic regions, store-bought food imported from more southern regions makes up a great part of people's diets. As a result of a poorly developed infrastructure and unreliable connections, including inadequate storage facilities, and the associated high import costs, bought food is often expensive, of inferior quality and/or simply unavailable at times. These challenging conditions have are particularly pernicious effect on fresh, perishable foods, which is why supermarkets in isolated Arctic communities heavily rely on non-perishable, highly processed foods of low nutritional value (Chen and Natcher, 2019). These are some of the reasons why food insecurity is prevalent in Arctic communities that are easily cut off from imported supply (De Schutter, 2012; Nobel, 2013), an issue also raised by the northern Canadian grassroots advocacy group 'Feeding Nunavut'. In 2014, 60 and 29%, respectively, of children in Nunavut and the Northwest Territories lived in food insecure households (Tarasuk *et al.,* 2016).

Producing food locally can make the supply more dependable and reduce food insecurity. The risk of damage during transport or storage can also be greatly diminished through faster harvest-to-shelf cycle of locally produced vegetables, resulting in better food quality (e.g. Friedrich, 2018; Schreier, 2019). Loring and Gerlach (2010: 183) argue that even though not an exclusive means of subsistence in the past, small-scale so-called outpost gardening in Alaska was nevertheless 'one of many equally important components in a flexible and diversified subsistence strategy.' According to Lamalice *et al.* (2018), complementing the current food supply with the local cultivation of crops increases the resilience of Arctic food systems.

Therefore, gardening initiatives help to lessen the dependence on imported food and external economic actors and processes, as well as the associated vulnerability, which is a motivating factor for many communities (Loring and Gerlach, 2010). However, striving towards increased food security must take into account and prevent possible trade-offs with environmental issues and happen in a sustainable, climate-friendly way, while also alleviating poverty (Poppy *et al.,* 2014). Some initiatives pay attention to these interconnections and take care of the environmental impact of cultivating vegetables, as well as reducing poverty and food insecurity within the community by distributing vegetables to people in need (e.g. Varga, 2014).

## Health and well-being

Health is another paramount reason for many communities to start gardening initiatives, in order to steer away from unhealthy food in Arctic supermarkets and counter the rise of obesity and type 2 diabetes (Loring and Gerlach, 2010). For the community of Arviat, the push came from finding out about the children's poor eating habits. To address this concern, the community decided to focus on two areas of nutritional promotion – improving access to nutrient-rich country foods and 'eating green' – which is why Arviat started looking at the local potential for producing food, such as a community greenhouse. In addition, knowledge about the potential benefits of climate change for plant growth, such as warmer and longer growing seasons, combined with local observations of a 'significant change in the quantity and diversity of plants' encouraged the community on this path (S. Tagalik, personal communications).

For Rainey Hopson in Alaska, health issues also gave the impulse to improve her diet and go on a quest to find out how to grow vegetables in the Arctic, as she recounts on her website (Gardens in the Arctic n.d.): 'In 2008 or thereabouts I was confronted with my age and health. Like most residents on the North slope [sic] I was on a path to diabetes, lung issues and other health problems stemming from my diet and lifestyle. At the time I (…) ate a weird mix of about 15% wild caught food and about 85% pre-packaged food from the store or take out.' As she moved from North Slope to Anaktuvuk pass, her decision to quit smoking and the helpful change of scenery led to a series of realizations, including that she should 'eat better'. However, she 'found almost instantly that it was near impossible to eat a healthier diet here in the village' (Gardens in the Arctic n.d.). From there, she started researching how to grow vegetables and experimenting with different growing techniques.

Another issue that should not be underestimated is the psychological well-being and advantages that people can gain from growing their own food. In their study of community greenhouses in Nunavik, Canada, Lamalice *et al.* (2018) found that for many people (27% in one study site, 14% in another), gardening is relaxing and pleasant and in some cases a family activity. In line with this, gardening and farming programmes can enhance well-being and mental health and offer a safe communal space. They can also be seen as an opportunity for training revolving around food cultivation, preparation, storage, and use. Thus, several researchers (e.g. Chen and Natcher, 2019; Skinner *et al.,* 2014) argue that gardening initiatives' provision of community social services plays a crucial role, even beyond community gardens' and greenhouses' production of food. Skinner *et al.* (2014: 15) further point to the possibility of greenhouse projects as 'avenues for building individual and community empowerment, whereby program champions and community members are able to take control over initiatives that they feel are worthwhile.'

## Climate change and food procurement in the Arctic

The Arctic is warming twice as fast as the global average, which is why climate change in the Arctic deserves special consideration (Hassol and Corell, 2006). There are several ways in which climate change affects the procurement of food in the Arctic.

Firstly, the changing climate impacts the availability and access to some country foods, for instance by altering the growing conditions and range of wild plants and the migration routes of game animals (Seebacher and Post, 2015). Moreover, some transportation modes that rely on secure ice and that are required to hunt seals or other animals become more dangerous and unreliable (Berkes and Jolly, 2001; Ford, 2009).

Secondly, more frequent extreme weather events, including droughts, floods, and unpredictable rainfall, are potentially challenging for plant growth (see, e.g. Bardalen, 2016; Hansen *et al.,* 2014; Nobel, 2013). In addition, warmer temperatures allow plant diseases and animals considered as pests to expand further north (Bardalen, 2016). According to Willsrud (co-director and farm manager of Calypso Farm in Alaska, personal communications), more extreme conditions and especially the unpredictability of weather and seasons is a challenge for agriculture and renders farmers vulnerable.

Thirdly, a warmer climate does present opportunities and makes some Arctic regions more conducible to growing plants and a wider variety of them. The long hours of sunlight during the summer are an advantage that can result in several harvests, if temperatures are high enough (Friedrich, 2018; Gewin, 2016). The melting of permafrost frees soil that can be used for agricultural purposes. For these reasons, Bardalen (2016) estimates Arctic regions' potential for food production at 20% or more in the next 30 years.

Conversely, the cultivation of vegetables can have a positive impact on the environmental (and economic) sustainability of Arctic communities. With an environmentally friendly choice of materials and techniques used for gardening and agriculture, such as fertilized and limed local soil instead of imported soil, Arctic communities can reduce their carbon footprint that stems from the transport of food items, which often have amassed many kilometres on their way north (Lamalice *et al.,* 2018). What is more, it allows for more diversified economies that are less dependent on the oil and gas industry, even though it is questionable how big these effects are and can be.

In addition, the focus on a sustainable, local food production can also shed light on other aspects of the economy and their potential for improvement when it comes to being environmentally friendly. In Svalbard, Norway, the establishment of the world's northernmost greenhouse, 'Polar Permaculture', led to the rethinking of the local waste management, recycling practices, resource management, and pollution (B. Vidmar, personal communications; see also Friedrich, 2018). This is also the case in other initiatives (see, e.g. Skinner *et al.,* 2014). In Arviat, the community organization Aqqiumavvik, which offers programmes aiming amongst others at increasing food security, has published a Community Climate Change Manual to inform people about additional ways to act environmentally friendly (Aqqiumavvik n.d.).

The changing climate, the mitigation of its negative effects and the exploitation of its advantages for food procurement has been the reason behind some initiatives, such as Iqaluit's Piruqsiavut greenhouse in Canada (George, 2008). Another example is Arviat's research, which led to the initiative 'Arviat Goes Green' and successful experiments with greenhouses and growboxes. On the other hand, the reduction of communities' environmental impact, among others their greenhouse gas emissions, was the driving factor of a number of projects focused on growing food in the Arctic (e.g. Skinner *et al.,* 2014), including 'Polar Permaculture'.

## Case study: food insecurity in northern Canada and 'Arviat Goes Green'

Colonialism had and still has a profound impact on many isolated Arctic Indigenous communities. Some of the changes accompanying colonial processes were the intertwined transitions of energetic and nutritional systems (Martin, 2003). In Arctic Canada, many developments affected the Indigenous peoples' access to food, among them the depletion of important wildlife species due to their commercialization, the forced settlements, and changes in infrastructure, including the development of air traffic. While this facilitated access to foods items imported from southern regions, many of them highly processed, the access to country foods was hampered and became more expensive, effectively 'westernizing' Indigenous diets. As a result, a substantial part of the nutritional intake in many communities in northern Canada nowadays comes from store-bought, low-quality, high-sugar, high-fat foods (Blanchet and Rochette, 2008), thereby contributing to a severe food insecurity that is even more pronounced among Indigenous peoples (De Schutter, 2012; FNIGC, 2018; Tarasuk *et al.,* 2016). The First Nations Regional Health Survey recognizes this link between the food insecurity in Canada's Arctic and colonial structures (FNIGC, 2018), which adds an ethical and justice dimension to food procurement. Following a United Nations Report that highlighted the rampant food insecurity (De Schutter, 2012), Canada established a number of programmes focusing on northern agriculture and agricultural training (Chen and Natcher, 2019). A recent example is the Indigenous Agriculture and Food Systems Initiative, whose goal is to support Indigenous communities and entrepreneurs in the agricultural or agri-food sector (AAFC, 2020).

In 2018, Chen and Natcher (2019) counted 36 community gardens and 17 greenhouses across northern Canada, the majority of them in small communities in the Northwest Territories. One of the initiatives in northern Canada is the programme 'Arviat Goes Green', which has a strong focus on the local cultivation of produce in Arviat, a community situated in Nunavut at 61° North. The driving force behind the programme was the concern about the unhealthy eating habits of the community's youth, as well as the search for ways in which to adapt to a changing climate. Together with observations on the shifting quantity and diversity of plants, these factors encouraged the community's research on and experimentation with techniques to locally produce vegetables and other food (S. Tagalik, personal communications). The initiative had a positive impact on the community's food security and intake of healthy food. In addition to an increased consumption of country foods, which include wild plants, growboxes already planted with a variety of vegetables and accompanied by generous information and support gained a long-lasting popularity with families. Eventually, the community realized the potential stemming from the high demand of fresh vegetables and established two successful commercial hydroponic greenhouses (S. Tagalik, personal communications).

## Conclusion

Vegetable farming and gardening in the Arctic thus have a longer and more varied history and hold more promise than usually assumed. In addition to their positive outcomes for Arctic residents' health, well-being, and food security, they can further support Arctic communities in their adaptations to climate change while potentially mitigating their contribution to it by cutting down on greenhouse gas emissions. Despite the growing interest in gardening and vegetable farming in the Arctic in the last two decades, research on the reasons for this trend and its impacts on Arctic communities is severely lagging behind, a gap that this contribution seeks to reduce.

While limited through its reliance on the few academic studies available and through a lack of empirical research with the exception of interviews conducted with three projects leaders, this contribution can serve potential project initiators, policy makers, as well as customers as an introduction to the potential of vegetable farming in the Arctic and its ramification on communities' well-being, sustainability, and climate change. It further shows that the local production of vegetables has a number of benefits and can foster the development of other valuable ramifications, setting in motion a virtuous circle.

## References

AAFC (Agriculture and Agri-Food Canada) (2020). Indigenous Agriculture and Food Systems Initiative. Available at: https://www.agr.gc.ca/eng/agricultural-programs-and-services/indigenous-agriculture-and-food-systems-initiative/?id=1542835055742.

Aqqiumavvik (n.d.). Community Climate Change Manual. Available at: https://b4be1162-391a-4a89-9354-530c0ff9b928.filesusr.com/ugd/1f7032_829bfba6a0454e23848d246190d3f07f.pdf.

Bardalen, A. (2016). Arctic Agriculture: Producing More Food in the North. Arctic Deeply November 2, 2016. Available at: https://deeply.thenewhumanitarian.org/arctic/community/2016/11/02/arctic-agriculture-producing-more-food-in-the-north.

Berkes, F. and Jolly, D. (2001). Adapting to climate change: social-ecological resilience in a Canadian western Arctic community. Conservation Ecology 5(2): 18.

Blanchet, C. and Rochette, L. (2008). Nutrition and food consumption among the Inuit of Nunavik. Nunavik Inuit health survey 2004, Qanuippitaa? How are we? Quebec: Institut national de santé publique du Québec (INSPQ), Nunavik Regional Board of Health and Social Services (NRBHSS).

Chen, A. and Natcher, D. (2019). Greening Canada's Arctic food system: Local food procurement strategies for combating food insecurity. Canadian Food Studies/La Revue canadienne des études sur l'alimentation 6(1): 140-154.

De Schutter, O. (2012). Report of the special rapporteur on the right to food. Olivier De Schutter: Mission to Canada, United Nations General Assembly: 1-21.

FNIGC (First Nations Information Governance Centre) (2018). National Report of the First Nations Regional Health Survey Phase 3: Volume One. Ottawa: FNIGC.

Ford, J.D. (2009). Vulnerability of Inuit food systems to food insecurity as a consequence of climate change: a case study from Igloolik, Nunavut. Regional Environmental Change 9(2): 83-100.

Friedrich, D. (2018). Vegetable Farms 'Mushrooming' Across the Arctic. High North News January 3, 2018. Available at: https://www.highnorthnews.com/en/vegetable-farms-mushrooming-across-arctic.

Gardens in the Arctic (n.d.). Why do we grow our own food? Available at: https://gardensinthearctic.com/2019/04/19/why-do-we-grow-our-own-food/.

George, J. (2008). Ignoring earlier failures, 'pioneering; gardeners vow to bring green revolution to Iqaluit. Nunatsiaq News September 4, 2008. Available at: https://nunatsiaq.com/stories/article/Ignoring_earlier_failures_pioneering_gardeners_vow_to_bring_green_revolutio/.

Gewin, V. (2016). Arctic Agriculture: Norway's Ambitious Green Shift. Arctic Deeply November 1, 2016. Available at: https://deeply.thenewhumanitarian.org/arctic/articles/2016/11/01/arctic-agriculture-norways-ambitious-green-shift

Hansen, B.B., Isaksen, K., Benestad, R.E., Kohler, J., Pedersen, Å.Ø., Loe, L.E., Coulson, S.J., Larsen, J.O. and Varpe, Ø. (2014). Warmer and wetter winters: characteristics and implications of an extreme weather event in the High Arctic. Environmental Research Letters 9(11): 114021.

Hassol, S.J. and Corell, R.W. (2006). Arctic Climate Impact Assessment. In: Schellnhuber, H.J., Cramer, W., Nakicenovic, N., Wigley, T. and Yohe, G. (eds) Avoiding Dangerous Climate Change. Cambridge University Press, New York, USA, pp. 205-213.

Lamalice, A., Haillot, D., Lamontagne, M.A., Herrmann, T.M., Gibout, S., Blangy, S., Martin, J.L., Coxam, V., Arsenault, J., Munro, L. and Courchesne, F. (2018). Building food security in the Canadian Arctic through the development of sustainable community greenhouses and gardening. Écoscience 25(4): 325-341.

Loring, P.A. and Gerlach, S.C. (2010). Outpost gardening in interior Alaska: food system innovation and the Alaska native gardens of the 1930s through the 1970s. Ethnohistory 57(2): 183-199.

Martin, T. (2003). De la banquise au congélateur: Mondialisation et culture au Nunavik. Les Presses de l'Université Laval, Saint-Nicolas, Québec, Canada.

Naumov, A., Sidorova, D. and Goncharov, R. (2020) Farming on Arctic margins: Models of agricultural development in northern regions of Russia, Europe and North America. Regional Science Policy and Practice: 1-13.

Nobel, J. (2013). Farming in the Arctic: It Can Be Done. Modern Farmer October 18, 2013. Available at: https://modernfarmer.com/2013/10/arctic-farming/.

Poppy, G.M., Jepson, P.C., Pickett, J.A. and Birkett, M.A. (2014). Achieving food and environmental security: new approaches to close the gap. Philosophical Transactions of the Royal Society B: Biological Sciences 369(1639): 20120272.

Schreier, F. (2019). Feasibility Study – Local Vegetable Production in Longyearbyen, Svalbard.

Seebacher, F. and Post, E. (2015). Climate change impacts on animal migration. Climate Change Responses 2(1): 5.

Skinner, K., Hanning, R., Metatawabin, J. and Tsuji, L.J. (2014). Implementation of a community greenhouse in a remote, sub-Arctic First Nations community in Ontario, Canada: a descriptive case study. Rural and Remote Health 14: 2545.

Tarasuk, V., Mitchell, A. and Dachner, N. (2016). Household food insecurity in Canada, 2014. Toronto: Research to identify policy options to reduce food insecurity (PROOF).

Varga, P. (2014). Iqaluit greenhouse society ready to set new record yields. Nunatsiaq News May 27, 2014. Available at: https://nunatsiaq.com/stories/article/65674iqaluit_greenhouse_society_ready_to_set_new_record_yields/.

# 6. A Lockean approach to justice for food security under global climate change

*A. Inoue*
*Department of Advanced Social and International Studies, Graduate School of Arts and Sciences, University of Tokyo, Japan; inoue@waka.c.u-tokyo.ac.jp*

## Abstract

This paper argues that the Lockean proviso can be utilised as a relevant principle of justice for food security under global climate change. Greenhouse gas (GHG) emissions directly or indirectly affect food systems. There can be no doubt that GHG emissions ought to be reduced to enhance food security. This would seem to suggest a global food security scheme that apportions GHG emission quotas among all states. To consider such a global scheme for dealing with GHG emissions relating to food systems, it is important to provide a principle of justice for food security under global climate change. A relevant principle of justice applying to the scheme must be one that provides for food systems to meet the basic needs of the global population. Furthermore, the principle of justice should incorporate the value of fairness. In this context: (1) the parties concerned under the global food security scheme are states; (2) as part of the ideal of justice, fairness does not undermine the requirement that the basic needs of all people must be met; when determining the fair burdens, a fair allocation of GHG emission quotas should be sensitive to (3) each state's responsibility for its (past) GHG emissions and (4) its (potential) effort to reduce the emissions in practice. Given these pointers, the Lockean proviso – or, more precisely, its egalitarian version – is a relevant principle of food security justice under global climate change in four corresponding respects. Firstly, the Lockean proviso can provide legitimate normative guidance for each state. Secondly, the Lockean proviso suggests that a state has: (1) a right to a food system that secures its citizens basic needs; and (2) a duty to promote the basic needs of other people. Thirdly, the Lockean proviso can be deployed not only as a principle of global justice, but also as a principle of intergenerational justice in the context of global climate change. Finally, the Lockean proviso enables us to count the reduction of GHG emissions by each state as 'the fruits of its labours': this reduction in financial burdens imposed on people can be regarded as 'savings.' Specifically, a state's financial burden for supporting the global food security scheme, determined by its required contribution to the restitution, is mitigated by the corresponding value that the reduction achieves. These points demonstrate that the Lockean proviso is not merely an ideal of justice for food security under global climate change, but offers tangible benefits as well.

**Keywords:** fairness, GHG sinks, global food security scheme, left-libertarianism, Lockean proviso

## The relevance of food systems when debating greenhouse gas emissions

This paper argues that the Lockean proviso can be construed as a relevant principle of justice for food security under global climate change.

According to an IPCC Special Report on climate change and land, on the one hand, climate change negatively affects food security through changes in temperatures, precipitation patterns, and other aspects of weather; pastoral systems and vegetable production necessary for the entire globe are vulnerable to climate change. On the other hand, food systems contribute to the increase of greenhouse gas (GHG) emissions through, for example, deforestation for agricultural development or the excretion of animal faeces during stock raising. Since, arguably, there are limits to direct improvements with soil, crops, and


EurSafe 2021 – DOI 10.3920/978-90-8686-915-2_6, © Wageningen Academic Publishers 2021

livestock breeding, those GHG emissions that bear directly or indirectly on food systems ought to be reduced for the sake of food security (Pimentel *et al.,* 1992).

## The global food security scheme and justice

A reduction in GHG emissions is linked to global food security. This urges me to propose 'a global food security scheme' that apportions GHG emission quotas among all states in terms of food systems in general. Note that this does not cover all the GHG-related quotas; it aims only to deal with food insecurity owing to the increase in GHG emissions. For example, displacement caused by tropical storms and rising sea-levels is also an important issue in the determination of GHG emission quotas, but is not covered here. However, we must note that the global food security scheme is a *pro tanto* determinant of emission quotas. Since the effects of global climate change disrupt food production that results in global hunger and inequalities, this deserves more attention when considering the relevant distribution of GHG emission quotas.

To consider such a global scheme for dealing with GHG emissions relating to food systems, it is important to provide a principle of globally applicable 'justice'. Justice requires that people enjoy basic moral rights. No plausible theories of justice deny that the fundamental right to basic needs is the most important one. Indeed, some may view this right as a target of humanitarian morality rather than justice. However, this does not alter the fact that protecting this right should be afforded the highest priority; justice does not operate prior to the satisfaction of basic needs (Rawls, 1993: 7; Tan, 2012: 102). As a first approximation, the relevant principle of justice that applies to a global food security scheme must be such that food systems meet the basic needs of all people around the globe. This is a minimum standard of justice that must be satisfied to determine GHG emission quotas for all states (Shue, 2014: 98-9).

However, the standard at issue only provides a minimum condition for food security justice under global climate change. To be plausible, a principle of justice should incorporate the value of fairness. Fairness denotes that the benefits and burdens of maintaining a just scheme, under which people have basic moral rights, is appropriately apportioned among them. Note that the question of fairness concerning the distribution of burdens is often overlooked in theories of justice. Since the benefits apportioned to people must have their source, a relevant principle of justice should come to terms with the question of fair burdens that unavoidably involve transfers among the parties (Shue, 2014: 58-9). In this context: (1) the parties are states; (2) as part of the ideal of justice, fairness does not undermine the requirement that the basic needs of all people must be met; furthermore, and more specifically, when determining the fair burdens, a fair allocation of GHG emission quotas should be sensitive to (3) each state's responsibility for its (past) GHG emissions and (4) its (potential) effort to reduce the emissions in practice.

Since (1) and (2) are quite obvious, let me explain (3) and (4). We cannot ignore the cumulative effects of past GHG emissions on food systems by developed countries; if this were implemented, (future) emission quotas would be unfairly imposed on developing countries (and future generations). Nor can we disregard the importance of incentivizing a reduction in emissions. A relevant principle of justice should require each state to reduce their food system emissions, given that this effort of reduction would affect the well-being of future generations (Parfit, 1986: Part IV).

Using these pointers, I will show the Lockean proviso to be a relevant principle of justice for food security under global climate change.

## The Lockean proviso

While the Lockean proviso was originally proposed by Locke (1960), it was developed into its present iteration by Nozick (1974), a renowned libertarian, as a necessary and sufficient condition for entitling one to possess previously unowned things through one's mixing of labour with them. Provided that the labour-mixing presumes that each person can exercise their robust right of self-ownership, we should firstly notice the important role of this right in the theory of justice incorporating the Lockean proviso. It goes without saying that the right of self-ownership is that of using one's mind and body. This right does not deny that third parties may be just in enforcing their rights only insofar as they have not reasonably dissented, and the enforcement in question echoes their fundamental interest. The robustness of the right of self-ownership thus does not entail the absoluteness of this right. The right of self-ownership may be overturned if exercising it would violate a fundamental interest to life (Nozick, 1974: 178-179). Since a fundamental interest has much to do with basic needs, the right of self-ownership does not allow us to forgo the importance of meeting basic needs.

With the premise of self-ownership in mind, we can see that the Lockean proviso constitutes justice in acquisition. But the proviso also provides an evaluative standard for the whole system of justice including the principle of compensation; satisfying the proviso guarantees that people rightfully own the natural resources at their disposal. The original Lockean proviso says that we can acquire unowned natural resources if and only if we leave 'enough and as good left in common for others' (Locke, 1960: II Sec. 27). Nozick interprets this proviso as a principle of justice that permits one to appropriate original resources if, and only if, one leaves enough resources, or compensates others, so that nobody else is made worse off than he or she should be if the acquired resources were used in a state of nature. Since this understanding does not forbid people from appropriating unowned things in a contemporary society, Nozick regards it as an apt refinement of the Lockean proviso.

However, there is a problem with Nozick's Lockean proviso: under this proviso, a single person can acquire all natural resources 'if she compensates all others by hiring them and paying a wage that ensures that they end up no worse off than they would have been if they had continued to live the meagre hand-to-mouth existence of hunters and gatherers on non-private land' (Otsuka, 2003: 23). This seems unfair because the single person is undoubtedly exploiting the others. To deal effectively with this problem, 'left-libertarianism' presents a different interpretation of the Lockean proviso.

Left-libertarianism holds that, while people have a robust right of self-ownership, natural resources belong to everyone in an egalitarian manner (Vallentyne *et al.,* 2005). The latter phrase bears upon the interpretation of the Lockean proviso that requires an appropriator to leave enough, or provide compensation, so that everyone else can enjoy an equal division of natural resources. The metric of equality comes in several variations, such as the resourcist metric, according to which the value of natural resources is determined by supply and demand in accordance with all peoples' preferences toward them at a given time (Steiner, 1994: 235-236), and the equal-opportunity-for-advantage metric that pushes divisors to attend to each person's accessibility to, and usability of, natural resources whose value is determined by supply and demand at any given time (Otsuka, 2003: 24-25; Vallentyne, 2007: 200-201). I will not discuss the metric issue here. Suffice to say that the egalitarian Lockean proviso enables us to avoid the counterintuitive implications of Nozick's Lockean proviso (hereafter, the term Lockean proviso should be construed as referring to the egalitarian Lockean proviso).

It becomes clear that the Lockean proviso satisfies the fundamental requirements of fairness. It can reasonably answer the standard question of fairness, such that the initial allocation of entitlements to a certain equal portion of natural resources under the proviso signifies the fair share of benefits. Also, it can reasonably answer the question of fair burdens, in that the proviso requires the exploiters to compensate responsibly the exploited for the loss that rendered them worse off than they would have been had they possessed an equal portion of natural resources.

## The Lockean proviso as a principle of justice for food security under global climate change

The Lockean proviso can be a relevant principle of justice for food security under global climate change. A related suggestion is seen in Boven (2011), according to whom the capacities of the atmosphere, the land, and oceans – the so-called 'GHG sinks' – can be regarded as natural resources that ought to be left in sufficient supply and quality for all people, not only synchronically but also diachronically. This can reasonably be adapted to the context of food security under global climate change.

Firstly, the Lockean proviso can provide a legitimate normative guidance for each state. As seen above, the question of fair burdens is of enormous importance for developing the relevant principle of justice. In this context, each state's responsibility strongly matters for sustaining the global food security scheme in fair terms. With this in mind, let me focus on what makes the Lockean proviso special. According to the Lockean proviso, any community has an exclusive right to use natural resources required for sustaining a food system only if appropriators and their successors constituting the community have agreed on the shared use of the natural resources. This constitutes a relevant criterion for each state because it legitimatises a state's exclusive use of the natural resources, provided, as a condition, that appropriation of a portion of the GHG sinks meets the proviso. Admittedly, this is an ideal picture of states that deviate (radically) from current states in the world. However, this is exactly the point: this ideal picture plays a normative role within the purview of justice in that the deviations prescribe the real states to reinstate responsibly the world to a situation where, as the Lockean proviso stipulates, GHG sinks can be equally assigned to all states on a per capita basis. Although we cannot precisely identify the trajectories regarding how far the existing states have deviated from the ideal, a rough estimate of these deviations can reasonably be hoped for and would be sufficient to guide us to correct many past wrongs, such as colonisation, in establishing the current food systems (Simmons, 2016).

Secondly, the Lockean proviso suggests not only a state's right to a food system that secures its citizens basic needs but also its duties to (directly or indirectly) aid in the protection of the basic needs of other people. This is because, as seen above, the Lockean proviso is not only a principle of justice in acquisition, but also a principle of compensation that defines the degree to which unjust appropriators and their successors ought to compensate others. To illustrate this point, recall that the principle of justice that accommodates the proviso is founded on the exercisability of the right of self-ownership, which requires that the basic needs of all people be met. Furthermore, that the proviso clarifies a reasonable demand of indemnification that involves transfers from the exploiters to the exploited. As such, the operation of the Lockean proviso in the global food security system has three strengths. Firstly, with the Lockean proviso, we can say how stringently the duty to assist ought to be incurred by those existing states whose record involves unjust appropriation. Secondly, there is no need to evict unjust users of the land and oceans in order to deliver justice; the Lockean proviso imposes on them a reasonable requirement to indemnify for their overuse of GHG sinks. This is hardly an extreme measure. Thirdly, on this argument, given the Lockean proviso applies only to the global food security scheme that apportions GHG emission quotas, the proviso may, very likely, require exploiters to redress the socioeconomic conditions of other people 'only' by the extent to which they can enjoy the equal division of natural resources required for a food system where their basic needs are met. Since the global food security scheme under the proviso does not demand further equalization, the duties owed to the exploited are less demanding and therefore feasible.

Thirdly, the Lockean proviso can play a key role not only as a global justice principle but also as an intergenerational justice principle in the context of global climate change. As seen above, the Lockean proviso can also be regarded as a principle of compensation that defines how much unjust appropriators and their successors ought to compensate others. This implies that, in actual circumstances, the developed countries which have exploited GHG sinks for years are held accountable to the developing countries which have been deprived of effective opportunities to use their assigned portion of GHG sinks. Moreover, the Lockean proviso can apply to an intergenerational context where the ownership of natural resources by earlier generations are questioned. For example, when the second generation comes into existence, the first generation may lose some of their right over GHG sinks if they would disadvantage the second generation by taking more than their apportioned share (Steiner and Vallentyne, 2009: 65).[1] This application of the Lockean proviso as a principle of compensation gives fair treatment both to developing countries and also to future (next) generations as 'others'.

Finally, we are now in a position to see that the Lockean proviso gives states an incentive to reduce GHG emissions in a normatively relevant manner. As a principle of justice for food security under global climate change, the Lockean proviso enables us to count the successful reduction of GHG emissions for maintaining a food system by a state as 'the fruits of its labours' if it does not disadvantage others. Importantly, under the global food security scheme defined by the Lockean proviso, a state's achievement of successful reduction of GHG emissions would improve the global situation of GHG sinks in that it would lead to a greater proportion of the sink being left for others. This gives the state its due, such that it mitigates a responsibility of the state for restitution of the current shape of GHG emissions for food systems to equally divisible sinks, based on the degree to which the state successfully reduced GHG emissions. Specifically, a state's financial burden for supporting the global food security scheme, determined by its required contribution to the restitution, is mitigated by the corresponding value that the reduction achieves. This shows that the Lockean proviso gives a normatively relevant incentive for states to reduce GHG emissions.

## Objections and replies

There may be (at least) two objections to my proposal. First, one might object that my proposal is too demanding because leaving good and sufficient GHG sinks for others in today's world seems unachievable.[2] Response: My proposal does not aim strictly to apply the Lockean proviso to the global food security scheme. Rather, my proposal is quite moderate, such that the Lockean proviso may require indemnification only to the extent to which other people can enjoy the equal division of natural resources required for a food system where their basic needs are met; nor does it constitute an absolute moral imperative: it aims to give states an incentive to reduce GHG emissions for food security under global climate change. This implies that, even if there are few GHG sinks to be left in the physical sense, the proviso can reasonably function as a relevant criterion for the fair distribution of the cost of coping among states.

Second, any cut-off point that the Lockean proviso denotes seems morally arbitrary in such a way as to render any starting point dysfunctional in the long run; the proviso apparently appeals to a cut-off point that operates as an initial standard of applying justice to all states (Gosseries, 2005: 297; Moellendorf,

---

[1] However, it might seem that the choices of people in earlier generations play a decisive role in the birth of new agents. Given that people have a right (not a duty) to procreate through contracts, on this argument, procreating agents may bear responsibility for the existence of the procreated and, thus, should provide an equal share of natural resources for them (Steiner and Vallentyne, 2009: 67-68). However, as I see it, the responsibility at stake should, at most, be partial, because many features with respect to procreation cannot boil down to simple choices. Hence, unlike Steiner and Vallentyne, I contend that the procreated can largely count as 'others' in the proposed manner and, thus, require the recalculation of the equal value of natural resources to which they are entitled under the Lockean proviso.

[2] I owe this objection to an anonymous reviewer.

2009: 117). Response: The Lockean proviso need not be taken as applicable only once: it can, or, more precisely, 'should' apply consecutively; because otherwise, no newcomers can be entitled to GHG emission quotas (Knight, 2013: 415). An objection to this might be that this stultifies any initial appropriation. However, such an objection ignores the fact that, in my proposal, the Lockean proviso is presented only as a principle of justice for food security under global climate change. Put differently, it does not exhaust all relevant considerations of determining GHG emission quotas. This, I believe, counters the view that any initial acquisition of natural resources would easily be nullified under the Lockean proviso.

## Conclusion

The Lockean proviso is not merely an ideal of justice for food security under global climate change. Under the Lockean proviso, the developing countries can utilise emission quotas for development, and attach importance to the satisfaction of the basic needs of their citizens; nor does the Lockean proviso deny the duty to contribute to historical restitution and the compensation of others for their losses.

Admittedly, this is a rough sketch of the global food security scheme under the Lockean proviso. Arguing how to institutionalise this scheme is an important task for another day.

## Acknowledgments

This work was partly supported by JSPS KAKENHI Grant Number 18H00602 and 20H01446. I would like to thank Linc Science (https://lincscience.com/) for English language editing.

## References

Boven, L. (2011). A Lockean defense of grandfathering emission rights. In Arnold, D.G. (ed.) The ethics of global climate change. Cambridge: Cambridge University Press, UK, pp. 124-144.

Gosseries, A. (2005). Cosmopolitan luck egalitarianism and the greenhouse effect. Canadian Journal of Philosophy, Supplementary Volume 31: 279-309.

Knight, C. (2013). What is grandfathering? Environmental Politics 22: 410-427.

Locke, J. (1960). Two treatises of government. P. Laslett (ed.) Cambridge: Cambridge University Press, UK, 478 pp.

Moellendorf, D. (2009). Global inequality matters. Basingstoke: Palgrave Macmillan, UK, 195 pp.

Nozick, R. (1974). Anarchy, state and utopia. New York: Basic Books, USA, 400 pp.

Otsuka, M. (2003). Libertarianism without inequality. Oxford: Oxford University Press, UK, 174 pp.

Pimentel, D., Brown, N., Vecchio, F., La Capra, V., Hausman, S., Lee, O., Diaz, A., Williams, J., Cooper, S. and Newburger, E. (1992). Ethical issues concerning potential global climate change on food production. Journal of Agricultural and Environmental Ethics 5: 113-146.

Parfit, D. (1984). Reasons and persons. Oxford: Clarendon Press, UK, 560 pp.

Rawls, J. (1993). Political liberalism. New York: Columbia University Press, USA, 401 pp.

Shue, H. (2014). Climate justice: Vulnerability and protection. New York: Oxford University Press, USA, 353 pp.

Simmons, A.J. (2016). Boundaries of authority. New York: Oxford University Press, USA, 263 pp.

Steiner, H. (1994). An essay on rights. Oxford: Blackwell, UK, 305 pp.

Steiner, H. and Vallentyne, P. (2009). Libertarian theories of intergenerational justice. In Gosseries, A. and Meyer, L. (eds) Intergenerational justice. Oxford: Oxford University Press, UK, pp. 50-76

Tan, K-C. (2012). Justice, institutions, and luck: The site, ground, and scope of equality. Oxford: Oxford University Press, UK, 220 pp.

Vallentyne, P. (2007). Libertarianism and the state. Social Philosophy and Policy 24: 187-205.

Vallentyne, P., Steiner, H. and Otsuka, M. (2005). Why left-libertarianism is not incoherent, indeterminate, or irrelevant. Philosophy and Public Affairs 33: 201-215.

# 7. Animals and climate change

*C.E. Blattner[1]* and E. Meijer[2]*
*[1]Institute for Public Law, University of Berne, Schanzeneckstr. 1, 3001 Berne, Switzerland; [2]Department of Social Sciences, University of Wageningen, Hollandseweg 1, 6700 EW Wageningen, the Netherlands; charlotte.blattner@oefre.unibe.ch*

## Abstract

The climate crisis is often presented as a crisis for humans. It is, however, also a crisis for other animals with whom we share this planet. This raises the question of what we owe to animals in this crisis, which is not merely an ethical one. Indeed our relationships with other animals are distinctively political, raising new demands for policy making and change in political institutions and practices. In this process, it is key to recognize nonhuman animal agency, in order to do justice to them and to be able to overcome the human exceptionalism that led to the ecological crises we are facing.

**Keywords:** multispecies society, mitigation and adaptation, Anthropocene, political turn in animal ethics, climate justice

## Introduction

In politics, culture, and academia, the climate crisis is often presented as a crisis for humans. Global warming, heat waves, rising sea levels, and mounting environmental disasters threaten habitability, food security and economic prosperity of humans. The climate crisis is, however, also a crisis for other animals with whom we share this planet. Wild, domesticated, companion, liminal, sedentary, migratory and many other animals are struggling for food and water, habitable land, suitable temperatures, etc. as we speak. The fact that countless nonhuman animals are profoundly affected by the climate crisis does not just raise ethical questions. Rather, as we show in this paper, since our relationships with other animals are distinctively political, animals should be considered in climate crisis debates as a matter of justice, meaning their interests should be taken into account in policy making and political institutions. Because of the scope of this paper, we cannot present a full argument for why nonhuman animals' interests should be taken into account in ethical and political decision-making. Rather, we wish to explore what it would mean to take seriously their interests and agency in thinking about addressing climate change, and sketch ways in which this could be realized. In this paper we investigate how anthropogenic climate change affects nonhuman animals and how we can develop a multispecies account of climate justice.

## Animals affected by climate change

Climate change denotes the process by which increased greenhouse gas concentrations alter the climate's energy balance, leading to an abrupt (and rapid) increase in global mean temperature. Relative to pre-industrial times, globally averaged combined land and ocean surface temperatures reached +1 °C in 2017 (IPCC, 2018: 51). With economic activity, lifestyle, energy use, land use patterns, technology, and climate policy continuing at business as usual level, we must expect exponential temperature increases of up to +4.8 °C by 2100 (IPCC, 2014: 10). Given the multi-layered, non-linear, interconnected, and sensitive nature of the climate system, its precise trajectory and effects are difficult to predict. Still, there is broad agreement about key risks, i.e. potentially severe impacts across sectors and regions (due to high hazard or high vulnerability) that manifest themselves in physical, biological, and human systems (IPCC, 2014: 64 ff).


EurSafe 2021 – DOI 10.3920/978-90-8686-915-2_7, © Wageningen Academic Publishers 2021

According to the 2019 IPCC Special Report on Climate Change and Land Systems, the land surface air temperature has risen nearly twice as much as the global average temperature, leading to more intense temperature extremes and more variable weather (IPCC, 2019). This exacerbates existing stresses on land systems (such as forest destruction, land degradation, and freshwater consumption), with direct negative consequences for wild-living animals, who depend on these resources. All animals, whether they live in the wild or not, will suffer from high ambient heat, causing them to suffer (and, often, die) from cardiovascular and respiratory illnesses, heat strokes, hyperthermia, preterm births, kidney disorders, and mental health issues (Gaughan *et al.,* 2009; Green, 2019: 166; Rojas-Downing *et al.,* 2017). Thanks to tropical diseases now spreading to previously temperate zones, new pests, pathogens, and parasites will spread to animals and plants (IPCC, 2014: 17). Many animals will flee and migrate to more temperature zones (indeed, a majority of land animals are now on the move: Pecl *et al.*, 2017), however, many will also face death. At 1.5 °C warming, 6% of insects, 8% of plants, and 4% of vertebrates would disappear – compared to 18% of insects, 16% of plants, and 8% of vertebrates at 2 °C warming (IPCC, 2018: 8). While these figures consider species only, they give us some indication about the countless individual animals threatened to suffer and die from a warming planet.

Marine ecosystems and their inhabitants are among the most affected by climate change. The oceans have so far absorbed about one-third of anthropogenic GHGs, which comes at a high price, namely ocean acidification, a decrease in oxygen levels, and an increase in global ocean temperature. This has fatal consequences for calcium skeleton-building creatures (such as mussels), corals, and the marine life, more generally, which will suffer from significant die-offs and migration to higher latitudes (i.e. cooler waters) (IPCC, 2018: 8).

As the temperature rises, polar ice caps and glaciers begin to thaw, causing sea levels to rise (by up to 82 cm on average by 2100: IPCC, 2014: 13). Coastal areas, in which a great number of animals live and depend on for important life functions (like nesting), are rendered uninhabitable due to flooding and erosion. Rising sea water threatens freshwater habitats and causes aquifer and agricultural soil contamination, affecting many more inland-living animals (CBD, 2013). With a drier climate comes limited availability and access to water resources. In the Mediterranean region, for example, a drier climate is spreading, increasing the risk of droughts and forest fires.

Upsetting the relatively sensitive climate system through unprecedented GHG emissions leads to more extreme weather events (e.g. heavy precipitation, floods, and droughts), and more numerous and intense natural disasters (e.g. hurricanes, cyclones, and tornadoes). The 2019-2020 bush fires that burnt 18.6 million hectares in SE/SW Australia, killing and/or seriously injuring an estimated 3 billion animals (143 million mammals, 2.46 billion reptiles, 180 million birds, and 51 million frogs: WWF, 2020) shows how devastating natural disasters can be for animals.

Given the complex nature of climate systems and their heightened sensitivity to even the smallest changes, its precise breadth and implications for animal life on this planet are difficult to predict. Some animals will be more affected than others, depending on the magnitude of impacts, the region they inhabit, their vulnerability, and adaptive capacities. Some animals will experience changes in behaviour, such as through altered mobility, migration, resource-switching, or by seeking refuge; other will undergo changes in life history traits, such as survival, aging, longevity, recruitment, reproductive restraint, and breeding; yet others will change morphology, such as rapid evolution through size-dependent mortality and metamorphosis; some will experience changes in physiology, such as altered body condition, or altered stress responses to emergency conditions; and yet others will be subject to changing genetic structures, such as changes in frequency of polymorphic variants (Sergio *et al.,* 2018). For any of those changes, it will be difficult to determine whether they emerge exclusively as a result of a climate change, or whether they are merely exacerbated by it (see e.g. Almiron and Faria, 2019). Despite these variables, overall, the pervasive effects of climate change on animals – be they wild, domesticated, companion, liminal, sedentary, or migratory – is undisputed.

As we're nearing a +2 °C world, the likelihood of triggering tipping points, i.e. critical thresholds resulting in severe, widespread, long-lasting and often irreversible change to the climate with catastrophic consequences for humans, animals, and the environment) is significantly amplified (the degradation of the Amazonian rainforest being one of them: IPCC, 2014: 72). Research using even very conservative assumptions shows that climate change is already outpacing animals' ability to adapt, causing a 'sixth mass extinction' that is already under way (Ceballos *et al.*, 2015; UNFCC, 2014: 67).

## Animals in climate change action

While climate change is a pervasive force, affecting all animals irrespective of their species, action in response to climate change is almost exclusively focused on benefitting human animals.

The United Nations Framework Convention on Climate Change (UNFCCC), adopted in 1992 at the Earth Summit in Rio as the basis for a global response to tackle the challenge and enjoying near-universal membership with 197 members, recognizes as 'adverse effects of climate change' 'changes in the physical environment or biota resulting from climate change which have significant deleterious effects on the composition, resilience or productivity of natural and managed ecosystems or on the operation of socio-economic systems or on human health and welfare' (UNFCCC, 1994: article 1(1)). Animals, themselves, are *a contrario* not recognized as being adversely affected by climate change, only insofar as their compromised lives negatively affect the ecosystem they inhabit. UNFCC aims at the 'stabilization of greenhouse gas concentrations at a level that would prevent dangerous anthropogenic interference with the climate system' (UNFCCC, 1994: article 2) and lays down that parties 'should protect the climate system for the benefit of present and future generations of humankind' (UNFCCC, 1994: article 3(1)). Animals are not mentioned either as direct beneficiaries of climate action or as agents whose interests determine at what level GHG stabilization is 'safe'. UNFCCC concretizes that said level 'should be achieved within a time frame sufficient to allow ecosystems to adapt naturally to climate change, to ensure that food production is not threatened and to enable economic development to proceed in a sustainable manner' (UNFCCC, 1994: article 2). Animals, many of which are an integral part of ecosystems, are expected to adapt naturally and without human aid, and as a means to secure food and ensure economic development for human society.

The Kyoto Protocol (adopted in 1997), which operationalized UNFCCC, committed industrialized countries and economies in transition to limit and reduce GHGs. Though the protocol does not refer to animals, successful mitigation will also benefit them (Pepper, 2019: 596). The Kyoto Protocol's successor, the Paris Agreement, which recognizes climate change as 'a common concern of humankind' (2015, preamble, recital 11), aims to keep the increase in global average temperature to 'well below 2 °C above pre-industrial levels' and to pursue 'efforts to limit the increase to 1.5 °C, recognizing that this would substantially reduce the risks and impacts of climate change' (Paris Agreement, 2015: article 2(1)(a)). Here, too, mitigation will benefit animals, as well. Unlike UNFCCC and Kyoto, the Paris Agreement also aims to increase the ability of parties to adapt to the adverse impacts of climate change, '(...) to protect people, livelihoods and ecosystems (...)' (Paris Agreement, 2015: article 7(2)). The Agreement mentions ecosystems at several instances alongside people, communities, and places, departing from the narrow focus on humankind that UNFCCC maintains. Animals inhabiting said ecosystems could in turn benefit from respective adaptation action.

When looking at the Intergovernmental Panel on Climate Change's (IPCC) reports that assess and summarize scientific knowledge about climate change, its natural, political, and economic impacts and risks, and possible response options, frequently factor in animals (both as victims and drivers of climate change: IPCC, 2019). The IPCC understands animals predominantly as exchangeable members of a broader group (a certain species, e.g. or part of an ecosystem) that – alone – is worthy of protection. However, 'animals are more than parts of a whole, like drops of water or grains of sand. They are living, breathing, thinking, feeling individuals' (Sebo, 2021: 28). Individual hazards faced by animals, be it loss of life, pain and suffering experienced due to diseases, habitat loss, hunger, food insecurity, loss of one's family and community, vulnerability to temperature and weather extremes, have not received the attention in international climate policy they deserve. It may not be surprising that climate change is almost exclusively seen through human eyes, yet, given the interconnected effects of climate change faced by humans and animals alike, the benefits of tackling the two in tandem (Blattner, 2021), and the many states committed to ameliorating the lot of animals (Blattner, 2019), suggest we should begin to work toward a multi-species approach to climate justice.

## Climate justice for other animals

Different authors have argued that the interests of nonhuman animals should be taken into account in climate justice. Pepper (2019) for example draws on human rights literature to argue for a cosmopolitan approach to climate justice. She argues that the rights to life, health, and subsistence, which have been presented as focal points for human climate justice, should also apply to other animals because they are sentient beings with the capacity for self-willed action, who have lives that can go better or worse for them. Pepper develops what she calls the 'radical equality thesis', which builds on the view that all sentient animals are equal subjects of cosmopolitan climate justice. From the perspective of radical equality, duties of mitigation are not enough to address climate justice for nonhuman animals (Pepper, 2019: 596). We should, similar to how this works in the human case, also consider duties of adaptation.

Pepper formulates a general duty of adaptation: 'The relevant moral agents have a duty to facilitate adaptation whenever rights to life, health, and subsistence are violated or threatened by anthropogenic climate change' (2019: 597). From this principle follow derivative duties, notably that we have a duty to ensure that human-built settlements adapt in ways that protect all of the sentient animals that dwell there and that they do not violate or jeopardise the rights of wild animals; to remove man-made barriers to animals' autonomous adaptation; and to facilitate the adaptation of wild animals unable to autonomously adapt to climate change.

Pepper recognizes the relevance of animal interests for establishing an account of climate justice. She also acknowledges that more-than-human animals are active agents with their own perspective on life, and that this is something we should take into account for normative reasons (2019: 594). However, she does not explore in detail the political consequences of this position, nor does she challenge human exceptionalism. This is unfortunate. Human relationships with other animals are political, and other animals act politically. Bringing to light the political aspects of climate justice for nonhuman animals not only clarifies what is at stake for them: it is intertwined with decentring the human. This is necessary for tackling the climate crisis. In what follows we therefore first briefly show the benefits of a political approach to relationships with nonhuman animals and then explore why we need to learn to think beyond the human in order to be able to confront the current climate crisis, as well as the other crises of the Anthropocene.

## Political relations between human and nonhuman animals

In recent years, there has been attention for the political relationships between humans and other animals. Drawing on insights from animal rights theory, authors of the 'political turn' in animal ethics offer a distinctly political approach to assessing the relations between humans and other animals, giving rise to a new type of ontological and normative framework. The most important proponents of this approach are Sue Donaldson and Will Kymlicka (2011), who develop a group-differentiated theory of political nonhuman animal rights. Donaldson and Kymlicka propose to supplement negative rights of non-human animals (such as the right not to be killed, captured or tortured), as developed in existing animal rights theories, with a relational political theory which is based on the ways that groups of animals relate to human political groups. They argue that the lives of humans and animals are interconnected in many ways, politically, but also historically, culturally and geographically, which leads to different rights and responsibilities.

Importantly, in these relations other animals are not passive recipients of human care or harm. They are actors who may form their own communities and in certain instances co-shape interspecies communities of which humans can also be part (2011: Chapter 3). By emphasizing this aspect of relations with other animals, Donaldson and Kymlicka not only do justice to animal subjectivity but also open up new pathways of change which are not shaped by humans alone. They show that insights from political philosophy are helpful in addressing common challenges because it can provide the conceptual tools to translate ethical and moral insights to institutions and practices, in which concepts such as democracy and citizenship can play a key role.

The current climate crisis is political in different ways. The questions at stake are in part about power, justice, and freedom: topics political philosophy has explored in much detail. The crisis asks for political decision-making, and change in political institutions and practices. In thinking about duties of mitigation and adaptation it is from this perspective not enough to simply broaden views about justice in the human case to include the interests of nonhuman animals. Instead, we must develop a multispecies approach to climate justice (Sebo, 2021: 13), which recognizes the fact that human and more-than-human animals are active agents who are connected to others in a multitude of relations, sharing the need for a sustainable future. This is not just relevant for nonhuman other animals. It can also help to reimagine the position of the human, as part of a bigger whole: a necessary ingredient for tackling the problems of the Anthropocene.

## Developing a non-anthropocentric attitude towards the climate crisis

Our current geological epoch has been named the Anthropocene, because of the impact of human acts on the planet. Human exceptionalism – the idea that humans are not only better than other animals but also categorically separated from them – lies at the heart of many of the systems and practices that led to the climate crisis, loss of animal lives, biodiversity, and ecosystems. For addressing these problems, developing new technological solutions within this system is not enough. Human behaviour in the Northern hemisphere needs to change, and for this transformation to begin, we need to move beyond human exceptionalism. In 'Pluriversal Politics' (2020), Arturo Escobar argues that we cannot address the current challenges from within the system that caused them. He argues we need to work towards a 'pluriverse', a world view that accepts there are different world views, in which different types of wisdom are articulated. Escobar draws on indigenous thinking from different peoples from Latin America to investigate how we can move away from a world view that centres 'reason' and 'the human' as formed in the Enlightenment. He proposes a multiversal understanding of knowledge and the world which is based on 'sentipensar' – feel-thinking. This can make possible different types of futures and bring about the social transformations that are needed to address the current planetary crises.

Nonhuman communities have their own forms of wisdom and insight, their own cultures and languages (Brakes *et al.*, 2019; Meijer, 2019). Learning from these can help with finding practical solutions for sharing the world in new ways with the multitude of others that are there and help us navigate our own position as humans. The name Anthropocene makes it seem as if humans are all knowing and capable of solving the current crisis alone. But we are not. In order to tackle the problems ahead, we do not need more human control – which was what led us to the problems we are facing in the first place. Taking responsibility should start from the understanding that we are part of a web of relations with other animals, in a natural world upon which we all depend.

## Conclusion

Given the breadth and unprecedented nature of the climate crisis, it seems tempting to focus on 'saving' humans only. This approach is self-defeating as we are in a deeply entangled web of relations with animals and neglecting them often directly hurts people. But even if this were not the case, there is a strong political case for taking a multispecies approach to climate justice, as we showed, which recognizes nonhuman animals as agents with their own perspective on life and their own forms of wisdom. Sketching the multi-level changes needed to move toward a full account of climate justice for all animals is beyond the scope of this paper. Nonetheless, in an attempt to prompt further research on this topic, we would like to offer suggestions we consider critical to ensure climate justice will become more 'multispecies.' In particular, we should:

- increase our scientific understanding of the effects of climate change on animals;
- build knowledge to better understand and aid animals' to recover from and become more resilient to climate change;
- stop blaming animals for climate change harms and instead consider and acknowledge our exploitation of animals as drivers of climate change (e.g. factory farming, wildlife trade);
- ensure that mitigation and adaptation action itself does not threaten animals but instead respects animals' rights;
- consider animals in mitigation and adaptation action – worthy of consideration on their own, in order to increase their resilience against direct effects of climate change and indirect effects (access to food, water, shelter, and other basic goods) (Sebo, 2021: 7);
- include animals in infrastructural changes and adaptations (Sebo, 2021: 148);
- establish principles to reconcile potentially conflicting goals, policies, and measures;
- recognize animals as political agents (Donaldson and Kymlicka, 2011);
- establish public assemblies or special climate committee, with – human and nonhuman – representatives for animals, building on insights from political animal philosophy (e.g. Donaldson and Kymlicka, 2011; Meijer, 2019).

## References

Almiron, N. and Faria, C. (2019). Climate change impacts on free-living nonhuman animals. Studies in Media and Communication 7(1): 37-48.

Blattner, C. (2019). Protecting animals within and across borders: Extraterritorial jurisdiction and the challenges of globalization. OUP, New York NY.

Blattner, C. (2021). Global migration crises, non-human animals, and the role of law. In: Khazaal, N. and Almiron, N. (eds) Like an animal: Critical Animal studies approaches to borders, displacement, and othering. Brill, Leiden, pp. 1-49.

Brakes, P. *et al.* (2019). Animal cultures matter for conservation. Science 363.6431: 1032-1034.

Ceballos, G, *et al.* (2015). Accelerated modern human-induced species losses: Entering the sixth mass extinction. Science Advances 1(5): 1-5.

Center for Biological Diversity (CBD) (2013). Deadly waters: How Rising seas threaten 233 species. Available at: https://www.biologicaldiversity.org/campaigns/sea-level_rise/pdfs/Sea_Level_Rise_Report_2013_web.pdf.

Donaldson, S. and Kymlicka, W. (2011). Zoopolis: A political theory of animal rights. OUP, Oxford UK.

Escobar, A. (2020). Pluriversal politics: The real and the possible. Duke UP, Durham NC.

Green, D. (2019). Animals in disasters. Butterworth-Heinemann, Oxford UK.

Guaghan, J., *et al.* (2009). Response of domestic animals to climate challenges. In: Ebi, K.L., Burton, I., and McGregor, G.R. (eds) Biometeorology for adaptation to climate variability and change. Springer, Cham, pp. 131-170.
Intergovernmental Panel on Climate Change (IPCC) (2014). Climate change 2014 synthesis report. IPCC, Geneva.
Intergovernmental Panel on Climate Change (IPCC) (2018). Global warming of 1.5°C. Summary for Policymakers. IPCC, Geneva.
Intergovernmental Panel on Climate Change (IPCC) (2019). Climate change and land. Summary for Policymakers. IPCC, Geneva.
Kyoto Protocol to the United Nations Framework Convention on Climate Change (1998). December 10, 1997. 37 I.L.M. 22 (1998); 2303 U.N.T.S. 148; U.N. Doc FCCC/CP/1997/7/Add.1.
Meijer, E. (2019). When animals speak: Toward an interspecies democracy. NYU Press, NY.
Paris Agreement (2015). Adoption by the Conference of the Parties. December 12, 2015. U.N. Doc. FCCC/CP/2015/L.9/Rev/1.
Pecl, G., Vérges, A., Popova, E. and McDonald, J. (2017). Climate-driven species on the move are changing (almost) everything. The Conversation April 5, 2017.
Pepper, A. (2019). Adapting to Climate Change: What We Owe to Other Animals. Journal of Applied Philosophy 36(4): 592-607.
Rojas-Downing, M.M., *et al.* (2017). Climate change and livestock: Impacts, adaptation, and mitigation. Climate Risk Management 16: 145-63.
Sebo, J. (2021). Animal ethics in a human world: Why animals matter for pandemics, climate change, and other global threats. OUP: Oxford. Manuscript in press (with authors).
Sergio, F., Blas, J., and Hiraldo, F. (2018). Animal responses to natural disturbance and climate extremes: a review. Global and Planetary Change 161: 28-40.
UN General Assembly (1994). United Nations Framework Convention on Climate Change: Resolution adopted by the General Assembly. January 20, 1994. A/RES/48/189.
World Wide Fund (WWF) (2020). Impacts of the unprecedented 2019-2020 bushfires on Australian animals. November 2020.

# 8. Food security and the moral differences between climate mitigation and geoengineering: the case of biofuels and BECCS

*H. Schübel* and I. Wallimann-Helmer*
*University of Fribourg, Environmental Sciences and Humanities Institute, UniFR_ESH Institute, Chemin du Musée 4, 1700 Fribourg, Switzerland; hanna.schuebel@unifr.ch*

## Abstract

Both biofuels and BECCS serve the purpose of reducing the concentration of carbon in the atmosphere, biofuels by reducing the quantity of $CO_2$ newly added and BECCS by removing the $CO_2$ already emitted. Both rely on the large-scale growth of biomass and hence compete with food production for arable land. Consequently, the implementation of both at large scales potentially endangers food security. Given this conflict and the need for climate action, this paper discusses whether there are differences between the ethical implications of mitigation by biofuels and geoengineering with BECCS. We apply a principlist approach, specifying and weighting four key principles of climate ethics: the polluter-pays principle, the ability-to-pay principle, the equal-per-capita principle, and the procedural involvement principle. Because the ethical and practical implications of BECCS and biofuels have many parallels, applying these four principles identifies relevant moral differences to decide between these two technologies.

**Keywords:** biofuels, BECCS, principlism, polluter-pays principle, procedural justice

## Introduction

Reaching the goal of the Paris Agreement to keep global mean temperature well below 2 °C above pre-industrial levels, requires steep reductions in carbon emissions while ensuring enough energy for economic development. Because of this, biofuels become an important part of the menu of renewable energy transitions (Fulton *et al.*, 2015). At the same time, most modelling scenarios for keeping the global mean temperature increase to 1.5 °C above pre-industrial levels assume the large-scale availability of geoengineering technologies to remove carbon from the atmosphere (Fuss *et al.*, 2018). To this goal, the use of bioenergy with carbon capture and storage (BECCS) is assumed most often (IPCC, 2018; Minx *et al.*, 2017). BECCS removes carbon from the air and stores it permanently underground. This is done by binding carbon in large plantations of biomass that is then combusted for energy production. The carbon produced in this process is captured and liquefied, and later stored by using parts of this energy (Bui *et al.*, 2018).

The large-scale planting of biomass is also necessary when scaling up the production and use of biofuels to reduce emissions in order to meet the 1.5 °C target from the Paris Agreement. However, if industrialized countries grow this biomass for BECCS and biofuels themselves, they tend to increase their imports of grain, rice, and other staple foods, which can increase global food prices. In consequence, implementing both these technologies at large scale may well endanger food security in the poorest regions of the world (Kortetmäki and Oksanen, 2016; Renzaho *et al.*, 2017).

In this paper, we compare the ethical implications of biofuels and BECCS in light of the risks they pose to food security. To assess their ethical implications, we apply a principlist approach by specifying and weighting four key principles of climate ethics we explain in more detail below (Gardiner *et al.*, 2010): the polluter-pays principle (PPP), the ability-to-pay principle (APP), the equal-per-capita


EurSafe 2021 – DOI 10.3920/978-90-8686-915-2_8, © Wageningen Academic Publishers 2021

principle (EpC), and the procedural-involvement principle (PIP), also sometimes called procedural justice. Because the ethical and practical implications of biofuels and BECCS have many parallels, the application of these four principles will indicate moral parallels and differences between the two strategies. We argue that the main difference between the two lies in the specification of the PPP, because biofuels can bring the burdens of climate action closer to the polluting parties, whilst deploying BECCS allows historical emissions to be taken into account. However, we argue that which of these specifications of the PPP should be prioritized must be decided by involving all those concerned most directly and hence means relying on the PIP.

## A disturbing conflict of goals

The starting point of our investigation is the challenge that the target of the Paris Agreement to reduce the negative impacts of climate change for the poorest regions of the world may demand measures that threaten the livelihoods of these very same regions. In the following, we first explain why climate protection is important to ensure food security in poor regions of the world. We then investigate in more detail the similarities and differences between biofuels and BECCS that are relevant for our ethical analysis.

In the poorest regions of the world, many people directly depend on agricultural food production for their livelihood. For the agricultural sector, dangerous climate change results in significant changes in the distribution of rainfall, longer droughts, and extreme weather events such as flooding and heavy rainfall. Hence, climate protection becomes a key factor to foster food security in these regions (Myers *et al.*, 2017). Even though some adaptation to changing climatic conditions is possible in the agricultural sector, at some point, the transformation of traditional ways of livelihood may become necessary. If these transformations mean leaving ancestral lands, such transformational adaptation may increase the risk of poverty and hunger (Preston *et al.*, 2013). Hence, keeping the current climate as stable as possible significantly contributes to food security in these regions of the world. Both biofuels and BECCS can support this goal.

The way we will consider biofuels and BECCS here, both require the large-scale plantation of biomass. In the case of biofuels, such as biodiesel or bioethanol, this biomass is processed into energy. In the case of BECCS, biomass is also processed into energy but the carbon released in this process is captured, liquified (using some of the energy released in the process of burning it) and stored in geological formations (Bui *et al.*, 2018). The plantation of biomass is most efficient if agricultural land is used for this purpose, as transforming other areas into farmland results in emissions and loss of natural carbon sinks (Field *et al.*, 2020). In terms of infrastructure, BECCS and biofuels may be combined (Field *et al.*, 2020). Biofuels demand facilities and energy for their production, and they must be transported and stored until they are used (Fulton *et al.*, 2015). Likewise, BECCS involves facilities for producing energy to capture and liquefy carbon, but additionally also for transporting the liquefied carbon and storing it. This storage of carbon is the main difference between two the technologies: BECCS removes carbon from the atmosphere whereas biofuels do not. Instead, biofuels mitigate emissions by displacing fossil fuels (Field *et al.*, 2020).

Because both technologies require the growth of vast quantities of biomass, both tend to threaten food security in poor regions of the world. Studies indicate that an important consequence of planting of biomass at large scale will be a massive increase in grain, rice and staple food prices on the global market, to the disadvantage of poorer regions of the world (Brinkman *et al.*, 2020; Hasegawa *et al.*, 2018). Thus, it is necessary to balance food security with implementing biofuels and BECCS at the scale assumed in climate models for avoiding dangerous climate change. To do so is challenging and both technologies are not easily differentiated in an ethical evaluation. In the next two sections, we apply a principlist approach to analyse the ethical conflicts presented.

## Specifying principles to understand the conflict of goals

The methodology suggested to analyse the conflict of goals identified in the last section is principlism, well known from medical practice and ethics. Principlism assumes that for a given practice there exists a core set of principles that need to be specified and weighted differently depending on the moral issue at stake (Beauchamp and Childress, 2009). Whereas the core principles are well established in medical ethics, such core principles have not yet been defined for environmental issues. For the purpose of this investigation, we assume that they are the core principles of climate ethics and the principle of procedural involvement as explained in Wallimann-Helmer (2019): the polluter-pays principle (PPP), the ability-to-pay principle (APP), the equal-per-capita principle (EpC), and the procedural-involvement principle (PIP). In the following, we specify each principle for the given case before proposing a weighting of them and offering concluding remarks.

### *Specifying the polluter-pays principle*

The PPP is usually understood as assigning responsibilities for climate action in proportion to contributions to anthropogenic climate change. Parties emitting more greenhouse gases should shoulder heavier burdens in avoiding dangerous climate change. In a variant, this principle also concerns those who benefit proportionally more from the emissions of others. Historical emissions are relevant here because current welfare most probably depends on emissions made in the past. Both biofuels and BECCS can make proportional contributions to climate protection. The more someone contributes or has contributed to anthropogenic climate change, the more they should reduce their production of emissions or remove emissions from the air.

However, the PPP identifies two crucial differences between these technologies. First, BECCS demand centralized facilities for producing energy and liquefying carbon from the biomass grown nearby. Storing liquefied carbon requires a transportation infrastructure (e.g. pipelines) and appropriate geological sites (Bui *et al.*, 2018). Although biofuels must be produced and stored in centralized facilities as well, their negative impacts will most probably be more decentralized. To lower transportation costs, biofuels most probably will be produced geographically closer to those who emit and hence bring negative impacts closer to those contributing more to climate change. This is not the case for BECCS, whose scale effects may favour large facilities, especially for storage. Second, biofuels only possess the potential to reduce currently produced carbon because the carbon they capture will be released again once they are burnt for energy production. By contrast, BECCS capture whatever quantity of carbon from the atmosphere that is processed by facilities. This also enables the capture and storage of high emissions produced in the historical past.

According to our specification, the PPP clearly assesses the potential negative impacts of both biofuels and BECCS on food security as an injustice. The poorest regions of the world did not contribute the most to climate change but most probably have negative impacts on their food security. Consequently, the PPP requires that the implementation of both biofuels and BECCS is realized in a way that minimizes their negative impacts on the poorest regions of the world. Beyond this, the PPP remains rather neutral regarding a choice between biofuels or BECCS. Biofuels allow negative impacts to be more closely aligned with individual pollution. By contrast, BECCS has the advantage of accounting for all emissions, both those currently produced and those in the past. Hence, it is difficult to fully dismiss one of the two technologies from the perspective of the PPP.

### *Specifying the ability-to-pay principle*

Most commonly, the APP does not assign responsibilities for climate action in proportion to current and past contributions to climate change but relies on agents' capacities. Those more capable of remedying environmental threats are ascribed heavier burdens for acting. Parties commanding more financial resources or better expertise are seen as having a duty to take the lead and assist those less well equipped. The APP entails a crucial further implication: Demanding action from those not able to act is considered illegitimate. Only those capable of acting can be ascribed responsibilities.

In our context this means that those parties financially and technologically able to foster research on biofuels and BECCS and to implement these technologies must take the lead. They are responsible for providing the finance, technology, and managerial skills for their implementation where needed and possible. Hence, the APP differentiates between those parties with financial capacities and those with the skills necessary to implement these two technologies. These skills might further differentiate responsibilities because they might not be equally distributed between the technologies. Because biomass must be planted where it can grow and liquefied carbon must be stored in appropriate sites, the APP may also be interpreted as defining where biomass should be grown and BECCS should be realized.

However, the APP has an important limitation when applied to both BECCS and biofuels: with the APP, tropical regions might have a duty to make land available to grow biomass if this is more efficient than growing it in colder climates. But arguments like these do not reflect any differences in the economic and social conditions of those affected by the implementation of these technologies. This is problematic if the negative impacts of their implementation worsen conditions that are already poor. However, realizing such facilities might also be of advantage for these regions because these technologies may also mean economic gain for those involved. If economic gain can outweigh the negative impacts on food security, implementing plantations and facilities might be justifiable.

### *Specifying the equal-per-capita principle*

In its classical interpretation, the EpC assigns rights to equal quantities of emissions to all human beings. This quantity is usually calculated from an emissions budget like the one defined by the climate target of the Paris Agreement. An entitlement to an equal quantity of emissions is usually justified by two arguments. First, no morally relevant differences exist between human beings regarding emission entitlements. Second, if this is the case then the only pattern of distribution that requires no justification is an equal distribution. However, if there are legitimate reasons to distribute unequally, such as the need to heat or the limited availability of low-carbon technology, then unequal distributions of emission entitlements might be justified.

The promise of biofuels is to produce similar amounts of energy like fossil fuels while releasing less carbon. This entitlement to an equal amount of energy seems relevant when specifying the EpC to decide between the two technologies. While biofuels reduce emissions released in energy production, BECCS delivers less energy and removes fossil emissions from the atmosphere. Deciding which technology is more closely in line with the EpC depends strongly on technological details and on the degree to which emissions of energy production can be balanced. The more energy that can be produced with a technology, the more favourable this technology becomes. In consequence, the decision depends on the degree to which each technology can help reduce emissions while at the same time ensure a high level of equal amounts of energy for all.

The EpC sheds some critical light on the consequences of biofuels and BECCS for food security. As a principle that defines rights to equal entitlements, the EpC demands food security for all to an equal degree. In whatever way such equality of food security is defined, deviation from it would have to be justified. Because human beings do not seem to differ in their need for food security, such inequalities can hardly be justified. Thus, unequal distribution of the negative impacts of biofuels and BECCS on food security need to be avoided unless they can be justified. Many egalitarians argue that an important exception to equal distribution is voluntary choice. This would be the case, for example, if regions accepted biomass plantations in exchange for economic advantages.

### *Specifying the procedural-involvement principle*

The procedural involvement on fair terms of all affected demands that environmental policy decisions are taken in a collective endeavour. This principle, more common in empirical environmental justice literatures, demands collective decision-making because not all environmental questions are strictly speaking questions of fair distribution. Important examples include the evaluation of risks and compensatory measures for climate loss and damage. In both cases, it can be argued that assessment is only legitimate if it is achieved by involving those who have morally relevant interests at stake (Wallimann-Helmer, 2015).

Biofuels and BECCS both involve negative impacts that might require compensation for unfair disadvantages. This is the case if the unequal distribution of their negative impacts cannot be justified. In such cases, those facing negative impacts have to be involved in decision-making under fair conditions to decide what compensation is appropriate for the unfairness faced. Because biofuels and BECCS might also have economic advantages for those living near these facilities, such processes might also legitimize unfairly heavier burdening. However, one of the most serious challenges is to ensure appropriate recognition of indigenous peoples, ethnic minorities, and other marginalized social groups in such decisions.

Special emphasis must be given to the food security of indigenous peoples and more generally to poor communities and regions of the world. Not only is their livelihood endangered due to climate change; there is also a high risk that their voices will not be heard when it comes to the threats to their food security. Hence, to ensure food security for the poor, it will be of key importance to involve these peoples and communities in decision-making about the implementation of biofuels and BECCS and to ensure appropriate recognition.

## Weighting principles to decide the conflict of goals and concluding remarks

According to the principlist approach of Beauchamp and Childress, once principles are specified, they need to be weighted to decide the given case. Principlism assumes that the principles building the core set of ethical principles in a given practice have no predefined hierarchy. The hierarchy of principles might differ by case and context. However, once a hierarchy of principles has been established, it justifies a decision. The principle or the principles that are at the top of this hierarchy decide the ethical conflict at scrutiny. What action these principles demand in a given context is the morally right thing to do.

Following our analysis, we deem the PIP to be the most important principle when comparing biofuels and BECCS. The reason for this is that the specifics of the efficiency and negative side-effects of biofuels and BECCS are both still to be determined, which makes it difficult to identify exactly which technology the other three principles would propose. Under such conditions, we believe that it is most important to involve those who are most directly affected by the large-scale implementation of biofuel and BECCS facilities in deciding how to prioritize them.

We take the PPP to be the second most important principle because it most clearly identifies the injustices in food security at stake. Those who have contributed the least to climate change will suffer the greatest threats to their food security. However, what the PPP recommends depends on how much weight is given to bringing the additional burdens of new technologies closer to those emitting, as is possible with biofuels, or to accounting for historical emissions, as BECCS allows.

We argue that both the APP and the EpC could serve as safeguards. The APP ensures that those disadvantaged are not carrying heavy burdens due to the implementation of both technologies, especially if their food security is endangered. Further context is needed to fully specify the EpC because its demands depend strongly on the technology used and its efficiency. However, the EpC demands that equal energy be ensured for all and thus that no one is left behind when implementing new technology for climate protection.

## References


Beauchamp T.L. and Childress J.F. (2009) *Principles of biomedical ethics*. New York, NY: Oxford Univ. Press.

Brinkman M., Levin-Koopman J., Wicke B., *et al.* (2020) The distribution of food security impacts of biofuels, a Ghana case study. *Biomass and Bioenergy* 141: 105695.

Bui M., Adjiman C.S., Bardow A., *et al.* (2018) Carbon capture and storage (CCS): the way forward. *Energy & Environmental Science* 11(5): 1062-1176.

Field J.L., Richard T.L., Smithwick E.A.H., *et al.* (2020) Robust paths to net greenhouse gas mitigation and negative emissions via advanced biofuels. *Proceedings of the National Academy of Sciences of the United States of America* 117(36): 21968-21977.

Fulton L.M., Lynd L.R., Körner A., *et al.* (2015) The need for biofuels as part of a low carbon energy future. *Biofuels, Bioproducts and Biorefining* 9(5): 476-483.

Gardiner S.M., Caney S., Jamieson D., *et al.* (eds) (2010) *Climate ethics*. Oxford: Oxford University Press.

Hasegawa T., Fujimori S., Havlík P., *et al.* (2018) Risk of increased food insecurity under stringent global climate change mitigation policy. *Nature Climate Change* 8(8): 699-703.

Kortetmäki T. and Oksanen M. (2016) Food Systems and Climate Engineering: A Plate Full of Risks or Promises? In: Preston C.J. (ed.) *Climate justice and geoengineering: Ethics and policy in the atmospheric Anthropocene*. London: Rowman & Littlefield International Ltd, pp. 121-135.

Myers S.S., Smith M.R., Guth S., *et al.* (2017) Climate Change and Global Food Systems: Potential Impacts on Food Security and Undernutrition. *Annual review of public health* 38: 259-277.

Preston B., Dow K. and Berkhout F. (2013) The Climate Adaptation Frontier. *Sustainability* 5(3): 1011-1035.

Renzaho A.M.N., Kamara J.K. and Toole M. (2017) Biofuel production and its impact on food security in low and middle income countries: Implications for the post-2015 sustainable development goals. *Renewable and Sustainable Energy Reviews* 78: 503-516.

Wallimann-Helmer I. (2015) Justice for climate loss and damage. *Climatic Change* 133(3): 469-480.

# 9. Ecofeminism, afforestation & *La Via Campesina*

*E. Woodhouse*
*University of Sheffield, Philosophy, Department of Philosophy, Victoria Street, Sheffield, S37QB, United Kingdom; ewoodhouse2@sheffield.ac.uk*

## Abstract

This paper uses a materialist ecofeminist analysis to compare two forms of carbon dioxide removal (CDR): Afforestation for carbon capture and La Via Campesina's proposals for soil carbon storage through a transition to 'food sovereignty'. It argues that adopting an ecofeminist perspective allows us to see why afforestation for carbon capture fails to tackle the root causes of climate change. Additionally, this ecofeminist perspective can give us insight into the contrasting position of small farmers between both forms of CDR.

**Keywords:** geoengineering, food sovereignty, carbon capture

## Introduction

In this paper I engage in a critical comparison of two forms of carbon dioxide removal (CDR): Afforestation for Carbon Capture, and in particular those afforestation projects which have been encouraged under the REDD+ Scheme, and La Via Campesina's (LVC) novel proposals for increasing soil carbon storage through a widespread shift to a 'Food Sovereignty' model of world agriculture (La Via Campesina, 2009). I consider these proposals on two points of comparison: how they appear to conceptualise of the 'root cause' of climate change, and how this manifests in their differing treatment of peasant farmers. I use a materialist ecofeminist analysis to examine and contrast the CDR proposals. I aim to demonstrate that this framework gives us a useful tool to explain the differences in each proposal's approach to root causes; and their relationship with small farmers.

## What is ecofeminism?

Ecofeminism is a broad theoretical tradition which emerged from the intersections of feminist and environmentalist theory in the 1980s and 90s. All ecofeminists believe that environmental problems/degradation are a feminist issue, and at a minimum this belief may stem from recognising that women bear the brunt of environmental problems worldwide. However, most ecofeminists go further and argue that the oppression of women and the domination of nature are connected and co-constituted. While some ecofeminists, such as Collard (1988) see a women/nature connection coming from a biological, spiritual or essentially feminine cultural connection with the natural world, as mothers, nurturers and carers, these 'affinity' (Mellor, 1992) positions are widely side-lined in favour of a 'social constructionist' account (Mellor, 2000). The social constructionist school view the women/nature connection as the result of women's and nature's position in a patriarchal society, which has also historically devalued nature in comparison with social, cultural, and otherwise essentially human characteristics. For social constructionist ecofeminists the study of women and nature begins with the material realities in which they are situated, particularly capitalism and patriarchy. In particular, the social constructionists have drawn our attention to the 'ideological devaluation of nature'. By this, I mean the ways in which the dominant forms of social and philosophical thought tend to view anything constructed as being natural as both essentially separate from, and inferior to, the human, the cultural or the societal. According to Merchant (1980) new ideas of objective scientific knowledge championed during the scientific revolution by were seen as beneficial for giving humankind a kind of 'power over nature'. At the same


EurSafe 2021 – DOI 10.3920/978-90-8686-915-2_9, © Wageningen Academic Publishers 2021

time, revolutions in the scientific ability to measure and record the natural world led to a 'mechanistic' vision of nature as predictable and inanimate (Merchant, 1980). These developments provided the ideological justifications for greatly expanding the scope and scale of extractive practices; providing a moral license to transform the natural world for humanity's benefit.

While some ecofeminists see the root cause of environmental degradation, including climate change, situated within this ideological turn, 'materialist ecofeminists' emphasise the importance these changes had in the development of the capitalist world system (Mellor, 2000). In particular, they note that capital is reliant on this systemic devaluation of nature's 'work'. Nature 'works' for capitalism in many ways: through the natural (re)producing of fertile soils, combustible fossil fuels, and other raw materials, of new human beings through the 'natural' work of women as mothers, as well as crucially for our sakes, by providing an atmospheric dump for the excess $CO_2$ (Mies, 2014; Moore, 2015; Sallah, 2017). The society/nature dualism results in environmental degradation not inevitably, but because of the role it plays in enabling capitalism's particular ecological regime. Just like a new machine or mining technique, the ideological devaluation of nature forms part of the productive forces of capital by allowing it to access nature's work with greater efficiency and lower cost. The social/natural dualism becomes harmful to nature, that is, causes environmental degradation, because it justifies natures exploitation.

Adopting the materialist ecofeminist position which I have outlined here shows us that a particular ideological construction of nature, as separate and inferior from the social, is the root cause of climate change. This occurs through the role it plays in justifying nature's exploitation. Presuming briefly that it is preferable to treat the causes, rather than just the symptoms of a problem (all else being equal), then ecofeminists suggest that neither greenhouse gas removal, nor emissions mitigation will be sufficient to address climate change 'unless' they also challenge the social, economic and ideological circumstances which allowed these environmental problems to come to be. This means planning for a carbon drawdown that does not see the job of removing carbon as separate from changing the economic system that put it there in the first place, since conceptualising it in this way would be reproducing the society and nature dualism that lies at the root of environmental problems. Instead, ecofeminists require environmentalism that moves beyond, rather than reproduces, the social/nature dualism seeing the calls for economic reform *as* environmentalism, not separate from or only related to it.

## Carbon stock indicators and the mechanistic vision of nature

The ecofeminist concept of the 'mechanistic vision of nature' provides a useful tool for considering the difference between afforestation and LVC's proposals for carbon capture. As I have suggested above, ecofeminists are likely to see forms of environmentalism that reproduce the mechanistic vision of nature as fundamentally unable to tackle the root cause of climate change. I argue that contemporary afforestation projects have tended to employ this mechanistic vision, particularly through their reliance on Carbon Stock Indicators (CSIs) to make assessments on land-use changes. Additionally, I argue that the displacements of forest-edge peasant farmers which have resulted from these projects will not be surprising for ecofeminists, because of the historic role the materialist vision of nature has played in expanding the frontiers of capitalised production into previously uncapitalised realms.

Afforestation is a well-known strategy for capturing and storing atmospheric carbon through increasing the total amount of carbon stored in forest biomass. In its simplest form this just means planting new trees, however for the purposes of this analysis when I refer to afforestation I am particularly referring to the types of afforestation projects which have been promoted by the UNFCCC Clean Development Mechanisms (UNFCCC, 2013) and the associated REDD+ schemes where the development of new, as well as the maintenance of existing, forests is financially incentivised through the trading of carbon credits. Sites for these new or expanded forests are typically selected through the use of CSI assessments:

estimates of the potential volume of carbon which could be stored in a given land-use system. For projects to attain REDD+ certification, they must adhere to standard methodological guide for conducting these carbon stock assessments published by the Intergovernmental Panel on Climate Change. While the use of CSIs is commonplace they are not without methodological controversy (Schiedel, 2018). Fox *et al.* (2014) argues that these assessments have been used to disqualify forest-edge peasant farmers 'almost automatically' from keeping their lands. The effect that the REDD+ afforestation program has had displacing farmers on the forest/agricultural boundary is additionally reported by Griffiths (2008), Enri (2009) and Lyons *et al.* (2014).

The ecofeminist perspective allows us to put the use of Carbon Stock Indicators within historic and ideological context of the mechanistic vision of nature. The development and expansion of capitalism in the sixteenth century occurred in tandem with the development of myriad new scientific ways of measuring, charting and ultimately conceptually simplifying the natural world. The Mercator projection allowed for an easier route to the American colonies, the mechanical clock broke the day into work hours, standard weights and measures allowed kings and states to accurately survey their kingdom's resources (Scott, 1998; Merchant, 1980). For ecofeminists, each of these developments, and many more, worked to order the chaos of nature and simplify it for easier exploitation. Carbon Stock Indicators fit this broad historical trend. They give us a means for 'seeing' the world, simplified down to potential tonnage of carbon storage biomass per hectare. Looking through this lens, the problem of climate change becomes simple: too much carbon is in the atmosphere and not enough of it is being stored as biomass. Climate Change is reduced to a management problem, where we only need to fiddle with the inputs and outputs like the dials of a machine. This is the mechanistic vision of nature in essence.

The problem for ecofeminists with reducing climate change to a problem of carbon management is that it reflects a misunderstanding of the problem. Climate change is a problem for ecofeminists because of the harms a changing temperature will have to both the natural world and to the people who will be displaced and have their lives altered by its effects. By employing CSIs as their means of seeing the world, the REDD+ afforestation projects have given themselves a kind of tunnel-vision, where climate change has become separated from its social and economic context and narrowed down to a parts per million count of $CO_2$. This is not to argue that the information gained from CSI evaluation is not important for preventing climate change, rather for ecofeminists it is going to be 'necessary' but not 'sufficient' information (Sheidal, 2018). Within this context, ecofeminist will be unsurprised that the side-effects of REDD+ afforestation has been the displacement and disruption of peasant farmers since these schemes only aimed to manage carbon concentration. This is the social/nature dualism in effect: environmental problems are seen as separate from, and solvable apart from, the social circumstances the occur within. Ironically, instead of preventing the harms of climate change, the REDD+ projects have brought the disruption and displacements to forest-edge farmers themselves!

## Root causes in LVCs proposals

I argue that by tying carbon sequestration to their program of social and economic reform, LVC offer a CDR proposal that, unlike the REDD+ afforestation schemes, do no reproduce the mechanistic vision of nature. Notably, and in sharp contrast with the REDD+ projects, LVC's proposals put the needs of peasant farmers first and foremost in their considerations: defending their interest in staying on their land and in their livelihoods, and by championing peasants as agents who can play a pivotal role in facilitating a sustainable carbon sequestration. I argue that LVC's CDR proposals will be seen favourably by ecofeminists.

La Via Campesina are a global conglomeration of local and national peasants' organisations and unions, ultimately representing the interests of over 200 million peasant farmers, small-holders and fisherfolk worldwide. Since the mid-1990s they have promoted their concept of 'Food Sovereignty' as a radical alternative to the neoliberal food system. Food sovereignty is understood as:

> A true agrarian reform, that strengthens small scale farming, promotes the production of food as the primary use of land, and regards food as a basic human right that should not be treated as a commodity. (La Via Campesina, 2009)

The scope and intentions of Food Sovereignty makes it, as Bernstein (2014) notes, a project with 'world historic' ambitions since its calls for local and democratic control over food production, as well as a widespread switch to the labour-intensive growing techniques of small farmers, should be seen as antithetical to the current world agricultural system, which relies on highly capital-intensive mechanised production techniques and growing food predominantly for export. In addition to being a program for economic and food justice, in a 2009 position statement, LVC propose that their Food Sovereignty program would have a carbon sequestration effect (La Via Campesina, 2009). This would be achieved firstly through a reduction of the carbon inputs into food production: through intensive use of pesticides, fossil fuelled machinery and transport to foreign markets. However, they additionally propose that the low-till agricultural methods which are typical of peasant farmers will increase the amount of carbon which can be stored in soil biomass, resulting in a subsequent $CO_2$ drawdown. LVC's proposals are therefore a hybrid form of CDR, straddling the line between carbon sequestration and emission mitigation, while all at the same time being an expansive program of economic and social reform for a new world agricultural system that secures the rights of peasant farmers.

LVC explicitly locate the root cause of climate change in:

> Current global modes of production, consumption and trade... a development model based on capital concentration, high fossil energy consumption, overproduction, consumerism and trade liberalization. (LVC, 2009)

This is a noticeably different from the REDD+ afforestation schemes described above, which I characterised as reducing climate change to merely a problem of atmospheric $CO_2$ concentration, divorced from its social and economic origins. This is not to say that the designers of REDD+ and the people working on these projects are 'personally' ignoring that climate change has social and economic dimensions, and the members of LVC do not. Rather LVC's proposals incorporate this belief into the design of their CDR project. This distinction is important in marking LVC's CDR as distinctly ecofeminist. By positioning their carbon sequestration program within their program for food and economic justice, LVC are proposing a form of CDR that moves beyond the social/nature dualism. This is because the environmental aspect of their proposal, the CDR, is indistinguishable from the social and economic side, food sovereignty. Nature is no longer seen as something that can be handled separate from society: nature 'is' society and society 'is' nature. While this distinction may only seem slight, for ecofeminists it is of paramount importance, since the conceptual separation of society/nature is what causes nature's conceptual devaluation, and vis a vis enables its exploitation.

The other, and perhaps most notable, difference between the two programs is the position of peasant farmers within them. The REDD+ afforestation schemes have enabled a new wave of 'green-grabbing': new appropriations of agricultural land of environmental purposes (Fairheld *et al.*, 2014), which have displaced peasants. Conversely, LVC champion the specific skills and knowledges of peasants to enact a carbon drawdown by remaining on their lands. For materialist ecofeminists the differences between the two schemes are not accidental but are a result of the different ways the two programs conceive of the society/nature dualism. For ecofeminists, the conceptual devaluation of 'the natural' also present in the devaluation of 'natural' reproductive labour in comparison with social/cultural productive, waged and capitalised labour (Mies, 2014). The fact that forest-edge farmers are engaging in primarily subsistence agriculture, that is, reproductive/natural rather than productive/waged work, is for ecofeminists an explanatory factor in why they are displaced in favour of capitalised land-usages.

Within this context the fact that LVCs proposals are 'peasant-led' is not incidental; they do not simply 'add peasants and stir'. Instead, by reimagining subsistence producers as the agents in their carbon drawdown is part of the way LVC challenge the society/nature dualism. For LVC, Food Sovereignty has always been a positioned as an alternative to neoliberalism. It should not be understated that the highly labour intensive and low capital input agricultural methods LVC argue will cause a carbon sequestration, are antithetical to neoliberalism's world agricultural model of highly capitalised production for export (Moore, 2015). The CDR that LVC propose is therefore not opposed only in theory to the economic forces that have resulted in both $CO_2$ being put in the atmosphere and peasants being displaced, but it cannot exist alongside it. In this way we can see that it is through challenging the social/natural dualism that LVC are able to build security for peasants into their CDR proposals.

## Conclusions

Throughout this paper I have demonstrated the potential of ecofeminism as an evaluative tool for CDR. Ecofeminist analysis shows us the dangers of trying to 'solve' environmental problems separately from their social and economic context. The ecofeminist evaluation of the REDD+ afforestation project shows us how reducing the climate change to an adverse concentration of atmospheric $CO_2$, resulted in the displacement of forest-edge peasant farmers. Conversely, I have argued that LVC's CDR proposals will be seen favourable for ecofeminists because of how they situate carbon drawdown as part and parcel of a broader struggle for social and economic justice. By doing so, they offer a vision of a CDR that moves beyond the society/nature dualism; they do this by refusing to believe that environmental problems are different, separate to, or only part-of social problems. Food Sovereignty IS their environmentalism, and their environmentalism IS Food Sovereignty. This move to seeing climate change as a problem of social organisation 'as well as' a carbon problem is essential for ecofeminists, since the idea that climate change could be solved simply by removing $CO_2$ from the atmosphere relies on the conceptual devaluation of nature which enables the atmosphere to become constructed as a waste dump in the first place.

If the tendency to see the solution to climate change as a lower atmospheric $CO_2$ count, 'is in itself part of the problem', we might reasonably conclude that CDR development is ultimately a flawed endeavour. However, this is not the argument I am seeking to make. Rather, ecofeminists show us how this kind of 'mechanistic thinking' contributes to environmental degradation as an essential component of capitalism. What I have hoped to show through my analysis of 'LVCs' program for $CO_2$ sequestration that there are alternatives. I therefore conclude that their proposals merit further attention from those in search of models for future development of an equitable and a just carbon drawdown.

## References

Arnim Scheidel (2019) Carbon stock indicators: reductionist assessments and contentious policies on land use, The Journal of Peasant Studies, 46:5, 913-934.

Bernstien, H. (2014). Food Sovereignty via the Peasants Way: A Sceptical View. Journal of Peasant Studies. 41:6, 1031-1063.

Collard, A., and Contrucci, J. (1988). Rape of the Wild. London: Women's Press.

Enri, C. 2009. 'Shifting the Blame? Southeast Asia's Indigenous People and Shifting Cultivation in the Age of Climate Change.' Indigenous Affairs I (9): 38-49.

Fox, J., J.-C. Castella, and A.D. Ziegler. 2014. 'Swidden, Rubber and Carbon: Can REDD+ Work for People and the Environment in Montane Mainland Southeast Asia?' Global Environmental Change 29: 318-326.

Griffiths, T. 2008. 'Seeing 'REDD'? Forests, Climate Change Mitigation and the Rights of Indigenous Peoples and Local Communities: Update for Poznan (UNFCCC COP 14).' (December).
Hall, D., P. Hirsch, and T.M. Li. 2011. Powers of Exclusion: Land Dilemmas in Southeast Asia. Singapore: NUS Press.
La Via Campesina (2010). Position Paper: Small Scale Sustainable Farmers are Cooling Down the Earth. La Via Campesina Views.
Mellor, M. (1992). Breaking the Boundaries: Towards a Feminist Green Socialism. London: Virago.
Mellor, M. (2000) Feminism and Environmental Ethics: A Materialist Perspective. Ethics and the Environment. 5:1, 107-123.
Merchant, C. (1980) The Death of Nature. Harper, San Franscisco.
Mies, M. (2014). Patriarchy and Accumulation on a World-Scale. Zed Books, London.
Moore, J.W. (2015). Capitalism in the Web of Life. Verso, London.
Plumwood, V. (1993) Feminism and the Mastery of Nature. Routledge, New York.
Sallah, A. (2017). Ecofeminism as Politics, 2nd ed. Zed Books, London.
Scott, J. (1998) Seeing Like a State. Yale University Press, New Haven.

# Section 3.
# Adapting agriculture to sustain food security

# 10. Who owns the taste of coffee – examining implications of biobased means of production in food

*Z.H. Robaey*[1*] *and C. Timmermann*[2]
[1]*Philosophy Group, Wageningen University, the Netherlands;* [2]*Institute of the History, Philosophy and Ethics of Medicine, Ulm University, Germany; zoe.robaey@wur.nl*

## Abstract

Synthetic foods advocates offer the promise of efficient, reliable, and sustainable food production. Engineered organisms become factories to produce food. Proponents claim that through this technique important barriers can be eliminated which would facilitate the production of traditional foods outside their climatic range. This technique would allow reducing food miles, secure future supply, and maintain quality and taste expectations. In this paper, we examine coffee production via biobased means. A start-up called Atomo Coffee aims to produce synthetic coffee with the aim of saving 'the taste of coffee' from the effects of climate change. This decontextualisation of coffee production ignores the current and historical contributions of coffee farmers in two ways: the traditional varieties in taste of coffee and their cultural significance, and the potential shade-grown coffee plantations have in capturing carbon. In addition, synthetic coffee may lead to the loss of agricultural biodiversity and the removal of resources away from production systems that provide a safe space for tropical flora and fauna. How should the 'taste of coffee' be owned? We investigate the property regimes under which we could consider owning the taste of coffee as a 'synthetic' agrobiodiversity to help identify rights and responsibilities. Building on this analysis, we consider dimensions of responsible innovation and social justice to help guide synthetic foods as an agricultural innovation.

**Keywords:** cellular agriculture, bioeconomy, ownership, responsible innovation, biodiversity

## Introduction

Synthetic foods advocates offer the promise of efficient, reliable, and sustainable food production. By engineering microorganisms (via conventional non-GM processes, gene insertions or artificial synthesis), these become factories for the production of food products, presenting a new trend called cellular agriculture. Proponents claim that through this technique important barriers can be eliminated which would facilitate the production of traditional foods outside their climatic range: In other words, anywhere a bioreactor could be placed with the engineered microorganisms, and with considerably less environmental impact. This technique would allow reducing food miles, secure future supply, contribute to circular processes, and maintain nutritional quality and taste expectations. In addition, this technique can tap into waste streams or use renewable resources and can therefore be an essential part of a new bioeconomy for food production.

In this paper, we examine a case of cellular agriculture to produce coffee, a biobased process. Biobased means processes typically employ microorganisms to transform inputs like sugar, or $CO_2$, to beneficial compounds. At this moment, a start-up called Atomo Coffee produces synthetic coffee, that they call atomic coffee. This means that the coffee itself does not come from a bean, but from the transformation of other products into elements that, together, make coffee. The mission of Atomo Coffee is to save the taste of coffee from the effects of climate change (Atomo Coffee, n.d.). While their aims are laudable, they also embody a central concern that the rising trend of cellular agriculture presents: a complete social and ecological decontextualization of coffee production. Indeed, what cellular agriculture produces are


EurSafe 2021 – DOI 10.3920/978-90-8686-915-2_10, © Wageningen Academic Publishers 2021

not coffee beans, but the taste of coffee with equivalent, or adapted nutritional properties. While the production of synthetic coffee itself might be decontextualised, it is hard to argue that a synthetic coffee that appropriates properties of coffee agrobiodiversity for the purpose of taste and nutritional qualities is devoid of connection and independent of commitments to this agrobiodiversity.

The new method of production afforded by cellular agriculture ignores current and historical contributions of coffee farmers in at least two ways: the traditional varieties in taste of coffee and their cultural significance, and the potential shade-grown coffee plantations have in capturing atmospheric carbon. In addition, synthetic coffee may lead to the loss of agricultural biodiversity and the removal of resources away from production systems that are currently providing a safe space for tropical flora and fauna. This is particularly unfortunate when synthetic coffee is presented as sustainable, as it is likely that it will take a cut from the market share of organic fair-trade producers while seeking for concerned customers. Because cellular agriculture is not producing coffee beans, but captures the taste of coffee, this invites asking: who owns the taste of coffee? Establishing such a claim has normative implications with regard to the conservation of farmed coffee beans and the contributions of farmers to agrobiodiversity and carbon capture.

Re-creating taste, or flavour, is not a new endeavour. Chemically, famous flavours to be synthesized are vanilla with vanillin, and the flavour of the now defunct Cavendish banana, that remains the staple for banana synthetic flavours. What is then the difference with flavours and food produced through bio-based processes instead of chemicals? One is that they make strong claims about being sustainable, and two, the emergence of new processes on multiple fronts in the food sector makes that such processes do more than re-creating flavour, or taste, but recreate the actual food item. Together, these two differences present a long-term challenge to farming and the conservation of agrobiodiversity.

In the following, we first consider the historic labour efforts of farmers globally to establish the taste of coffee as we know it today. Then, we consider what the taste of coffee consists of in cellular agriculture. Next, we investigate the property regimes under which we could consider owning the taste of coffee (Timmermann and Robaey, 2016), and finally, we sketch how dimensions of responsible research and innovation (RRI) and social justice (Robaey and Timmermann, 2018) might help guiding synthetic foods as an ethical agricultural innovation.

## Cultivating coffee, what led to its taste

Coffee is one of the oldest commodified agricultural goods. The early limitations to seek appropriate new cultivation sites outside its native Ethiopian Highlands and the Mountains of Yemen made controlling the trade with the beans very lucrative. After coffee first began to be consumed in Yemen in the 15$^{th}$ century, it was diffused through ample trade networks throughout the Arabic peninsula and later on to Turkey. Ensuring an adequate taste experience was from early on a central feature in the diffusion of coffee. Merchants introduced coffee to a new public in the Muslim world and Western Europe by establishing prestigious coffee houses that served as places of gathering and social exchange (Topik, 2009).

Nowadays coffee is grown in highly different production systems. Some regions, particularly in the Spanish-speaking Latin America, India and Africa, maintain highly biodiverse systems that grow coffee under the trees canopy, providing a refuge for wildlife and delivering important ecosystem services, from which farmers also draw immediate benefits. Shade-grown coffee farms vary in the number of trees per hectare, but even scant-shade coffee systems (including 1-3 tree species) sequester an additional 53-55 tons of carbon per hectare aboveground compared to sun grown coffee systems without counting the extra carbon stored in much richer soils (Jha *et al.*, 2014). Unfortunately, traditional production sites such as Brazil and new coffee areas in East Asia are concentrating on non-shaded production systems (idem.).

A major challenge for shaded coffee growers is to find support and financial compensation for their ecologically superior production methods. A strategy that has been followed is to secure additional revenues through labelling. Yet here too consumers may find it tricky to navigate around the large variety of labels, supporting goals such as fair trade, carbon footprint reductions, bird sanctuaries, agroforestry, gender justice and child labour prevention. Consumers are left on their own to figure out which label guarantees the fairest and most sustainable coffee (Ingenbleek and Reinders, 2013). An over-emphasis on ecological sustainability may leave labour issues unaddressed by giving the impression that coffee comes out effortlessly and without hazards from an idyllic forest (Jimenez-Soto, 2020). Fair trade also needs to be guaranteed in organic agriculture.

There is some indication that shaded coffee farms produce also a more tasteful coffee (Jha *et al.*, 2014). Here shaded coffee farmers' cooperatives could take advantage of an additional instrument provided by intellectual property: geographic indications. Here farmers have the advantage of securing additional income by offering added value, through special regional preparation methods, or agroecotourism services that provide a full coffee experience.

Due to the extensive commodification of coffee, much of the effort to add specialty value has concentrated on the roasting process, as it allows coffee traders additional control over the quality and thereby reduces the bargaining power of coffee growing regions. The fierce competition among major coffee brands is however opening up new marketing opportunities. Coffee taste varies with factors such as altitude, humidity, degrees of shade, plant age and maturation (Seninde and Chambers, 2020). An interesting example is offered by the Colombian coffee growers Juan Valdez. To highlight the diversity of geographic regions within the country and the differences in taste, the coffee growers launched their *Café deorigen* series branding coffee produced in the Colombian regions of Huila, Cauca, Antioquia, Santander, Nariño, among others (Juan Valdez, 2020). This series has allowed coffee growers to market a particular regionally specific taste, for example by offering an extra strong high-altitude variety from the highest Coastal mountain range of Sierra Nevada de Santa Marta, at the Caribbean Coast of Colombia.

## Synthesizing the taste of coffee

What does synthetic coffee mean? A literature search for synthetic coffee in scientific literature databases reveals a great interest in coffee, which can be broadly categorized in four groups: health effects of coffee, molecular composition of coffee, the role of coffee waste in the bioeconomy, and farming practices of coffee. Interestingly, we could not find a paper describing the phenomenon of bio-based synthetic coffee.

In a review of sustainable production of food flavours (Bel-Rhlid *et al.*, 2018), authors point to different techniques that can help reproducing coffee aroma, with enzymes, or fermentation. Interestingly, the very process of fermentation on coffee beans also seems to be a topic of research and an ambition for standardisation in order to maintain a more predictable flavour that can be related to a particular brand (Elhalis *et al.*, 2021; Lee *et al.*, 2015). What makes coffee flavour has therefore been a long endeavour, a paper from 2014 lists 72 compounds that play a role in Arabica coffee flavour with the ability to link sensory perception to composition (Sunarharum *et al.*, 2014). One flavour compound, alkyl pyrazines for instance, seems to be important to recreating flavour in many food items, including coffee (Mortzfeld *et al.*, 2020). In this review, they highlight bio-based synthesis options, in other words, synthetic options from microorganisms. Also, the articles cited all refer to the complexity of identifying coffee taste, precisely due to the complex process from farming, to fermenting, to roasting described in the previous section. So, while we did not find one paper that indicates how to make synthetic coffee, there is certainly knowledge and technology available to know how to produce it.

If we return to the start-up, Atomo Coffee, that sparked our interest in the matter of investigating the ethical implication of claiming benefits on the taste of coffee, the start-up explains on their website that they reverse engineer the taste of coffee, sourcing the needed components from waste streams (Atomo Coffee, n.d.). They call it up-cycling from plants and describe their proprietary technology to recreate a coffee bean that is then roasted and brewed. As described in the previous paragraph, they tap into the chemical properties of coffee to recreate aroma, body, colour, taste and caffeine. In 2020, Atomo raised 9 million USD in addition to a kick-starter and earlier venture seed funding (Brown, 2020). In a nutshell, Atomo Coffee is one of the first, and maybe many more to come to make coffee, without coffee.

## Property regimes and the taste of coffee

Establishing what property regime best fits 'owning the taste of coffee' is not a descriptive endeavour. Indeed, property regimes are inherently socially constructed systems of ownership to help manage resources (Thompson, 2007). The question here should be specified to which property regime best captures the connection between synthetic coffee, and farmed coffee beans? We consider each with a focus on conservation of agrobiodiversity of coffee beans, for the advantages it provides as highlighted previously. In Timmermann and Robaey (2016) we present five alternatives of property regimes: no property, commons, private goods, state sovereignty, and common heritage. In this section, we explore each for the taste of coffee produced via bio-based means (what we call synthetic in this paper) and their implications for agrobiodiversity conservation in a *long-term* perspective.

### *No property*

In the case that there are no property rights that can be established on the taste of coffee and its varieties, this means that the labour and ecological contributions of coffee farmers that are currently rewarded through selling coffee beans, might find themselves replaced by other ways of producing coffee. Should this replacement occupy a large portion of the coffee market in the future, there would be no incentive left to develop and conserve varieties of coffee *in situ*.

### *Commons*

In the case that the taste of coffee is considered as a commons, this means that there would be some norms developed to regulate the behaviour of stakeholders involved in coffee production. Coffee taste as a commons could organize around different varieties of coffee beans, and include other ways of producing coffee in the governance of a particular taste. A commons conception for the taste of coffee would favour a reciprocal relationship between stakeholder benefiting from the taste of coffee, and would likely create strong incentives for at least the conservation of popular coffee varieties. A potential risk would be that such a property regime might not encourage the development of new varieties, unless explicitly part of the governance norms for the taste of coffee as a commons.

### *Private goods*

In the case that the taste of coffee should be considered a private good, this would not be primarily linked to the way coffee is produced, via farming, or via bio-based means, or chemical synthesis, as the ultimate profile of a taste could be the object of property claims. At the moment, there are already patent claims on enhanced flavour profiles to be found. If a flavour profile would result of a specific process only, then a utility or process patent could also play a role. Either of these cases offer little incentives to preserve and develop agrobiodiversity in coffee in the long term, essentially pitting modes of production against each other in a market where the most efficient and affordable product will dominate. In addition, this might move farmed coffee to the realm of luxury goods, and therefore represent a much smaller portion of the market for this beverage.

#### *State sovereignty*

State sovereignty over the taste of coffee presents an interesting opportunity that is fraught with pitfalls (for full discussion see Timmermann and Robaey, 2016). A notion of property linked to the state could strengthen the geographical link to the taste of coffee. Coffee is consumed worldwide but only produced in the Global South at the moment. With the advent of new technologies to recreate the taste of coffee, this property regime could strengthen the requirement for conservation of agrobiodiversity. However, major pitfalls include the lack of political continuity in sovereign states, as well as the unwillingness of industrialized countries to fairly share in conservation costs where agricultural genetic resources are already stored in seed banks. State sovereignty therefore offers little incentives for preserving and developing coffee agrobiodiversity.

#### *Common heritage*

The notion of common heritage of humankind was adapted to manage crop genetic resources in the 1980s. The concept never gained ground due to existing global inequalities that would render implementing the principle nearly impossible. What this principle does underline is the need of a recognition of global efforts in creating genetic diversity. Considering the taste of coffee as a common heritage of humankind, without going into the details of how to implement the different demands of the principle, calls for recognizing that the taste of coffee, and its varieties, comes from the agrobiodiversity of coffee, as well as the multiple fermentation and roasting practices. What this concept offers to the taste of coffee is the opportunity to keep it contextualised, independently of its method of production.

## Moving forward responsibly with synthetic coffee

In this paper, we take a critical look at developments in the bioeconomy for food production, specifically looking at coffee production through cellular agriculture, or through bio-based processes. We ask, who owns the taste of coffee and why does this discussion matter when considering historical efforts of farmers in the Global South, that contribute not only to taste, but to creating carbon sinks, and agrobiodiversity. Building on previous research on how property regimes incentivize agrobiodiversity in terms of conservation and development of new varieties, we explore that these property regimes can mean for the taste of coffee.

Currently, the taste of coffee is under no property regime. Our analysis suggests that in the long-term, this could lead to market dominance of bio-based processes, where agrobiodiversity of coffee beans could come to be threatened.

As explained in the introduction, such techniques decontextualize coffee production. In order to recontextualize coffee production, we suggest framing it within the context of a property regime. Our analysis suggests that none of the existing property regimes are without pitfalls. However, having a commons governance that specifies conservation and agrobiodiversity goals could step up to the challenge. Likewise, a concept of common heritage of humankind can come to strengthen the link between coffee beans and the taste of coffee.

Our aim is not to stigmatize these new developments, but to reflect on how to move forward responsibly. There is a considerable amount of learning to be done in the bioeconomy (Asveld and Stemerding, 2017) and for doing responsible innovation (Van de Poel *et al.*, 2020). For specific concerns relating to agricultural innovations, in Robaey and Timmermann (2018) we present five key social challenges that need to be addressed: availability, accessibility, participation in science and governance, arbitration and long-term sustainability.

Taking into account these challenges, the need to learn, and the opportunities of property regimes such as commons and common heritage of humankind, ethicists would invite the cellular agriculture community to keep a close link to the geographical origins of coffee, in order to recontextualize it, continue technological development in a responsible manner, and recognize that synthetic coffee is inextricably linked to coffee beans. Without the historic work of coffee farmers, synthetic coffee would simply not have come to exist.

## References


Asveld, L. and Stemerding, D., 2017. Social learning in the bioeconomy: The Ecover case. In: I. van de Poel, L. Asveld and D.C. Mehos (eds), Experimentation Beyond the Laboratory: New Perspectives on Technology in Society. Routledge, Oxon, pp. 103-124.

Atomo Coffee, n.d. Our Tech | Atomo! Molecular Coffee. Available at: https://atomocoffee.com/our-tech.

Bel-Rhlid, R., Berger, R.G. and Blank, I., 2018. Bio-mediated generation of food flavors – towards sustainable flavor production inspired by nature. Trends in Food Science & Technology 78: 134-143.

Brown, N., 2020. Atomo Coffee Raises $9 Million More, Plans Seattle Production Facility. Available at:https://dailycoffeenews.com/2020/08/11/atomo-coffee-raises-9-million-more-plans-seattle-production-facility/.

Elhalis, H., Cox, J., Frank, D. and Zhao, J., 2021. Microbiological and biochemical performances of six yeast species as potential starter cultures for wet fermentation of coffee beans. LWT 137: 110430.

Ingenbleek, P.T. and Reinders, M.J., 2013. The development of a market for sustainable coffee in the Netherlands: Rethinking the contribution of fair trade. Journal of Business Ethics 113: 461-474.

Jha, S., Bacon, C.M., Philpott, S.M., Ernesto Mendez, V., Läderach, P. and Rice, R.A., 2014. Shade coffee: update on a disappearing refuge for biodiversity. BioScience 64: 416-428.

Jimenez-Soto, E., 2020. The political ecology of shaded coffee plantations: conservation narratives and the everyday-lived-experience of farmworkers. The Journal of peasant studies, https://doi.org/10.1080/03066150.2020.1713109

Juan Valdez, 2020. Café de origen. Available at: https://www.juanvaldezcafe.com/es-co/productos/nuestros-cafes/cafe-de-origen/.

Lee, L.W., Cheong, M.W., Curran, P., Yu, B. and Liu, S.Q., 2015. Coffee fermentation and flavor – an intricate and delicate relationship. Food chemistry 185: 182-191.

Mortzfeld, F.B., Hashem, C., Vranková, K., Winkler, M. and Rudroff, F., 2020. Pyrazines: Synthesis and industrial application of these valuable flavor and fragrance compounds. Biotechnology Journal 15: 2000064.

Robaey, Z. and Timmermann, C., 2018. Fair agricultural innovation for a changing climate. In: E.C. Gilson and S. Kenehan (eds), Food Justice, the Environment, and Climate Change. Rowman & Littlefield, Lanham, pp. 213-230.

Seninde, D.R. and Chambers, E., 2020. Coffee flavor: A review. Beverages 6: 44.

Sunarharum, W.B., Williams, D.J. and Smyth, H.E., 2014. Complexity of coffee flavor: A compositional and sensory perspective. Food Research International 62: 315-325.

Thompson, P.B., 2007. Conceptions of Property and the Biotechnology Debate, Food Biotechnology in Ethical Perspective. Springer, Dordrecht, pp. 233-259.

Timmermann, C. and Robaey, Z., 2016. Agrobiodiversity under different property regimes. Journal of Agricultural and Environmental Ethics 29: 285-303.

Topik, S., 2009. Coffee as a social drug. Cultural Critique: 81-106.

Van de Poel, I., Asveld, L., Flipse, S., Klaassen, P., Kwee, Z., Maia, M., Mantovani, E., Nathan, C., Porcari, A. and Yaghmaei, E., 2020. Learning to do responsible innovation in industry: six lessons. Journal of Responsible Innovation 7: 697-707.

# 11. Food security and cultural identity in agricultural adaptation – a trade-off?

*T. Carrillo** *and I. Wallimann-Helmer*
*Environmental Sciences and Humanities Institute, University of Fribourg, 1700 Fribourg, Switzerland; tanja.carrillo@unifr.ch*

## Abstract

This paper analyses the impact of climate change on the food security and cultural identities of indigenous peoples. The rising intensity and frequency of extreme weather events and changes in rainfall patterns are increasing the pressure on agricultural systems and food availability. Indigenous communities are particularly affected by these negative impacts because they often rely on subsistence farming for their livelihoods and maintain a deep connection with the environment through their cultural practices. We claim that a potential trade-off exists between food security and cultural identity, notably when it comes to climate adaptation strategies. Despite the clear need for agricultural adaptation to manage climate-related risks, enhance food security, and protect the livelihoods of communities, most adaptation strategies do not consider the importance of traditional foods to indigenous communities. We argue that this neglect entails an ethical conflict and suggest two related approaches to addressing the emerging injustices: First, for the planning and implementation of adaptation to be fair, indigenous peoples and their relationships to nature need to be recognized. Second, such recognition needs to translate into the fair procedural involvement of the communities concerned, so that they acquire the power to actively shape adaptation strategies, influence policy decisions, and thus protect their interests.

**Keywords:** climate change, indigenous communities, procedural involvement, recognition

## Introduction

Food security exists when 'all people, at all times, have physical and economic access to sufficient, safe, and nutritious food that meets their dietary needs and food preferences for an active and healthy life' (FAO, 2020). Originally, food security was used to describe whether a country had enough food available to meet the dietary energy requirements of its citizens, but the focus now lies at the individual and household levels. International agencies, governments and civil society organizations alike stress the importance of eradicating hunger and malnutrition worldwide to achieve the Sustainable Development Goal 2 of the UN Agenda 2030 (Blesh *et al.,* 2019).

Food security maintains a bidirectional relationship with climate change. More frequent extreme weather events such as floods, droughts, and storms and altered hydrological cycles and rainfall variance are increasing the pressure on agricultural systems and heavily affect food security (Carter *et al.,* 2018). Conversely, global food production needs to increase significantly to satisfy the food security demands of a growing population by 2050 (FAO, 2020). However, conventional agriculture is one of the most significant sources of greenhouse gas emissions, and its intensification is likely to worsen global warming (Clark and Tilman, 2017). Thus, maintaining global food security without exacerbating climate change requires that agricultural practices are adapted and transformed.

We argue not only that climate change jeopardizes food security but that essential agricultural adaptation threatens the cultural identity of indigenous communities. We define cultural identity as the shared traits of a group of people, such as religion, language, food, rituals, art, and music. For indigenous peoples,


EurSafe 2021 – DOI 10.3920/978-90-8686-915-2_11, © Wageningen Academic Publishers 2021

cultural identity is often closely connected to traditional forms of agriculture: many communities depend on subsistence farming to sustain their livelihood, and for millennia, land and natural resources have provided the foundations of their cultures and identities (Whyte and Cuomo, 2017). However, adaptation strategies aimed at increasing overall food security often alter traditional livelihoods and implement new methods of farming that potentially lead to a loss of traditional knowledge and critically change the relationships of communities to their foods and environment (Carter *et al.,* 2018).

This paper discusses the impact of climate change on the food security and cultural identities of indigenous peoples and shows why an ethical trade-off can arise between the two. It suggests two pathways that may overcome this struggle. First, we claim that justice as recognition opens up a particularly large adaptations space that allows the cultural interests of indigenous peoples to be considered. Second, we argue that such consideration needs to translate into effective practices of political participation, providing communities with real opportunities to influence adaptation decisions that protect both their food security and their cultural identity.

## The impact of climate change on indigenous food systems

Rising food insecurity threatens the entire global population. However, the harvest of local natural resources for human consumption is particularly important for indigenous peoples. Many of the approximately 476 million indigenous people worldwide pursue a combination of subsistence and commercial harvesting activities to maintain their livelihoods (World Bank, 2020). Furthermore, traditional agriculture is intricately connected to the cultural, spiritual, and emotional health of tribal communities. Traditional foods are used in storytelling, rituals, and ceremonies. For example, the Wabanaki people, a Native American tribe, rely on berry plants as food medicine (Lynn *et al.,* 2013). Depending on the season, the Wabanaki's purpose, and the age of the plant, traditional knowledge guides which parts of the plant should be collected and how they should be prepared. The Strawberry Ceremony, for example, is of great importance for Wabanaki women because it initiates the transition from adolescence to womanhood. However, as climate change alters environmental conditions and harvest times, such traditional knowledge loses its usefulness, and cultural practices cease to be viable (Michelle, 2012).

Empirical evidence shows that indigenous forms of food production most often are sustainable and low carbon, conserve biodiversity, and produce nutritious food for the communities (IIED, 2020). Additionally, they tend to be climate resilient, because indigenous peoples are experienced in adapting to varying environmental conditions. However, the negative impacts of climate change are progressively decreasing these peoples' adaptive capacity (Maldonado *et al.,* 2016). For the Wabanaki, climate change affects the diversity, quality, and availability of their berries and thus threatens not only the tribe's health but also its cultural foundation (Michelle, 2012). Similarly, climate change significantly impacts the temperature and availability of water. In many communities, water is considered a sacred good and plays an important role in cultural identity. However, the lack of water is also one of the main reasons for the declining ability of indigenous peoples to cultivate traditional crops. Consequently, climate change threatens the food security of many communities and undermines their deep identification with traditional crops (Cozzetto *et al.,* 2013).

These examples illustrate that the negative impacts of climate change endanger both indigenous communities' access to nutritious and culturally appropriate food and the cultural practices that define their identities. However, large-scale climate adaptation strategies tend to prioritize overall food security over traditional agriculture. As non-forest land areas that are not used for agricultural purposes are often considered as unproductive by non-indigenous cultures, adaptation strategies repeatedly do not acknowledge their importance as sources of tribally valued bulbs and roots (Lynn *et al.,* 2013).

Consequently, climate adaptation itself sometimes represents an injustice (Heyward, 2017). Hence, we claim that realizing climate adaptation under fair conditions requires that a wide range of interests and both goals of ensuring food security and protecting cultural identity are integrated in governance structures. In the following section, we present two approaches to achieving such an integration.

## Recognition and procedural involvement – a possible way forward?

As the negative impacts of climate change begin to materialize, researchers, policy makers, and civil society adopt frameworks of climate justice to assess the relationship between the effects of global warming and conceptions of fairness. However, the majority of climate ethics researchers focus on questions of distributive justice (Gardiner *et al.,* 2010; Wallimann-Helmer, 2019a) or follow a rights-based approach (Caney, 2012; Vanderheiden, 2017). By contrast, this paper focuses on two aspects of environmental justice that have been somewhat neglected in the climate ethics literature but which we claim can provide potential pathways to achieving justice for indigenous communities in climate adaptation: justice as recognition and procedural justice.

### *Pathway 1: recognition*

First, we want to discuss how lack of recognition of individuals, communities, and circumstances often leads to fundamental injustices (Schlosberg, 2012). According to Young (1990), it is institutional processes and contexts and social status that determine the distribution of resources and power. Consequently, she claims that distributive injustices often arise from a lack of recognition and that political exclusion and cultural disrespect are sources of inequality and vulnerability. Similarly, Fraser (1997) also sees a strong correlation between the lack of recognition and maldistribution. She distinguishes three forms of misrecognition: The first stems from the broader system of cultural domination. The second concerns nonrecognition, understood as being ignored or rendered invisible. The third type involves open disrespect by regular stereotyping and maligning in cultural representations.

Indigenous peoples often experience all three forms of misrecognition. Communities regularly argue that they perceive the stereotyping and misrepresentation of their cultures as a form of abuse that is directly linked to distributive injustices (Whyte, 2013). It is striking that, although the link between environmental threats and cultural identity is recognized in many UN declarations, such as the Universal Declaration on the Rights of Indigenous people, the UNFCCC process is still dominated by Western knowledge, and indigenous cultures and practices are largely ignored in climate negotiations (Etchart, 2017). Furthermore, as explained above, traditional foods and their production are often not considered relevant to climate adaptation strategies. This shows a lack of recognition not only of these peoples and their cultures but also of these communities' reliance on the natural environment to fulfil their basic needs.

Consequently, planning and implementing climate adaptation strategies need to recognize how they affect the cultural practices, identities, and everyday life of indigenous peoples. According to Fraser (1997), such recognition calls for both deconstructive and affirmative practices, which include dismantling the norms and social practices that neglect traditional relationships with the environment. Furthermore, attention must be given to the direct experiences of vulnerable communities. This requires integrating indigenous people into the political processes of adaptation planning.

### *Pathway 2: procedural involvement*

In the previous section, we argued that adaptation planning must recognize the cultural identities of indigenous communities and their special relationship with the environment. Such recognition must also translate into practices of political participation that give these communities control over their destinies (Adger, 2016). In the literature on environmental ethics, procedural justice generally refers to the conditions of fairness in processes of collective decision-making that ensure that marginalized social groups can influence decisions about, for example, climate adaptation (Holland, 2017; Wallimann-Helmer, 2019b).

However, although the importance of addressing the procedural fairness of adaptation planning is increasingly recognized, its implementation is still rather vague. In practice, an understanding of procedural justice often amounts to financial assistance and capacity-building, which is unlikely to provide the affected communities with real power to shape any decisions (Maldonado *et al.,* 2016). Likewise, ethicists have yet to theorize much about what a procedurally just approach to climate adaptation should entail. We argue that rendering indigenous communities politically capable and treating them as equals in the decision-making processes means engaging them in the discussions and the mapping of their own vulnerabilities and in designing adaptation strategies that protect their interests in ways they believe they must be protected (Wallimann-Helmer, 2016).

We suggest that this means integrating indigenous communities and their traditional knowledge into climate change research in order to shape the common understanding of the negative impacts of climate change and design culturally appropriate adaptation strategies (Huggel *et al.,* 2016; Williams and Hardison, 2013). Moreover, indigenous peoples may enter government-to-government relationships within a state to increase their political power in local adaptation management, maintain access to environmental areas to harvest traditional foods, and lease and buy agricultural land to sustain their food security (Whyte, 2013). For such interactions between indigenous peoples, research communities, and policy makers to be fair, they must comply with the demanding conditions of procedural justice, such as transparency, accountability, and free, prior, and informed consent (Golay, 2018).

## Conclusion

In this paper, we have discussed the consequences of ongoing climate change for the food security and cultural identities of indigenous communities. Climate change is a global challenge that particularly threatens the cultural practices and traditional knowledge of indigenous peoples. We discussed how altered environmental conditions and a decline in berry plants affect the ceremonies of the Wabanaki people and how the decreasing availability of water impedes communities from cultivating traditional crops. Besides the evident challenges for both food security and cultural identity, we have argued that there is a potential ethical conflict between the two, notably when agricultural adaptation strategies prioritize overall food security over the cultural aspects of traditional food. To overcome this trade-off and prevent injustices, we suggest two possible, interconnected approaches: First, indigenous people and their cultural identities must be recognized in adaptation planning by means of practices that deconstruct current social norms and which foregrounds the direct experience of vulnerable communities. Second, such recognition needs to result in the procedural involvement of indigenous communities to give them the power to actively shape adaptation strategies and participate in political decisions. Future work, both theoretical and in practical, should thus pay more attention to recognition and procedural justice and strive to define successful ways of integrating indigenous communities in the planning and implementation of climate adaptation strategies.

## References

Adger, W.N. (2016). Place, well-being, and fairness shape priorities for adaptation to climate change. Global Environmental Change. 38: A1-A3.

Blesh, J., Hoey, L., Jones A. D., Friedmann, H. and Perfecto I. (2019). Development pathways toward 'zero hunger'. World Development 118: 1-14.

Caney, S. (2012). Global Justice, Climate Change, and Human Rights, in: Hicks, D.A. and Williamson, T. (eds), Leadership and Global Justice. Palgrave Macmillan, Basingstoke, pp. 91-111.

Carter, R., Ferdinand, T. and Chan, C. (2018). Transforming agriculture for climate resilience: A framework for systemic change. World Resources Institute, Washington D.C. Available at: https://www.wri.org/publication/transforming-agriculture-climate-resilience-framework-systemic-change

Clark, M. and Tilman, D. (2017). Comparative analysis of environmental impacts of agricultural production systems, agricultural input efficiency, and food choice. Environmental Research Letters 12: 064016.

Cozzetto, K., Chief, K., Dittmer, K., Brubaker, M., Gough, R., Souza, K., Ettawageshik, F., Wotkyns, S., Opitz-Stapleton, S., Duren, S., Chavan, P., Koppel Maldonado, J., Pandya, R.E. and Colombi, B.J. (2013). Climate change impacts on the water resources of American Indians and Alaska Natives in the U.S. Climatic Change 120: 569-584.

Etchart, L. (2017). The role of indigenous peoples in combating climate change. Palgrave Communications 3: 17085.

Food and Agriculture Organization of the United Nations (2020). The State of Food Security and Nutrition in the World 2020. Available at: http://www.fao.org/3/ca9692en/online/ca9692en.html.

Fraser, N. (1997). Justice Interruptus: Critical Reflections on the 'Postsocialist' Condition. Routledge, New York, 241 pp.

Gardiner, S.M., Caney, S., Jamieson, D. and Shue, H. (2010). Climate Ethics: Essential Readings. Oxford University Press, Oxford, New York, 368 pp.

Golay, C. (2018). The rights to food sovereignty and to free, prior and informed consent. The Geneva Academy, Geneva. Available at: https://www.geneva-academy.ch/research/publications/detail/355-the-rights-to-food-sovereignty-and-to-free-prior-and-informed-consent.

Heyward, C. (2017). Ethics and Climate Adaptation, in: Gardiner, S.M. and Thompson, A. (eds), The Oxford Handbook of Environmental Ethics. Oxford University Press, Oxford, pp. 474-486.

Holland, B. (2017). Procedural justice in local climate adaptation: political capabilities and transformational change. Environmental Politics 26: 391-412.

Huggel, C., Wallimann-Helmer I. and Cramer W. (2016). Reconciling justice and attribution research to advance climate policy. Nature Climate Change 6: 901-908

International Institute for Environment and Development (2020). Indigenous Peoples' food systems hold the key to feeding humanity Available at: https://www.iied.org/indigenous-peoples-food-systems-hold-key-feeding-humanity.

Lynn, K., Daigle, J., Hoffman, J., Lake, F., Michelle, N., Ranco, D., Viles, C., Voggesser, G. and Williams, P. (2013). The impacts of climate change on tribal traditional foods. Climatic Change 120: 545-556.

Maldonado, J., Bull Bennett, T.M., Chief, K., Cochran, P., Cozzetto, K., Gough, B., Redsteer, M.H., Lynn, K., Maynard, N., Voggesser, G., Jacobs, K., Moser and S., Buizer, J. (2016). Engagement with indigenous peoples and honoring traditional knowledge systems. Climatic Change 135: 111-126.

Michelle, N. (2012). Uses of plant food-medicines in the Wabanaki bioregions of the Northeast; a cultural assessment of berry harvesting practices and customs. University of Maine, Orono.

Schlosberg, D. (2012). Climate Justice and Capabilities: A Framework for Adaptation Policy. Ethics and International Affairs 26: 445-461.

Vanderheiden, S. (2017). Human Rights and the Environment, in: Gardiner, S.M. and Thompson, A. (eds), The Oxford Handbook of Environmental Ethics. Oxford University Press, Oxford, pp. 301-310.

Wallimann-Helmer, I. (2016). Differentiating responsibilities for climate change adaptation. Archiv für Rechts-und Sozialphilosophie 149: 119-132.

Wallimann-Helmer, I. (2019a). Justice in managing global climate change, in: Letcher, T.M. (ed.) Managing Global Warming: An Interface of Technology and Human Issues. Academic Press, London, San Diego, Cambridge, pp. 751-768.

Wallimann-Helmer, I. (2019b). The ethics of waste policy, in: Lever, A., Poama, A. (eds), The Routledge Handbook of Ethics and Public Policy. Taylor & Francis, New York, pp. 501-512.
Whyte, K.P. (2013). Justice forward: Tribes, climate adaptation and responsibility. Climatic Change 120: 517-530.
Whyte, K.P. and Cuomo, C. (2017). Ethics of Caring in Environmental Ethics: Indigenous and Feminist Philosophies, in: Gardiner, S.M. and Thompson, A. (eds), The Oxford Handbook of Environmental Ethics. Oxford University Press, Oxford, pp. 234-247.
Williams, T., Hardison, P. (2013). Culture, law, risk and governance: Contexts of traditional knowledge in climate change adaptation. Climatic Change 120: 531-544.
Young, I.M. (1990). Justice and the Politics of Difference. Routledge, London. 304 pp.
World Bank (2020). Indigenous people. Available at: https://www.worldbank.org/en/topic/indigenouspeoples.

# 12. Inclusive biobased value chains: building on local capabilities

*L. Asveld[1]*, Z.H. Robaey[2] and S. Francke[3]*
*[1]Delft University of Technology, Department of Biotechnology, Van der Maasweg 9, 2629 HZ Delft, the Netherlands; [2]Wageningen University, Department of Social Sciences, Postbus 8130, 6700 EW Wageningen, the Netherlands; [3]DSM, Baden-Würtemberg, Germany; l.asveld@tudelft.nl*

## Abstract

Uncertainties about how to achieve sustainable and reliable biobased value chains can be remedied by inclusion of local biomass producers. Such inclusion implies that the knowledge, values, interests and skills of these local producers are integrated into the set-up, design, development and associated distribution of risk and benefits of the specific value chain. To make sure that this inclusion is both fair and effective, capabilities of relevant actors need to be taken into account, i.e. the capabilities of biomass producers and of companies and local governments. Building on these capabilities can lead to context sensitive design of biorefineries that allow for true inclusion of local stakeholders. We support this claim with reference to the case of sugarcane production in Jamaica.

**Keywords:** inclusion, biorefinery, Jamaica, context sensitive design

## Introduction

Biomass such as crops and animal waste is increasingly used as the primary resource for products like bioplastics and biofuels, which provide an alternative to fossil-based energy and materials. Biobased applications hold the promise of reducing climate change and the plastic soup, while at the same time providing an incentive for socio-economic development and sustainable agriculture.

However, there are many uncertainties about how to organise biobased value chains in a sustainable and economically fair way (Asveld and Stemerding, 2017), especially because these value chains usually have a global set-up, while the production of biomass takes place in widely diverging contexts (Meckenstock *et al.*, 2016). Including local stakeholders in the set-up or design of biobased value chains – taking into account their needs, values, wishes and knowledge – can be a way to deal with these uncertainties (Rist *et al.*, 2007); local producers often have valuable knowledge about land management, their natural environment and the associated biomass (Sumane *et al.*, 2018). This knowledge is indispensable for the sustainable supply of biomass and reducing economic inequalities (IPCC, 2019). Including local biomass producers can thus be expected to have both epistemic and moral benefits. Also, approaches to creating sustainable agriculture that focus mainly on technological aspects while neglecting local stakeholders are moreover often unsuccessful from a business perspective (Hounkonnou *et al.*, 2012).

However biomass producers are often located in developing countries where conditions for inclusion can be suboptimal (Postal *et al.*, 2020). Additionally, cultural values and practices between producers and other actors in the biobased value chain may vary widely. Although inclusion can be to the benefit of both producers of biomass as well as other actors in the value chain, these circumstances complicate the effort of building inclusive value chains (*ibid.*).

In this paper we want to address the question: how to achieve meaningful inclusion in global biobased value chains?


EurSafe 2021 – DOI 10.3920/978-90-8686-915-2_12, © Wageningen Academic Publishers 2021

We argue that for meaningful inclusion the actors that are being included have to be able to 'choose' to be included, otherwise they are not being included, but co-opted. We see the Capability Approach as a useful framework to identify whether actors are able to choose to be include, since the Capability Approach has freedoms or opportunities at its core (Sen, 1990; Robeyns, 2017).

Furthermore we argue that such inclusion can only be achieved when the actors setting-up the value chain take the responsibility for the inclusion of other actors. However we will also argue that such responsibility should be assigned in way that is feasible and respects the relevant limits that a given context bestows on those enacting their responsibilities (cf Van de Poel and Sand, 2018). We will illustrate our arguments a case-studies in Jamaica, where actors producing biomass are relatively poor and have relatively little resources.

## Inclusion

Although the need for inclusive biobased value chains may be clear, what inclusion implies exactly is not yet clear. What does it mean to include someone? For inclusion to have a positive effect on biobased value chains, we posit that it needs to impact the processes that surround the building of these value chains. These processes encompass the invention, design, development and distribution of benefits of biobased products (Heeks *et al.*, 2014). Inclusion implies that the values, knowledge, interests and skills of biomass producers are taken into account in these processes.

In a context of economic, social and institutional inequalities, such as with global value chains, the issue of capabilities is crucial. In the context of innovation, distributive justice cannot be achieved by merely giving marginalised individuals access to a specific innovation or even to the process leading to that innovation (*ibid.*). Distributive justice also requires that inequalities are identified at a more structural level. To achieve inclusive value chains, the question of responsibility is of paramount importance. Who should make sure that a value chain is indeed inclusive?

Many of the existing global inequalities can be captured by the notion of capabilities and the lack thereof. Actors in Western countries typically have more capabilities to shape and market innovations compared to actors in developing countries. To develop truly inclusive global biobased value chains, capabilities should therefore be taken into account. We want to argue that actors with more capabilities to influence the set-up of an inclusive value chain should also take more responsibility for this. In this paper we aim to apply a rudimentary framework for inclusive biobased value chains that takes capabilities into account.

## Capabilities as opportunities

Distributive justice and agency are central to the capability approach. To conceptualise these values, the approach attaches central importance to individual human capabilities, or a person's real 'freedoms or opportunities' to do and be what (s)he has reason to value (Robeyns, 2017; Sen, 1990). Examples are capabilities to get educated, live a healthy life and grow one's own food. Realisation of such capabilities depends on a range of interconnected inputs, such as institutions, public goods, social practices, resources and skills (Robeyns, 2017). When people exercise their agency, they make choices on which of their capabilities to turn into so-called functionings (e.g. actually becoming educated, living healthfully, growing food).

## Capabilities for inclusion

Inclusion can only be meaningful if those who are to be included have a real choice. Unwanted inclusion is not inclusion – it is co-option (Kleine *et al.*, 2012). Inclusion thus requires the capability to be included to begin with (Simpson and Basta, 2018). This capability depends on actors having specific skills and access to resources, reliable infrastructure and education. For example, the capability of handling sophisticated technologies may be a prerequisite for the capability to be included in an advanced biorefinery.

## Responsibility for inclusion

Can private companies developing a biobased value chain be expected to improve the opportunities of the producers of the relevant biomass? The societal expectations concerning the responsibilities of companies have increased in recent years, which is shown by the increased attention for corporate social responsibility (CSR).

Responsibility can be divided in two types: forward looking and backward looking responsibility (Van de Poel and Sand, 2018). Backward looking responsibility concerns questions about liability and accountability: if things go awry, who is to blame? Forward looking responsibility concerns responsibilities to see to it that something comes about. For instance an individual in a company might be responsible for seeing to it that the 10 principles for Corporate Sustainability are implemented within the company.

With regard to socio-economic issues, the ten principles of the UN Global Compact state that companies should respect human rights and avoid forced labour and discrimination. Such responsibilities can be interpreted as falling under backward looking responsibility. They stipulate that companies comply with specific governance regimes (such as the UN Charter of Human Rights). The principles do not require an active care for the socio-economic circumstances of individuals employed by them, which makes sense since such circumstances are usually considered to be the responsibility of governments.

However in the case of inclusive value chains we can expect companies to go beyond backward looking responsibilities and actively fulfil their forward looking responsibilities. Van de Poel and Sand (2018) have argued that assigning forward looking responsibilities to companies is unfair and ineffective because of the many uncertainties surrounding innovations. However in the case of value chains, it is likely that companies are able to exercise at least some control over the set-up of the value chains, as exemplified by companies such as Unilever and initiatives such as Fair Trade and the Dutch Initiative Sustainable Trade. Of course this control will differ between varying contexts. Also the capabilities of relevant companies will play a role. A big multinational such as Unilever will have more resources to influence the set-up of a value chain as compared to a small start-up. So in order to fairly and effectively assign responsibilities, this context as well as the capabilities of the company involved will have to be taken into account.

Another argument why it is fair to assign responsibility for inclusive value chains to biobased companies is that such inclusive value chains are in their direct interest. Inclusion can reduce uncertainties about sustainability as well as about availability of biomass. Additionally, the creation of inclusive value chain can only be done effectively when companies involved take an active stance in realising this. Other actors such as local governments might not have the knowledge or the resources to achieve such value chains, and such involvement might even be undesirable because it doesn't fit with free market principles. The question is then how commercial companies might go about achieving inclusive biobased value chains and what the extent of this responsibility is exactly.

## The case of Jamaica

To explicate how the concepts of capabilities and responsibility for capabilities might be applied, we refer to a case-study on Jamaica. We visited Jamaica in January 2018 as part of a research project called

IBIS: Inclusive biobased innovation: Securing sustainability and supply through farmers' involvement. We conducted about 10 site-visits and 15 interviews. Interviewees included government representatives, farmer representatives, agricultural extension officers and local scientists. For the purpose of this paper we will mostly rely on a rather general description of the case, because it serves mainly to bring out the challenge of building inclusive value chains and to shed light on what this implies for using the capability approach to conceptualize inclusion and how to assign responsibilities for this.

## Local context

Jamaica is an island in the Caribbean Sea, with a tropical climate. Jamaica is categorised as a upper middle income country and is ranked in the 'high human development' category of the UN's 2016 Human Development Index (UNDP, n.d.). Jamaica ranks 43 out of 198 on the corruption index of transparency.org in 2020. Jamaica has a large sugarcane industry. It is one of the main crops grown on the island. However many farmers are struggling because they own only a very small piece of land. On top of that, the preferential pricing agreement with the EU for sugar from sugarcane ended (Burrell, 2016).

Within this context, the Jamaican Sugar Authority along with co-operatives of sugar cane farmers were looking for a way to diversify their income. This caught the attention of some commercial actors looking for new opportunities in the bioeconomy. Although none of them had very concrete plans, these parties supported us with access to their contacts. We, as a research team, tried to find the conditions under which a sugar cane biobased enterprise might be inclusive.

## Relevant local capabilities

We found that the farming community is aging rapidly with little access to funds to invest in new technologies or even in new plants. Farming is heavily dependent on government funding, while at the same the farmers distrust the government. The community is desperate for new opportunities but at the same time is conservative about innovations. There is a strong tendency to rely on proven results shown in the field. On top of that, the available scientific research did not match with the actual needs of the farmers in the field. A lot of sugar is imported, because it is cheaper and has more reliable quality. Contextually, infrastructure is not very reliable. The roads are not taken care of very well. Climate change is posing a lot of challenges. Harvests are failing because the weather conditions change very quickly. The climate has become too humid to have a good sugar cane yield.

This implies that for local capabilities for inclusion are limited. Farmers have very limited opportunities. They do not have the capability to access and apply new technologies. They do not have the capability to invest in new farming practices. They also do not have the capability to transport their sugarcane over long distances across the island. Additionally they do not have the capability to adapt to the new climate conditions or the capability to compete with imported sugar.

The capabilities the farmers do possess are the capability to organize, the capability to grow sugar cane and, for some of them, the capability to provide sugar cane to one of the still functioning sugar mills on the island.

## Responsibilities of commercial actors

What does this context imply for the responsibility of commercial actors for the inclusion of sugar cane farmers in a prospective biobased value chain? In the complex situation of Jamaica, it appears that their main responsibility is to build on locally available capabilities. That means that we consider the option of investing in local capabilities to be outside of the scope of the responsibilities of commercial actors. Investment in local capabilities could include investment in infrastructure, in agricultural extension services or local microloan schemes. We consider such investments as outside of the scope of the responsibilities of commercial actors because it is unclear whether the actors would indeed benefit from such investments in the long run. Capabilities are opportunities and as such they provide freedom of choice. Local beneficiaries might use their freedom for other purposes than a sustainable biobased value chain.

However we do consider it the responsibility of commercial actors to acknowledge local capabilities and take them into account when developing their value chain. This implies a biorefinery that uses low tech that can easily be operated and maintained by local, low-schooled people. For instance, the use of bagasse for local energy generation could fit very well in this concept because this is done by widely available technologies that are well known in Jamaica, and it can provide huge environmental benefits. The biorefinery should use sugar-cane instead of another crop that might be more advantageous in terms of yield or sustainability impact but that might not be accepted by the rather conservative farming community. Also the end products of the biorefinery should be tailored for local consumption. Producing for export will be too demanding at the initial stage of a biorefinery. This is an ambition that might be strived for at later stages.

## Conclusion

Inclusive value chains require a careful inventory of local capabilities. Such capabilities can be turned into design requirements for inclusive biorefineries and the associated value chains. Commercial actors can be held responsible to take local capabilities into account, because this will benefit these actors in the end, as well as the local population. Further investment in local capabilities cannot be expected from commercial actors.

## References

Asveld, L. & Stemerding, D. (2017). *Social learning in the bioeconomy: The Ecover case.* In Van de Poel, I., Asveld, L. & Mehos, D.C. (eds). (2017). New Perspectives on Technology in Society: Experimentation Beyond the Laboratory. London: Routledge.

Burrell D. (2016) The Decline of Preferential Markets and the Sugar Industry: A Case Study of Trade Liberalization in Central Jamaica. In: Beckford C., Rhiney K. (eds) Globalization, Agriculture and Food in the Caribbean. Palgrave Macmillan, London. https://doi-org.tudelft.idm.oclc.org/10.1057/978-1-137-53837-6_5

Heeks, R., Foster, C. & Nugroho, Y. (2014). New models of inclusive innovation for development. *Innovation and Development, 4*(2), 175-185. https://doi.org/10.1080/2157930X.2014.928982

Kleine, D., Light, A. & Montero, M.J. (2012). Signifiers of the life we value? – considering human development, technologies and Fair Trade from the perspective of the capabilities approach. *Information Technology for Development, 18*(1), 42-60.

Meckenstock, J., Barbosa-Póvoa, A.P. & Carvalho, A. (2016). The wicked character of sustainable supply chain management: evidence from sustainability reports. *Business strategy and the environment, 25*(7), 449-477.

Postal, A.M., Benatti, G., Palmeros Parada, M., Asveld, L., Osseweijer, P. & Da Silveira, J.M.F. (2020). The Role of Participation in the Responsible Innovation Framework for Biofuels Projects: Can It Be Assessed?. *Sustainability, 12*(24), 10581.

Rist, S., Chidambaranathan, M., Escobar, C., Wiesmann, U. & Zimmermann, A. (2007). Moving from sustainable management to sustainable governance of natural resources: The role of social learning processes in rural India, Bolivia and Mali. *Journal of rural studies, 23*(1), 23-37.

Robeyns, I. (2017). *Wellbeing, freedom and social justice: The capability approach re-examined.* Open Book Publishers.

Sen, A. (1990). Development as capability expansion. *The community development reader*, 41-58.

Simpson, N.P. & Basta, C. (2018). Sufficiently capable for effective participation in environmental impact assessment?. *Environmental Impact Assessment Review*, *70*, 57-70.

Šūmane, S., Kunda, I., Knickel, K., Strauss, A., Tisenkopfs, T., des Ios Rios, I., ... & Ashkenazy, A. (2018). Local and farmers' knowledge matters! How integrating informal and formal knowledge enhances sustainable and resilient agriculture. *Journal of Rural Studies*, *59*, 232-241.

Van de Poel, I. & Sand, M. (2018). Varieties of responsibility: two problems of responsible innovation. *Synthese*, 1-19.

# 13. The loss of food sovereignty in synthetic meat transition: a critique from eco-republican justice

*C. Moyano Fernández*
*Department of Philosophy, Institute of Environmental Science and Technology (ICTA), Universitat Autònoma de Barcelona (UAB), 08193 Bellaterra, Barcelona, Spain; cristian.moyfe@gmail.com*

## Abstract

In a socio-ecological context of accelerated climate change, environmental pollution, ecosystems depletion, loss of biodiversity and growing undernutrition in an increasing number of countries, our societies are at a crossroads that requires food systems conversion. Synthetic meat has been achieving popularity during the last years because it is presented as one of the fairest food productions for the forthcoming decades. It could be an apparent solution for the ecological and food crisis which tendency is to worsen throughout the 21st century. It promises to ensure food security while environmental impacts and animal suffering are dramatically reduced. Hence, it seems reasonable that synthetic meat transition is defended from an utilitarian theory of justice. However, in such transition there are still moral deep-rooted issues in the food systems not addressed primarily by lab-grown meat, like our relationship with animals, the land-use change, the food sovereignty, or the interdependencies between fauna and flora. This paper aims to explore the challenges of synthetic meat in societies guided by a non-utilitarian justice framework. In particular, it will be focused on the food sovereignty loss due to the top down practices developed in high-tech labs with artificial interventionism. The production of cultivated meat in order to satisfy the growing protein demands is a supply process which prioritizes food security entitlements. Nevertheless, some other important values are missing from industrialized food chains and hubristic utilitarianism. The recognition of nonhuman nature from interspecific ethics or the caring for resilience and integrity from ecological ethics imply essential considerations in the protection of food sovereignty entitlements. In sum, the anthropocentric and consequentialist approach of synthetic meat is discussed in this contribution providing some arguments regarding an ecological justice approach based on a republican perspective.

**Keywords:** synthetic meat, utilitarianism, ecological justice, republicanism, food sovereignty

## Introduction

Since the second half of the 20th century, intensive livestock farming has profoundly transformed the socio-ecological metabolisms of our environments, and produced great changes in our lifestyles and in the ecosystem rhythms of the biosphere (FAO, 2013). Furthermore, after the recent Covid-19 pandemic, concern about zoonotic diseases is growing. This places the criticism of industrial livestock in the spotlight, whose conditions of massive enclosure and overcrowding with poor hygiene can favour these mutations and spread of viruses that are also harmful to our societies (Bryony *et al.*, 2013). In this way, any person aware of these causal chains whose ethics is minimally based on protecting himself or his relatives, should wonder whether it is worthwhile continuing to participate in this industrialized food system.

There are other food alternatives: extensive livestock farming is another option dedicated to satisfying the taste for meat. But there is an unresolved underlying problem in it: there is not enough surface on the planet for extensive livestock rearing to sustainably supply current trends in diets rich in meat proteins (Hayek *et al.*, 2020). If this practice is chosen to replace intensive livestock farming, a drastic reduction


EurSafe 2021 – DOI 10.3920/978-90-8686-915-2_13, © Wageningen Academic Publishers 2021

in meat consumption must be implicit. And this would lead to the other alternative of changing dietary preferences towards vegetarianism or veganism, because any of them could free up about 76% of the land dedicated to agriculture and livestock (Poore and Necemeck, 2018).

However, what about those people who resist giving up meat on their weekly menus but quite reasonably from an anthropocentric perspective are concerned with maintaining good planetary health? Many people, including environmentalists, despite knowing the detrimental effects of meat on the biosphere, find it difficult to reduce its consumption sufficiently. This is because they maintain a defeatist perception of individual responsibility and consider that great systemic changes will not be achieved by reducing individual consumption. Instead, it should be politicians who facilitate a structural transition towards more fair societies (Scott *et al.*, 2019).

Faced with this complex scenario, where food can be conceived as an essential leverage point for a just transition and at the same time a difficult habit to transform, the option of cultivated meat is being presented as a possible solution. This is because it promises to be the long-awaited magic formula that manages to maintain our expensive lifestyles without entailing a cost to the planet and our future generations. But is it certainly ethical?

## Cultivated meat from utilitarian point of view

Since the first laboratory-created hamburger, developed by the University of Maastricht in the Netherlands, was presented and tasted in 2013, the funding and public expectation of research with animal cells to create this type of synthetic food has greatly increased. Artificial meat is created in laboratories from stem cells extracted from muscles and other organic animal elements. These cells are harvested by biopsy and then reproduced under control in a synthetic culture that would mimic the animal's body, so that they grow and form new muscle tissue without the need for genetic intervention. It is not a simple or cheap investigation, but it is expected that it will be considerably cheaper if its consumption becomes widespread.

And why so much confidence in this possible future diet? It is often argued that the production of cultivated meat would significantly alleviate some of the major environmental problems associated with intensive livestock farming, such as animal suffering, overexploitation of land and water, methane emissions, deforestation, or fertilizers, pesticides, and fossil fuels abuse (Post, 2012). In theory, it presents these advantages of minimizing injustices towards animals, environment and people from a utilitarian point of view. Fewer animal sacrifices are needed to produce more meat products. Its processes of technological intensification in cell culture laboratories would entail a 'land sparing' strategy, through which much more land would be released than is currently maintained by agriculture and livestock. In addition, its efficient in vitro reproduction would supposedly generate enough meat food for the entire global population, helping to guarantee the right to food security. These potential benefits are often considered sufficient to at least support the cultivated meat process from a consequentialist point of view.

## Some epistemological challenges from utilitarianism

However, not all are advantages even from the utilitarian approach. Some of its promises to reduce animal suffering, environmental pollution and guarantee food safety can be discussed even from the point of view of consequentialism.

Cultivated meat requires foetal bovine serum for food growth in the laboratory, a culture based on calf stem cells. After a mother cow has been slaughtered and gutted, her uterus, which contains the foetus, is removed. Only foetuses older than three months are used, otherwise the heart is too small to perforate. No anaesthesia is administered during the process (Van der Valk *et al.*, 2010), which opens up a controversial debate about animal welfare. Although there are already tests with other culture media that do not require animal embryonic cells, such as algae, the option with the best results and most adopted remains the use of mammalian foetuses (Hocquette, 2016).

The key issue in a utilitarian argument is to weigh how much suffering is saved by generating cultivated meat. And here the answer depends basically on two conditions: to whom do we recognize basic capacities for suffering and at what time scale do we calculate the trade-offs. First, if the foetuses used during in vitro meat processing are understood as sentient beings, with intrinsic value and with a whole life ahead of them to potentially flourish, then the cost-benefit scale of cultivated meat may not be as advantageous as is often presumed. Second, although on a short time scale cultivated meat could reduce the global amount of animal suffering, on a longer scale it could perhaps find a ceiling from which it could no longer significantly reduce suffering. And in this case, it would also be necessary to make a food leap towards diets with less meat.

Regarding the supposed environmental benefits of cultivated meat, it must be taken into account that since 2015 the scientific literature seems to predict that cultivated meat could have less environmental impact than beef and perhaps pork, but more than chicken and plant-based proteins (Mattick *et al.*, 2015; Smetana *et al.*, 2015). Although the technology used in the process has significant scope for innovation that could reduce energy requirements below these assessments and offer better environmental results in the long-term (Stephens *et al.*, 2018), currently it cannot be said that it is an environmentally better solution than directly advocating for reducing the consumption of meat proteins.

In addition, another possible consequentialist challenge that the transition to artificial meat may bring should be taken into account: the Jevons paradox. According to this theory of environmental economics, policies that stimulate technological efficiency in the use of resources can lead to greater consumption of these, cancelling by doing so energy and environmental savings (Polimeni *et al.*, 2009). In other words, expanding cultivated meat could cause a 'rebound effect' in which total energy consumption would boost due to increased demand as a result of cheaper products and greater efficiency. If synthetic meat convinces the public and succeeds in the market, this may lead to fewer people opting for plant-based foods (even if it is more sustainable). And in this hypothetical scenario where the demand for in vitro meat would increase above the demand for vegetable proteins, would the net environmental impacts really be reducing? Again, some of the advantages from a utilitarian point of view are controversial if a long-term calculation is adopted.

## Preserving the value of food sovereignty

The consequentialist analyses focus mainly on how much collective well-being is gained with the cultivated meat procedure, assessing the extent to which it is worth promoting this food transition even if it implies accepting some trade-offs. The key argument of utilitarianism is that the benefits outweigh the costs, from a basically quantitative approach.

However, I differ with the view that to morally evaluate cultivated meat it is enough to do numerical calculations and under the cost-benefit mantra. There are values whose loss cannot be offset by the gain of other values. In the case of human food, I consider the right to sovereignty to be one of these non-compensatory values. On the one hand, we should be able to know how food is produced throughout its production chain. This would imply understanding in greater depth what is our relationship with nonhuman animals used in this process, as well as the resulting impacts on the land and the environment. On the other hand, we should at least have the option of being able to participate in the food-making process, in order to be self-sufficient. I think that the recent Covid-19 situation has awakened in many people a legitimate interest and concern to acquire resilience and not have to depend on transporters and supermarkets to be able to have enough food. Hence, there are quite a few people who began to set up small gardens and grow their own food in the gardens and balconies of their homes.

Seeking this food autonomy and recognizing the interdependencies generated during food production are not targets contemplated in the transition to cultivated meat. Nevertheless, they are deep-rooted issues in any food chain that we should be able to rescue and address. For this reason, I argue that in addition to the interesting criticisms of cultivated meat that may proceed from the same utilitarian perspective, there are other controversial points that deserve to be addressed from a republican and ecological justice approach.

## Eco-republicanism to face the oppression of cultivated meat

Power and domination should also be studied to determine only which collectives are privileged and which remain under dependency through cultivated meat production. In this direction, there are both approaches they are concerned about: the republican and the ecological.

### *Republican perspective*

The republican perspective differs from distributive justice because it studies the relations of power and domination with greater sensitivity than the latter. Thus, it is committed to determine which collectives are privileged and which remain under dependency through cultivated meat production. Cultivated meat is a product of our high-tech world and restricting production to this technology alone perpetuates the separation of key players who are aiming to create a sustainable food future. Farmers and land sovereigntists should be important players in our food systems (Borras *et al.*, 2015) since they have strong local and traditional knowledge of the land (Engdawork and Bork, 2015) and how to grow food while keeping the soil healthy (Rhodes, 2012). Nevertheless, shifting food systems to cultivated meat and high-tech processes could aggravate the loss of traditions based on respectful interactions with ecosystems and experience in obtaining food in a sustainable and regenerative way.

Industrializing systems and enhancing our reliance on technologies often cause further damage to rural collectives and people who decide to live without so many technological dependencies or who cannot afford to live according to this expensive system of machines, devices and ultra-processed products. In this developed model, cultivated meat would contribute to marginalize some collectives and make them more vulnerable.

While there is still domination among nations, communities and species, and power distribution is managed by a reduced group of people, externalities will keep being invisibilised and there will be no empowerment or food sovereignty. Not everyone is able to develop cellular agriculture, because prior specific knowledge of biotechnology, laboratories full of machinery, competent teams and industrialized systems are all required to this end. Consequently, the conditions to produce and obtain food would be extremely limited to a privileged sector of society and distributive procedures would broadly depend on this. The technological, technical and economic restriction to participate in the in vitro processes of food generation is far from a real democratization of food systems and close to a discrimination that further disconnects people from the land, nonhuman animals and self-sufficiency.

### *Ecological perspective*

The loss of food sovereignty and the different forms of social oppression that could derive from cultivated meat is a republican critique to take into account in order to evaluate the moral validity of its development and production. Another criticism from ecological justice that is not usually considered from utilitarianism consists of applying the concept of justice to the nonhuman world (Dobson, 2006) and recognizing which entities are being instrumentalized during the food production chain. Distributive theories of ecological justice broaden the spectrum of beings considered as beneficiaries of the fair distribution of resources. Here, animals, plants, ecosystems and species, previously considered merely as objects of distribution by classic anthropocentric approaches to justice, become legitimate recipients of political distribution activity (Baxter, 2005; Low and Gleeson, 1998). Recognition and participatory theories of ecological justice delve deeper and ask how beings and communities should be included in justice and ask how humans might listen to them without imposing our own voice (Dryzek, 1995; Schlosberg, 2007). From this perspective, it is often assumed that nonhuman nature and even ecosystems have a certain agency worthy of mention. The use of agency as a morally relevant characteristic is somewhat more problematic because its scope can be very narrow if it is linked to ideas such as cognitive intentionality or self-awareness.

However, agency does not imply assuming superstitious personification regarding nonhuman nature: even lacking the capability for rationality it could hold moral rights if classic reciprocity between rights and duties is abandoned (Curry, 2000). Many animals and ecosystems do not express themselves via cognitive strategies and do not make reasonable claims like humans. They evolve in line with the laws of physics, chemical reactions and biological interactions. Rationality or sentience are not a necessary condition for an entity to become a victim of injustice, but other capacities, like for example resilience, can be measured to discover how some entities change patterns when under stress conditions (Kortetmäki, 2017).

The central issue in evaluating whether cultivated meat is permissible from an ecological perspective essentially boils down to a question of whether nonhuman entities form part of food decision-making and in what way. This requires learning how to listen to other voices in food regeneration processes and including their sort of autonomy. Thus, the ecological perspective is engaged in caring for nonhuman entities involved in food chains and trying to represent their sovereignty about themselves. In accordance with this approach, rather than a final product, food may be considered an integral part of an interdependent chain. This suggests that cultivated meat should be analysed by taking into account how humans consider nonhuman nature during the food generation process.

## Conclusion

The cost-benefits from utilitarian point of view and consequentialist justice deserves to be taken into account. It is difficult to imagine a scenario where industrial meat based on intensive farming is completely abandoned and still human nutrition requirements are completely supplied. But utilitarian approaches aimed to analyse cultivated meat should consider long-term consequences and potential epistemological biases.

Furthermore, despite cultivated meat might trigger more benefits than costs compared to the current agribusiness models, there are still some missing values which should be cared for. Here, the objections provided from a theory of justice based on eco-republicanism are significant. Domination and power relations are concerns not properly dealt by the cultivated meat agenda, but they become crucial for an ethical assessment of food systems. Preserving food sovereignty, participatory processes and recognizing the integrity of beings and environment affected by food production, should be targets addressed by cultivated meat.

## References

Baxter, B. (2005). *A Theory of Ecological Justice.* London: Routledge.

Borras, S.M., Franco, J.C., and Monsalve, S. (2015). Land and food sovereignty. *Third World Quarterly*, 36(3): 600-617. https://doi.org/10.1080/01436597.2015.1029225

Bryony, J., Grace, D., Kock, R., *et al.* (2013). Zoonosis emergence linked to agricultural intensification and environmental change. *PNAS*, 110(21): 8399-8404. https://doi.org/10.1073/pnas.1208059110

Curry, P. (2000). Redefining Community: Towards an Ecological Republicanism. *Biodiversity and Conservation*, 9(8): 1059-1071. https://doi.org/10.1023/A:1008970518564

Dobson, A. (2006). Ecological citizenship: a Defence. *Environmental Politics*, 15(3): 447-451. https://doi.org/10.1080/09644010600627766

Dryzek, J. (1995). Political and Ecological Communication. *Environmental Politics*, 4(4): 13-30. https://doi.org/10.1080/09644019508414226

Engdawork, A., and Bork, H.R. (2015). Farmers' Perception of Land Degradation and Traditional Knowledge in Southern Ethiopia – Resilience and Stability. *Land Degradation & Development*, 27(6): 1552-1561. https://doi.org/10.1002/ldr.2364

FAO (2013). *Tackling climate change through livestock – a global assessment of emissions and mitigation opportunities.* Rome.

Hayek, M.N., Harwatt, H., Ripple, W.J. *et al.* (2020). The carbon opportunity cost of animal-sourced food production on land. *Nature Sustainability*. https://doi.org/10.1038/s41893-020-00603-4

Hocquette, J.F. (2016). Is *in vitro* meat the solution for the future?. *Meat Science*, 120: 167-176. https://doi.org/10.1016/j.meatsci.2016.04.036

Kortetmäki, T. (2017). Applying the Capabilities Approach to Ecosystems in advance: Resilience as Ecosystem Capability. *Environmental Ethics*, 39(1): 41-58. https://doi.org/10.5840/enviroethics20179263

Low, N., and Gleeson, B. (1998). *Justice, society and nature: An exploration of political ecology.* London: Routledge.

Mattick, C.S., *et al.* (2015). Anticipatory life cycle analysis of in vitro biomass cultivation for cultured meat production in the United States. *Environmental Science & Technology*, 49(19): 11941-11949. https://doi.org/10.1021/acs.est.5b01614

Polimeni, J.M., *et al.* (2009). *The Myth of Resource Efficiency. The Jevons Paradox.* Nueva York: Routledge.

Poore, J., and Necemeck, T. (2018). Reducing food's environmental impacts through producers and consumers. *Science*, vol. 360(6392): 987-992. https://doi.org/10.1126/science.aaq0216

Post, M. (2012). Cultured meat from stem cells: Challenges and prospects. *Meat Science*, 92: 297-301. https://doi.org/10.1016/j.meatsci.2012.04.008

Rhodes, C.J. (2012). Feeding and healing the world:through regenerative agriculture and permaculture. *Science Progress*, 95(4): 345-446. https://doi.org/10.3184/003685012X13504990668392

Schlosberg, D. (2007). *Defining Environmental Justice: Theories, Movements, and Nature.* Oxford: Oxford University Press.

Scott, E., Kallis, G., and Zografos, C. (2019). Why environmentalists eat meat. *PLoS ONE*, 14(7). https://doi.org/10.1371/journal.pone.0219607

Smetana, S., *et al.* (2015). Meat Alternatives: Life cycle assessment of most known meat substitutes. *International Journal of Life Cycle Assessment*, 20: 1254-1267. https://doi.org/10.1007/s11367-015-0931-6

Stephens, N., *et al.* (2018). Bringing cultured meat to market: Technical, socio-political, and regulatory challenges in cellular agriculture. *Trends Food in Science & Technology*, 78: 155-166. https://doi.org/10.1016/j.tifs.2018.04.010

Van der Valk, J., *et al.* (2018). Fetal bovine serum (FBS): Past – present – future. *Alternatives to animal experimentation (ALTEX)*, 35(1): 99-118. https://doi.org/10.14573/altex.1705101

# 14. Climate adaptation limits and the right to food security

*I. Wallimann-Helmer[1*], L.M. Bouwer[2], C. Huggel[3], S. Juhola[4], R. Mechler[5] and V. Muccione[3]*
*[1]University of Fribourg Environmental Sciences and Humanities Institute, Chemin du Musée 4, 1700 Fribourg, Switzerland; [2]Climate Service Center Germany, Helmholtz-Zentrum Geesthacht, Hamburg, Germany; [3]Department of Geography, University of Zurich, Switzerland; [4]Faculty of Biological and Environmental Science, University of Helsinki, Finland; [5]Risk and Resilience Program, International Institute for Applied Systems Analysis, Laxenburg, Austria; ivo.wallimann-helmer@unifr.ch*

## Abstract

Avoiding severe impacts from anthropogenic climate change requires not only substantial reductions in greenhouse gas emissions but also further implementation of adaptation measures. In many regions with smallholder farming systems adaptation can help ensure food security despite significantly changing climatic conditions. However, the space for adaptation measures has limits. In this paper, we investigate hard and soft adaptation limits and discuss their relevance to food security in smallholder farming food systems. We argue that soft adaptation limits can be defined by sufficiency levels of justice like basic needs, capabilities, human rights or levels of well-being. Depending on social, cultural, political and economic conditions different communities realize different sufficiency levels of justice. Furthermore, as normative claims of justice, these sufficiency levels not only allow the social, cultural and political and economic conditions of social systems to be described but also define entitlements. We show that for smallholder farming systems with low social and economic standards these entitlements might mean that soft adaptation limits can define entitlements to social and economic improvements despite transformational adaptation to changing climatic conditions.

**Keywords:** climate adaptation, smallholder farm, sufficientarian justice, food security, climate justice

## Introduction

Avoiding severe impacts from anthropogenic climate change requires substantial reductions in greenhouse gas emissions. But as several impacts are already occurring, and some further warming cannot be avoided, further upscaling of adaptation measures in agriculture will be required (IPCC, 2018). In many regions, adaptation of smallholder farming systems can help ensure food security despite significantly changing climatic conditions (Juhola *et al.*, 2017; Harvey *et al.*, 2018). However, the space for adaptation measures has limits (Mechler *et al.*, 2020). Once an adaptation limit is reached, transformation al adaptation becomes key. This paper analyses from an ethical perspective the entitlements of subsistence farmers depending on food systems that are forced into transformational adaptation. We investigate whether these entitlements entail a right to identical conditions for food production or whether and under what conditions the need for transformation only involves either the right to the same standards of food security or improvements in this regard.

The literature on climate adaptation limits distinguishes between hard and soft adaptation limits (IPCC, 2014). A hard limit is usually defined as the point at which objectives cannot be secured from intolerable risks. Hard adaptation limits are often associated with biophysical conditions such as temperature, water availability, and salinity beyond which adaptation is physically no longer possible (McNamara and Jackson, 2019). By contrast, a soft limit is determined by social, cultural, political and economic factors. A soft limit is a limit that can be overcome if the necessary adaptation options and capacities are available (IPCC, 2014).


EurSafe 2021 – DOI 10.3920/978-90-8686-915-2_14, © Wageningen Academic Publishers 2021

Like every social system, the functioning of a smallholder farming food system depends on biophysical, social, cultural, political and economic factors. Whilst the first kind of factors is set by the geographical location and biophysical requirements of a food system, it is the social and economic factors that can be adapted to secure food production. Social, cultural, political and economic factors define the relevant conditions under scrutiny if a food system enters the transformation space. We argue that the factors defining soft adaptation limits can be described by different sufficientarian standards of justice. As normative standards they do not only allow describing the characteristics of different smallholder farming systems but also define entitlements.

Our argument proceeds as follows. First, we explain in more detail how food systems depend on factors related to hard and soft adaptation limits. This allows definition of the conditions under scrutiny when a smallholder farming food system enters the space of transformational adaptation. Second, we discuss various thresholds of justice to define whether the conditions to be secured after transformation of a food system must remain the same as before transformation or simply have to ensure the security of food production or entail entitlements to improvement. Finally, we argue that, depending on local social, cultural, political and economic conditions, the entitlements range from minimal food security to improving the circumstances of communities in terms of justice and sovereignty.

## Climate adaptation limits in smallholder framing

It has been estimated that globally about 84% of farms are smallholder farms cultivating no more than 2 hectares of land. In low and middle-income countries, about 70-80% of farms are of this size (Lowder *et al.*, 2016). Typically, climate change has severe impacts on the food security of smallholder farming food systems with regard to in all four dimensions defining food security: food availability, access to food, its utilization, and the stability of food production (FAO, 2018). In this section, we explain how food security is linked to adaptation limits and under what conditions breaching limits might necessitate transformational adaptation, in many cases probably leading to abandoning formerly cultivated farmland and searching for other ways of livelihood.

Many smallholder farms depend directly on specific climatic and biophysical conditions that allow the production of the amount of nutrients needed to ensure livelihood (Lit). Climate Change can significantly undermine these conditions through changes in precipitation, heat and $CO_2$ significantly affecting crop growth and consequently the nutrients available (Challinor *et al.*, 2014). Similarly, climate change often increases the frequency and magnitude of extreme weather events such as windstorms, heavy rainfalls and floods, or droughts potentially destroying or heavily reducing harvest. In addition, the occurrence of pests and diseases can increase, threatening crops as well as livestock (Myers *et al.*, 2017). These consequences from climate change put food security at risk. Changing biophysical conditions and extreme weather events threaten cultivation conditions and can lead to significantly lower crop yields. These changes impact both smallholder farms themselves and those living in cities who depend on smallholder food production because they do not cultivate crops themselves. Hence, climate change in smallholder farming food systems most probably is one of the key factors threaten to undermine not only food availability but also economic and physical access to food (Sandstrom and Juhola, 2017).

The literature presents three prominent strategies for dealing with changing climatic conditions and extreme weather events, each involving trade-offs (Fedele *et al.*, 2019). Up to a certain magnitude of negative impacts, extreme weather events can be addressed with coping strategies and incremental adaptation. In the case of flooding, for example, coping means just replanting crops after they have been lost while not changing the social and ecological traits of a food system. By contrast, incremental adaptation means small adjustments to the food system in order minimize the risk of future harvest loss. It could mean building dykes to meet flooding risks or develop and expand irrigation systems to address the risks of rainfall variability or drought. Whilst the socio-ecological characteristics of a food system remain the same or similar with coping strategies and incremental adaptation, as a third strategy transformational adaptation fundamentally alters the social and economic characteristics of an existing system (IPCC, 2018). This becomes necessary if extreme weather events become too threatening or if the biophysical conditions permanently change so that food security can no longer be ensured. This could require changing how cultivation is organized, the relocation of farms or even shifting ways of livelihood.

In smallholder farming food systems, hard adaptation limits are typically faced when biophysical conditions permanently change, making transformation necessary. Crops such as wheat and maize are very sensitive to changes in precipitation and temperature. Once the environmental conditions no longer allow their cultivation, the agricultural system must be transformed (Käyhkö *et al.*, 2020). At a minimum, this means changing the crops cultivated, which might involve altering social and economic interactions within the food system. Further transformational adaptation strategies with heavier second-order consequences for social and economic conditions especially at the household level include relocating farms and abandoning farming altogether. However, such hard adaptation limits might also be reached if the frequency and impacts of extreme weather events become intolerable and demand relocation or abandoning farming. Such adaptation requires new ways of livelihood, and this change probably risks poverty due to the lack of opportunities or competition among farmers for new income.

Analysing soft adaptation limits is more complex because they are closely interlinked not only with ensuring food security but also with respecting food sovereignty. Soft adaptation limits are typically defined as limits that exist because the financial or technological means are not yet given or because the necessary decisions cannot be taken for social or political reasons (Warner, 2016). If building a dyke is not possible because engineering expertise or the necessary finance or equipment and material is lacking, this is a soft adaptation limit. Similarly, smallholder farmers sticking to traditional ways of cultivation face a soft adaptation limit since in principle adaptation might be possible. What is crucial about soft adaptation limits is that in principle they can be overcome, but if no action is taken for too long, they may turn into hard limits, because adaptation options are not available anymore. However, overcoming such limits by imposing measures in a top-down approach probably undermines food sovereignty: the right of communities to decide themselves how they want to produce their nutrition (Fairbairn, 2010).

## Adaptation limits as thresholds of justice

Engaging with the concept of food sovereignty raises why food security cannot be simply translated into a minimum level of nutrients available and accessible to households in smallholder farming systems in the face of climate change and transformation. We argue that it depends on the social, cultural, political and economic conditions of such food systems what sufficiency level can be considered the minimal acceptable standard. Defining such a level as appropriate and legitimate establishes soft adaptation limits. Here, we introduce several metrics of sufficientarian justice to illustrate how soft adaptation limits can be defined by such standards and to explain why adaptation limits in different food systems might differ depending on social, cultural, political and economic conditions.

Food sovereignty is a concept that has been developed to counter top-down approaches to ensuring food security without properly involving those affected (Noll and Murdock, 2020). Food sovereignty is often presented as concept in conflict with food security because its main aim is collective self-determination of food production by communities. In the case of smallholder farming, this would mean letting members of such systems decide themselves how to and what crops to cultivate as well as decide how to adapt to changing climatic conditions. However, food sovereignty is not totally disconnected from food security.

Rather, it is often argued that food sovereignty allows more sustainable food production and hence more stable availability of nutrients.

As a sufficientarian standard of justice, food sovereignty is highly demanding, especially if understood as a democratic standard. A democratic standard demands not only fair procedures and institutions of participation but also substantial conditions of justice to ensure that all can participate on an equal footing (Wallimann-Helmer, 2019). If democratic food sovereignty defines a smallholder farming food system, then this would be the sufficientarian standard to be ensured despite changing climatic conditions. Once biophysical, social, cultural, political and economic conditions change so that these conditions can no longer be secured, a social system has to undergo radical transformation. This is why sufficientarian standards of justice such as democratic food sovereignty define adaptation limits.

These limits are soft limits because they are defined by social, cultural, political and economic conditions that translate into institutional structures and distributions of resources. Given appropriate means to protect the social, cultural, political and economic structures or to re-establish the necessary institutions and other conditions of justice by geographically moving communities, they could be overcome. However, the more demanding sufficientarian standards of justice become, the more difficult it is to push limits and maintain a community without radically transforming its social, cultural, political and economic structures. This seems most obvious when the standard of justice is defined by a level of welfare as suggested by egalitarian thinkers of justice (Lippert-Rasmussen, 2016). The higher the welfare standard, the more demanding will be the measures to be taken to ensure that this limit is not breached. By contrast, if a soft adaptation limit is defined by a less demanding subsistence level, such as satisfying basic nutrient needs or even bare survival, then maintaining such a limit is much less challenging (Schuppert, 2011). The same might be true for ensuring basic human rights to food as sufficientarian standard of justice (Caney, 2010).

However, which standards of justice define adaptation limits within smallholder farming systems depends on the social, cultural, political and economic structures in these communities and on how their conditions are perceived by its community members (Whyte, 2018). We argue that the more demanding these soft adaptation limits are, the more demanding soft adaptation limits they will entail. For example, satisfaction of basic nutrient needs is usually understood as part of the more demanding standard of human rights, and the human right to education is crucial for democratic decision-making. This means that the higher the standards of justice realized in a community, the higher is the number of soft adaptation limits that can be at risk due to changing climatic conditions. It also means that communities that have realized more demanding sufficiency standards of justice might less quickly be pushed into the transformational adaptation space, probably deeply altering their social and economic structures.

## Socio-economic conditions defining transformational entitlements

From an ethical point of view, sufficientarian standards of justice define the entitlements to be secured for communities and individuals. Because we argue that soft adaptation limits are defined by such standards, any kind of measure taken must ensure that these standards can be preserved despite changing climatic conditions. By definition, coping action and incremental adaptation would be such measures, but it is less clear how to deal with these entitlements if communities are pushed beyond soft adaptation limits into the transformational adaptation space. It is unclear whether standards of justice must be maintained or improved or can justifiably be lowered. Here, we argue that standards for smallholder farming food systems most probably need to be improved, if not for the system as a whole then for individual households, because the standards of justice realized in terms of economic conditions are most probably very low anyway.

In the literature, sufficiency standards of justice define the conditions to which each individual is entitled (Herlitz, 2019). Food security's four dimensions of food availability, access to food, its utilization and the stability of food production could be understood as such a standard of justice. However, hardly any sufficientarian standard of justice accepts that food security alone defines the sufficiency conditions for human lives. Beyond food shelter, relations with the environment and friendship certainly require mention as well (Nussbaum, 2000). In the literature on sufficientarian justice, what are seen as the necessary conditions for a human life depend on an understanding of human dignity or of human self-understanding. But it is a matter under scrutiny in ethical research exactly which social and economic conditions must be provided. Some define them by examining the history of human cultures, others rely on human rights as they evolved, and still others require detailed interpretations of the concept of human dignity (Shields, 2016).

In whatever way the sufficiency conditions of a human life are defined, they certainly extend beyond the four dimensions of food security. Hence, food security as a standard of justice will most probably be part of a more encompassing sufficientarian standard. Due to their poor economic conditions, such an encompassing standard of justice might define the most crucial soft adaptation limit for smallholder farming systems (Hansen *et al.*, 2019). If food sovereignty is not provided by such a system, it seems appropriate to improve conditions in transformational adaptation such that food sovereignty can be ensured. However, it is also possible that food sovereignty is given even though other crucial aspects of sufficiency are not secured. In such circumstances, improving standards of justice when a community transforms might be appropriate because despite food sovereignty the sufficiency conditions for human lives seem not to be given in a full sense. By contrast, communities and households with very high standards of wellbeing might not be entitled to the high standards they enjoyed before transformation became necessary. Their standards might justifiably be lower than before climatic conditions changed.

These considerations indicate that the social and economic conditions of communities define their entitlements in terms of sufficientarian standards of justice once they need to transform due to changing climatic conditions. Whether these conditions can and must be maintained for all individual households in a community depends on their various social and economic conditions. It might well be that some households of a community are entitled to improvements whilst others are not independently of whether the community as a whole is entitled to improved sufficiency standards of justice. For smallholder farming food systems, this most probably means entitlements to improvements of living conditions for most households, whereas less favourable conditions might also be justifiable for those communities and households enjoying higher standards of living. However, exactly which standards of sufficiency are minimally acceptable for communities and households and how exactly they must be defined from an ethical perspective is a matter of investigation and cannot be decided here.

## Conclusion

In this paper, we analysed the relation between smallholder farming systems, food security, and climate adaptation. We first discussed that such food systems face not only biophysical limits to adaptation but also soft limits defined by social, cultural, political and economic conditions. They can be overcome if the economic means, technology, and know-how are available. We then argued that soft adaptation limits can be defined by sufficientarian standards of justice like basic needs, human rights or levels of well-being, if not democratic conditions of food sovereignty. We claimed that more demanding sufficiency levels require more extensive adaptation action but communities that have realized higher standards will most probably less quickly enter the space of transformational adaptation.

From an ethical perspective, sufficiency standards of justice help not only to describe the social and economic conditions of communities and households but also to establish entitlements. We argued that a minimal objective standard of justice that needs to be maintained for all communities and households despite transformational adaptation has yet to be defined and cannot be derived from ethical literature. For smallholder farming communities, such a standard most probably means entitlements to improvements of living conditions despite transformational adaptation whereas communities and households with higher standards of living might need to accept less favourable living conditions.

## References

Caney, S. (2010). Climate Change, Human Rights, and Moral Thresholds. In: Gardiner, S.M., Caney, S., Jamieson, D., Shue, H. (eds) Climate Ethics. Essential Readings. Oxford University Press, Oxford, New York, pp. 163-177.

Challinor, A.J., Watson, J., Lobell, D.B., Howden, S.M., Smith, D.R., Chhetri, N. (2014). A meta-analysis of crop yield under climate change and adaptation. Nature Climate change 4: 287-291.

Fairbairn, M. (2010). Framing Resistance: International Food Regimes and the Roots of Food Sovereignty. In: Desmarais, A.A., Wiebe, N., Wittman, H. (eds) Food sovereignty. Reconnecting food, nature & community. Fernwood Publ, Halifax, pp. 15-32.

FAO (2018). The state of food security and nutrition in the world: Building climate resilience for food security and nutritionFAO, Rome, 181 pp.

Fedele, G., Donatti, C.I., Harvey, C.A., Hannah, L., Hole, D.G. (2019). Transformative adaptation to climate change for sustainable social-ecological systems. Environmental Science & Policy 101: 116-125.

Hansen, J., Hellin, J., Rosenstock, T., Fisher, E., Cairns, J., Stirling, C., Lamanna, C., van Etten, J., Rose, A., Campbell, B. (2019). Climate risk management and rural poverty reduction. Agricultural Systems 172: 28-46.

Harvey, C.A., Saborio-Rodríguez, M., Martinez-Rodríguez, M.R., Viguera, B., Chain-Guadarrama, A., Vignola, R., Alpizar, F. (2018). Climate change impacts and adaptation among smallholder farmers in Central America. Agric & Food Secur 7.

Herlitz, A. (2019). The indispensability of sufficientarianism. Critical Review of International Social and Political Philosophy 22: 929-942.

IPCC (2014). Climate Change 2014: Synthesis Report. Contribution of Working Groups I, II and III to the Fifth Assessment Report of the Intergovernmental Panel on Climate Change [Core Writing Team, R.K. Pachauri and L.A. Meyer (eds)]., IPCC, Geneva, Switzerland, 151 pp., 151 pp.

IPCC (2018). Global Warming of 1.5°C: Summary for Policy Makers: In IPCC Special Report on the impacts of global warming of 1.5°C above pre-industrial levels and related global greenhouse gas emission pathways, in the context of strengthening the global response to the threat of climate change, sustainable development, and efforts to eradicate povertyWorld Meteorological Organization, Geneva, Switzerland.

Juhola, S., Klein, N., Käyhkö, J., Schmid Neset, T.-S. (2017). Climate change transformations in Nordic agriculture? Journal of Rural Studies 51: 28-36.

Käyhkö, J., Wiréhn, L., Juhola, S., Neset, T.-S. (2020). Integrated framework for identifying transformative adaptation in agri-food systems. Environmental Science & Policy 114: 580-586.

Lippert-Rasmussen, K. (2016). Luck egalitarianism. Bloomsbury, London, 259 pp.

Lowder, S.K., Skoet, J., Raney, T. (2016). The Number, Size, and Distribution of Farms, Smallholder Farms, and Family Farms Worldwide. World Development 87: 16-29.

McNamara, K.E., Jackson, G. (2019). Loss and damage: A review of the literature and directions for future research. Wiley Interdisciplinary Reviews: Climate Change 10.

Mechler, R., Singh, C., Ebi, K., Djalante, R., Thomas, A., James, R., Tschakert, P., Wewerinke-Singh, M., Schinko, T., Ley, D., Nalau, J., Bouwer, L.M., Huggel, C., Huq, S., Linnerooth-Bayer, J., Surminski, S., Pinho, P., Jones, R., Boyd, E., Revi, A. (2020). Loss and Damage and limits to adaptation: recent IPCC insights and implications for climate science and policy. Sustain Sci 15: 1245-1251.

Myers, S.S., Smith, M.R., Guth, S., Golden, C.D., Vaitla, B., Mueller, N.D., Dangour, A.D., Huybers, P. (2017). Climate Change and Global Food Systems: Potential Impacts on Food Security and Undernutrition. Annual review of public health 38: 259-277.

Noll, S., Murdock, E.G. (2020). Whose Justice is it Anyway? Mitigating the Tensions Between Food Security and Food Sovereignty. J Agric Environ Ethics 33: 1-14.

Nussbaum, M. (2000). Women and Human Development: The Capabilities Approach. Cambridge University Press, Cambridge.

Sandstrom, S., Juhola, S. (2017). Continue to blame it on the rain? Conceptualization of drought and failure of food systems in the Greater Horn of Africa. Environmental Hazards 16: 71-91.

Schuppert, F. (2011). Distinguishing basic needs and fundamental interests. Critical Review of International Social and Political Philosophy: 1-21.

Shields, L. (2016). What is Sufficientarianism? In: Shields, L. (ed.) Just Enough: Sufficiency as a Demand of Justice. Edinburgh University Press, pp. 15-43.

Wallimann-Helmer, I. (2019). The ethics of waste policy. In: Lever, A., Poama, A. (eds) The Routledge handbook of ethics and public policy. Taylor & Francis, Oxon / New York, pp. 501-512.

Warner, B.P. (2016). Understanding actor-centered adaptation limits in smallholder agriculture in the Central American dry tropics. Agric Hum Values 33: 785-797.

Whyte, K.P. (2018). Food Sovereignty, Justice and Indigenous Peoples: An Essay on Settler Colonialism and Collective Continuance. In: Barnhill, A., Budolfson, M., Doggett, T. (eds) The Oxford Handbook of Food Ethics, vol. 1. Oxford University Press, Oxford, New York, pp. 345-366.

# 15. How to techno-fix climate change problems in winemaking – and legitimately get away with it

*R. Nydal*
*Programme for applied ethics, Department of philosophy and religious studies, Norwegian University of Science and Technology, Dragvoll, 7491 Trondheim, Norway; rune.nydal@ntnu.no*

## Abstract

Efforts are currently made to preserve traditional viticulture by finding ways to adapt to ongoing climate changes. A key challenge for wine making is global warming causing increasing levels of sugar in grapes. This makes it difficult to preserve the quality of wine due to the corresponding and unwanted increase in alcohol level in wine. This work is part of a European research project aiming to modify yeast to produce less alcohol – and thereby offer a powerful climate adaption tool of critical importance for winemakers in areas like Spain. This paper discuss how to avoid the techno-fix critique of being a technological solution that suppresses symptoms while amplifying the root causes like our lifestyles, our consumption, and our mind-set. To avoid this techno-fix critique, this paper argues, we need to undo sociotechnical opacity. We need to uncover the sociotechnical world created by using yeast technology to preserve traditional viticulture. This involves exposing and acknowledging ways in which technological components are part of the identity of the wine, winemaking and wine culture. The visibility of technology thus becomes a test of the technologies moral viability: a test of whether, or under what conditions, living with certain technologies support our concept of the good life and the good society.

**Keywords:** viticulture, sociotechnical opacity, a new normal

## Wine industry as a test case for how to respond to climate change

The quality of wine is particularly sensitive to climate changes since wine plants are one of the most affected plants. A key challenge is to respond to the way in which global warming affects the ripening process of wine grapes. This maturing process include two separate processes in need of being balanced; the development of the sugar level (sugar ripeness) on the one hand and the colour, flavour and aroma (physiological ripeness) on the other hand. Finding the optimal window for harvesting grapes is of critical importance and involves careful monitoring of the two ripening processes. The climate conditions are the most important factor. Decisions on when to harvest would have to be made locally and include various options for adapting to climate variations like the monitoring and control of growing conditions (like temperature, sunlight exposure and water supply) or intervening in wine stocks biology through cutting or covering of wine stocks, or perturb the ripening process more directly through chemical additives (Goode, 2019; Robinson, 2015).

Climate changes have created serious challenges to winemakers. They have constantly been searching for ways to expand their tool case for climate adaption during the last decades (Mozell and Tach, 2014). The current limitations of the tools at hand have created a sense of urgency. The uncertain future of wine worries producers (Famila Torres, 2019), energize stakeholder conferences (The Porto Protocol, 2019) and scientific conferences (Science and Wine, 2019) and capture the attention of the media (Daniels, 2020; Lensom, 2020). The steady unwanted growth of alcohol percentage of wine is a key indicator of how climate changes affect the wine production process. The alcohol percentage has risen to a level where it seriously threatens the quality of wine. The geographic distribution of established wine regions may over time drift towards cooler locations. This would not only raise a number of quality concerns of


EurSafe 2021 – DOI 10.3920/978-90-8686-915-2_15, © Wageningen Academic Publishers 2021

its own, but also imply the loss of highly valued viticultures. Some traditional viticultures, like the ones found in Spain, is in particular vulnerable to climate changes (Santos *et al.,* 2020).

The place of production is essential for wine, expressed in the valued notion of Terroir. Wine making exhibits an intimate relationship between humans and nature: cultures of fine-tuned interplay of climate, soil, plants, microbes and human practices. Viticultures represent important cultures worth preserving due to its economic, social and cultural value. The possible loss of old viticultures forcefully brings into life our concept of the good life – and hence also the tragedy of root causes of climate change – like unsustainable habits and political choices.

Winemaking becomes in this situation an interesting site for discussing technological fixes. It is a case where the effects of climate changes are clear to view calling for immediate actions to counteract climate change. Winemaking reminds us of how fragile human-nature relationships are, and what we may lose. A powerful techno-fix of viticultures however – risk becoming yet another case masking root causes of climate change.

## Technology for the preservation of viticultures

This work is part of a European research project coordinated from Spain. Researchers in the project reached out to philosophers and anthropologist for help to assess the projects technological climate adaption strategy. The rationale for our participation, like any other participants in the project, could be understood as the one of contributing to the realisation of the overall end goal of the project: to preserve Spanish viticulture. This comprises the ethos of the project – for which reason the group is assembled and accepted as a project worth funding.

The project aims to engineer yeast cultures capable of producing less alcohol without compromising the characteristic colour, flavour and aroma of the wine. Winemakers would be offered a powerful tool of particular importance in areas where viticultures are in jeopardy.

Contemporary laboratories may, one could say, extend and enhance human capacity of coexistence and adaption to natural environments. Socio-humanist research may help tailor the process of modifying wines to the larger social context in which the wines are to be produced, sold and consumed. For instance, the technological solutions are constrained by the projects commitment to avoid GMO strategies. It is taken to be a shared concern among stakeholders that wine needs to be maintained as a consistent GMO free product. A social impact assessment is conducted in the project to help identify, clarify and assess such social bottlenecks and acceptability criteria. Social impact studies may inform strategies of winemaking, marketing as well as future research projects.

Even if we assume all this work help to preserve Spanish viticulture, the project could still be vulnerable to the critique of techno-fix. How to avoid this critique is the topic of this paper, claiming it is a matter of undoing sociotechnical opacity.

## Techno-fix and technological opacity

The nuclear physicist Alvin Weinberg introduced the notion of techno-fix in the 1960s in promoting technological fixes to societal problems. Technical solutions may provide simple, solvable and stable alternatives to social solutions as they factor out human unpredictability (Johnston, 2018). Compare for instance seatbelt alarms in cars to legal measures or moral campaigns.

Techno-fix is often used as a critical concept, describing a technocratic mind-set favouring technological solutions even if the problem primarily calls for social solutions. Technological solutions may in such cases target wrong problems or create more problems than they solve. We should in particular be sceptical to simple technological solutions to climate changes as it may undermine social action (Huesemann and Huesemann, 2011). But as Dane Scott notes, the techno-fix notion has little analytical value if it simply refers to a 'mind-set' of naive techno-optimism. The moral concerns that the notion of techno-fix expresses are important sources for understanding the role technologies play in contemporary societies.

> The fact that 'technological fix' has become a dismissive term in public rhetoric hides its significance as an important concept for understanding the roles technologies play in contemporary life. Through analysing the technological fix criticisms, insights can be gained into the ambiguous relationship between modern industrial societies and science and technology. (Scott, 2011: 208)

One way to understand the ambiguity, the moral tensions we often experience when adapting new technologies, is to pay attention to the interconnected nature of technological and social orders. New technologies restructure these orders, a dynamics of essential importance for understanding the ethics of technology (Winston, 1999). As Weinberg argued, technological solutions may offer powerful alternatives to social solutions by factoring out the messiness of social dynamics. This simplification is at the same time at the heart of the techno-fix critique. The blessing and curse of techno-fix follows from a disentanglement of technical and social orders where socio-technical complex problems are reduced to simple technological problems.

We should not be concerned with promotion of technological fixes to social problems as such – but rather what we could call the blindness – or the forgetting of the social dimension of technology. This concealing of the social is part of a normal technological innovation process, what Bruno Latour (1979) referred to as 'black boxing'. Technology tends to hide its sociotechnical conditions – because that is when it functions at its best. Technologies function efficiently when experienced as a tool – as a technical artefact – allowing users to not pay attention to them. It is an integral part of engineers training to find ways to background, or black box their technology by integrating them into our daily habits and practices. Engineers succeed when their products are no longer noticed, i.e. when the technology they produce become so user friendly, domesticated and backgrounded that they are taken to be a natural part of the practice. It is an integral part of engineers training to find ways to hide them by integrating technological solutions into users' daily habits and practices. The more engineers make themselves and their products invisible – the more they succeed.

The hiding matters more than ever as technologies have come to permeate modern societies – causing what Shannon Vallor describe as a 'growing technosocial blindness, a condition that I will call "acute technosocial opacity"' (Vallor, 2016: 6). What is at stake here are the hiding, or forgetting of the ways in which technologies take part in the formation of our moral selves individually and collectively through our appropriation of new technologies.

When moral questions of techno-fix are raised, the social dimensions are called into question. Weinberg's own call for techno-fix illustrates the point, although with a surprising twist as Weinberg himself did not coin the term techno-fix as a critical term.

Weinberg's (1961) agenda was in part to protect university research from trends towards 'big science' exemplified by high energy physics. Such large scale efforts were bound to be more bureaucratic, dependent on funding sources and mass media coverage. The trends where in many respects unavoidable following an exponential growth of scientific activity creating institutional changes one simply had to 'learn to live with'. In this situation Weinberg called for an assessment of the social value of research. Weinberg criticized Big Science efforts in physics for not being oriented towards social needs. For all its prestige, high energy physics was only oriented towards itself, often expressed as a pursuit of answering big philosophical topics. Why not put all that money, effort and prestige into something useful – like biotechnology (that at the time had far lower prestige). Biotechnology expressed a more viable alternative ethos for Weinberg: Researchers would be oriented towards fixing social problems that matters for citizens and could make a real difference for them.

Weinberg's call for technological fixes to social problems then, does not express a naive mind-set where complex sociotechnical problems are reduced to simple technical problems. On the contrary, the problem Weinberg addressed in his notion of techno-fix is a mind-set that does *not* question the social value of technology – where high prestige priority areas may turn out to be indifferent to or in conflict with the common good.

In cases where we risk amplifying root causes of the climate crisis a given technology aims to fix (like yeast engineering) this paper suggests, it is particularly important to find ways to undo sociotechnical opacity. We need to live with the technology and the problem it aims to fix in plain sight. Such transparency is tricky since technology resists it – we need to imagine a new normal in times of such crisis. This will be discussed further by drawing on two other examples of climate adaption.

## Legitimate techno-fix – imagining a new normal

Norwegians experience the effect of climate change every winter. The skiing season has in many areas been reduced by one month during the last thirty years – and the climatic snow line is expected to be pushed towards higher altitudes in the coming decades.

These changes have been accompanied by a number of climate adaption strategies. In Trondheim, large amount of snow are stored outdoors under chopped wood during summer to allow skiing in November. Outdoor ski resorts are regularly maintained by artificial snow during season. In January 2020, an indoor ski resort where even opened in Norway – one of the largest in the world. All this help preserve Norwegian ski cultures. This aim is explicitly stated at the webpage of a research project in Trondheim that aims to develop energy efficient ways to produce artificial snow even at temperatures above freezing point.

> Skiing is part of the Norwegian identity and innately defines Norway culturally in a global context. With the future challenges of climate change and lack of snow in people's communities, the foundation of skiing and other traditional winter activities is being threatened. (Snow for the future, 2020)

In this case the climate adaption strategy reminds us of the tragedy of climate change. We are reminded of what may be lost, in this case highly appreciated Norwegian ski cultures as the technological fixes are plain to sight.

It is not equally evident that efforts to preserve Spanish viticultures would keep reminding us about the tragedies of climate change – simply because the technological solution could be more powerful and thereby also more opaque. While it seems evident that a technological fix of Norwegian ski cultures could help call attention to responsible action of climate change from the industry, politicians and the general public – it is not equally evident in the case of the technological fix of Spanish viticultures. In this case we may need to find imaginative and creative ways to counteract the forgetting.

Another example may be illustrative. Ross Harding (2019) advocates a need for 'a new normal' directing our attention to ways in which we normally organise our cities. The problem of sustainability is not restricted to issues like circular economy or zero emission systems. Part of the problem is that our daily habits and mind-set are by design disentangled from the technological systems that conditions our comfortable lives. Our energy, water and waste management systems are efficiently brought out of sight. The opacity of the technological systems disconnect technology from our conceptions of the good lives lived comfortably in the cities. Harding suggests a new normal where we make our technological mediated life conditions visible by building the vital infrastructures in plain sight in the middle of our town. Such deliberate design provides means to 'connect culture with technology'. We need find ways to be able to live comfortably with the technology that enable our comfortable lives.

It appears to be a particularly risky approach to expose the technology involved in the work of preserving viticultures. This becomes evident in the discussion on labelling, as wine is one of the few products that have not been subjected to mandatory listing of wine ingredients and nutritional information on labels. Mandatory labelling could help expose and acknowledge ways in which technological components in various degrees already are part of the identity of the wine, winemaking and wine culture. But labels are primarily designed for the purpose of listing additives not accounting for dilemmas and choices of wine making. Labelling would need to be part of a larger task of creating a narrative calling attention to the complexity of the hard work invested in preserving the quality of wine and cultural heritage of wine making.

The work of creating such narratives is to explore ways to test the moral viability of the technological fix of global warming in this case. It is a test of whether, or under what conditions, living with these technologies support our concept of the good life and the good society. It is also a way to re-think well-established standards of well-constructed technological solutions, considering when and how to counteract technological opacity.

## References

Daniels, S. (2020). Climate change is rapidly altering wine as we know it. Wine Enthusiast Magazine. February 3. Available at: https://www.winemag.com/2020/02/03/wine-climate-change/.

Familia Torres (2019). Familia Torres brings together international experts to present urgent actions to combat climate change. Available at: https://www.torres.es/en/we-are/news-item/familia-torres-brings-together-international-experts-present-urgent-actions-combat.

Huesemann, M. and Huesemann, J. (2011). Techno-Fix: Why technology won't save us or the environment. New Society Publishers, Philadelphia, USA.

Goode, J. (2019). Ripeness in wine. Wineanorak. Global Wine Journal. Available at: https://wineanorak.com/2019/10/20/ripeness-in-wine-part-1/.

Johnston, S. (2018). Alvin Weinberg and the promotion of the technological fix. Technological Cultures 59: 620-651.

Latour, B. and Woolgar, S. (1979). Laboratory life. The construction of scientific facts. Beverly Hills, Sage Publications, USA.

Ledsom, A. (2020). How climate change is altering your favorite glass of wine. Forbes. September 12. Available at: https://www.forbes.com/sites/alexledsom/2020/09/12/how-climate-change-is-altering-your-favourite-glass-of-wine/?sh=7c28eb14491b.

Mozell, R.M., and Tach, L. (2014). The impact of climate change on the global wine industry: Challenges & solutions. Wine Economics and Policy 3: 81-89.

Robinson, J. (ed.) (2015). The Oxford Companion to Wine. Oxford University Press, Oxford, UK.

Harding, R. (2019). A new normal. Available at: https://www.youtube.com/watch?v=KA_9fWVm1AE&fbclid=IwAR0eDfU7sKpxKf4FLh1q_7mIDqrk1cbP_2Unj1r7zcJlr5wurhkSJSxNTcA.

Scott, D. (2011). The technological fix criticisms and the agricultural biotechnology debate. Journal of Agricultural and Environmental Ethics 24: 207-226.

Santos, J., Fraga, H., Malheiro, A.C., Moutinho-Pereira, J., Dinis, L-T., Correia, C., Moriondo, M., Leolini, L., Dibari,.C., Costafreda-Aumede, S., Kartschall, T., Menz, C., Molitor, D., Junk. J., Beyer, M. and Schultz, H.R. (2020). A review of the potential climate change impacts and adaptation options for European viticulture. Applied Sciences 10: 3092.

Science & Wine (2019). 1st Science & Wine world congress. The wine of the future. Available at: https://www.ciencia-e-vinho.com/last-and-next-conferences/1st-science-wine-world-congress-the-wine-of-the-future/.

Snow for the future (2020). Snow for the future. Efficient and climate-friendly snowmaking. Available at: https://www.sintef.no/en/projects/2019/snow-for-the-future-will-produce-artificial-snow-s/.

The Porto protocol (2019). Climate change leadership. Solutions for the wine industry. Available at: https://climatechange-porto.com/.

Vallor, S. (2016). Technology and the virtues. A philosophical guide to a future worth wanting. Oxford University Press, New York, USA.

Weinberg, A. (1961). Impact of large-scale science on the United States. Science 134:161-64.

Winston, M. (1999). The scope of technology. In Winston, M. and Edelbach, R. (eds) Society, ethics and technology. Wadsworth Publishing, Belmont, USA.

# 16. Policy and institutional responses to climate change and food security in Bangladesh

*M.H.I. Khan and M. Das Gupta**
*Supreme Court of Bangladesh and Centre for Climate Justice-Bangladesh (CCJ-B), 14/B, 11, Poribag, Dhaka, Bangladesh; dgmoumita2014@gmail.com*

## Abstract

Bangladesh is widely recognized as one of the most climate vulnerable countries in the world taking into account both existing vulnerabilities and future climate change impacts. Historically it experiences frequent natural disasters having adverse impacts on lives and livelihoods, especially of poor people. Climate change impacts are going to exacerbate many of the current problems and natural hazards the country faces and those impacts and vulnerabilities are already evident in Bangladesh, which is reducing inter alia, agricultural production and threatening food security of the country. The Government of Bangladesh (GoB) is considering relentlessly these challenges through taking relevant measures and policy approaches to ensure food security and to achieve continuous higher economic growth to eradicate poverty at an expected pace. This paper has identified initially the causes of threat to food security associated with climate impacts and explored some effective adaptation measures to reduce and manage the risk of reducing agricultural products in Bangladesh. It thereafter has scrutinized the national legislations, policies, plans, programs and institutions aligning with the international policy regime on climate change and food security and analysed the scopes and gaps to make some recommendations for required effective national policy and institutional structures. Such policy investigations, at the national level as a bottom up approach can also provide some required policy guidance for international regulatory regime.

**Keywords:** climate vulnerability, national responses, threat to food security

## Introduction

Bangladesh got wide recognition to be one of the most climate vulnerable countries of the world by considering both existing vulnerabilities and future apprehensions. Historical experiences of devastating natural disasters with adverse impacts on lives and livelihoods, especially of the poor people, are anticipated to be increased due to climate change impacts. Predicted and evident adversities include frequent and severe tropical cyclones, more erratic rainfall (heavier/lower), higher river flows, riverbank erosion, increased sedimentation, melting of the Himalayan glaciers, sea level rises, warmer and more humid weather, etc., reducing inter alia, agricultural production and threatening food security of Bangladesh. The Government of Bangladesh (GoB) is relentlessly taking relevant measures and policy approaches to ensure food security for higher economic growth and eradication of poverty (MoEF, 2009). Flash flood at haor basins is considered to be the main disaster of south-east Bangladesh threatening productivity and food security of this area (BHWDB, 2012).

The Fifth Assessment Report (AR5) of Intergovernmental Panel on Climate Change (IPCC) has identified that current temperatures and heat stress are already approaching critical levels during the susceptible stages of rice plant in Bangladesh during the months of March to June. Sea water inundation is also identified as a major problem for agriculture. It has further reported that Bangladesh will experience a net increase in poverty of 15% by 2030 in context of low crop productivity. Adaptation measures including salinity-tolerant crop cultivation, hydroponics, embankment cropping aquaculture, early flood

 Hanna Schübel and Ivo Wallimann-Helmer (eds.) ***Justice and food security in a changing climate***
EurSafe 2021 – DOI 10.3920/978-90-8686-915-2_16, 

warning systems, risk insurance mechanisms are identified as effective adaptation measures to manage the risk of reducing food production (AR5, WG2; Chapters 5 & 24).

Literature review supporting AR5 reveals that climate change adversities in Bangladesh are projecting cut of Gross Domestic Products (GDP) by -0.11% by 2030, lowering agricultural production by -1.23%, hiking imports of goods by 1.52%, limiting consumption of total calories by 17% and also leaving some households to be underfed as a result of uneven distribution of foods (Banerjee *et al.*, 2015). By emphasizing on urgency of proper adaptation strategies struggling to ensure growth of annual yield through necessary technical progress is desired to fight with this situation adequately (Kobayashi and Furuya, 2011). Besides advocating international negotiations developing adequate legal and policy framework at national level is a prerequisite to protect the citizens from both existing and projected loss and damage situations arising out of hitting consequences of climate change (Al Faruque and Khan, 2013).

In this position, this study tends to address three questions. 1. What is the current state of legal and institutional responses to food security and climate change in Bangladesh? 2. What are the gaps of the existing system? 3. What are the appropriate legal and institutional frameworks required for dealing with threats towards food security due to climate change impacts?

## Legal, policy and institutional responses in Bangladesh

The Constitution of the People's Republic of Bangladesh has imposed duties on the State to make endeavour for improvement and protection of natural environment and biodiversity for both the present and future citizens (Article18A) and recognized basic necessities including food by Article 15(a) which obliges the GoB to ensure the food security for the citizens. However, the relevant legal enactments as existing in Bangladesh, namely, the Environment Conservation Act, 1995, the Environment Conservation Rules, 1997, the Environment Court Act, 2010 (repealed and re-enacted the Environment Court Act, 2000), have basically defined the offences against environment, punishments for such offences along with procedural and institutional mechanisms to enforce those laws containing no direct concerns about public good or food security. The Environment Conservation Act, 1995 has established the Department of Environment under the than Ministry of Environment and Forest (MoEF), presently renamed as the Ministry of Environment, Forest and Climate Change (MoEFCC), for implementation of relevant laws and regulations.

The Bangladesh Climate Change Strategy and Action Plan (BCCSAP), 2009 identified 24 ministries, and their responsibilities regarding 44 programs and it also identified the role of the concerning development partners. Specifically, there are 9 relevant ministries adopted different legislations, policies, and plans and also currently implementing different projects and programmes related to climate change and food security. Hereby relevant ministries and their initiatives in compliance with these issues are scrutinized to identify the scopes and gaps to ensure food security in the context of climate change impacts and vulnerabilities in Bangladesh.

### *Ministry of Environment, Forest and Climate Change*

The MoEFCC (previous MoEF) as to the focal point of the United Nations Framework Convention on Climate Change (UNFCCC) is leading global climate change negotiations and also providing necessary guidance to develop national policy regime. MoEF has prepared the National Adaptation Programme of Action (NAPA), 2005, updated, 2009 and the Bangladesh Climate Change Strategy and Action Plan (BCCSAP), 2008, reviewed, 2009. Both of these have stepped forward towards food security by emphasizing on agricultural research to develop crop verities which are tolerant of climate change impacts like flooding, drought and salinity, based on indigenous and other varieties suited to the needs of resource. The Bangladesh Climate Change Trust Fund (BCCTF) was established under MoEF through legislation, namely, the Climate Change Trust Fund Act, 2010 to implement the six thematic areas of BCCCSAP, 2009 and this fund is financed from the national budget of Bangladesh. Also a multi-donor trust fund, the Bangladesh Climate Change Resilience Fund (BCCRF) was created in 2009 and came operational in 2010 with a view to implement projects of both government agencies and Non-Government Organizations (NGOs) to facilitate the thematic areas of BCCSAP including food security. Intended Nationally Determined Contribution (INDC), submitted to UNFCCC by GoB in 2015, which is now NDC as to the Paris Agreement, 2015, specifically identified the need for investment for adaptation measures related to food security and livelihood and health protection including water security (MoEF, GoB, 2015). The Bangladesh Climate Change and Gender Action Plan (BCCGAP), 2013, prepared by MoEF, also addressed the issue of integration of gender considerations in the thematic area related to food security, social protection, and health. Both of the Fourth National Report to the Convention on Biological Diversity submitted by Bangladesh in the year 2010 headed as the National Biodiversity Assessment and Programme of Action 2020 and the Bangladesh Country Investment Plan for Environment, Forestry and Climate Change (2016-2021) (BCIPEFCC) as a living document have focused on the issues relating to management of natural resources and adaptation and those are some key points to ensure national food security.

#### *Ministry of Disaster Management and Relief*

The Ministry of Disaster Management and Relief (MoDMR) adopted number of plans, policies and legislations to fight with the adversities of climate change aiming to Disaster Risk Reduction (DRR). The Disaster Management Act, 2012 has provided guidance to set up institutional mechanisms to fight with the disasters and adverse climate change impacts of which the MoDMR is the authority of implementation. The National Disaster Management Policy, 2015, and the National Plan for Disaster Management (NPDM), 2010-2015; 2016-2020 and 2021-2025 are adopted by the MoDMR. The MoDMR adopted the Standing Orders on Disaster (SoD), 2019, revising the Standing Orders on Disaster (SOD), 2010 to coordinate between national policies and international frameworks for DRR by identifying duties and responsibilities of the relevant agencies, institutions, bodies of central and local government and NGOs to reduce vulnerabilities, rehabilitate, and provide humanitarian assistance to the victims of both disasters and climate change impacts (MoDMR, 2019). However, though disasters have severe adverse impacts on national food security, only NPDM, 2021-2025 is concerned about creating food stock at family level merely and not about local or national food security in crucial times, like disasters, as a whole.

#### *Ministry of Agriculture*

The Ministry of Agriculture (MoA) has identified development of climate resilient crop variety and technology as one of the priority areas by the National Agriculture Policy, 2018. Also the MoA has developed National Agricultural Extension Policy, 2020 by addressing the challenges of climate change aiming to extension of good agricultural practices as a strategy to produce safe and nutritious food.

#### *Ministry of Food*

The Ministry of Food developed the National Food Policy Plan of Action 2008-2015, which identified climate change that poses an additional burden on food security, especially in areas where agriculture and water resources are already under stress due to adverse meteorological conditions. A monitoring report is submitted by assessing this plan by the year 2010.

### *Ministry of Health and Family Welfare*

The Ministry of Health and Family Welfare (MoHFW) adopted the Health Policy in 2011, which provided guidance to adapt with climate impacts and relevant diseases. The MoHFW also adopted the National Nutrition Policy in 2015 and this policy provide provision to adapt for ensuring food security, employment and disease management strategies in line with the situation related to climate change in Bangladesh.

### *Ministry of Water Resources*

The Ministry of Water Resources (MoWR) is responsible to regulate the water resources in Bangladesh. The MoWR developed a Master Plan of Haor Areas in 2012, which addressed climate impacts and vulnerabilities in water resources in Haor basins in its various sections. This Master Plan of Haor Areas, 2012 also recognized the climate change implications and the loss of fisheries biodiversity, which is evident in the haor areas. Though water is an essential part of food security being inevitable for both agricultural production and drinking purpose yet these matters are not focused in the Bangladesh Water Act, 2013.

### *Ministry of Fisheries and Livestock*

Fisheries and Livestock is identified as one of the climate vulnerable sector in BCCSAP, 2009, but this sector regulating agency. The Ministry of Fisheries and Livestock (MoFL) took little effort to develop relevant policies, plans, and programs. The MoFL recently adopted the National Shrimp Policy in 2014 and this policy considered the climate change impacts and vulnerabilities in shrimp sector.

### *Ministry of Planning*

The Ministry of Planning (MoP) is responsible for developing overall planning of the country and for implementation of such plans. National Sustainable Development Strategy (NSDS), 2010-2021; Perspective Plan of Bangladesh 2010-2021; Sixth Five Year Plan and Seventh Five Year Plan as developed by the MoP have identified the development priorities for ensuring food security taking into account the climate impacts. The MoP also developed the National Social Security Strategy (NSSS), 2015 addressing the issue of climate change and food security.

### *Ministry of Finance*

The Ministry of Finance (MoF) adopted the Bangladesh Climate Fiscal Framework (CFF) in 2014 which, identified the principles and tools for development of Climate Fiscal Policy (CFP) of which 'food security, social protection and health' is considered to be the first thematic area following BCCSAP. The Economic Relations Division (ERD) of the Ministry of Finance is acting as the National Designated Authority (NDA) of Bangladesh to the Green Climate Fund (GCF) (Finance Division, 2014).

## Key findings and way forward

Being one of the worst victims of climate change admittedly, food security of Bangladesh has headed forefront as a burning issue due to both sudden and slow onset environmental disasters. The GoB is playing active role in climate negotiations and international climate policy development processes besides developing national legislations, policies, plans, programs and institutions aligning with the international policy regime on climate change and food security to protect its vulnerable communities from the adversities of climate change. However, different ministries of Bangladesh are taking different relevant steps by focusing on their own agenda and avoiding the interdisciplinary approach which is inevitable to fight with the challenges of climate change and food security adequately. Even the same ministry has created different guiding documents sometimes which created ambiguities about de facto focus of the ministry. Though the BCCSAP has empowered the focal points of climate change in all associated ministries to make plans as well as implement those exsisting, it is still missing the urgency of a single and coordinated plan to be implemented by all the relevant ministries hands in hands. Without creating different documents it could be wise for MoEFCC to adopt a single document by addressing all the concerns as focused by NAPA, BCCSAP, BCCGAP and BCIPEFCC. Also the MoA could adopt a single policy with combination of the focal points of agriculture and agricultural extension. Matters relating to agriculture, food, nutrition, water fisheries and livestock are so much inter-connected that dealing with these aspects separately is not preferable in any way. Also the MoP and MoF should work together as without proper finance planning cannot be accurately implemented. In spite of numbers of sectoral legislations, policies and plans have been adopted by different ministries of GoB in connection with climate change impacts and food security yet the urge for a consolidated approach to deal with this phenomenon continued to remain unheard and unattended. No scope is provided for integrating the community people within these procedures to explore and utilize their indigenous knowledge to tackle this situation in a bottom up approach. As such, further comprehensive studies and steps are required to explore scopes for promoting coordinated and integrated policy and institutional approaches to address the relevant issues all together by developing an action plan for sectoral coordination and integration in consonance with all the stakeholders by applying bottom up approach for setting a positive example to be followed by the world community. The BCCSAP is a plan for ten years of which the targeted period is already over (MoEF, 2009). It will be highly appreciated if the second version of this document can take an integrated approach based on realistic research studies by accommodating all the relevant institutions and sectors within the purview of this prime instrument.

## References


Al Faruque, A., and Khan, M. H. I. 2013. Loss *& Damage Associated with Climate Change: The Legal and Institutional Context in Bangladesh*. Available at: http://asiapacific.domain50.com/sites/default/files/resource/attach/loss-n-damage-associated-w-climate-change-legal-n-institutional-context-in-bangladesh.pdf.

Banerjee, O., Mahzab, M., Raihan, S., and Islam, N. 2015. An economy-wide analysis of climate change impacts on agriculture and food security in Bangladesh. *Climate Change Economics*, *6*(01), 1550003.

BHWDB, 2012. *Master Plan of Haor Area.* Dhaka: Bangladesh Haor and Wetland Development Board, Ministry of Water Resources, Government of the People's Republic of Bangladesh.

Disaster Management Bureau, 2010. *National Plan for Disaster Management 2010-2015.* Dhaka: Disaster Management Bureau, Disaster Management and Relief Division, Ministry of Food and Disaster Management, Government of the People's Republic of Bangladesh.

Finance Division, 2014. *Bangladesh Climate Fiscal Framework, 2014,* Fiscal Division, Ministry of Finance, Government of the People's Republic of Bangladesh.

FPMU, 2008. *The National Food Policy Plan of Action (2008-2015).* Dhaka: Food Planning and Monitoring Unit, Ministry of Food and Disaster Management, Government of the Peoples Republic of Bangladesh.

FPMU, 2010. *The National Food Policy Plan of Action: Monitoring Report, 2010.* Dhaka: Food Planning and Monitoring Unit, Ministry of Food and Disaster Management, Government of the People's Republic of Bangladesh.

General Economics Disvision, 2012. *Perspective Plan of Bangladesh 2010-2021.* Dhaka: General Economics Division, Planning Commission, Government of the People's Republic of Bangladesh.

General Economics Division, 2015. *7th Five Year Plan FY2016-FY2020.* Dhaka: General Economics Division, Planning Commission, Government of the People's Republic of Bangladesh.

IPCC, 2014: Climate Change 2014: Impacts, Adaptation, and Vulnerability. Part A: Global and Sectoral Aspects. Contribution of Working Group II to the Fifth Assessment Report of the Intergovernmental Panel on Climate Change [Field, C.B. *et al.* (eds)]. Cambridge University Press, Cambridge, United Kingdom and New York, NY, USA, 1132 pp.

IPCC, 2014: Climate Change 2014: Impacts, Adaptation, and Vulnerability. Part B: Regional Aspects. Contribution of Working Group II to the Fifth Assessment Report of the Intergovernmental Panel on Climate Change [Barros, V.R. *et al.* (eds)]. Cambridge University Press, Cambridge, United Kingdom and New York, NY, USA, pp. 688.

Kobayashi, S., and Furuya, J. 2011. Comparison of climate change impacts on food security of Bangladesh. *Studies in Regional Science, 41*(2), 419-433.

MoA, 2018. *National Agriculture Policy, 2018,* Ministry of Agriculture, Government of the People's Republic of Bangladesh.

MoA, 2020. *National Agricultural Extension Policy, 2020,* Ministry of Agriculture, Government of the People's Republic of Bangladesh.

MoDMR, 2017. *National Plan for Disaster Management (2016-2020).* Dhaka: Ministry of Disaster Management and Relief, Government of the People's Republic of Bangladesh.

MoDMR, 2021. *National Plan for Disaster Management (2021-2025).* Dhaka: Ministry of Disaster Management and Relief, Government of the People's Republic of Bangladesh.

MoDMR, 2019. *Standing Orders on Disaster, 2019,* Ministry of Disaster Management and Relief, Bangladesh Secreteriate, Dhaka – 1000.

MoEF, 2009 a. *Bangladesh Climate Change Strategy and Action Plan 2009.* Dhaka, Bangladesh: Ministry of Environment and Forests, Government of the People's Republic of Bangladesh.

MoEF, 2009 b. *National Adaptation Programme of Action.* Dhaka: Ministry of Forests and Environment, Government of the People's Republic of Bangladesh.

MoEF, 2010. *Biodiversity National Assessment and Programme of Action 2020.* Dhaka: Ministry of Environment and Forests, Government of the People's Republic of Bangladesh.

MoEF, 2013. *Bangladesh Climate Change and Gender Action Plan.* Dhaka: Ministry of Environment of Forests, Government of the People's Republic of Bangladesh.

MoEF, 2015. *Intended Nationally Determined Contributions (INDC).* Dhaka: Ministry of Environment and Forests, Government of the People's Republic of Bangladesh.

MoEF, 2016. *Bangladesh Environment, Forestry and Climate Change Country Investment Plan.* Dhaka: Ministry of Environment and Forests,Government of the People's Republic of Bangladesh, Bangladesh Secreteriate, Dhaka – 1000.

MoFL, 2014. *National Shrimp Policy, 2014,* Fisheries – 2 and Law Section, Ministry of Fisheries and Livestock, Government of the People's Republic of Bangladesh.

MoHFW, 2015. *National Nutrition Policy.* Dhaka: Ministry of Health and Family Welfare, Government of the People's Republic of Bangladesh.

MoHFW, 2015. *National Health Policy.* Dhaka: Ministry of Health and Family Welfare, Government of the People's Republic of Bangladesh.

The Bangladesh Water Act, 2013 (Act No. 14 of 2013).

The Climate Change Trust Fund Act, 2010 (Act No. 57 of 2010).

The Constitution of the People's Republic of Bangladesh, 1972 (As Amended Up To Date).

The Disaster Management Act, 2012 (Act No. 34 of 2012).

The Environment Conservation Act, 1995 (Act No. 1 of 1995).

The Environment Conservation Rules, 1997.

The Environment Court Act, 2010 (Act No. 56 of 2010).

World Bank, 2016, *BCCRF Annual Report,* Dhaka: Bangladesh Climate Change Resilience Fund and Ministry of Environment and Forest, Government of the People's Republic of Bangladesh.

# 17. A water culture perspective for food security

*S. Meisch[1,2*] and S. Bremer[1]*
*[1]Centre for the Study of the Sciences and the Humanities (SVT), University of Bergen, Parkveien 9, 5020 Bergen, Norway; [2]International Centre for Ethics in the Sciences and Humanities (IZEW), University of Tübingen, Wilhelmstr. 19, 72074 Tübingen, Germany; simon.meisch@uni-tuebingen.de*

## Abstract

The changing climate is disrupting freshwater infrastructures and this affects food security. Dominant technocratic systems of knowledge and practice supporting water infrastructures are being undermined by rapid changes, which introduce new uncertainties and contest old securities. Novel ways of mobilising knowledge and practice for water infrastructures are needed that readmit communities' diverse cultural interactions with water and promote adaptive decision-making. The paper introduces the project 'Changing Water Cultures' (CANALS) claiming: (1) that current understandings of water, climate and infrastructures struggle to address emerging uncertainties due to climate change; (2) that a proper regard for water cultures is needed to overcome these obstacles; particularly by (3) extending the 'peer communities' of diverse knowledge-holders. The paper's water culture perspective understands cultures as both practices and systems of symbols and meanings. It recognises the hybrid character of water, which exists independently of humans and at the same time is culturally enacted through human practices. From this perspective, the paper looks beyond infrastructures physical manifestation to include the social practices connecting people and objects in the world in socio-material relations. A water-cultural perspective thus better captures the complex processes around water infrastructures and food in the context of challenges posed by climate change. The paper first argues for a novel way of analysing water infrastructures and second explains its relevance for understanding food security in a changing climate.

**Keywords:** climate change, water infrastructures, social practices

## Introduction

The fast-changing global climate is altering precipitation patterns, thus posing major challenges to freshwater infrastructures – be it that sewage and stormwater systems can no longer cope with the amount of rain, or that it becomes difficult to create spaces to accommodate flood waters in densely populated areas. Dominant technocratic and managerial systems of knowledge and practice supporting water infrastructures in Europe are being undermined by these rapid changes, which introduce new uncertainties and contest old securities. In particular, a narrow focus on engineering 'climate-proof' infrastructures can ignore the multifarious ways communities culturally relate to, know and use water. This narrow focus can potentially have maladaptive consequences, where a flood-proof river loses its other functions as a swimming hole or habitat for example. So, novel ways of mobilising knowledge for water infrastructures are needed – from sewage, pipes or urban flooding areas to new customs and traditions – which readmits communities' diverse cultural interactions with water, in order to promote adaptive decision-making. The project 'Changing Water Cultures' (CANALS), located at the Centre for the Study of the Sciences and the Humanities (Bergen, Norway) set itself precisely the task of mobilizing such knowledge by developing a water-cultural approach.

In this paper, we will present key assumptions of the CANALS project and discuss its relevance for food security. We will firstly introduce the concept of modern water, which we then contrast with a water cultural perspective (around water infrastructures). Thirdly, we will introduce the CANALS project, before concluding by discussing its implications for food security.


EurSafe 2021 – DOI 10.3920/978-90-8686-915-2_17, © Wageningen Academic Publishers 2021

## Modern water

Dominant modern frameworks of freshwater governance in Europe are criticized for failing to acknowledge the cultural embeddedness of water (Böhme, 1988; Swyngedouw, 2009; Bruns *et al.*, 2021; Wesselink *et al.*, 2017). Steered mainly by engineering and planning professionals, water governance has adopted a highly techno-scientific and reductionist focus on establishing, using and maintaining hard infrastructure like pipes and reservoirs; reinforcing the separation of nature and culture, and the exclusion and objectification of water (quantification, commodification). Water has become a resource managed by technoscientific means, in isolation from its social and cultural contexts, with resulting injustices through the deprivation of rights and the destruction of the environment (Linton, 2013; Meisch, 2019; Swyngedouw, 2009). For instance, Bremer *et al.* (2019a) show how calls to climate-proof the Vosso river in Norway against floods ignores the rivers functions for fish-spawning, white-water rafting and its intrinsic natural value.

The prospect of rapid climate change makes starkly apparent the weaknesses of a narrow techno-scientific focus. These changes are undermining infrastructures design parameters, which are often based on a statistically defined 'normal' precipitation regime that is increasingly unrecognisable. New precipitation patterns are stressing these systems to their limits, with sometimes severe consequences, e.g. through compromised dams or overburdened sewage systems. And this is arguably challenging water professionals' privileged role as water managers.

## A water culture perspective

Against this background, there are calls for new approaches for both the analysis and governance of water, which take seriously waters complex linkages to the natural, social and cultural world (Bremer *et al.*, 2019b; Bruns *et al.*, 2021; Linton and Budds, 2014; Meisch, 2019; Orlove and Caton, 2010). This includes the cultural practices, artefacts and representations that intimately link the ways we live with water and climate (Donner, 2011; Hulme, 2015; Rudiak-Gould, 2013). It invites a perspective that assumes that water, climate and infrastructures 'co-produce' each other (Table 1).

Key to the CANALS project is its 'water culture perspective' understanding cultures as both practices and systems of symbols and meanings (Sewell, 1999). This perspective allows three challenges to be met in order to better perceive water infrastructures in times of climate change.

First, it recognises the hybrid character of water, which exists independently of humans and at the same time is culturally appropriated and enacted through human practices.

*Table 1. Definitions of water, climate, and infrastructure as entangled in water cultures.*

| | | |
|---|---|---|
| WATER 'is neither purely "natural" nor purely "social" but simultaneously and inseparably both [... Water] exists apart from human influence [...], and simultaneously is produced and enacted through human labour and social action within a given mode of production [...].' (Perreault, 2014) | CLIMATE is 'a tacit idea held in the human mind or in social memory of what the weather of a place "should be" at a certain time of year. [...] climate allows humans to live culturally with their weather.' (Hulme, 2015) | An INFRASTRUCTURE is 'a practice of connecting people and things in socio-material relations that sustain urban life. It is not just a context or a noun, but a verb: social infrastructure is made and held stable through work and changing ways of connecting.' (McFarlane and Silver, 2017) |

Second, it understands infrastructures as social practices connecting people and objects in the world in socio-material relations. Consequently, CANALS chose to focus on infrastructures as 'heterogeneous infrastructure' configurations (Lawhon *et al.*, 2018). This focus 'enables a clearer analysis of infrastructural artefacts not as individual objects but as parts of geographically spread socio-technological configurations'; hard works like pipes, entangled with soft practices, from riverbank maintenance to traditional water allocation (Lawhon *et al.*, 2018, 720). These infrastructures connect people who live in and with them and interact with their water. CANALS studies how formal and informal institutional practices and water cultures co-produce each other and how water cultures can be mobilised to adapt water infrastructures to climate change. A water-cultural perspective can thus capture these complex processes around water and water infrastructures more appropriately and better address more and different challenges posed by climate change. A water culture perspective means acknowledging the many connections between people and their waters and opening up for the multiplicity of institutions and practices emerging with and structuring these connections. For example, we can consider municipality engineers and their consultant counterparts as one water culture; composed of a network of professionals, conducting technical practices towards defined ends (e.g. the efficient and effective supply of water) based on a set of scientific water models and values. A conflicting culture could be seen in a network of river-care NGOs and schools, who look to reinstate the riparian habitats, motivated by environmental and educational values, and steered by horticultural practices.

Third, a water cultural perspective better captures how water and climate mutually produce particular water cultures and are produced from these water cultures (on climate co-production, see Bremer and Meisch, 2017). People have expectations that their normal weather provides, at different times, particular kinds and amounts of water. And these expectations create formal and informal practices and connections around objects in the world like swimming pools, sewage, rainwater retention basins, horticulture, car washes, or (supra-) regional food production. In this vein, the water cultural perspective is an analytical lens to better grasp the social nature of tangled water and climate in infrastructures (Castree, 2017). It studies the 'systems of symbols and meanings' that determine and govern the manifold relationships between people and their water as well as the 'practices' of people living within these systems, maintaining and changing them (Bremer *et al.*, 2019b).

## The CANALS project

The challenges discussed above have spurred a field of social and cultural sciences critically engaging with heterogeneous water infrastructures as 'water cultures' (Bichsel, 2016). A broad range of studies about water infrastructures especially from critical geographies, political ecologists, economists and STS scholars, unravel how the social processes and metabolic ecological processes (i.e. the transformation of 'nature' by means of different technologies and practices) constantly produce and reproduce each other. These studies, in the tradition of critical theory and poststructuralist approaches, critically explore the meanings, symbols, and power structures that are inherent in seemingly technical water infrastructures. Yet most of this scholarship stops at providing a critical perspective (on implicit power relations or injustices) but does not go on to produce directly action-guiding knowledge or practices (Floyd, 2013; Meisch, 2018).

The CANALS project will extend the boundaries of this field of water cultures by going beyond the diagnosis of problems with the technoscientific paradigm, and asking people engaged in water cultures to actively mobilise new ways of co-creating knowledge for policymaking on water. The project will first make visible the whole spectrum of knowledges and practices used by different social groups for creating and maintaining infrastructures. Going further, the project will second develop ways of convening an 'extended peer community' (Funtowicz and Ravetz, 1993) of water stakeholders in these cultures, to together appraise the quality of their water knowledges for policymaking under climate change. CANALS begins from three main claims; (1) that current understandings of water, climate and infrastructures struggles to address the emerging uncertainties due to climate change; (2) that a proper regard for water cultures is needed to overcome these obstacles; and (3) that new ways of knowledge production in a water culture opens up approaches to extend peer communities to diverse knowledge-holders, introducing transdisciplinary perspectives and problem solving mechanisms.

Addressing the limitations of the literature, CANALS asks 'how can a water culture perspective improve the quality of knowledge production for water infrastructures facing rapid climate change?' This overarching research question will be broken up into two parts, addressed by two sub-questions, which in turn translate into two research objectives that each steer a work package (Table 2).

*Table 2. Guiding questions and objectives of CANALS.*

| Overarching research question: How does a water culture approach improve the quality of knowledge production for water infrastructures facing rapid climate change? | |
|---|---|
| Sub-question A: How is knowledge about water and climate culturally co-produced in water infrastructures? | Sub-question B: How does a water cultures perspective extend the peer community for assessing the quality of knowledge for water infrastructures? |
| Objectives 1: Advance understanding of how knowledge is co-produced in water cultures. | Objective 2: Mobilise an extended water cultures peer community for assessing current forms of knowledge for water infrastructures and coproducing new forms better adapted to climate change |
| Work package 1: Analysing Water Cultures | Work package 2: Mobilising Water Cultures |

## Implications for food security

The CANALS project deals with both the different co-productions of water, climate and infrastructures and the concrete, situated configurations of water cultures. It thus follows on from the critical literature on the Water-Energy-Food (WEF) Nexus (Leese and Meisch, 2015; Völker *et al.*, 2019; Wiegleb and Bruns, 2018; Williams *et al.*, 2019) and at the same time goes beyond it.

According to Bruns *et al.* (2021):

> the Water-Energy-Food (...) Nexus has been employed in discussions as a conceptual approach, an analytical framework, and a governance perspective to address the relationships between vital resource flows and basic supply systems (...). Originating from a development and resource management discourse (...), centre stage in WEF-Nexus thinking is that water, energy, and food are interlinked sectors that can neither be fully appreciated nor indeed be governed independently of each other.

Bruns *et al.* acknowledge the heuristic added value of the nexus perspective. Yet, they criticise the generalising focus of most approaches ('the one nexus') and their techno-managerial ways of approaching the issue, which are also predominantly concerned with networked infrastructures. In contrast, Bruns *et al.* stress the relevance of a perspective dealing with 'heterogeneous infrastructure' configurations and support an analytical lens

> that focusses on infrastructure in-the-making as well as in-the-use [which in turn] discloses the socio-material entanglements characteristic of any one individual Nexus. We reveal how and by which 'matters of practices' (Shove, 2017) a particular Nexus is crafted, how it is stabilised, and under what circumstances any specific Nexus configuration turns into another. (Bruns *et al.*, 2021)

They expect their approach not only to provide a deeper understanding of how different resource flows are interrelated, but also to address numerous normative issues at different levels. At the macro and meso level, it is about the socio-ethical questions which scientific and political pre-decisions determine whose values, norms and knowledge are heard and considered in decision-making processes. It also concerns the recognition of informal and invisible contributions to the flow of resources in a society. Last, but not least, it is about the recognition of injustices when people are pushed into informal and often harmful social domains or their livelihoods are disrupted by formal nexus configurations. At the micro level, it is about the perception and empowerment of agency.

So, when CANALS deals with water cultures in times of climate change, looking at heterogeneous infrastructures that necessarily include both social and material aspects, it will invariably also have to engage with the interactions of water with other resources. Water, agriculture and food production are closely intertwined. By adopting a water-cultural perspective, CANALS can broaden the view of the interconnectedness of resource flows and thus initiate communal discussion on how people want to be connected to each other and their natural environment in times of climate change.

## Acknowledgement

Meisch acknowledges funding for the 'CANALS' project from the European Union's Horizon 2020 research and innovation programme under the Marie Skłodowska-Curie grant agreement No 895008. Bremer had support from the 'CALENDARS' project, which has received funding from the European Research Council (ERC) under the European Union's Horizon 2020 research and innovation programme (Grant agreement 804150).

## References

Bichsel, K. (2016). Water and the (Infra-)Structure of Political Rule: A Synthesis. Water Alternatives 9(2): 356-372.

Böhme, H. (1988). Umriß einer Kulturgeschichte des Wassers. Eine Einleitung. In Böhme, H. (ed.) Kulturgeschichte des Wassers. Suhrkamp, Frankfurt/Main, pp. 7-42.

Bremer, S., Wardekker, A., Dessai, D., Sobolowski, S., Slaattelid, R. and van der Sluijs, J. (2019a). Toward a multi-faceted conception of co-production of climate services. Climate Services 13: 42-50.

Bremer, S.; Schneider, P. and Glavovic, B. (2019b). Climate Change and Amplified Representations of Natural Hazards in Institutional Cultures. In: Oxford Research Encyclopedia of Natural Hazard Science. https://doi.org/10.1093/acrefore/9780199389407.013.354

Bremer, S. and Meisch, S. (2017). Co-production in climate change research: reviewing different perspectives. WIREs Climate Change, 8: e482. https://doi.org/10.1002/wcc.482

Bruns, A., Meisch, S., Ahmed, A., Meissner, R. and Romero-Lankao, P. (2021). Nexus disrupted: Lived realities and the Water-Energy-Food Nexus in the urban and beyond. Geoforum (accepted).

Cairns, R. and Krzywoszynska, A. (2016). Anatomy of a buzzword: The emergence of 'the water-energy-food nexus' in UK natural resource debates. Environmental Science & Policy 64, 164-170.

Castree, N. (2017). Nature, in: The International Encyclopedia of Geography. https://doi.org/10.1002/9781118786352.wbieg0522

Donner, S. (2011). Making the Climate a Part of the Human World. Bulletin of the American Meteorological Society 92(10): 1297-1302.

Floyd, R. (2013). Analyst, theory and security: A new framework for understanding Environmental Security Studies. In: Floyd, R. and Matthew, R. (eds) Environmental Security: Approaches and Issues, Routledge, London, pp. 1-20.
Funtowicz, S. and Ravetz, J. (1993). Science for the post-normal age. Futures 25(7): 739-755.
Hulme, M. (2015). Climate and its changes: a cultural appraisal. Geo 2: 1-11.
Linton, J. and Budds, J. (2014). The hydrosocial cycle: Defining and mobilizing a relational-dialectical approach to water. Geoforum 57: 170-180.
Lawhon, M.; Nilsson, D., Silver, J., Ernstson, H. and Lwasa, S. (2018). Thinking through heterogeneous infrastructure configurations. Urban Studies 55(4): 720-732.
Leese, M. and Meisch, S. (2015). Securitising Sustainability? Questioning the 'Water, Energy and Food – Security Nexus'. Water Alternatives 8(1): 695-709.
McFarlane, C. and Silver, J. (2017). Navigating the city: Dialectics of everyday urbanism. Transactions of the Institute of British Geographers 42(3): 458-471.
Meisch, S. (2019). *I Want to Tell You a Story*: How Narrative Water Ethics Contributes to Re-theorizing Water Politics. Water 11(4), 631. http://doi.org/10.3390/w11040631
Meisch, S. (2018). Water Ethics – Lessons from Post-Normal Science. In: Springer, S. and Grimm, H. (eds) Professionals in Food Chains: Ethics, Roles and Responsibilities. Wageningen Academic Publishers, Wageningen, pp. 459-464.
Perreault, T. (2014). What kind of governance for what kind of equity? Towards a theorization of justice in water governance. Water International 39(2): 233-245.
Rudiak-Gould, P. (2013). 'We Have Seen It with Our Own Eyes': Why We Disagree about Climate Change Visibility. Weather, Climate, and Society 5(2): 120-132.
Sewell, W. (1999). The Concept(s) of Culture. In: Bonell, V. and Hunt, L. (eds) Beyond the Cultural Turn. New Directions in the Study of Society and Culture. University of California Press, Berkeley, pp. 86-94.
Strang, V. (2004). The Meaning of Water. Berg Publishers, Oxford, New York.
Orlove, B. and Caton, S. (2010). Water Sustainability: Anthropological Approaches and Prospects. Annual Review of Anthropology 39: 401-415.
Shove, E. (2017). Matters of Practice. In: Hui, A., Schatzki, T.R. and Shove, E. (eds) The nexus of practices. Connections, constellations and practitioners. Routledge, London, New York, pp. 155-168.
Swyngedouw, E. (2009). The political economy and political ecology of the hydrosocial cycle. Journal of Contemporary Water Research & Education 142(1): 56-60.
Voelker, T., Blackstock, K.; Kovacic, Z.; Sindt, J.; Strand, R. and Waylen, K. (2019). The role of metrics in the governance of the water-energy-food nexus within the European Commission. Journal of Rural Studies. https://doi.org/10.1016/j.jrurstud.2019.08.001
Wesselink, A., Kooy, M. and Warner, J. (2017). Sociohydrology and hydrosocial analysis: toward dialogues across disciplines. WIREs Water, 4: e1196.
Wiegleb, V. and Bruns, A. (2018). What Is Driving the Water-Energy-Food Nexus? Discourses, Knowledge, and Politics of an Emerging Resource Governance Concept. Frontiers in Environmental Science 6, 870.
Williams, J., Bouzarovski, S. and Swyngedouw, E. (2019). The urban resource nexus: On the politics of relationality, water-energy infrastructure and the fallacy of integration. Environment and Planning C: Politics and Space 37: 652-669.

# 18. Agricultural resilience and wine production: a value analysis

*G. De Grandis*
*Department of Philosophy and Religious Studies, Norwegian University of Science and Technology, Dragvoll, 7491 Trondheim, Norway; giovanni.grandis@ntnu.no*

## Abstract

Climate change presents the agricultural and agro-industrial sectors with formidable challenges in meeting the food demands of the world population. What could be the role of those agricultural productions with no or negligible nutritional value but high economic value, like floriculture and wine production? We look at the case of the wine industry and the role it may play in an agroindustrial sector that needs to tackle the challenge of feeding the world under climate change circumstances. The wine industry may be accused of competing with other more important agricultural products for land, water (at least in some countries), fertilisers, etc. while producing a luxury product with limited nutritional value, as well as producing considerable GHG emission per bottle (including the various stages of farming, winemaking and marketing). Yet, viticulture and winemaking may contribute to the resilience of local agricultural systems and provide inspiring examples, know-how and capital resources. The wine sector has a deep connection with locality, is considerably vulnerable to global warming and has repeatedly shown the capacity to be a fly-wheel for the local economy and its agricultural and tourist sector. These features may be important assets in boosting the resilience of local agricultural systems under global warming stress. I put forward some ethical commitments and conditions that concerned wine producers and consumers need to meet in order to contribute to the sustainability and resilience of local agricultural systems. Since the high economic value of wine is so closely intertwined with its cultural and symbolic value, success needs the joint effort of producers, consumers and wine critics, which share responsibility in reorienting the sector towards supporting agricultural resilience.

**Keywords:** food ethics, climate change, food scarcity, viticulture, sustainability

## Background and problem statement

> We always need to remember that wine does not represent a fundamental good for human survival like protein, carbohydrates, lipids, therefore it needs a more advanced ethics of production. (Lorenzo Corino)

It is now widely acknowledged that climate change will have significant impacts on food security, although these impacts will be different according to the entity of climate change, to the timescale considered (less dramatic and widespread in the short term, potentially much more widespread and radical, but also more uncertain, in the long run), and to the different regions of the world (CCAFS, 2010; FAO, 2008). Instead of working under one specific scenario that would need accurate scientific backing, I will make some general assumptions that are fairly weak and seem in line with the scientific literature. The first assumption is that in the short term (2-3 decades) the most severe impacts will be felt by people working in the first sector in low income countries and by the urban poor, starting with those living in cities within low income areas. The second assumption is that in the longer term, as the impact of climate change become deeper and more severe, food security will require considerable shifts in food production patterns, new knowledge, methods and capabilities. The third assumption is that at some point in the future, uses of agricultural land that do not contribute to food security will need to provide convincing justification, at least from an ethical point of view. The last assumption suggests that a number of farmed products without significant nutritional value may come under scrutiny, and


EurSafe 2021 – DOI 10.3920/978-90-8686-915-2_18, © Wageningen Academic Publishers 2021

that if the food security crisis reaches critical points, producers and consumers of these products may themselves feel that their production and consumption choices are ethically problematic or perhaps even just wrong and indefensible. This could be the case, for instance, for flowers, for spices and for beverages like wine and beer (although these are not without any nutritional value and in the past have been not negligible sources of calories and safe drinks when clean water was difficult to obtain). How strong the pressure on these products will be at different points in time is very difficult to foresee, for it depends on how adaptation to and mitigation of climate change will unfold and on political, economic and cultural factors. It is clearly possible that at some point the situation will become so critical that the question will not arise and conversion to different cultivations will be inescapable, as it is possible that other changes (e.g. dramatic improvements in food production efficiency, or sharp declines in population) will prevent the issue to emerge. However, this paper adopts the perspective of what may be called the conscientious wine producer and wine drinker: i.e. people who are interested in producing or consuming wine, but are very aware of the impact that climate change will likely have on food security and believe that the wine industry has a role to play in tackling this issue. So, the question that I ask is what will make possible for this fictional and ideal typical wine producer and wine consumer to feel at peace with their conscience. Answering this question is not only relevant for wine producers and consumers who identify with our ideal conscientious specimen but can also be relevant for agricultural policy making. The approach adopted here bypasses two other positions that I consider rather extreme: (1) the moralistic stance of those who may claim that in the face of the threat of malnutrition and starvation any use of land that does not maximise food production and availability is unjustifiable; and (2) the laissez-faire perspective of those who may claim that one should leave it to the market to decide how producers will use land in response to consumers demands (and hence that as long as demand for luxury goods will be there any interference with the free market is wrong or counter-productive).

## Method

The aim of the paper is to offer some normative suggestions on how viticulture and wine making should respond to the possible impacts of climate change on food security. The normative recommendations are the result of working out how a set of values can be sustainably honoured and realized within the current and prospective circumstances in which viticulture and wine making operate. Therefore, two steps are needed: one is to identify the set of relevant values, the other is to understand the relevant empirical circumstances in which wine producers and consumers operate currently and in the prospective future.

The first task is performed through a theoretical construct, namely the outline of the ethical perspective of two hypothetical moral agents: what I call the conscientious wine producer and the conscientious wine consumer. The conscientious wine producer (henceforth CWP) is constructed by attributing to her a set of professional goals and moral concerns. The professional goals are kept to a simplified minimum, namely the CWP wants to make its wine producing activity economically viable – in the sense of generating and preserving the resources needed for it to continue in the future (as long as natural and socio-cultural enabling conditions obtain) – and profitable. The moral concerns are also kept to a few simple but broad principles. The first is that the CWP wants to act in respect of laws, regulations and of the principle of offering to consumers a product which is safe and enjoyable. The second is that CWPs want to work in such a way that contributes to the mitigation and adaptation to climate change, as long as this is compatible with preserving wine quality and economic sustainability of their activity. This aspiration is seen as an ongoing task, not as something that implies a sudden and radical transformation, nor as something that can be achieved simply by making one change and thinking that having done this satisfies once and for all the task. On the contrary, every successful step leads to the question: what can I do next (presenting this principle as a linear temporal progression is of course another simplification to make the point clear: the CWP is not satisfied with a token achievement)?

The conscientious wine consumer (henceforth CWC) too is constructed by attributing to him a goal and a moral concern. The goal is to continue to drink wines that give him satisfaction (hedonic and convivial value) while being prepared to adapt his consumption pattern, for instance through educating himself to different standards of wine appraisal, to reduce consumption and to be prepared to pay more. The moral concern is to support and encourage viticultural and wine making practices that contribute to climate change mitigation, adaptation and resilience.

It remains to understand the empirical circumstances in which wine producers and consumer operate now and in the prospective future. This could be an almost impossible task if interpreted as setting very detailed and accurate requirements, especially when it comes to future circumstances. What can be done in the context of this paper is to rely on the author's knowledge of the reality and main trends of vine growing, wine making and global wine markets acquired through some relevant literature (research done mostly while working on a biotechnology project on wine yeasts) and through his own two decades of experience as an interested and curious wine consumer. Moreover, I do not provide the empirical background at the outset. Rather, I take from the FAO (2008) framework document the three main dimensions of a response to climate change related food security and describe the implications for the wine sector on the basis of relevant knowledge of the circumstances of the sector.

## Responding to climate change related food security

According to an important document produced by the Food and Agriculture Organization of the United Nations (FAO) in 2008, a response to climate change threat to food security includes adaptation to changed climatic circumstances, increasing resilience to climate change impact and mitigating the driving causes of climate change, most notably through reducing greenhouse gas emissions. I will now try to illustrate what can be demanded to the wine sector in relation to these three dimensions.

### *Mitigation of climate change*

I start with mitigation of greenhouse gases emissions. At all three stages of vine growing, wine making and wine marketing substantial amounts of greenhouse gases are produced and can be measured, for instance, in terms of carbon footprint (CF) (Scrucca *et al.*, 2018). Studies and forward-looking producers have shown that actions are possible to reduce these emissions. Hence it is easy to conclude that CWPs ought to work out a transition plan that could enable them to reduce the CF at all stages of the lifecycle of the product: from using less fossil fuel in the vineyard (e.g. by limiting the use of tractors) to packaging the final product in lighter and easily recyclable containers and privileging nearby markets. Parallel to reducing emission could be the effort of increasing carbon capture, a goal that can be achieved by adopting viticultural practices that abandon a monocultural approach to the vineyard that eliminates any other form of vegetation and by embracing instead a more biodiverse approach to vineyard management and a more promiscuous land use. This enables both an increase in carbon capture and a more sustainable type of agriculture that preserves the quality and productive potential of the soil, this latter is a result important also for climate change adaptation and resilience (Corino, 2019).

What are the implications of mitigation for the CWC? CWCs have a very important role to play. First, they need to demand transparent, accurate and truthful information about the CF of wine producers so that they can make informed choices that reflect and are consistent with their values. Second, they should be prepared to accept higher production costs and hence higher final prices, for mitigation may bring additional costs (a least in the short period). Third, they should be prepared to change their expectations when it comes to the packaging of wine and also to the taste of wine. However, it should be noted that more sustainable viticultural and wine making practices may be expected to have interesting if not downright positive effects on the organoleptic qualities of wine. But consumers may have to accept less standardised taste and consistency. This is probably not something that would displease the wine connoisseurs, but may be more challenging for the less experienced, casual drinkers with a limited budget, who may not welcome the risk of spending money on a product that does not meet their expectations. In this respect, casual CWCs may need to adapt their wine taste and preferences to honour their ethical commitment. Yet, this taste adaptation can be consistent with an aesthetic appreciation of wine that is more subtle, authentic and aware of the connection between wine and the environment and climate that enable its production.

### *Adaptation to climate change*

Adaptation to climate change was defined by IPCC as 'Adjustment in natural or human systems in response to actual or expected climatic stimuli or their effects, which moderates harm or exploits beneficial opportunities' (IPCC, 2001). In food production an important element of this adjustment is to change practices so as to prevent or mitigate the decline of food production and supply, as well as to protect ecosystems and preserve natural resources (FAO, 2008). These principles have some important implication for CWPs. The first is that they should not plant new vineyards on fertile farmland that is better suited to crops with high nutritional value. Vineyards do best in poor hillside soils that are often not ideal for other crops and in the face of the climate change related challenges to food security, an attitude of restraints towards planting new vineyards is required. Restraint that should not be limited to fertile farmland, but also to land whose current use provides high carbon capture, like woodland (while grapevines do capture some greenhouse gases, tending a vineyard produces more GHG than it is captured by the grapevine) or that have great biological value in that they provided the habitat to either endangered species or to species that play a fundamental role in the preservation of the local ecosystem. Another important implication is that the use of scarce or non-renewable resources in viticulture should be avoided or minimised. The most important and telling example is water for irrigation. CWPs should try to avoid irrigation as far as possible (the *Vitis vinifera* is a plant that copes well with water scarcity stress) and when this is not possible the most efficient systems of irrigation should be adopted, and no groundwater should be used. Moreover, CWPs should farm their vineyards so as to avoid soil deterioration and erosion, in order for the land to retain its productive potential and to be converted to food production should the necessity arise. Finally, CWPs should commit to minimize their GHG emissions, not only because they are responsible of climate change, but also because $CO_2$ is also very likely to be responsible for lowering the nutritional value of many crops and foods (Ebi and Loladze, 2019; Loladze, 2016).

Furthermore, viticulture and winemaking have so far benefited from warmer climate, but are now reaching the tipping point in many wine producing regions. This is pushing many producers to develop new approaches and techniques that can be inspirational examples for other production. Thanks to the high commercial value of wine and profitability of premium wines, wine growers have the resources to lead in this experimentation and adaptation practices. And the fact that so far, they have reaped benefits from climate change puts an extra moral demand on them to make a special effort to contribute to adaptation.

The implication for CWCs is that they should demand clear standards and certifications that enable consumers to be informed about wine producers land use, resource management and vineyard practices.

### *Agricultural resilience*

Viticulture and wine production can be economically very rewarding because premium wines command high prices on the market. This, together with the *V. vinifera* high productivity make wine an important resource for many farmers and for many agricultural districts. Furthermore, it has been shown that producing quality wines with a strong connection with the local *terroir* can promote tourism and the development of other high quality and profitable agricultural products. This can significantly enhance the resilience of rural areas and hence contribute to preserve adequate supply of food.

What are the implications for CWPs and CWCs? CWPs who have good profits should accept the responsibility of contributing to the economic sustainability of the whole farming district to which they belong. When their products achieve the level of attention and interest that attract visitors interested in quality products, they should try to promote the sustainable development of the whole local community, striving to use the new sources of income to promote more sustainable and resilient farming practices within their district. Since this kind of development brings with it the risk that a district enjoying increased opportunities to sell premium products may switch from basic crops with great nutritional value to luxury production that do not contribute to satisfying basic food supply, it is important that specialized luxury productions do not displace basic crops. This could be achieved by adopting the principle that the district output of basic crops is not reduced. In practice this means that it is acceptable to allow luxury production to take up land that was previously devoted to basic crops, only if the income generated by these production can support better farming practices that enable the total production of basic crops to either increase or at least not to fall below previous levels. These principles can be embedded in a concept of responsible local farming transition, whose ideal would be to reinforce the resilience, sustainability and adaptive capacity of farming districts while honouring the duty of providing a sufficient supply of agricultural products needed to ensure local and global food security. This duty needs to be interpreted against the background of a global division of responsibility in food production, so that every region is expected to produce its fair share according to its productive potential and the global needs.

CWCs have an important role to play. The principles laid out for producers would be much easier to achieve if consumers are not too subject to fads and ephemeral fashion, but try instead to appreciate deserving but less famous and fashionable areas, so that the benefits of a precious and profitable product like good wine are distributed more widely among the regions that have the potential to produce good and fine wines. A distribution of the benefits based more on the intrinsic value of the wine and less on the successful branding of few fortunate areas would also contribute to avoid generating the wrong incentive to plant too many vineyards in certain regions with a negative impact on sustainability and biodiversity. CWCs should also be prepared to accept that global supply of wine, and of cheap wine in particular, may decline because it is not compatible with sustainable farming practices.

Clearly other stakeholders, like wine critics and traders have important responsibilities in supporting a transition towards a more sustainable and responsible viticulture and wine making. Since wine has always had complex cultural and symbolic value, changes in production and consumption need to be accompanied and supported by steering wine culture in the desired direction, a task to which many stakeholders need to contribute. However, for the sake of simplicity I have here focused only on the role of producers and consumers.

## References


CCAFS (2010). Agriculture, food security and climate change: Outlook for knowledge, tools and action. https://cgspace.cgiar.org/bitstream/handle/10568/3090/ccafs_report_3-low-res_final.pdf.

Corino, L. (2019). *L'essenza del vino e della viticoltura naturale*. Amazon.

Ebi, K.L. & Loladze, I. (2019). Elevated atmospheric CO2 concentrations and climate change will affect our food's quality and quantity. *The Lancet Planetary Health*, 3(7), e283-e284.

FAO (2008). Climate change and food security: a framework document. *Food and Agriculture Organization of the United Nations*, http://www.fao.org/3/k2595e/k2595e00.htm.

IPCC (2001). Climate Change 2001: Impacts, Adaptation, and Vulnerability. Contribution of Working Group II to the Third Assessment Report of the Intergovernmental Panel on Climate Change. *Cambridge University Press*, Cambridge UK.

Loladze, I. (2016). Reaping what we sow. *New Scientist, 230*(3068), 18-19.

Scrucca, F., Bonamente, E. & Rinaldi, S. (2018). Carbon footprint in the wine industry. In Muthu, S.S. (ed.). *Environmental Carbon Footprints. Industrial Case Studies* (pp. 161-196). Butterworth-Heinemann, Oxford UK and Cambridge Ma.

# 19. Producing food in a fragile food system – a case study on Isle of Skye, Scotland

*A. Bruce[1*], D.M. Bruce[2], I. Fletcher[1] and C. Lamprinopoulou[1]*
*[1]Science Technology and Innovation Studies, The University of Edinburgh, Old Surgeons' Hall, High School Yards, Edinburgh EH1 1LZ, United Kingdom; [2]Edinethics Ltd., 11/6 Dundonald Street, Edinburgh EH3 6RZ, United Kingdom; ann.bruce@ed.ac.uk*

## Abstract

The current food system, emphasising efficient food supplies to urban centres, works at the expense of providing food security to remote rural communities. We draw on a case study of a remote, rural location in Scotland (the Isle of Skye) to demonstrate that systemic factors in the food system disadvantage both the provision of food locally, and the contribution that agricultural activity in the rural area can make to wider markets. Skye's agriculture consists mostly of sheep and beef cattle production, the soil and climate being poorly suited to other types of activities. The production systems in turn contribute to maintaining landscapes and high nature value biodiversity. However, pathways to consumers, which require continual supplies of consistent product, fail to reward these producers. On the other hand, citizens in these remote, rural areas, mainly purchase food from supermarket chains, allied to global supply chains. Because they are at the extreme end of supply chains, they are vulnerable to perturbations in the system. Where local food is available, it tends to be an expensive, premium product, often aimed at tourists. Producing food in a fragile food system does not easily secure food security for locals in food production areas.

**Keywords:** food security, beef, lamb, retailers, butchers, shopping

## Introduction

The security and resilience of our food systems to external events has been an issue for some time, put into even stronger focus by the Covid-19 pandemic. Considerable attention has been paid to urban food security (e.g. Milan Urban Food Policy Pact), but the impact that current food production systems have on food resilience in remote, rural areas has been less studied. Securing adequate nutrition (food security) and food production that gives priority to local consumption (food sovereignty) are concepts that have been widely debated (Murdock and Noll, 2015). This paper argues that the current food system, emphasising efficient food supplies to urban centres (satisfying food security), works at the expense of providing food for local consumption (food sovereignty) to remote rural communities. We take the example of the Isle of Skye, in the Scottish Highlands, as a case study.

Paradoxically, while beef and lamb is produced on the island, residents rely heavily on supermarket-based supply chains to buy their food. Such remote locations are at the extremities of supply chains, which are prone to failure, for example in shortages of milk and vegetables experienced even before Covid 19. Whereas one might expect that a rural area would be food secure by definition, the evidence suggests that the current organisation of food production and supply can leave such areas as being to some degree food insecure.

After describing the context and the research method, this paper examines two aspects; the production of food in the area, and the provision of food in the area, highlighting a mis-match between the two.

 Hanna Schübel and Ivo Wallimann-Helmer (eds.) ***Justice and food security in a changing climate***
EurSafe 2021 – DOI 10.3920/978-90-8686-915-2_19, 

## Context

The Isle of Skye is located off the North West coast of mainland Scotland with a population of only 12,000 but with a very large influx of tourists. For example in 2019 the island hosted an additional 650,000 visitors (Moffat Report, 2020). Figure 1 provides a map of the island in its UK context. Although linked to the mainland by a bridge, it is geographically remote, with a three hour drive to the nearest large town. However, it is well served by public transport and road infrastructure. It also forms the primary supply route serving the Outer Hebrides.

Most farming in Skye is done as crofting. Crofting is a scheme originally introduced into the northwest Highlands of Scotland in the late 19th century, to protect the rights of tenant subsistence farmers from the then frequent abuses and evictions by land-owners. Crofters are tenants (in some cases now owners) of small farms (typically around 5 ha[1]) with the right to pass on the croft in perpetuity to their family or a person of their choice. Originally intended for the subsistence for their own families, most crofters farm part-time and gain other income by a wide range of additional employment. Many croft as a way of life, rather than for economic reasons; but for some it is their main economic enterprise, often made possible by joining several crofts together. The crofters maintain land areas which are both of very high nature value and are also integral to the landscape and culture of Skye which attracts and pleases so many visitors.

The area has traditionally kept beef cattle and sheep, which are best suited to Skye's maritime climate and environment, often on poor soils and in challenging topography. Traditionally a few cereal crops and potatoes were grown, mostly for own consumption, but little is grown today. More recently, sheep have become more prominent, although their number has also declined over recent years. Few crofts are

[1] https://www.crofting.scotland.gov.uk/crofting.

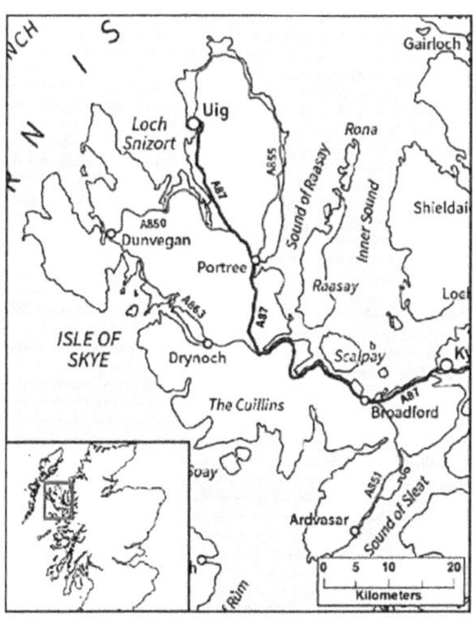


*Figure 1. Isle of Skye.*
*MiniScale® [TIFF geospatial data], Scale 1:1000,000, Tiles: GB, Updated: 18 November 2019, Ordnance Survey (GB), Using: EDINA Digimap Ordnance Survey Service, https://digimap.edina.ac.uk, Downloaded: 2021-02-16 14:48:45.741. GB Overview [TIFF geospatial data], Scale 1:5,000,000, Tiles: GB, Updated: 19 August 2013, Ordnance Survey (GB), Using: EDINA Digimap Ordnance Survey Service, https://digimap.edina.ac.uk. © Crown Copyright and Database Right, OS Licence number 100025252, 2021. Ordnance Survey (Digimap Licence) © SRUC 2021*

able to rear many animals to slaughter weight, and so the animals are sold in the autumn as 'stores' for continued rearing in more favourable, lowland conditions. With the lack of an abattoir on the island, even lamb and beef originally reared on Skye will have come from a mainland abattoir three hours drive away. Despite having cattle, there is no dairy on the island, so all milk has to be imported. The production of vegetables and fruit on Skye is challenging due to the poor soils, and wet, salty climate due to prevailing strong winds off the sea. Production is mostly only practical if protected by expensive plastic tunnels.

## Method

As part of wider concerns about the resilience of food systems, the challenges to farming practices based largely in upland livestock production are being researched. In a project examining resilience in upland and island beef and sheep production in Scotland and Northern England, we undertook individual face-to-face interviews with 16 people in the Isle of Skye, consisting of 8 crofters, and 8 associated with agriculture more broadly, several of whom were also part-time crofters. The interviews explored issues around the resilience of the crofting activities. We also undertook 12 face-to-face or telephone interviews with people involved in the wider meat chain in the UK, outside Skye. These included representatives of 4 supermarkets, 2 butchers, 1 intermediate processor and 5 people in associated roles.

To inform our understanding of food provisioning in Skye, we conducted an additional 18 interviews amongst food shoppers on the island, posing questions on four key topics: where interviewees got their food from; their purchase of locally produced foods; whether they experienced problems in shopping for food; and their opinions on how food shopping had changed in the area or might change in the future. The most successful method of recruitment for shoppers was standing outside the shop (with the permission of the manager/proprietor) and approaching customers to ask if they would be willing to be interviewed by phone at a convenient time for them in the next few days. We also undertook a shopping basket exercise, examining prices for a limited number of basic products, such as eggs and potatoes, from a range of outlets. All interviews were conducted during 2019, before the Covid-19 outbreak and at a time when Brexit negotiations were ongoing.

## Food production in the Isle of Skye

As noted earlier, the primary food produced from Skye is beef and lamb meat. The current UK meat chain is very complex, and has a number of different routes to the consumer. Most pertinent for this research is the supply of meat through supermarkets and butchers.

Our meat chain interviewees (including both supermarkets and butchers) strongly emphasised the need for consistency in supply: in volume, size, shape, fatness and eating quality. According to them, their customers demand a good quality eating experience and expect to be able to get a broadly similar product whenever they want it, as indicated by the quote:

> I want consistency, and I get that consistency when it's being picked out of 100 cattle, I can't get that consistency out of fifteen. (interview with butcher)

Supermarket retailers similarly referred to the need for consistency. In the UK as a whole, only 50-55% of the carcasses are reported as falling within their desired specification. This is a challenge when the 'product' is biological, coming from different animals, dependent on weather and other unpredictable factors, and seasonal in its basis (e.g. sheep only lamb in the spring). Because farms in Skye are small and a range of different breeds and breed-crosses are kept, it is challenging for them to provide the volumes of meat of the consistency which retailers and butchers say that they require. Thus, the food supplied from the island is somewhat problematic even for local retailers.

In such a context, crofters seeking to sell store animals into mainstream markets find that prices for larger, meatier animals from continental breeds (such as Simmental, Limousin or Charolais) gain much better prices than smaller, slower growing native breeds, such as Highland and Aberdeen Angus. There is therefore an economic incentive to use continental breeds rather than native ones. Against this, the native breeds are generally found to be more suited to the local environments in an area such as the Isle of Skye, and to require less external inputs such as high protein or cereal feeds. There is thus a mismatch between the centralised demands of the retail markets, and the products that are best suited for Skye.

Yet the meat from the island has the characteristics that many modern consumers consider ethical, such as being produced under more natural conditions, pasture fed and free-range. However, the infrastructure for local sales is lacking. As observed, the absence of an abattoir on the island means that animals must be transported 100 miles for slaughter, and the meat returned the hundred miles back in order to be sold locally. The small volumes produced in Skye are also insufficient to attract special branding as 'Skye meat' unless a number of farmers can work together and commit to produce to a common production protocol. Additional challenges with branded meat include the need to be able to gain a sufficient income from all parts of the carcase, not just the high-value cuts. An alternative, which came to the fore during Covid-19 outbreak, is selling on-line, but the costs to the farmer of distributing their own meat in this way are very high. A small number of crofters are able to supply mainly high-end restaurants, but these outlets are inevitably limited. So although local, high quality meat is produced on Skye, the rest of the food chain is not set up to enable it to be provided where it is wanted.

## Food provision in the Isle of Skye

Many farmers may consume their own meat, and there is considerable backyard egg production and some vegetable production mostly for personal use. Nonetheless, there was a very limited amount of local food available on Skye and much of that seems to be aimed at the sizeable tourist market. 'Locally' produced foods included chocolate and chutney made on the island, based around imported raw materials. Some venison from a local estate was available and lots of 'backyard' eggs but little in the way of beef and lamb, fruit and vegetables or even fish or the shellfish caught by local fishermen which are largely sold to foreign outlets. The Co-op supermarket in Portree, which might be expected to have the greatest demand from tourists, had prominent displays of smoked salmon and venison skirlie (an oatmeal based mixture) all produced in the Highlands, but not on Skye itself. Most food produced locally, such as fish and shellfish, appears to be luxury production for the hotel and catering trade. Everyday consumption is dominated by food brought onto the island. Interviewees acknowledged the importance of income generated by tourism but several expressed frustration at the strain increased numbers put on supply chains and the higher prices of goods and services targeted at tourists. Interviewees identified periods of milk shortages. Short-term disruptions to fruit and vegetable supplies were observed during the interview period, sometimes due to inclement weather preventing supermarket deliveries, but also other failures within the delivery process (e.g. failure of refrigeration during transport).

Supermarket chains dominate food retailing in the UK: Tesco, Sainsbury's, Asda and Morrison's account for 70% of all food purchases[2]. This finding was echoed by our interviews in Skye, where the predominant source of food was either the one supermarket chain on the island (not one of the big four listed above), a limited number of local general stores, or from limited direct supermarket deliveries to the island. Despite the importance of smaller outlets, supermarkets form a key part of food provision in Skye. In common with the findings of Marshall *et al.* (2018), our consumers interviewees often had to travel long distances, both within the island to access the single supermarket, or to the mainland in order to

[2] https://www.kantarworldpanel.com/en/grocery-market-share/great-britain.

shop for food in other supermarkets. This meant planning their trips carefully to fit in with their other commitments.

Our shopping basket exercise data has to be treated with caution as there is a great deal of variability in what is included in each product. However, this exercise was repeated in four case study areas (Skye, Orkney, Scottish Borders and Yorkshire Dales), and what was clear was that the highest prices were demanded for local products, for example beef mince cost between £5.78/kg and £10.99/kg, but a local product was priced at £11.23. Six eggs could be purchased for £0.90 to £1.65 but locally produced eggs could cost up to £2.50. Outside Skye, Cheddar cheese could be purchased for £8.54/kg to £9.20/kg but locally produced could be purchased for £12.50/kg, and even plant products, such as oats, became more expensive when produced locally, with standard prices ranging from £1.40/kg to £2.00/kg, and local oats coming in at £2.89/kg. These four diverse regions each had different local products which might have been expected to be cheapest, but turned out to come with a premium price compared with products brought in from outside. Thus, the purchase of locally produced food is somewhat problematic.

## Discussion

Current ethical values in the livestock food system put stress on such qualities as local production, knowing the sources, a 'circular' economy, pasture-fed largely outdoor animals which have as natural a life as possible, breeds well adapted to the natural environment, high animal welfare standards, low environmental impacts and reduced net carbon emissions. The current food system has led to a paradoxical situation for the remote rural area of the Isle of Skye, that produces food which should score highly according to many of these criteria, nonetheless finds itself unable to market a range of affordable, locally-produced food products for its local community. Furthermore, the wider food system does not recognise and reward the products from Skye for their ethical value.

In the way the food supply system is currently set up, the produce from these areas can only be sold as premium as added-value and branded products, such as from a specific breed or 'primarily grass-fed' identity, which can only be gained by selling to high-end hotels or similar outlets. Offering these animals to the commodity market, as mainly happens now through sale of store animals, leaves them uncompetitive in comparison to production from more geographically and climatically favoured areas.

In the UK context, the main way a farmer can contribute to the global economy is if they can produce something that cannot be produced cheaper somewhere else. This tends to favour certain types of agriculture. Farms in areas of physical disadvantage (such as geology, weather, farm size or ownership, or lack of infrastructure) are at a disadvantage, and yet what they produce has value to ethical consumers. The product of the local producers is not given its true value in the current, mainstream food system. What the food system values, is not sufficiently weighted to the values embodied in a local product. However, this outcome is not a knowing, unethical act of particular actors so much as the unintended, unethical consequences of how the system has developed over multiple decisions and interactions. There can be individuals who make decisions in it; individuals, or corporate and policy decision-makers, that can be seen as contributing to the ethical problem, but none of them have sole responsibility for the system.

The food system of mainly centrally supplied supermarkets does provide consumers in remote areas access to a range of foods that would not be available locally. However, locations like Skye are at the end of a long supply chain and to that extent are vulnerable to unexpected perturbations in the chain. Local shops provide particular services to communities that extend beyond food supply, including provision of social services such as bank ATMs. During covid-19, the local village stores on Skye provided an outstanding service in food provision to their localities communities and were able to be very flexible and innovative in sourcing food.

One of the question this food system raises is why consistency of product has apparently become so important? This perhaps reflects on an unintended consequence of a reliable supply of varied foodstuffs available on the shelf on any day, in that is that food becomes viewed as a commodity, like electricity or water, that is available on-tap immediately when required. Just as the centralised services for electricity have increasingly become questioned, and more distributed electricity supplies are being considered, so perhaps the need to have the same, consistent, year-round supplies of foods can be questioned.

One should be careful not extrapolating too far from a single case study. The specific case of Skye has a particular history and a specific modern context of extraordinary levels of tourism. The historical context is still influential to much of the local population. Yet the paradox and challenges identified in Skye in this study have proven to be reflected in the other three areas of the wider study, notwithstanding their different histories, geographies and contexts.

The current ethical emphasis of modern food production systems on the efficient and cheap delivery of food to cities, is thus at the expense of providing food security to remote rural communities, often where food is produced, leaving neither consumers nor producers in a satisfactory and secure food situation.

## Conclusions

It would be easy to imagine that a rural area is in a position to eat the food it produces, but a more detailed examination of this case study demonstrates that this is not necessarily the case. Furthermore, the ethically, high quality meat produced in Skye does not fit the industrial paradigm that requires consistent, year round supplies of meat that is substantially the same all the time. It seems that food security for rural populations, via supermarket chains or even butchers, have systematic barriers to products from small farms in remote areas. Infrastructural barriers to local production, such as lack of abattoirs and dairies, further exacerbate this conundrum. Policy incentives to move from a centralised food system to a distributed system, by supporting provision of infrastructure, would not only provide a more resilient food system but also limit the systematic barriers to production in remote areas.

## Acknowledgements

This work was funded by the Global Food Security's 'Resilience of the UK Food System Programme' with support from BBSRC, ESRC, NERC and Scottish Government: project BB/R005796/1.

## References

Marshall, D., Dawson, J. and Nisbet, L. (2018). Food access in remote rural places: consumer accounts of food shopping. Regional Studies 52(1):133-144. https://doi.org/10.1080/00343404.2016.1275539

Milan Urban Food Policy Pact. Available at: https://www.milanurbanfoodpolicypact.org/.

Moffat report (2020). Isle of Skye and Isle of Raasay Tourism Economic Impact Study commissioned by SkyeConnect and undertaken by Glasgow Caledonian University's Moffat Centre as reported by the Press & Journal. Available at: https://www.pressandjournal.co.uk/fp/news/islands/2343962/year-long-economic-study-finds-skye-visitors-boosted-economy-by-211-million-pre-lockdown/.

Murdock, E. and Noll, S. (2015). Beyond access: integrating food security and food sovereignty models for justice. Chapter 49. In: Dumitras D.E., Jitea, I.M. and Aerts, S. (eds) Know your food: food ethics and innovation. Wageningen Academic Publishers, Wageningen, the Netherlands, pp. 327-332. https://doi.org/10.3920/978-90-8686-813-1_49

# 20. Interpreting 'ethics' in IPCC AR5: insights for agricultural adaptation to climate change

*A. Voisard*
*Université de Fribourg, Environmental Sciences and Humanities Institute, Chemin du Musée 4, 1700 Fribourg, Switzerland; anthony.voisard@unifr.ch*

## Abstract

This research paper aims to provide a brief review of recent developments regarding the concept of 'ethics' in IPCC reports. We argue that reflection on the ethical dimensions of climate change in these reports helps to highlight explicit, and often implicit, evaluative judgements in climate change politics. However, ethical issues such as those related to climate change adaptation seem insufficiently discussed and constitute blind spots in this documentation. From an interpretative point of view, we seek to broaden the scope and understanding of the ethical implications of agricultural adaptation to climate change, using as an example this specific sector which is particularly affected by the impacts of climate change and where ethical considerations are little discussed in official documentation. We conclude that an increased focus on ethical reflection would provide additional insights to strengthen the work of the IPCC. In turn, these ethical insights would better equip the world's political decision-makers, to whom these state of knowledge reports on climate change are addressed, among others. Critical and empirical interpretation of 'ethics' in the work of the IPCC will support our pluralist reasoning.

**Keywords:** climate ethics, Intergovernmental Panel on Climate Change, interpretative analysis, environmental pragmatism, food security

## Introduction

The Intergovernmental Panel on Climate Change (IPCC) has been instrumental in shaping the research community in climate science and climate politics. The work of this UN expert body on climate change benefits from a high visibility and an important capacity to influence climate negotiation processes both at the regional and global levels. In 2014, Working Group III 'Mitigation of Climate Change', with the contribution of several experts, including philosophers John Broome and Lukas Meyer (Broome, 2020), published the first chapter of IPCC reports dealing substantially with ethics. We consider that the recent incorporation of ethics into the Fifth Assessment Report (AR5) of the IPCC is a significant step forward in the recognition of climate ethics as a disciplinary field. But what does the concept 'ethics' mean in the work of the IPCC? And what is the contribution of ethics to the reflection on adaptation to climate change in these reports? Based on a review of relevant passages from the official reports, we will address these two questions in this paper. We will first provide an overview on our methodological approach, as well as precision on data sources. The research results and their interpretation will then be presented. In our view, the ethical category should be understood more comprehensively than what is presented in the main IPCC chapter on ethics. Other chapters and sections of the AR5 that we have analysed provide nevertheless important insights into the meanings of 'ethics'. Finally, we close this article by deepening the discussion on climate change adaptation ethics, based on a targeted sector in IPCC reports, namely the field of food security.



Hanna Schübel and Ivo Wallimann-Helmer (eds.) ***Justice and food security in a changing climate***
EurSafe 2021 – DOI 10.3920/978-90-8686-915-2_20, 

## IPCC AR5 analysis method

Our method of analysis is informed by an interpretative approach in practical philosophy (Létourneau, 2019), but it is also rooted in a version of environmental pragmatism that combines conceptual analysis in philosophy with research techniques more frequently associated with social sciences (Voisard, 2019). In this case, the aim was to produce a critical synthesis of the IPCC's ethical perspectives in the AR5. Since these reports are authoritative, their understanding of ethical issues is necessarily of great importance. For the purposes of this analysis, we have focused in particular on two reports, one published by Working Group III (Mitigation of Climate Change; IPCC, 2014a, see in particular chapter 3), and the other, written in two volumes and produced by the second working group (impacts, adaptation and vulnerability; IPCC, 2014b; 2014c). This choice of data sources is explained by the fact that it is in these IPCC reports that we have identified substantial, recent and relevant developments of the ethical concept. *NVivo* content analysis software was used to efficiently target the relevant passages for our textual analysis. We validated a total of 196 occurrences relating to the targeted passages in the two IPCC reports.

## Results from WGIII and WGII reports

Throughout the 1,435 pages of the third working group report (AR5), we identified a total of 126 occurrences in the body of the document referring to the concept 'ethics', or one of its declinations ('ethical'; 'ethically'). We may note that these occurrences are mostly found in the chapter entitled 'Social, Economic, and Ethical Concepts and Methods'. This chapter alone contains 59% (74/126) of the occurrences relating to the concept 'ethics' in the body of the document. Developments in ethics are mainly located in this chapter. Moreover, a total of 70 occurrences of the term 'ethics' or one of its declinations can be identified in the body of the second working group report. These occurrences are mostly found in the chapters entitled 'Foundations for Decision Making' and 'Adaptation Opportunities, Constraints, and Limits'. Taken together, these chapters contain 60% (42/70) of occurrences in this report. The ethical dimension, when it is actually dealt with, is found to be integrated with a plurality of other considerations (Table 1).

*Table 1. Distribution of the occurrences of the word 'ethics' and its declinations in the body of IPCC's Working Group III and Working Group II reports (AR5).*

| Section title in WGIII report | Number of occurrences (n=126) | Section title in WGII report | Number of occurrences (n=70) |
|---|---|---|---|
| Chap. 3: *Social, Economic, and Ethical Concepts and Methods* | 74 (58.7%) | Chap. 2: *Foundations for Decision Making* | 27 (39%) |
| Chap. 4: *Sustainable Development and Equity* | 21 (16.7%) | Chap.16: *Adaptation Opportunities, Constraints, and Limits* | 15 (21%) |
| *Technical Summary* | 12 (9.5%) | Chap. 20: *Climate-Resilient Pathways: Adaptation, Mitigation, and Sustainable Development* | 5 (7.1%) |
| *Preface* | 4 (3.17%) | Chap. 1: *Point of Departure* | 3 (4.3%) |
| Chap. 6: *Assessing Transformation Pathways* | 4 (3.17%) | Chap. 6: *Ocean Systems* | 3 (4.3%) |
| | | Chap. 17: *Economics of Adaptation* | 3 (4.3%) |
| *Summary for Policymakers* | 3 (2.38%) | *Technical Summary* | 3 (4.3%) |
| Chap. 1: *Introductory chapter* | 3 (2.38%) | Chap. 4: *Terrestrial and Inland Water Systems* | 2 (2.86%) |
| Chap. 13: *International Cooperation: Agreements & Instruments* | 3 (2.38%) | Chap. 22: *Africa* | 2 (2.86%) |
| Chap. 2: *Integrated Risk and Uncertainty Assessment of Climate Change Response Policies* | 2 (1.59%) | Chap. 27: *Central and South America* | 2 (2.86%) |
| | | Other sections (chs. 13; 19; 21; 28; 29) | 5 (7.1%) |

## Comparative interpretation of the research results

We identify a total of five major meanings of the term 'ethics' across the two reports analysed in this research. We present our categorisation in Table 2. Although the meanings that we have identified may fit in both reports to varying extents, we draw from our analysis strong trends in the understanding of the word 'ethics' for each of the working groups.

### *A. Ethics as a philosophical standpoint on climate change*

The predominant ethical conception of these reports refers to a relatively narrow view of philosophical ethics. These developments are mainly presented in the third chapter of the AR5 WGIII report. Within this framework, the ethical discussion is divided into two main constituent branches, namely the theory of justice and the theory of value. This ethical posture can be criticised for marginalising or ignoring the contribution of a plurality of possible developments in philosophical ethics (pragmatist ethics, applied ethics, care ethics, ecofeminism, etc.). In addition, the authors of this chapter themselves acknowledge that this rather simplistic dichotomy is subject to debate in the field of moral philosophy. Indeed, is there not a conception of the Good within the theories of justice? And how can we think about this Good, if at least a preconception of justice is not acquired? Nevertheless, this division between justice and value aims to provide a basis for organising the discussion on the principles and concepts of philosophical ethics.

### *B. Ethics as a component of decision-making in the context of climate change adaptation*

These reports also focus on the ethical discussion from the point of view of responsible decision-making. This is specifically addressed in the report of the second working group. In this case, ethics is an integral component of decision-making in climate change adaptation. Ethics is then integrated into various

*Table 2. Key meanings of the word 'ethics' in AR5 reports.*

| Identified key meanings | Working group(s) most frequently associated with key meanings |
|---|---|
| A. Ethics as a philosophical standpoint on climate change | Working Group III (see in particular chapter 3) |
| B. Ethics as a component of decision-making in the context of climate change adaptation | Working Group II (see in particular chapter 2) |
| C. Ethics as a transversal reflection on climate equity | Working Group III (see in particular chapter 4) |
| D. Ethics as a dimension in the debates on the opportunities, constraints and limits of adaptation to climate change | Working Group II (see in particular chapter 16) |
| E. Ethics as a discourse on value judgements about climate change | Working Group III<br>Working Group II |

cross-cutting dimensions and several complementary disciplines (decision-making theory, cultural studies, organisational theory, behavioural science, risk governance, etc.). Under this framework, ethical decision-making is understood in a situationist and pluralist perspective. This is because climate change adaptation options cannot themselves be fully evaluated out of context and without considering the diversity of interests, objectives and values of the participants in the decision-making process. This ethical approach is in line with the posture of environmental pragmatism, which is cited several times in the report of the second working group.

### C. Ethics as a transversal reflection on climate equity

Within these two IPCC reports, ethical discourse can also be interpreted according to a transversal perspective of climate equity, i.e. a viewpoint that addresses the issue of sharing burdens and benefits in an international climate regime. A number of justice principles and frameworks for equitable burden-sharing are emphasised in the fourth chapter of the third working group report. Several elements are dealt with in this fourth chapter, including the issue of distributive equity, as well as the issue of procedural equity. Other considerations at the intersection of equity, climate policy and sustainable development are also addressed.

### D. Ethics as a dimension in the debates on the opportunities, constraints and limits of adaptation to climate change

According to a fourth point of view, ethical reflection can be integrated into the discussion on the opportunities, constraints and limits of climate change adaptation. More specifically, within the sixteenth chapter of the second working group report, a total of seven ethical dimensions emerge from these considerations on climate change adaptation. Here we recall these seven ethical dimensions of climate change adaptation: access to opportunities; distribution of constraints; differing attitudes to risks; rights and potentials of people to secure particular valued objectives; differing rates at which limits are reached; trade-offs in securing valued objectives; intergenerational and interspecies equity and adaptation limits. In this discussion, the ethical issue is developed from a contextual, interdisciplinary, multisectoral, and decision-making perspective.

#### *E. Ethics as a discourse on value judgements about climate change*

Finally, in these two IPCC reports, ethics also constitute a discourse on value judgements in relation to climate change. The transparency of ethical value judgements is emphasised, as well as the complex interrelationships that this specific type of judgement can have within climate politics and climate science. Both reports recognise that the climate issue is not only about factual judgements. It includes a normative component. Climate change issues refer to evaluative judgements, e.g. in terms of assessing and recommending climate policies. This level of discourse seeks to highlight explicit or implicit evaluative judgements specific to the discussion on climate change within the IPCC reports themselves.

This research has shown that IPCC contributors and authors recognise the relevance of at least five meanings attributable to the term 'ethics'. While each of these meanings provides insights into the issues of value and justice in climate politics, each of them has a number of characteristic specificities (see the categorisation presented above). The predominant approaches discussed by the third working group in the AR5 report have this tendency to ground ethical reflection in a perspective of moral universalism. That is to say, the authors of this working group aspire, more or less implicitly, to a 'universal' ethical approach to the articulation of rights and duties. This perspective is arguably a pertinent approach to addressing the problem of climate change mitigation in an international governance regime. On the contrary, the report of the second working group is more in line with an ethical conception of a particularistic nature. Indeed, it is argued in this report that localised climate change adaptation strategies are 'place and context-specific'. It emerges from our reading of the documentation that these two areas of climate action, in this case mitigation and adaptation, refer to distinct metaethical conceptions that can presumably be justified by the objectives, specificities and orientations of these areas of climate action. A final key meaning attributed to ethics complemented the universalist or situationist approach of the two working groups. This level of discourse, identified in both reports, provides clarity and nuance on the role of evaluative judgements in climate policy debates.

## IPCC's ethical perspective on food security and agriculture in adapting to climate change

The authors of chapter sixteen produced by Working Group II argued that the discussion on the opportunities, constraints and limits of adaptation should be understood in the context and specificities of the different sectoral and regional chapters (IPCC, 2014b: 908). Here we consider the ethical issues raised by one of these sectoral chapters, namely the chapter on food security and food production systems (IPCC, 2014b: Chapter 7). While the authors of this sectoral chapter do not discuss ethics directly, they do address issues that raise important ethical dimensions. Indeed, the notion of food security is based on a number of key elements that evoke underlying social and cultural values, as well as evaluative judgements to integrate them judiciously. In other words, food security is based on moral values, and not only on values, but also on possible standpoints from which the associated evaluative judgements can validate or invalidate the legitimacy of the concepts from the 'food security' conceptual framework. These include the concepts of healthy and sufficient food access for an active life and for the growth and development of children, the sustainability of the agri-food system, intra- and inter-generational food justice, and human dignity. The concept of food security refers in particular to a plurality of issues related to the equitable production, distribution and consumption of the world's food resources. These ethical issues are likely to raise a variety of interests, objectives, norms, values and potential trade-offs for the implementation of climate change adaptation options (see also chapter 16). Reading this IPCC report, one might get the impression that the scientific literature on food security and food production systems seems to be caught in a logic of intensive farming when it comes to the limits and constraints of adaptation. Are we dealing with a pure ethic of bargaining over advantages and disadvantages, which would not even evaluate the available options according to other values? At the moment, the ethical models of sustainable agriculture proponents seem to be relatively absent from the discussions in the scientific literature at the intersection of climate change adaptation and food security. The IPCC authors themselves acknowledge in the section dedicated to explaining the gaps in research related to food security that subsequent studies should include a broader range of systemic and transformational adaptation options that are available to agriculture (IPCC, 2014b: 520).

The ethical issue remains essentially in the subtext of the core content of the sectoral chapter under study (chapter 7). Our reading of this chapter of the AR5 report, as well as the more general study of the entire report of the second working group, leads us to believe that the ethical issue could be given greater prominence. On the one hand, we suggest that more space be given to ethical and normative issues in the report of the second working group. These types of issues could be dealt with at least as much as in the report of the third working group. The results of the IPCC's synthesis report can be considered to reflect the relative novelty of the domain of climate change adaptation ethics. This being said, we may go further, recalling that the choices and priorities of writing, as well as the scientific methodologies deployed, necessarily influence the results of this synthesis report on climate change adaptation. On the other hand, we suggest that the ethical discussion be integrated into the various topics discussed throughout the chapters of the full report, akin to what is done for the issue of equity in the report of the third working group. The effort to integrate ethical reflection in a transversal, contextual, interdisciplinary and multisectoral manner is already well under way with regard to the foundations of decision-making, as well as the opportunities, constraints and limits of adaptation to climate change. In our view, these developments should be pursued by the IPCC in subsequent reports.

## Conclusion

In this paper, we have carried out a textual analysis of recent developments of 'ethics' in the IPCC reports. We first briefly presented our method of analysis and our research results. We then proceeded to interpret our research results by identifying five key meanings of the concept of 'ethics' in the AR5 reports two and three. We have seen that the predominant ethical conception of these reports corresponds to a rather narrow view of philosophical ethics. Our research results showed, however, that there was not just one chapter on ethics in the AR5 report, but a plurality of sections and chapters dealing with a variety of ethical dimensions in the areas of climate change mitigation and adaptation. Some of these sections were more ethically developed than others. We note, for example, that the issue of climate change adaptation ethics is proportionally less developed than the ethical dimension in the domain of mitigation. Unexpectedly, we have found that the developments in ethics proposed by the second working group are in some respects consistent with a perspective of environmental pragmatism. On the other hand, the pragmatist intention expressed by the Co-Chair of the AR5's third working group, Ottmar Edenhofer, and his colleague Martin Kowarsch (Kowarsch and Edenhofer, 2016), was not entirely clear from a reading of the official documentation. Finally, we closed this paper by drawing on a sector targeted in the documentation, namely the area of food security. Reading these reports, we see that an ethical vision for inspiring transitional agriculture is missing (see for example Holmgren, 2011). Certainly, a pluralistic, interactionist and democratic ethical standpoint such as environmental pragmatism could in no way ignore adaptation options that respect agro-ecological environments, non-human organisms, and nature more broadly. We may argue that environmental pragmatism provides a legitimate input for the development of an adaptive ethics to address emerging issues in the different contexts of climate action. In this case, the incorporation of care ethics into the pragmatist meta-philosophical framework would, in our view, make it possible to avoid remaining a slave to a logic of 'performance' (Guérin, 2011), or to a productivist vision of agri-food systems. This perspective of environmental ethics would position itself in favour of a benevolent accompaniment in relations between humans and non-humans. The incorporation of another conceptual framework in ethics could, however, be more appropriate depending on the given situation. This conception of environmental pragmatism suggests methodically marshalling the resources of a variety of conceptual frameworks in order to best articulate our positions according to the requirements and specificities of the problem situation.

## References

Broome, J. (2020). 'Philosophy in the IPCC'. In A Guide to Field Philosophy: Case Studies and Practical Strategies, edited by Evelyn Brister and Robert Frodeman. New York; London: Routledge, pp. 95-110.

Guérin, S. (2011). Du care à la société accompagnante : une écologie politique du concret. *Écologie & politique* 42 (2): 115-34.

Holmgren, D. (2011). Permaculture: Principles and Pathways Beyond Sustainability. Australia: Permanent Publications.

IPCC (Intergovernmental Panel on Climate Change). (2014a). Climate Change 2014: Mitigation of Climate Change. Contribution of Working Group III to the Fifth Assessment Report of the Intergovernmental Panel on Climate Change. Cambridge; New York: Cambridge University Press.

IPCC (Intergovernmental Panel on Climate Change). (2014b). Climate Change 2014: Impacts, Adaptation, and Vulnerability. Part A: Global and Sectoral Aspects. Contribution to Working Group II to the Fifth Assessment Report of the Intergovernmental Panel on Climate Change. Cambridge; New York: Cambridge University Press.

IPCC (Intergovernmental Panel on Climate Change). (2014c). Climate Change 2014: Impacts, Adaptation, and Vulnerability. Part B: Regional Aspects. Contribution of Working Group II to the Fifth Assessment Report of the Intergovernmental Panel on Climate Change. Cambridge; New York: Cambridge University Press.

Kowarsch, M., and Edenhofer O. (2016). 'Principles or pathways? Improving the contribution of philosophical ethics to climate policy'. In Climate Justice in a Non-Ideal World, edited by Clare Heyward and Dominic Roser. New York: Oxford University Press, pp. 296-318.

Létourneau, A. (2019). L'approche interprétative dans un cadre de type éthique appliquée. Revue française d'éthique appliquée 8 (2): 113-128.

Voisard, A. (2019). Field Philosophy: Deweyan Inquiry on Climate Change Adaptation Perspectives. Dewey Studies 3 (1): 54-77.

# 21. Implications for animal welfare of food system changes towards circular agriculture

*F.L.B. Meijboom*[1,2]*, J. Staman*[1*]*, R. Pothoven*[1] *and I.J.M. de Boer*[1,3]
[1]*Council on Animal Affairs RDA, P.O. Box 20401, 2500 EK The Hague, the Netherlands;* [2]*Centre for Sustainable Animal Stewardship, Utrecht University and Wageningen University and Research, Yalelaan 2, 3584 CM Utrecht, the Netherlands;* [3]*Animal Production Systems group, Wageningen University & Research, P.O. Box 338, 6700 AH Wageningen, the Netherlands; info@rda.nl*

## Abstract

Dutch agriculture has become highly efficient and productive, but this has come at a cost. The quality of values, such as biodiversity, soil, air, water and nature has been compromised. Climate change and resource scarcity further complicate this situation and require farmers to reassess their production methods. A transition to circular agriculture seems a promising answer to these issues: closing cycles of nutrients and other resources in order to minimize losses and to end the impact on climate change. However, circular agriculture does not only answer questions, it raises also new ones. One set of questions focuses on animals and their health and welfare. On the one hand, current forms of livestock farming contribute to climate change and loss of biodiversity and, therefore, are part of the reason to find more sustainable forms of agriculture. On the other hand, animals can play an important role in circular agriculture by converting biomass that humans are unable or unwilling to eat and so helping to close circles. Although the role animals can play is not ignored, the animal welfare dimensions of circular agriculture seem to be a blind spot. In this paper we argue that this is a problem for both the success of circular agriculture and for the animals involved. To elaborate on this claim we analyse the background of the lack of attention. We claim that the lack of attention in the current discussion is embedded in the concept of animal welfare itself, in the current legal and economic possibilities and in the way circular agriculture is framed. Finally we present some steps that contribute to an animal welfare inclusive approach of circular agriculture.

**Keywords:** sustainability, food transition, livestock farming

## Introduction

Dutch agriculture has become highly efficient and productive, but this has come at a cost. The quality of values, such as biodiversity, soil, air, water and nature has been compromised. Climate change and resource scarcity further complicate this situation and require farmers to reassess their production methods. A transition to circular agriculture seems a promising answer to these issues: closing cycles of nutrients and other resources in order to minimize losses and to end the impact on climate change. However, circular agriculture does not only answer questions, it raises also new ones. This includes the questions: at what scale should we close cycles? and; does this change have an impact on our consumption patterns? (De Boer *et al.*, 2019, Poore and Nemecek, 2018). Additionally, a transition has impact on animals and their health and welfare. On the one hand, current forms of livestock farming contribute to climate change and loss of biodiversity and, therefore, are part of the reason to find more sustainable forms of agriculture. On the other hand, animals can play an important role in circular agriculture by converting biomass that humans are unable or unwilling to eat and so helping to close circles (Van Zanten *et al.*, 2018, 2019). Although the role animals can play is not ignored, the animal welfare dimensions of circular agriculture seems to be a blind spot (e.g. Ministry of Agriculture, Nature and Food Quality, 2019). This is remarkable given the increasing attention paid to animal welfare in recent years and in this paper, we argue that it also is a potential problem for both the success of circular agriculture and for the animals involved. Next we analyse the background of the lack of attention and present some steps to better embed the position of animal welfare in the transition towards a circular agriculture.

## Animal welfare as a blind spot, is it a problem?

The relative silence about animal welfare in the discussions on circular agriculture is not a self-evident problem. This especially holds if one claims that in the end sustainable agriculture has to be animal free (Kortetmäki and Oksanen, 2020). Nonetheless, we have two arguments that can justify more prominent attention to animal welfare (including health) in the context of circular agriculture.

First, as mentioned above, in the circular agriculture animals can play an important role. They can convert low-value plant-based biomass into high-value food that is suitable for human consumption. In addition, the role of animals in circular agriculture could consist of other providing ecosystem services besides food provisioning, such as landscape conservation (e.g. cows and sheep grazing on pastures), nutrient recycling, and other provisioning of leather, wool, etc. A second argument starts in the assumption that at least in the coming decades we are in a food system that asks for animal products. Therefore, animals still play and will play an important role in the discussion on circular agriculture.

These two arguments may stress that animals can play an important role in a future circular agriculture and will be present as a result of a demand for animal product. However, it does not yet show why animal welfare is an essential element for circular agriculture. To make this step, we refer to the combination of moral arguments to take the interests of animals into account that is also reflected in national and European legislation and the increase of public attention for animal welfare (EC, 2016). This entails that the current lack of attention to the animal welfare implications of the innovations towards circular agriculture is problematic. It is even an urgent problem, because circular agriculture raises new animal welfare questions. For instance, we still have relative limited knowledge about potential welfare issues related to insects although they can play quite a prominent role in circular agriculture. Also, for more 'traditional' animals, such as cows, sheep, pigs and chickens, closing cycles may pose risks to animal welfare due to changes in animal feed and housing conditions. To get further grip on this problem and find possible ways to include animal welfare more explicitly in the transition to circular agriculture, it is important to explore the background of the current silence.

## The background of animal welfare as a blind spot

The observation that animal welfare is not sufficiently on the radar in the discussions on circular agriculture is not easy to explain. In general attention to animals and their welfare is not as strongly embedded in politics and policy as the human equivalent. Nonetheless, in the European context animal welfare has been recognized as an important element in livestock farming (Schukken *et al.,* 2019). We think the lack of attention in the current discussion is embedded in the concept of animal welfare itself, in the current legal and economic possibilities and in the origin of the concept of circularity. These points we further elaborate on in this section.

First, animal welfare is a complicated concept that knows many definitions (RDA, 2018). However, they all share that welfare is linked to individual entities rather than to systems or collectives (Broom, 2010). This easily conflicts with developments towards sustainable farming that focuses on animal groups and collectives. Furthermore, it is not easy to quantify or calculate animal welfare by animal-based measures whereas more physical parameters, such as body weight and physical injuries, seem easier to translate in figures one can calculate. As a result, this easily leads to a situation in which animal welfare is largely left out as subjective and complicated and mainly some health related aspects of welfare are included such as mortality rates or use of antibiotics. This does not deny the improvements in animal welfare that have been achieved during the last decades, for instance in housing systems and transportation. However, it is a problem that reoccurs in many animal related innovations and that, for instance, can also be recognized in debates on formulating animal welfare related breeding goals.

Second, within the current legal and economic frames the options to take animal welfare on board are limited. It is beyond the scope of this short paper to expand in a way that do justice to this point, but for instance in many national laws, animals can be commodities that are less well protected than humans. Consequently, when trade-offs are necessary between human and animal interests, for instance, when measures to fight climate change have negative impact on animal welfare, the human interests easily trump. This can also be recognized from a market perspective. Animals clearly represent an economic value in the context of farming, but it is not easy to show and calculate the economic added value of animal welfare measures. Nonetheless there are opportunities for animal welfare (RDA, 2017) that can be applied in the context of circular agriculture.

Finally, the circularity concept has its origins in both agro-ecology and industrial ecology and is aimed at preserving and managing natural resources for future generations. In practice this entails that the circularity concept focuses on the ecological dimension of sustainability (De Boer en Van Ittersum, 2018). Therefore, circular agriculture is a form of agriculture aimed at producing food while maintaining ecological values, such as fertile soil, clean air, pure water, a healthy climate, preserving the quality of the landscape, nature and biodiversity. Although this account is promising in addressing urgent problems in today's agriculture, it has its shortcomings. Next to the ecological dimension of sustainability, circular agriculture should also consider the economic and social dimensions of sustainability. In other words, it should be economically viable and socially responsible (RDA, 2020, Figure 1). This broader approach

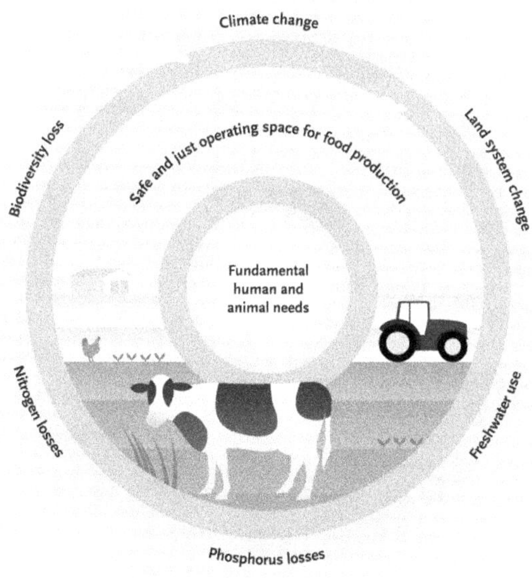


*Figure 1. Determining safe and just operating space for sustainable food systems (De Boer* et al.*, 2019). The operating space for agricultural entrepreneurs is determined by the ecological ceiling (such as nitrogen losses and loss of biodiversity), on the one hand, and by the social foundations, our social values (such as animal welfare and working conditions), on the other (RDA, 2020).*

of circular agriculture allows to include animal welfare as an element within the social dimension of sustainability rather than within the ecological dimension. This implies that animal welfare does not automatically form part of circular agriculture and that good animal welfare practices should be a separately formulated and monitored requirement for the development of circular agriculture.

## Towards an animal welfare inclusive approach of circular agriculture

The short presentation of the background of the limited attention to animal welfare in the transition towards circular agriculture shows that the problem will not solve itself.

It first requires innovation at the level of the concept and assessment of animal welfare. Although welfare is primarily linked to individuals, it is important to further reflect on the relation between the individual animal and its position as member of a group (e.g. Ohl and Putman, 2014). Furthermore, the challenge is not only to include attention to animal welfare as such. It requires also that animal welfare is taken into account in a way that can deal with both current and foreseen problems that come with circular agriculture and with the problems it aims to address. For instance, due to the current effects of climate change, such as more extreme weather, be it warm, cold, dry or wet, the living conditions of animals are affected. At this point it is important to stress that circular agriculture may have positive as well as negative implications for animal welfare. However, little is known to date about these specific implications. This requires innovation in animal welfare science and the development of improved measures to address and monitor these welfare challenges. Finally, a welfare inclusive approach requires innovations that aim to do justice to the complexity of the concept of animal welfare and incorporate this dimension in assessment models that enable to bring welfare on par with other relevant dimensions (e.g. Tallentire *et al.*, 2019). Only if these three dimensions of animal welfare are taken seriously a welfare inclusive approach of circular agriculture is feasible. Here we see that science has to play an important role in the development next to NGOs, government and private parties for the implementation.

Second, we already argued that next to the ecological dimension of sustainability, circular agriculture should take into account the social dimensions of sustainability. This entails that animal welfare can be embedded in the concept of circular agriculture rather than an add-on.

However, even with animal welfare innovations that are responsive to circular agriculture and an account of circular agriculture that encompass attention to animal welfare, we have not addressed the more institutional problems that are rooted in current legal or economic frames. If these remain unaddressed animal welfare will not be taken on board. Therefore, the third field of solutions has to focus on this institutional dimension. This is complex and cannot be the task and responsibility of one party only. However, we see possibilities. Circular agriculture calls for a redesign of our food system, in which all dimensions of sustainability (people, planet and profit) are taken into account. This requires the input from many parties including government, the business sector and NGOs. This includes those who are able to shape and redesign political, legal and market conditions of food production. Furthermore, this transition requires time. This is not only a risk; it also is an opportunity. Institutional frames do not change overnight. Therefore, it is important to utilise that time to take the institutional dimensions into account and embed the attention to animal welfare. This can result in making good animal welfare practices a requirement rather than a dependence of the goodwill of individual farmers or NGOs. This can be part of governmental policy but can also be embedded in a private certification system. Furthermore, a level playing field is required within Europe. This asks for strengthening the network and coordination in Europe in order to develop circular agriculture in a European context as well. As a first step this can result in the investment of a research agenda and provide multi-year financing and to establish (national) Centres of Expertise for Animal Welfare that can function as a source of information and knowledge exchange for entrepreneurs who want to embark on the transition to circular agriculture.

These three steps will not solve all problems regarding animal welfare in the context of circular agriculture. There still will be tensions between economic considerations and animal welfare, public health risks of animal welfare improvements or conflicts between animal welfare and nature that ask for a careful assessment. However, by taking animal welfare as one of the key concepts from the start in the redesign of the food system, one can better anticipate on potential challenges and look for ways to mitigate problems that may occur when animal welfare conflicts with other public values. This requires time and energy and an attitude of openness and transparency of all involved to critically discuss the animal welfare concept in the light of circular agriculture as well as critically reflect on how circular agriculture can encompass animal welfare. However, this can speed up the process in the long run and enable to develop a food system that can deal with the impact of urgent challenges of climate change, population growth and biodiversity loss *and* includes animal welfare as a key concept of circular agriculture.

## References

Broom, D.M. (2010), Animal welfare: an aspect of care, sustainability, and food quality required by the public. Animal welfare in education and research. Journal of veterinary medical education, 37:83-88.

De Boer, I.J.M. and Van Ittersum, M.K. (2018). Circularity in agricultural production, Scientific basis for Mansholt lecture 2018, Wageningen University & Research, The Netherlands, 72 pp.

De Boer, I.J.M., Van Der Linden, A. and De Olde, E.M. (2019). How to handle trade-offs and synergies in our search towards a sustainable food system? In: Book of Abstracts of the 70th Annual Meeting of the European Federation of Animal Science, Wageningen Academic Publishers, Wageningen, the Netherlands, p. 345.

EC European Commission. (2016). Special Eurobarometer 442: Attitudes of Europeans towards Animal Welfare. Available at: https://data.europa.eu/euodp/en/data/dataset/S2096_84_4_442_ENG.

Kortetmäki, T. and Oksanen, M. (2020). Is there a convincing case for climate veganism? Agriculture and Human Values. https://doi.org/10.1007/s10460-020-10182-x

Ministry of Agriculture, Nature and Food Quality (2019). Plan of Action. The Dutch government's plan to support the transition to circular agriculture, The Hague, The Netherlands, 10 pp.

Ohl, F. and Putman, R.J. (2014). Animal welfare at the group level: more than the sum of individual welfare?. Acta Biotheoretica 62, 35-45. https://doi.org/10.1007/s10441-013-9205-5

Poore, J. and Nemecek, T. (2018). Reducing food's environmental impacts through producers and consumers. Science 360: 987-992.

RDA (2017). Advisory Report – Animal welfare for sale!, Council on Animal Affairs (RDA), The Hague, The Netherlands, 16 pp.

RDA (2018). Conceptual Framework Animal Welfare. Council on Animal Affairs (RDA), The Hague, The Netherlands, 32 pp.

RDA (2020). Advisory Report – Animal Welfare in Circular Agriculture, Council on Animal Affairs (RDA), The Hague, the Netherlands, 40 pp.

Schukken, Y.H., van Trijp, J.C.M., van Alphen, J.J.M. and Hopster, H. (eds) (2019). The state of the animal in The Netherlands. Council on Animal Affairs (RDA), The Hague, The Netherlands, 202 pp.

Tallentire, C.W., Edwards, S.A., Van Limbergen, T. and Kyriazakis, I. (2019). The challenge of incorporating animal welfare in a social life cycle assessment model of European chicken production. The International Journal of Life Cycle Assessment 24, 1093-1104. https://doi.org/10.1007/s11367-018-1565-2

Van Zanten, H.H.E., Herrero, M., Van Hal, O., Röös, E., Muller, A., Garnett, T., Gerber, P.J., Schader, C. and De Boer, I.J.M. (2018). Defining a land boundary for sustainable livestock consumption. Invited review in Global Change Biology 24: 4185-4194.

Van Zanten, H.H.E., Van Ittersum, M.K. and De Boer, I.J.M (2019). The role of farm animals in circular food systems. Global Food Security 21: 18-22.

# 22. Developing principles and criteria for just transition in food systems: a transdisciplinary endeavour

*T. Tribaldos[1*] and T. Kortetmäki[2]*
*[1]Centre for Development and Environment, University of Bern, Mittelstrasse 43, 3012 Bern, Switzerland; [2]Department of Social Sciences and Philosophy, University of Jyväskylä, P.O. Box 35, 40014 Jyvaskyla, Finland; theresa.tribaldos@unibe.ch*

## Abstract

Modern industrialised food systems around the world suffer from well-documented shortcomings including damaging impacts on the environment, biodiversity, and climate, indefensible conditions of workers, as well as unhealthy dietary patterns. Significant system-level transitions to less harmful food systems in these respects are needed. Justice questions stand at the core of these issues and it is crucial to analyse who bears the costs and who enjoys the benefits of different solutions to these problems. But how can we decide if a solution is just and for whom? In this paper, we present the development of a framework with a preliminary set of principles and criteria for evaluating whether a transition towards more sustainable food systems is just and for whom and for improving the justness of the process. We establish the framework in a transdisciplinary approach suitable to capture normative components of food systems such as justice perspectives of different actors and contexts. Methodologically, we base our results on mid-level principles of social, food, and environmental justice, and sustainability criteria through literature review, interdisciplinary discussions with food system researchers and food system actors from Finland. We first present the preliminary set of principles and criteria for just transition, which is work in progress. We then reflect on the different purposes the framework serves: it can be used to assess and evaluate ongoing developments in food systems; to support future decision and policy making; to steer scenario work on acceptable future developments and desirable transition pathways; or to structure discussions about solutions in food system transition and to make trade-offs visible. We conclude by discussing some of the challenges faced during this work and give an outlook for the next steps.

**Keywords:** sustainable food systems, climate mitigation, food system assessment, food justice, transdisciplinary approach

## Introduction

Modern industrialised food systems around the world suffer from well-documented shortcomings including damaging impacts on the environment, biodiversity and climate, indefensible conditions of workers, as well as unhealthy dietary patterns (Tribaldos and Rist, in press). Significant system-level transitions, able to address interconnected and complex food system relations across time and scales, are needed in order to move towards less harmful food systems. Low carbon transitions mitigating climate change are of the utmost importance in this regard. But who benefits from transitions and who bears the costs? Income and social position determine to a large degree, whether people can afford healthy diets or not (Kousta, 2020) and participation in decision-making in food systems is more challenging for marginalised people than for others. In addition, marginalised people have less access to health care, intact environments and other goods, and are more affected by environmental degradation and pollution (Temper *et al.*, 2018). This is true both within countries and globally. Power dynamics and power concentration in food systems often inhibit actors outside the powerful networks to effectively defend their viewpoints or to compete against the financial force they face (Clapp and Purugganan,


EurSafe 2021 – DOI 10.3920/978-90-8686-915-2_22, © Wageningen Academic Publishers 2021

2020). This can be seen for example in oligopolistic settings with a large number of agricultural producers yet few processors and retail companies, or globally with concentration processes around transnational corporations. The disadvantaged food system actors – often low-paid food workers and farmers, some of whom have high investment debts and live on the verge of economic viability, and low-income or otherwise food insecurity prone households – have lower adaptive capacities for change, due to which the least well-doing groups are also often particularly vulnerable to systemic transitions.

Hence, justice questions stand at the core of food system transitions, and injustices in food systems are manifold (Kortetmäki, 2019, 2021). This raises the question how to determine whether a transition process is just and for whom. What questions does justice pose to thinking about policies towards more sustainable food systems? We address the issue by introducing a transdisciplinary process for developing principles and criteria for just transition. The following chapter explains the methodology for creating a framework that can guide policy processes and the use of individual parts of the framework with policy and food system actors. We then present the preliminary results of this process and discuss implications for research and policy making.

## Methodology

Developing a guiding framework for just transition in food systems involves on the one hand interdisciplinary knowledge of different dimensions in food systems, their drivers and outcomes, relations among food system actors, and associated dynamics. On the other, such a framework engages with fundamental principles of justice and normative goals mirroring conflicting and contradicting positions and expectations, common in questions of sustainability science (Boda and Faran, 2018). A transdisciplinary research approach is thus well suited to acknowledge the normativity of this endeavour and to facilitate a co-production process of the guiding framework made up of principles, criteria and indicators. The principles represent general justice or sustainability foundations, which ought not to be violated. Criteria are standards for judging, evaluating, and dealing with the impacts of the transition. Lastly, indicators operationalise the criteria by providing empirical or modelling-based information about the actual state of a situation or its changes. Accordingly, the elaboration of the framework consists of the following steps: (1) literature review of relevant justice, sustainability and food system literature; (2) interdisciplinary iteration rounds for selecting relevant principles and criteria; and (3) discussions with policy and food system actors to test the applicability of framework parts in policy and decision making (Figure 1). The development of the framework is part of a research project with a large, international research consortium on just transition in food systems, which started in 2019. The consortium provides us the opportunity to regular, interdisciplinary exchanges as well as engagement

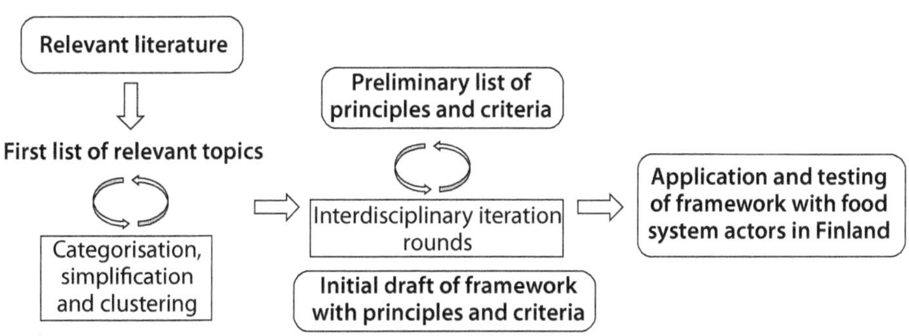


*Figure 1. Process of framework development.*

with food system actors in Finland. The exchange rounds in points 2) and 3) refer to these groups of actors. The interdisciplinary rounds took place between May and December 2019 and the first stakeholder exchange in October 2020.

To gather the first list of relevant concerns of justice for food system transition, we reviewed literature on environmental and food justice, food systems, food sovereignty and democracy, food sustainability, resilience and agroecology. After that, intense discussions among the author team followed on how to categorise and cluster the topics, and to eliminate overlaps. Iterative feedback rounds with the research consortium allowed us to come up with the first draft of the framework with principles and criteria for just transition. Several interdisciplinary iteration rounds involved exchanges with each work package on their specific expertise. Framework elements were first tested in the stakeholder discussions with Finnish food system policy and decision makers to reflect on the justice implications of different policy options.

## Results

We present here the preliminary framework of principles and criteria, but we do not discuss the indicators as a theoretically and methodologically distinct issue at the moment. In its current state, the framework consists of seven mid-level principles of justice with their associated criteria. The following paragraphs depict the principles and criteria, though details are omitted due to space limitations.

1. The right to food is the first and most important principle in our framework as it covers the access of all people to nutritious and adequate food at all times. The right to food as a fundamental human right addresses states' obligations in legislation and stands at the centre of what food systems are supposed to deliver. It covers the different dimensions of food security (availability, accessibility, utilisation, stability) but in addition includes age- and gender-specific requirements and considers cultural food preferences (De Schutter, 2014). Criteria for just transition under this principle include attending to the needs of food insecure and vulnerable groups (Dieterle, 2015); food safety considerations; and the maintaining of resilience in supply chains (e.g. Tendall *et al.*, 2015).
2. Fair distribution of income, value, and innovation in the food chain represents the distributive dimension of economic domains of justice in food systems (Dieterle, 2015). The focus lies on economic and livelihood aspects but also considers opportunities for innovation and change of those actors or regions affected by transitioning processes. Criteria in this principle therefore place fair treatment and payment of food workers at the centre (Gilson and Kenehan, 2019). An important criterion under this principle for just transition is the support for (and desirably enhancement of) the viability of farming as a profession.
3. Justice for the distant ones speaks to the universal and cosmopolitan aspects of justice. The associated criteria call the attention to distant interactions and demand that the right to food, the access and control over productive resources, livelihood opportunities of distant actors, or the prospects of future generations must not be violated, undermined or disregarded when taking decisions. Particular attention is paid to the right to food in faraway places, who are indirectly connected through trade and power relations (Boillat *et al.*, 2020), and to future generations (De Schutter, 2014) who will live with our legacy without being able to have a say in our decisions. Transparency in food systems is a crucial requirement for having a basis for decision making (Pereira, 2020).
4. The principle of human-nature relations has two interconnected sides to it – human and nonhuman. While the human side includes the benefits for humans from intact environments, recognition of diverse human-nature relations, participation in environmental decision making, and the capabilities to do so, the nature side touches upon protecting nature in its own right (Schlosberg, 2007). In terms of criteria, this means that transition must not approach nature simply as a resource to be harnessed for sustainable resource management and food production, but nature deserves protection for its own sake.

5. Procedural and epistemic justice includes the right of people to define their own food systems and not being imposed a system they do not want (Patel, 2009), as well as the availability of information about food systems and their dietary outcomes, which can enable and empower food system actors to actually participate in decision making (Pereira *et al.*, 2020) and climate actions in their own lives. This principle is essentially based on respectful treatment of others and democratic procedures for decision making (Loo, 2019). The associated criteria address these points.
6. Recognitive justice specifically addresses the respect for traditional, indigenous and local knowledge, cultural diversity and heritage, and equal rights for food system actors from those traditions (Cabell and Oelofse, 2012). In addition, it recognises all paid and unpaid work in food systems (Allen and Sachs, 2012) and covers the protection of gender, vulnerable groups and minorities from discrimination (Schlosberg, 2007). The criteria cover these points for all actors in the food system and include a respectful treatment of nonhuman beings.
7. Capacity development. The last principle deals with the capacities of food system actors and the enabling environment allowing them to actively respond to transition pressures and the need to change practices and or behaviour. Food literacy, participation in innovations and lifelong learning are important to make well-informed choices (Robaey and Timmermann, 2019), and governance can and should create enabling environments for sustainability transformations (Gliessman, 2016). The criteria demand that a just transition supports capacity development of food system actors to being able to sustain their livelihoods, eat well, and feed their families.

As mentioned, these principles and criteria for just transition are still work in progress. The interdisciplinary collaboration with the whole the research team and the food system actors in Finland, Switzerland, and Brazil is ongoing but we can already reflect on the different purposes the framework can serve. We see the main scope of these principles and criteria in four areas:

1. They present a toolset for assessing and evaluating ongoing sustainability developments in food systems in terms of justice impacts. Do transition developments respect and account for justice? Are all affected actors considered, including vulnerable and marginalised groups and actors that are only indirectly related to the developments?
2. They support decision-making and policy planning towards more sustainable food systems. Reflecting on, for example, the outcomes of different low carbon policies with relation to their justice implications will help decide on the desirability and acceptability of different policy options.
3. In research, the principles and criteria can steer scenario work on acceptable future developments and desirable transition pathways. Scenario work with stakeholders can benefit from guiding principles and criteria.
4. Related to the previous points, the principles and criteria are useful to structure discussions around solutions to food system challenges and to make trade-offs visible. Not every situation will be in need of the entire framework with all principles and criteria. Sometimes it will be enough to focus on certain parts of it – e.g. in relation to certain sector-specific policies or expert stakeholder discussions – but it can as well give indications about issues that have been neglected before.

## Discussion and conclusion

During the process of principle and criteria development, we have faced two major challenges. One of them is conceptual-methodological and the other relates to group dynamics and different disciplinary background of the involved experts. The first challenge was to define the relationship between just transition and food system sustainability and how to integrate them in the framework. Whereas just transition focuses on the process itself, food system sustainability is the desirable outcome. However, sustainability questions have to inform already the process itself – otherwise it would not be 'just transition' in the sense transition is defined here. It was helpful in this regard to ask what the concepts just transition and food system sustainability can learn from each other (Boone, 2010), and to have interdisciplinary group discussions. The latter brings us to the second major challenge. In discussions with researchers from different fields, finding the right balance for providing necessary background information but not overwhelming participants with it was a challenge. It helped to plan several rounds of discussions and to come back to certain issues while also feeding new questions to each round. Interdisciplinary collaboration was crucial for choosing the focal areas of concern, such as identifying groups who are most vulnerable to different types of transition policy impacts, and for seeing the problematic implications of ambiguously formulated criteria. In the coming year, we will test parts of the framework in stakeholder policy dialogues and in expert discussions in order to finalise the framework. Trans- and interdisciplinary work will endow us with deeper understanding about the question: what is just in just transition?

## References

Allen, P., Sachs, C. (2012) Women and Food Chains: The Gendered Politics of Food, in: Williams-Forson, P. (ed.), Taking food public: redefining foodways in a changing world. Routledge, New York, N.Y, pp. 23-40.

Boda, C.S., Faran, T. (2018) Paradigm Found? Immanent Critique to Tackle Interdisciplinarity and Normativity in Science for Sustainable Development. Sustainability 10, 3805.

Boillat, S., Martin, A., Adams, T., Daniel, D., Llopis, J., Zepharovich, E., Oberlack, C., Sonderegger, G., Bottazzi, P., Corbera, E., Ifejika Speranza, C., Pascual, U. (2020) Why telecoupling research needs to account for environmental justice. Journal of Land Use Science 15, 1-10.

Boone, C.G. (2010) Environmental justice, sustainability and vulnerability. International Journal of Urban Sustainable Development 2, 135-140.

Cabell, J., Oelofse, M. (2012) An indicator framework for assessing agroecosystem resilience. Ecology and Society 17, 18. https://doi.org/10.5751/ES-04666-170118

Clapp, J., Purugganan, J. (2020) Contextualizing corporate control in the agrifood and extractive sectors. Globalizations 17, 1265-1275.

De Schutter, O. (2014) Final report: The transformative potential of the right to food. UN General Assambly, Human Rights Council, Twenty-fifth session, A/HRC/25/57, New York.

Dieterle, J.M. (2015) Just food: philosophy, justice and food. Rowman & Littlefield International, London.

Gilson, E.C., Kenehan, S. (2019) Food, environment, and climate change: justice at the intersections. Rowman & Littlefield International, Lanham, Boulder, New York, London.

Gliessman, S. (2016) Transforming food systems with agroecology. Agroecology and Sustainable Food Systems 40, 187-189.

Kortetmäki, T. (2019) Nobody's fault? Structural Injustice, food, and climate change, in: Kenehan, S., Gilson, E.C. (eds), Food, environment, and climate change: justice at the intersections. Rowman & Littlefield International, Lanham, pp. 47-62.

Kortetmäki, T. (2021) Tensions between food justice and climate change mitigation, EurSafe 2021, Fribourg, Switzerland.

Kousta, S. (2020) The cost of a healthy diet. Nature Human Behaviour 4, 9-9.

Loo, C. (2019) Participation and Food Justice in Light of Global Climate Change in: Kenehan, S., Gilson, E.C. (eds), Food, environment, and climate change: justice at the intersections. Rowman & Littlefield International, Lanham.

Patel, R. (2009) Food sovereignty – what does food sovereignity look like? Journal of Peasant Studies 36, 663-706.

Pereira, L. (2020) Food System Transformation: Integrating a Political-Economy and Social-Ecological Approach to Regime Shifts. International journal of environmental research and public health 17, 1313.

Pereira, L., Frantzeskaki, N., Hebinck, A., Charli-Joseph, L., Drimie, S., Dyer, M., Eakin, H., Galafassi, D., Karpouzoglou, T., Marshall, F., Moore, M.-L., Olsson, P., Siqueiros-García, J.M., van Zwanenberg, P., Vervoort, J.M. (2020) Transformative spaces in the making: key lessons from nine cases in the Global South. Sustainability Science 15, 161-178.

Robaey, Z., Timmermann, C. (2019) Fair Agricultural Innovation for a Changing Climate in: Kenehan, S., Gilson, E.C. (eds), Food, environment, and climate change: justice at the intersections. Rowman & Littlefield International, Lanham, pp. 213-230.

Schlosberg, D. (2007) Defining environmental justice: theories, movements, and nature. Oxford University Press New York.

Temper, L., Demaria, F., Scheidel, A., Del Bene, D., Martinez-Alier, J. (2018) The Global Environmental Justice Atlas (EJAtlas): ecological distribution conflicts as forces for sustainability. Sustainability Science 13, 573-584.

Tendall, D.M., Joerin, J., Kopainsky, B., Edwards, P., Shreck, A., Le, Q.B., Kruetli, P., Grant, M., Six, J. (2015) Food system resilience: Defining the concept. Global Food Security 6, 17-23.

Tribaldos, T., Rist, S. (in press) Connecting human health to nature health? The bridging potential of sustainable food systems in: Scrimshaw, S.C., Lane, S.D., Rubinstein, R.A., Fisher, J. (eds) The SAGE Handbook of Social Studies in Health and Medicine, 2nd ed. Sage Publications Ltd., London, UK.

# 23. Eco-anthropological tools to improve food self-sufficiency through the use of wild edible plants

*L. Benoit*[1*]*, É. Olmedo*[2] *and B. Chen*[3]
[1]*L'Institut Agro – Agrocampus Ouest, Département d'Ecologie, 2 rue Le Nôtre, 49000 Angers, France;* [2]*Institute of Ethnic Studies (KITA), Universiti Kebangsaan Malaysia, The National University of Malaysia, 43600 Bangi, Selangor, Malaysia;* [3]*Independent researcher, 40 rue Edouard Aynard, 69100 Villeurbanne, France; lucie.benoit1@gmail.com*

## Abstract

Food security, or 'the ability for people to have at all times physical and economic access to sufficient, safe and nutritious food that meets their dietary needs and food preferences for an active and healthy lifestyle' (United Nations Food & Agriculture (FAO), 1996) is an issue that has been historically associated with the global South (Ahmadi *et al.*, 2018; Crush *et al.*, 2020). Being historically global South-centric, the FAO definition does not cater for the hidden hunger phenomenon, nor for preservation of indigenous food systems. The authors posit that the macrosocial construct of food security reflects only one aspect of empirical reality, the part where national economies rely on conventional agri-food systems (Maye and Kirwan, 2010). The research objective of this paper is a first methodological step into the process of producing an alternative paradigm: in lieu of food security, we suggest a wild edible plants (WEP)-inclusive food self-sufficiency framing. This new paradigm endows food self-sufficiency with a well-being social function: well-being is here translated as benefiting from a balanced diet incorporating WEP, whose intake is measured through ingested quantification and nutritional analysis. The mobilised theoretical framework draws from a post-modern stance and incorporates locavorism and Alternative Food Networks (AFN) as key-concepts. One indigenous social group from Malaysia is selected as empirical field for a pilot survey. The food system of this indigenous community's economy relies partly on foraged wild edible plants, thus legitimating an AFN framing. Targeted outcomes are scientific and social. Expected scientific outcome lies with the design of a methodological toolbox tentatively labelled as 'eco-anthropological tools'. The suggested methodology is mainly ethnographic and qualitative: subsequent data collection would enable a comprehensive reconstruction of a given community food system, thus inducing the most optimal 'toolbox'. The targeted social outcome is to identify 'tipping points' along the alternative food supply chain. Tipping points manifest a potential to activate individual and social change leading to new food consumption behaviour. Replicability and scalability tests of such an eco-anthropological toolbox are to be conducted in Europe with a population of gatherers. The use of comparison between population samples of both Europe and Malaysia enables us to qualify and quantify potential benefits and limitations of WEP systemic use within the framework of an alternative food self-sufficiency paradigm.

**Keywords:** alternative food networks, plant food, well-being, foraging, nutrition.

## Introduction: wild-edible-plants gathering as alternative food network

In 1997, Feenstra gave one of the most commonly cited definitions of Alternative Food Network (AFN): 'rooted in particular places, [AFN] aim to be economically viable for farmers and consumers, use ecologically sound production and distribution practices, and enhance social equity and democracy for all members of the community' (Feenstra, 1997: 31). Over the years, the definition expanded as the field of AFN research became more sophisticated but also more politicized, thus leading to controversy. The


EurSafe 2021 – DOI 10.3920/978-90-8686-915-2_23, © Wageningen Academic Publishers 2021

well-known definition of Goodman and Goodman coined in 2009 attempts capturing this complexity of the field's boundaries, while acknowledging its politicisation as well as its ambiguity:

> [AFN are] new and rapidly mainstreaming spaces in the food economy defined by – among other things – the explosion of organic, Fair Trade, and local, quality, and premium specialty foods. In these networks, it is claimed that the production and consumption of food are more closely tied together spatially, economically, and socially; however, the politics and practices of alternative food networks have recently come under critical scrutiny from geographers and others as a narrow and weakly politicized expression of middle- and upper-class angst. (Goodman and Goodman, 2009)

A consensual definition is hardly attainable in social science, as the targeted social objects are intertwined in the dynamics time and space; therefore, what can be captured is solely an instant picture of that particular social object embedded in the spatiality of a given social fabric, at a given point of time. In that sense, the tentative approach of AFN as an archetypal case of 'economy of qualities' by the French sociologist Michel Callon (2002) is much more astute, especially when he and his team at the 'Centre for the Sociology of Innovation' in Paris develop the phrase 'a gathering moment in market relations in which the conditions and competences of production, consumption and regulation become molten in the heat of intense social reflexivity'. Thereby, notion of economy of qualities becomes subject to reorganisation or/and qualification (Callon *et al.*, 2002). Referring to the praxis of gathering WEP, in identified and protected natural sites, as well as the sustainable techniques of plucking leafy plants – the latter being a mere illustration of a WEP subcategory –, without uprooting them and/or leaving the core of the flower untouched, notwithstanding the specific education and value system of the forager. We prefer to locate WEP gathering under the precept of an 'economy of care', rather than an economy of qualities, while subscribing to the necessary method of continuous social reflexivity.

## Beyond food security: alternative plant food network as pillar of food self-sufficiency

A part of the problem lies in the contextualised definition and instrumentation of the term 'food security'. Food security shares many similar points with agricultural development and the highlight is to increase food production and food consumption. Food sovereignty, on the other hand, shares many topics with rural development and the main goal for both is improving the quality of life for peasants, indigenous populations and, in general, rural inhabitants. A third concept, although not as often used in the literature, is highly important to take into account as it includes all the types of populations – not only covering peasants and their close network: food self-sufficiency. In this concept, lies also an intention of finding the solutions for a food autonomy (sometimes used as a synonym) of the citizens in a more balanced space: between the capitalistic approach of what food security might have become in the hands of international and national authorities and the politicized conceptualization of food sovereignty partly lead by La Via Campesina. In contrast, food sovereignty is an environmental, agricultural and rural developmental policy framework. It was first introduced in the 1980s in Central America by peasant activists as a complementary approach to food security. It was then separated from its original roots and presented as opposed to the capitalistic aspects of food security in 1996 at the World Food Summit in Rome by La Via Campesina (McMichael, 2014). LVC is a coalition of independent farmers, which consists of 182 organizations in 81 countries. Food sovereignty addresses the basic entitlements of every human to determine their food and agricultural systems and offset the ruling 'market economy' (Rosset and Martinez, 2010). Although useful, both concepts of food security and food sovereignty tend to be instrumentalised and politicized, the former serving economic liberalism and the latter taking its roots in Marxist theory. In the context of this research, we wish to propose food self-sufficiency as an alternative and more politically neutral construct. The Food & Agriculture Organisation of the United

Nations (FAO) defines food self-sufficiency (FSS) in broad terms: 'The concept of food self-sufficiency is generally taken to mean the extent to which a country can satisfy its food needs from its own domestic production.' (Thomson and Metz, 1999). Recognizing that WEP are a cheap and relevant source of nutritive food (e.g. Grivetti and Ogle, 2000), we posit (WEP)-inclusive food self-sufficiency as an operational concept acting as a mean of emancipation to acquire well-being for a given community. We therefore view FSS as the primordial characterisation of an ideal-typed indigenous food system. The notion of well-being calls for another subsequent definition, which is being addressed in the following section.

## Social functions of alternative plant food networks: a conceptual framework

### *Defining well-being*

There are various definitions of well-being used in the literature. These definitions can be broadly classified into three main groups: (1) General or global definitions that have no detail of the possible components of well-being; (2) Component definitions that break down well-being into its constituent parts, dimensions or domains, or identify key characteristics considered essential to evaluate well-being; (3) Focused definitions that either explicitly or implicitly refer to just one or a few components of well-being.

Despite the absence of a single definition of well-being, the Organisation of Economic Co-operation and Development (OECD, 2013) argues that most experts and ordinary people around the world would agree that it requires meeting various human needs, some of which are essential (e.g. being in good health), and includes the ability to pursue one's goals, to thrive and feel satisfied with their life. In the context of the present article, we posit that well-being is a composite construct, from which one feature may become more salient according to the social context where well-being is actualised. We also wish to emphasize that well-being is conceptualised here as a desired effect of food self-sufficiency.

## Contrasting case-study: the role of WEP for Malaysian Indigenous Communities

Drawing form anthropological comparative analysis methodology, but without its evolutionary undertones (Borofsky, 2019), we intend to use the cultural area of Southeast Asia as a mirror for Europe, in order to render cultural traits and behaviours of modern foragers in Europe more salient. Within Southeast Asia, we select Malaysia as empirical field, where one of the authors of this paper (Eric Olmedo) is based. His access to fieldworks with indigenous communities in West and East Malaysia shall provide useful information to design the desired eco-anthropological tools to analyse and make sense of WEP foraging practices.

### *Aborigines of the Malay Peninsula: wild edible plants at the core of their food system*

The Orang Asli is a social group of indigenous people that accounts for 13.9% of the 31 million people found in Malaysia (IWGIA, 2017). Their ethnic and historical identity have long been contested, though they are generally regarded as the first people to occupy Peninsular Malaysia (Minority Rights Groups International, 2018). The Orang Asli are caught in between the protectionist law and the hyper development policy where they are viewed as failures as a direct consequence of the reinterpretation of colonial law to fit the current political wave of the time. Hence, matters arising from their identity, way of life, education, to the houses they live in, economic pursuits, and the food they consume have caused them to suffer from the loss of identity, malnutrition, non-communicable diseases, social marginalization and poverty in recent times. Self-determining their own food systems might enable them to respond to their own needs, utilizing culturally-defined indigenous food – based mostly on game and WEP, as well as making decisions about the allocation of their resources.

## Methodology for designing eco-anthropological tools

The aforementioned eco-anthropological tools are meant to go beyond the econometric view of a self-sufficiency ratio by reconstructing the totality of a localised food indigenous system. In order to do so, we focus our investigation on the following components of the given indigenous food system: food availability, food access, food self-acquisition (gathering WEP) and production, eating habits, social actors in the food chain, value system and cosmology.

Food availability looks at the allocation and exchange of food for each habitant/population/community. It also revolves around issues such as land ownership, selection of crops and harvesting techniques, as well as the contribution and the proportion of WEP into the food system. Food access, in its economic sense, considers poverty levels, territory, income and inheritance of a family or individual. Food self-acquisition (gathering) and production are being compared in terms of their relative proportions and their nutritional participation within the supply chain. Eating habits shall be investigated through the degree of dependence on imported products, and specifically for this research, the proportion of WEP in the diet. Value system and cosmology related to food shall be analysed and made sense of using induction method.

### *Data collection method*

After the first phase of primary data collection through ethnographic and qualitative methods aiming at reconstructing communities' food systems, we intend to put in place focus groups, gathering key stakeholders in order to reach a group consensus on the characterization of the risks at stake in the absence of FSS. Gathered and known as consumed WEP samples shall be collected for identification and nutritional analysis.

### *Data analysis method*

Data analysis methods vary according to selected data collection techniques. Three (3) specific data collection techniques are mobilized for the purpose of this research: participant observation (aim: reconstruction of the food supply chain); food journal (aim: reconstructing eating habits); semi-structured interviews (aim: making sense of actions, choices and behaviours). (1) 'Participant observation' is recorded through field notes: data analysis shall focus on extracting two main variables: 'formal trading acts' alongside the food supply chain (purchasing, selling) and 'informal trading acts' (food exchange between households). (2) 'Food journal': thematic codification shall be conducted in order to segregate data entry subordinated to the following categories: food supply source; dining-in versus dining-out; intra-cultural variations of food intake. (3) 'Semi-structured interviews': thematic analysis is to be performed, deriving from five analytic themes mainly: (1) symbiotic relation between WEP and cosmovision; (2) inter- and intra-generational transfer of knowledge (WEP) and skills (foraging, producing); (3) gender roles variations in the light of the rising cash economy; (4) consciousness and/or revitalization of modern indigenous identity through the marketing of indigenous foods; (5) perceived level of food self-sufficiency.

### *Expected outcomes*

The reconstruction of the semantic corpus applied to food in the different communities is to be carried out alongside the logical sequence of the food supply chain, including WEP foraging. One of the main targeted outcomes of this research program remains the identification of tipping points alongside the whole supply chain of a community's' main food staples (including WEP). These tipping points are named after their capacity for triggering food consumption choices. Leveraging on these tipping points, one is then able to isolate the sociological, cultural and economic factors that act as levers or obstacles towards the incorporation of WEP into the community members' diet.

Narrowing down to a small number of preventive factors with universal value would enable reconstructing a programmatic scope of what could be the constraints for European foragers who wish to include WEP in their diet.

Secondly, nutritional analyses will allow us to understand the contribution of WEP in the diet, not only in terms of quantities but also in terms of quality, thus filling a gap of quality and reusable data on the topic in the literature (Crittenden and Schnorr, 2017). Thirdly, this study will look at how foraging knowledge, access to WEP and uses of WEP are related to perceived and actual food food self-sufficiency and how an introduction or an increase in WEP accessibility, harvesting and consumption is related to changes in perceived and actual food security. Finally, anthropological and ethnobotanical data put together will allow us to qualify and quantify the benefits and limits of WEP in terms of nutrition, economy, socio-cultural facilitators and taboos, education, etc.

## Conclusion: indigenous food systems as matrix to understand European WEP foraging

Indigenous food systems in Malaysia are well spatially framed, as indigenous communities are isolated and do not necessarily speak the same language, even across settlements of relative proximity. Such indigenous food systems would therefore be relatively easily reconstructed in comparison to the much broader and heterogenous European food systems. More interestingly, such a small scale might enable ethnographers and researchers to understand more clearly the interdependencies at stake between WEP and the other components of the food system. The beam of independencies induced from ethnographic research may then be used as a yardstick to measure the place, weight, symbolic meaning and social function of WEP foraging for European consumers.

### *Epistemological note*

We are used to read statistics about the diminutive size of our natural environment, due to deforestation. We are less familiar with the transformational journey of other types of biomes, such as WEP foraging sites. These foraging sites exist 'in the wild' and are known by professional WEP gatherers. Maxime Leloup (personal communications, 2018) (and other before him such as A.B Cunnigham, 2001) argues that WEP move on the species' continuum throughout the interactions with gatherers, progressively shifting from a 'genuine wild' to a 'domesticated' status. WEP gatherers become then agents of the eco-socio-sphere. The spontaneous foraging site co-evolves together with its gatherers to become a production system in itself, both modifying the original direction of the site's trajectory (M. Leloup, pers. com., 2018). A perfect illustration of such continuum shift is the ancestral practice of agroforests (Kumar and Nair, 2006; Leloup, 2020), or mimicking forest system functioning with various useful plants including tended WEP, a complex agroforestry system that ensures food sustainability. In turn, WEP range over a continuum defined by several parameters. WEP may fall into non-altered species as well as feral varieties categories. WEP can be – intentionally or not – enhanced by human practices – 'tended', favoured by removing plants around them for instance, or picking the edible leaves in a certain way to promote regrowth. The continuum paradigm if systematically observed and recorded as empirical phenomenon in Southeast Asian and European fieldworks may call for an alternative naming of our Wild Edible Plants, the first adjective reflecting no longer social and environmental realities. It may lead as well to propose sustainable ways to produce and use WEP on bigger scale, contributing to improvement of food self-sufficiency.

### *Factors for success*

In order to secure the effective application of the eco-anthropological toolbox, there is a need for participatory design by the targeted community at large (informal and professional WEP gatherers, local population, institutions): participatory design does not only enhance self-appropriation of the project but it also enables to incorporate an emic perspective into the design of our set of tools.

### *Further research*

The Malaysian case study represents a pilot survey, which will be tested for replicability and scalability in a different cultural area within the European Union, as well as in a more heterogenous group.

## References

Ahmadi, D. *et al.* (2018). The Impact of Remittances on Food Security in the Global South, Remittances Review, 1(3): 135-150.

Borofsky, R. (2019). Where Have the Comparisons Gone? Member Voices, Fieldsights, September 10. https://culanth.org/fieldsights/series/comparison.

Callon M, Méadel C., Rabeharisoa V. (2002). The economy of qualities. Economy and Society 3(1/2): 194-217.

Crittenden and Schnorr (2017). Current views on hunter-gatherer nutrition and the evolution of the human diet. American journal of physical anthropology. 162: 84-109.

Crush, J. & Si, Z. (2020). COVID-19 Containment and Food Security in the Global South. Journal of Agriculture, Food Systems, and Community Development, 9(4), 149-151.

Cunningham, A. B. (2001). Applied Ethnobotany: People, Wild Plant Use & Conservation. Earthscan Publications, VA, 300 pp.

FAO – World Food Summit (1996). Declaration on World Food Security. Rome.

Feenstra, G. (1997). Local food systems and sustainable communities. American Journal of Alternative Agriculture 12(1): 28-36.

Goodman D & Goodman M.K. (2009). Alternative Food Networks, in Kitchin R. & Thrift N. (eds). International Encyclopaedia of Human Geography, New York: Elsevier, 208-220.

Grivetti, L.E. and Ogle, B.M. (2000). Value of traditional foods in meeting macro-and micronutrient needs: the wild plant connection. Nutrition Research Reviews, 13(1): 31-46.

Kumar, B.M. and Nair, P.K.R. (2006). Tropical homegardens. Springer, Dordrecht, 377 pp.

Leloup, M. (2020). Propriété, règles, pouvoir: quelle place pour les agroforêts ? Masters thesis, Grenoble University. Zenodo.

Maye, D. and Kirwan, J. (2010), Alternative food networks. *Sociopedia.isa*.

McMichael, P. (2014). Historicizing food sovereignty. Journal of Peasant Studies, 41(6), 933-957.

Minority Rights Group International. (2018). World Directory of Minorities and Indigenous Peoples. Available at: https://minorityrights.org/minorities/orang-asli/.

OECD (2013). Economic well-being. In OECD Framework for Statistics on the Distribution of Household Income, Consumption and Wealth, OECD Publishing, Paris.

Pachón-Ariza, F.-A. (2013). Food sovereignty and rural development: beyond food security. Agronomía Colombiana. 31(3): 362-377.

Rosset, P. and Martinez, M.E. (2010). La Vía Campesina: the birth and evolution of a transnational social movement. Journal of Peasant Studies. 37 (1): 149-175.

Thomson A. and Metz M. (1999). Implications of Economic Policy for Food Security: A Training Manual. FAO. Available at: http://www.fao.org/docrep/004/x3936e/x3936e03.htm.

# 24. Is a thriving food sovereignty-based global future plausible?

*Y. Saghai*
*Department of Philosophy, University of Twente, Drienerlolaan 5, 7522 NB Enschede, the Netherlands; y.saghai@utwente.nl*

## Abstract

Because the global food system is complex, we cannot forecast whether by 2050 we will be able to adequately feed 9.6 billion people, while preserving the environment. Instead, we can imagine plausible scenarios whose storylines depict contrasted, challenging, relevant future worlds. To inform policy-makers and civil society actors, modelers quantify those narrative scenarios using computer simulations. Yet, none of the numerous quantified global food scenarios produced between 2000 and 2018 features a future governed by the principles that the transnational food sovereignty movement defends. Food sovereignty is the right to direct and participatory democratic control over small-scale, largely autonomous, just, and sustainable re-localized agri-food systems. In this paper, I investigate why food sovereignty is absent from quantified global food futures scenarios. I argue three points. First, it is absent because scenario-builders and modellers question the plausibility of a 'thriving' food sovereignty future, that is, a future in which food sovereignty principles would be universally adopted and food security ascertained without a large extension of farmland. Second, food sovereignty proponents are reluctant to take part in quantified scenario-building exercises since they claim dominant models are biased against their radically transformative vision, and therefore cannot be trusted in their plausibility judgments. Third, I argue that disagreements on the plausibility of a thriving food sovereignty future reside less in factual debates than in methodological, political, economic, and epistemic disagreements, some of which are easier to address than others.

**Keywords:** food ethics, food security, models, scenarios, futures studies

## Introduction

Will we be able to feed, by 2050, an estimated population of 9,6 billion with food that is healthy and sustainably grown, despite global warming, biodiversity loss, and the water crisis? This familiar question is difficult to answer since we cannot forecast the long-term evolution of a complex global food system. We can, however, create scenarios (storylines that depict a plausible future world) that explore a broad range of plausible futures to inform policymakers and civil society actors concerned with the future of food. The central elements of these narrative scenarios are then quantified, using computer simulation models to estimate how the global food system could evolve under different hypothesized conditions. To be helpful, future scenarios have to be contrasted, relevant, challenging and plausible; the scenario set must also cover the main solutions proposed by experts and civil society actors for addressing the food problem. For instance, most scenario sets feature a future governed by sustainable intensification principles that many experts promote. It is therefore surprising that none of the quantified global food futures scenarios produced between 2000 and 2018 includes the vision of the future defended by today's largest global agrarian social movement: food sovereignty. 'The food sovereignty movement stands for the right to direct and participatory democratic control over small-scale, largely autonomous, and re-localized agri-food systems based on: (1) sustainability (agroecology or organic farming); (2) social justice; (3) gender equity; and (4) respect for cultural diversity, nature, the value of food, and the peasant way of life' (Saghai in press). The question I address here is, 'why is food sovereignty absent from global food futures quantified scenarios studies?' The response is two-fold. On the one hand, scenario-builders and modellers question the plausibility of a 'thriving' food sovereignty future. Such


EurSafe 2021 – DOI 10.3920/978-90-8686-915-2_24, © Wageningen Academic Publishers 2021

a future would be governed by universally adopted food sovereignty principles, while food security would be ascertained without a large extension of farmland. On the other hand, from a food sovereignty perspective, models used to quantify scenarios are biased against their radically transformative vision and therefore cannot be trusted in their plausibility judgments. The presentation proceeds by first introducing the food sovereignty movement before an overview of the methods used in global food futures quantified scenarios studies. Next, I examine positions on the plausibility of a thriving global food sovereignty future in turn from a modellers' perspective and a food sovereignty perspective to determine the precise nature of the disagreements. I conclude by considering the types of disagreements at stake and ways of moving forward.

## The food sovereignty movement

Food sovereignty is a grassroots farmers' movement that emerged in Central and South America in opposition to neoliberal structural adjustment policies imposed by the International Monetary Fund and the World Bank in the 1980s and 90s. Deregulation, privatization and accelerated adoption of modern agricultural methods and monocropping led to the destruction of local markets and disrupted small farmers' way of life. In 1993, *La Vía Campesina* (LVC) was created to gather like-minded agrarian organizations from around the globe representing smallholding family farmers, farm workers, landless farmers, and medium-size farmers who practice low-input agriculture and are profoundly attached to nature, their land and cultures. Its members self-describe as 'peasants' (Claeys, 2015). Since 1996, LVC has become world-famous for opposing the World Trade Organization and advocating 'food sovereignty,' an umbrella term used to designate its demands, values, and vision of the future. Today, LVC has grown into an entity that groups 182 organizations in 81 countries and claims to represent 200 million peasants.

How exactly to characterize food sovereignty beyond its opposition to globalization is debatable. For our purpose, I propose the following characterization:

> Food sovereignty is the right to direct and participatory democratic control over small-scale, largely autonomous, and re-localized agri-food systems based on: (1) sustainability (agroecology or organic farming); (2) social justice; (3) gender equity; and (4) respect for cultural diversity, nature, the value of food, and the peasant way of life. It is also the process that leads to fully realizing that right and vision of the future. (Saghai in press)

Some issues are hotly disputed within the food sovereignty movement: (1) the scale at which food sovereignty must be implemented (strictly local or micro-regional below the national level); (2) the identification of the 'sovereign' (peasants, all farmers, peoples, all residents of nation-states); (3) the ideological foundations and specific shape of the alternative to capitalism (anarcho-syndicalism, eco-communalism, left-libertarianism, eco-socialism). Despite these indeterminacies, the food sovereignty movement stands for a vision of a desired food future in strong contrast with mainstream alternatives.

The food sovereignty movement has successfully lobbied for the adoption of its principles in laws and constitutions in Bolivia, Ecuador, Egypt, Mali, Nepal, Senegal, and Venezuela. In December 2018, United Nations' General Assembly adopted the Declaration on the Rights of Peasants and other People Working in Rural Areas (UNDROP) which features a right to food sovereignty. This new declaration builds on previous UN texts that recognized a right to food and nutrition security (such as, Article 11 of the International Covenent on Economic, Social and Cultural Rights (1966)) but goes beyond them by stressing a fundamental right to direct and participatory control over food systems. Thereby, it re-politicizes the debate on food.

## Global food futures quantified scenarios

A systematic review (Van Dijk *et al.*, 2020a) of fifty-eight global food security quantified scenarios studies (2000-2018) reveals that none of those hundreds of scenarios features a future based on food sovereignty principles. This is a surprising finding given the importance of food sovereignty in the debate on the future of food.

What are these quantified scenarios? As mentioned earlier, when a system is too complex to be predictable, the method of choice for anticipating its future states is scenario-building. According to futurist Jerome C. Glenn, a 'scenario is a story with plausible cause and effect links that connects a future condition with the present, while illustrating key decisions, events, and consequences throughout the narrative.' Scenarios are neither philosophical thought experiments designed to test our intuitions on conceptual possibility nor just-so stories disconnected from reality. A scenario-building process starts with an analysis of the system at hand to understand how it works, and to identity the main factors that influence future changes in the system. For instance, population size, income growth, technological change, dietary patterns, poverty and inequality, urbanization and climate change are some of the main drivers of change in the global agri-food system. Some of the drivers identified generate more uncertainty than others (because we do not know how they will evolve). Typically, two uncertain drivers are identified, and each is given two values (two 'future states') to generate a two-by-two matrix (and potentially more possibilities). The FOODSECURE project (Van Dijk *et al.,* 2020b), for example, builds a matrix based on sustainability (lifestyle and use of natural resources) and (in)equality to generate four futures labelled '1% World', 'Ecotopia', 'Food for all but not for ever', and 'Too little too late.' The study describes in detail each world in narrative terms and sketches a plausible pathway leading from the present to those futures. Participants are then asked to identify the main indicators to allow a computer model to simulate how different components of the system interact (e.g. interactions between population growth, income growth and education).

Importantly, we use computer models to test and refine the way we described our narrative of a future world: are all the features we intuitively ascribed to this future world compatible? Is the future we envision 'plausible'? Plausibility is not an easy concept to define and is the centre of a lively debate in futures studies, but in this context, it roughly means whether we have 'sufficient evidence' to believe a scenario element could obtain because 1. it already occurred in the past, 2. it presently occurs at another location, or 3. we have a strong 'proof of concept' that it could occur in the future (Wiek *et al.,* 2013). For instance, is it plausible that a world with 10 billion people, no income growth, low technological change, high international rivalry and hostility, and high climate change would have a low malnutrition rate? Although such a world is not 'logically' impossible, we don't have strong reasons to believe the conjunction of these scenario elements is occurable, and the credible evidence we 'do' have points in the other direction. Although not impossible and perhaps not 'entirely' implausible, its plausibility is low.

### Questioning the plausibility of a thriving food sovereignty future: a modeller's perspective

From a modeller's perspective, the plausibility of a scenario is determined by whether the elements of a scenario that interact are consistent and compatible (as determined by a validated model), and by the quality of the evidence in favour of knowledge claims. So, the selection of the model is key. Three main models are used in food futures studies: economic equilibrium models (predominant), biophysical models (a minority), and integrated models that combine models of different types, including climate models (on the rise). Models are expensive to create and run, this is why modelling teams are located in a handful of high-income countries (the United States, the Netherlands, Austria, France, Germany, and Japan).

Although global food futures studies do not explicitly discuss food sovereignty – with the exception of McIntyre (2009) and Le Mouël *et al.* (2018), see Saghai (in press) – we can surmise from existing scenarios and model analysis why modellers question the plausibility of a thriving food sovereignty future (in the above-mentioned sense), at the global scale.

Most food futures modellers use partial equilibrium models. A 'partial equilibrium model' is a fine-grained simulation of the agri-food sector in which global food demand and supply are determined by trade and price; a 'general equilibrium model', in contrast, represents the entire economy and the food sector more coarsely. As these models simulate the way international trade affects prices, a drastic reduction of international trade will always result in a price spike, make food less affordable for many populations, and therefore increase malnutrition and hunger. For these models, a world almost without global trade but with autonomous micro-regions (at a scale below the nation-state, in contrast with macro-regions gathering several countries, like the EU) cannot lead to food security. Thus, such models would deem a thriving food sovereignty future implausible.

Biomass balance models are the main biophysical models. 'They quantify the import, export and other uses of agri-food products (e.g. wheat, maize, dairy), estimate the maximum cultivable area in each region (for instance, to simulate whether an increase in demand for rice can be satisfied by increasing domestic production or requires imports), and estimates the difference between the actual per-hectare yield of each product in each region and its potential maximum yield according to local agro-climatic conditions' (Saghai, in press). Although these models do not take into account prices, their focus is still on international flows of food. In addition, they rely heavily on the idea of 'yield gap' (difference between actual and potential maximum yield) and assume that yield gaps should be reduced. In contrast, food sovereignty advocates envision a world in which each micro-region would produce enough food to feed its population, rather than optimize production capacity to reduce its yield gap for export. Here again, the model would not deem highly plausible a world in which 9,6 billion people would be adequately fed without strong international exchanges and yield gap reduction.

Both types of models would be sceptical of a thriving food sovereignty future for an independent reason: we have no strong evidence that either organic agriculture or agroecological practices are as productive as sustainable intensification. I will not dive into this familiar debate here, but simply note that measuring the productivity of a wide variety of content-sensitive agroecological practices is challenging, especially because the agroecology is aims at improving more holistic outcomes, such as increased resilience of communities. Few studies have compared the best organic agricultural practices and the best agroecological practices with the best sustainable intensification practices, in part because enduring, normatively significant disagreements on what counts as a relevant outcome and within what temporal horizon. If organic agriculture or agroecology turns out to be, *ceteris paribus*, less productive than alternative sustainable methods (as many would gauge) and if we should not expect their productivity to grow as much as more technologically driven practices, then to produce the same amount of food more land should be converted into farmland, aggravating biodiversity loss and climate change.

## The plausibility of a thriving food sovereignty future: a food sovereignty activist perspective

I have discussed at length the food sovereignty perspective elsewhere (Saghai in press). Here I will summarize the main points of disagreement between scenario-builders and modellers and food sovereignty advocates on the plausibility issue.

First, food sovereignty advocates reject equilibrium models *en masse*: they believe their assumptions are 'politically' biased in favour of international trade. Those models have simply no way to assess what would happen if trade were to take place at a scale below the national level, if prices were controlled by states or non-commercial food transactions were encouraged, or if the goal were to meet food 'needs' rather than food 'demand' of consumers based on their income level. As a consequence, rejecting the plausibility of a global food sovereignty future is going too fast.

Second, biomass balance models do not focus on market prices but still suffer from an excessive fixation on international trade. In addition, they are committed to a productionist 'political' assumption that yield gaps ought to be reduced.

Third, both types of models suffer from a lack of 'epistemic pluralism': their evidence basis is overly narrow. Food sovereignty activists claim that local and indigenous knowledge based on farmers' practices and success stories are on a par with the results of modern research methods in agronomy and agricultural economics. If a more pluralistic epistemology were adopted, then the plausibility of a food sovereignty future would be increased.

Forth, both types of models rely on past or present, quantifiable occurrence of an event or process to adjudicate plausibility and are not supportive of the possibility of a break from the past and the present, of a radical social, political, economic, ecological, and civilizational transformation. What is needed is a plausibility criterion less focused on the past and the present.

## Conclusion

To sum-up, the disagreement on the plausibility of a thriving food sovereignty future relies less on factual debates than on methodological, political, economic, and epistemic disagreements. How can we move forward from here? There are calls within the modellers' community to create truly transformative scenarios (e.g. de-growth and eco-communalist scenarios) and to invent new models to quantify them, given the limits of equilibrium and current biophysical models (Müller *et al.,* 2020). This is an area where mutual learning is possible, at least if both parties (1) desire to explore more transformative futures, (2) acknowledge the 'usefulness and the limits' of quantitative models and (3) have the political will to work together, despite pressures by lobbies or governments who may favour scenarios congruent with their own agendas. What seems more difficult to address is epistemic disagreements on what counts as good evidence. This is where recent philosophical work on deep disagreement, epistemic pluralism and indigenous knowledge might prove useful for clarifying the issues and perhaps finding common ground.

## Acknowledgements

I would like to thank Michiel van Dijk, Eric Kemp-Benedict, Kasper Kok, Olivier Mora, Antonio Onorati, Detlef van Vuuren, and audience members at the Departmental Colloquium Series of the Department of Philosophy of the University of Twente, the Netherlands, for helpful discussions. This research was partly funded by a grant from the Stavros Niarchos Foundation at the Johns Hopkins Berman Institute of Bioethics.

## References

Claeys, P. (2015). Human rights and the food sovereignty movement: Reclaiming control. Routledge, London, UK and New York, USA.

Glenn, J. C. (2009). Scenarios. In Futures Research Methodology version 3. The Millennium Project: Global Futures Research and Studies. Washington, DC. http://www.millennium-project.org/publications-2/#method.

Le Mouël C, Lattre-Gasquet M. de, and Mora O. (2018). Land use and food security in 2050: A narrow road. Editions Quae, Paris, France.

McIntyre, B.D. (2009). International assessment of agricultural knowledge, science and technology for development (IAASTD): synthesis report with executive summary: a synthesis of the global and sub-global IAASTD reports. No. E14-197.

Müller, B., Hoffmann F., Heckelei T., Müller C., Hertel T.W., Polhill J.G., Van Wijk M. *et al.* (2020). Modelling food security: Bridging the gap between the micro and the macro scale. Global Environmental Change 63: 102085.

Saghai, YI. (In press). Subversive future seeks like-minded model: on the mismatch between visions of food sovereignty futures and quantified scenarios of global food futures. Ethics and International Affairs.

Van Dijk, M., Saghai Y., Rau M.L., and Morley, T. (2020a). Global food demand projections. In *Feeding the World Well: A Framework for Ethical Food Systems*, edited by Goldberg A.M: pp. 98-124. Johns Hopkins University Press, Baltimore, MD, USA.

Van Dijk, M., Gramberger M., Laborde D., Mandryk M., Shutes L., Stehfest E, Valin H, and Faradsch K. (2020b). Stakeholder-designed scenarios for global food security assessments. Global Food Security 24: 100352.

Wiek A., Withycombe Keeler L., Schweizer V., and Lang, D.J. (2013). Plausibility indications in future scenarios. International Journal of Foresight and Innovation Policy 9: 133-147.

# 25. Exploring alternative food futures through critical design

*T. Antonsen[1,2*] and J.M. McGowan[1,3]*
*[1]GenØk – Center for Biosafety, Postboks 6418, 9294 Tromsø, Norway; [2]UiT The Arctic University of Norway, Institutt for filosofi og førstesemesterstudier, 9037 Tromsø, Norway; [3]UiT The Arctic University of Norway, The Academy of Arts, 9037 Tromsø, Norway; trine.antonsen@uit.no*

## Abstract

In this paper, we are experimenting with how design fiction might be used to speculate alternatives, foster debate and engage publics in current debates on emerging biotechnologies and future food. The debate concerning the governance of genetically modified organisms (GMOs) in food production has always been conflicted. With the emerging genome editing techniques, such as the CRISPR/cas-systems, GMOs are again topics of intense socio-political debate. What is more, decision-making about the use and release of biotechnologies is dominated by scientific and economic knowledge and rarely considers the cultural significance of the human/nature relations being envisioned through different technologies. Underlying the arguments and attitudes of both proponents and opponents of the regulation of genome editing as GMOs, we find not only scientific disagreement, but also a diverse range of future visions, narratives of science, food and nature, and views of the role of ethics and values in the debate. Within the context of a larger research project on these ethical and communicative aspects of the controversies around genome editing, we collaborate with designers to craft speculative interventions that prompt reflection on how human/nature relations are being reconfigured through agricultural biotechnologies. A particular aim for these interventions is to present alternative conceptions, narratives and visions for how these relations can unfold. The paper presents the preliminary finding of this practice-based research approach to the debate concerning the role of genome editing in the pursuit of sustainable food futures in particular. Our primary focus on methodology and the sharing of work-in-progress invites discussion on the potentials (and limitations) of creative practice compared to 'traditional' social science and humanities approaches.

**Keywords:** genome editing, design fiction, public engagement, ethics

## Introduction

In this paper we present and discuss the methodology and preliminary findings of a design-based research project that targets the ongoing debates concerning the development and use of genome editing techniques in agri- and aquaculture. We first introduce the Norwegian context of debate, as well as shortly introduce design fiction in order to show the relevance of this particular approach to the topic. We then move on to describe the methodology and the results of the first of a series of workshop, as well as present examples of expected output from the upcoming workshops. The purpose of the paper is to invite discussion on the potentials (and limitations) of creative practices compared to 'traditional' social science and humanities approaches used in contested and technological enterprises.

## Emerging techniques and regulation of GMOs

The background for our project is the debate concerning the governance of the use and release of genetically modified organisms (GMOs) in food production. This debate has long been conflicted. With the emerging genome editing techniques, such as the CRISPR/cas-systems, GMOs are again topics of intense socio-political debate.


EurSafe 2021 – DOI 10.3920/978-90-8686-915-2_25, © Wageningen Academic Publishers 2021

Genome editing techniques are claimed to offer the potential for future crops that are more productive, more resistant to heat stress (climate adaptation), and offer better protection from pests than both previous generations of GMOs and traditional crops (Ricroch *et al.*, 2017). Proponents also say the technique is more precise, cheaper and 'more natural' than earlier forms of GM techniques (Doudna and Charpentier, 2014; Jasanoff *et al.,* 2015; Ryffel, 2012). However, how these new tools for genome editing should be understood and regulated, are topics of public and scientific disagreement (Myskja and Myhr, 2020).

The Norwegian Gene Technology Act (GTA) demands that GMOs are sustainable, ethical justifiable and of societal benefit (these are the so-called non-safety criteria) (Act of 2 April 1993 No. 38). In addition, use and release must be considered for health or environmental risk. The demand for risk analyses in the GTA is given by international standards (such as EU Directive 2001/18/EC). The non-safety criteria are based on the precautionary principle and are specific for Norway. In 2017 the Norwegian Biotechnology Advisory Board (NBAB) initiated a public debate on the GTA as they suggested revisions to the Act. In 2021-22 a committee appointed by the Norwegian government will develop an official report. We expect the main questions to be elaborated by this committee to be: Should all the GM techniques continue to be regulated as the traditional GMOs, that is, still being given full assessment including the precautionary non-safety criteria? Or should there be (partly) exempts from the law, for instance for the use of genome editing techniques (such as CRISPR/cas)?

The debates concerning regulation of emerging biotechnologies is dominated by scientific and economic aspects. However, the non-safety criteria open the assessment of GMOs for broader considerations, including cultural issues. GMOs do and genome-editing organisms may change and disrupt our food systems and hence impact the culture associated to this. This may also reduce or have other adverse impacts on the present diversity in the way we produce, distribute and consume food. This, in and of itself, brings tension to the debate as (particularly) the criterium of ethical justification may be perceived to be based on less scientific grounds than the risk assessments.

Social scientists and humanists go beyond the scientific and economic discourses and consider the cultural significance of the human/nature relations being envisioned through different technologies. Underlying the arguments and attitudes of both proponents and opponents of the regulation of genome editing as GMOs, we find a diverse range of future visions, narratives of science, food and nature, as well as diverging views of the role of ethics and values in the debate. (e.g. De Graeff *et al.*, 2019; Halliwell, 2019) Furthermore, ethicists evaluate and develop normative concepts suitable for the normative assessment of genome edited food, with hopes of providing the flexibility necessary to accommodate different ethical, cultural and ecological priorities in decision making (e.g. Preston and Antonsen, 2020, in press; Preston and Wickson, 2016). And they develop models for the ethical assessment of GMOs, such as the Ethical Matrix (e.g. Forsberg *et al.*, 2020; Kaiser *et al.*, 2007).

In our project we aim to go further beyond the scope of the descriptive social scientist and the normative ethicist, and creatively explore and push norms, values and visions. *Design fiction* targets a disruption of the status quo, probes norms, ethics and values, and attempts to open spaces for other alternatives and scenarios that might work as a critical foil or counterpoint to dominant narratives, established histories and mainstream expectations.

## Design practice as critical practice

The past three decades has seen the emergence of 'art based', 'practice based' or 'practice-led' research, with the creative disciplines collectively working to legitimize, concretize and operationalize methods of working, modes of knowledge and types of research outputs that are distinctly different – and at times seemingly far removed – from the well-established norms and systems used within both the sciences and the humanities (Frayling, 1993). Within this terrain of 'research-into', 'research-through' and 'research-for' the creative disciplines, our specific approach links most closely to tendencies within design practice and research that can roughly be grouped under the rubric of 'critical design'. Most often associated with the work of Anthony Dunne and Fiona Raby (2009, 2013), 'critical design' takes an oppositional stance to the traditional roles, methods and outputs of design – and product design in particular – as a discipline. Through tactics such as defamiliarization, subversion, bricolage, relational ambiguity, fiction and counter factuality, amongst others, critical designers look to activate the potential of designed objects to mobilize debate, organize publics and instigate change.

At the same time, 'critical design' can be rather insular and bound up in the disciplinary limits and internal concerns of design only, perhaps running the risk of dead ending as either 'a kind of criticism from within' (Mazé and Redström, 2007), 'design for design's sake', or even 'design for critical designers' (Malpass, 2013). Yet, when handled carefully and strategically, there remains potential for deploying design approaches and thinking within specific interdisciplinary, scientific, political and socio-cultural contexts, as a means for critiquing and speculating around dominant narratives, established histories and mainstream values and expectations.

The complex, multi-layered and often charged context of current discussions in Norway around the use of emerging genome editing techniques in the pursuit of sustainable food futures seems a promising site for employing design fiction as a tool for critical inquiry, as it is an approach that can be used to actively avoid utopian/dystopian rhetoric and visions of the future, and focus instead on the everyday, the mundane and the more ambiguous grey areas of (possible) present or (near) future conditions.

Central to design fiction is the creation of speculative and at times intentionally provocative 'scenarios', or 'story worlds', which are narrated through the use of 'artifacts' or 'prototypes'. These 'objects' and 'things' might range from physical/digital products and (fictional) user manuals, to (fictional) videos, catalogues, magazines and newspapers, advertisements and marketing materials, research papers/abstracts and more. The point is to open the discursive spaces, and to bring into view and make momentarily tangible several other futures that were otherwise missing before.

## Methodology

Positioned in these contexts, the speculative interventions we are developing respond to the underlying difficulty with unfolding critical practice in design – that of locating and articulating the terms of criticality, or establishing the critical move through design. Our approach is to employ and sharpen the oppositional posturing common for the field of design to fit with the distinctive rhetorical and discursive challenges of the debates concerning the innovations based on emerging gene technologies in food production. This is made possible through a distinct turn toward the radical, contested potentiality of 'the everyday' and 'the agrarian' with the aim to highlight the ethical and cultural issues that are challenged.

In our case, this has further to do with a driving framework of geographic, political and socio-cultural specificity (Norway, the Nordics, Circumpolar North/Arctic). Grounding critical practice in design in a recognizable, identifiable and nuanced local context serves to differentiate our work from the main thrust to intervene in and comment on issues of overarching disciplinary or global concern. Indeed, while the basic operative move of most critical practice in design is to look to emergent or possible future technologies, in our project we explore how alternative, agrarian and craft-based knowledge frameworks – and linked associations to low-tech, folk practices and subaltern forms of expression more broadly – might function as a springboard and context for ideating and prototyping design artefacts and attendant narratives.

We hope such an approach might productively raise questions around not only mainstream design's motives for engaging with fieldwork strategies and participatory practices, and their attendant rhetoric of democracy, agency and inclusion amongst other things, but also shed some critical light on the ways in which design is currently being used – both formally and informally – by both the proponents and opponents of GMOs is shaping and articulating the 'visual culture' (and attendant value systems) of emerging biotechnologies.

## Workshops

As part of the project, we facilitate a series of workshops. Built on techniques and approaches borrowed from design thinking and practice, such as ideation and rapid prototyping, for example, our overarching focus is on interdisciplinarity, a multiplicity of perspectives and voice and the dynamics of co-creation. At the time of writing (January 2021), we have had one digital workshop that brought together the researchers who are part of the project's working group. The participants were from the humanities and the social sciences, several with an expertise in social and cultural aspects of emering biotechnologies. The first workshop was primarily an 'icebreaker' to establish common ground and gain familiarity with the overall goal of the design project. The project designer ran a structured 'ideation' exercise to start mapping the potentials and possibilities at stake, including types of interventions, sites of intervention, as well as the core thematics and problematics. For this exercise, we used Google Jamboard as a collaborative, real-time / live platform for co-creation. We aimed for both depth and breadth at this stage, getting as many genuine alternatives and variations on the table as possible – to be grouped, sorted, and further explored going forward.

The point of departure was a three-fold logic that approached biotechnology as a field of critical enquiry that is comprised of: (1) sites; (2) agents; and (3) practices (Table 1). The participants in the workshop were tasked with a series of rapid-fire brainstorming exercises that alternated between image-based and text-based thinking and imaginaries in order to map as broad and diverse a range of associations, connotations, consensus and discrepancies as possible. The result was a near-instant creation of three rich and varied image and word banks, each housing a rather remarkable spread of different perspectives and potential points of departure.

The potential next steps in activating these image and word banks are to explore – through a further series of structured design exercises – the different types of narratives, scenarios and storylines that different combinations of these 'sites', 'agents' and 'practices' might begin to 'auto-generate'. The next workshops will bring together a broader range of people. In addition to the participants in our researcher group, we will engage artists/designers, stakeholders and members of the public, with the aim of generating and progressively implementing a series of ideas ('fictions'), each with their own specific materiality, form, concept, target group and associated set of concepts, ideas, methods and theories.

## Preliminary results

A next step for the activation of these image and word banks we have generated might for example be the collective creation of 'storyboards' or 'plot lines'. Choosing one component from each of the three lists above (the choice might be random or determined) serves to immediately established a visual and textual starting point from which to begin a story or narrative. There is a character (agent), doing something (practice) in a particular place (site) – speculation and fiction can then be foregrounded and utilized to elicit and provoke responses, insights, perspectives and reactions that might otherwise be difficult or impossible to identify, isolate, approach and discuss together.

*Table 1. Image&Word-bank: result (sample) of Jamboard ideation exercise.*

| Sites (images) | Agents (words) | Practices (words) |
|---|---|---|
| The human body | Journalists | Activism |
| Waste bin | Activists | Workshopping with academics |
| Farms/Fields | Cells | Creating relevant fora |
| Parliaments | Food retailers | Feeding families |
| The laboratory | White men | Eating |
| Farmers' markets | 'Indigenous people' | Plant breeding |
| Supermarkets | Natual selection | Talking loudest |
| Demonstrations/Sites of protests | Influencers | Nudging consumer preferences |
| Backyard gardens | Scientists | DIY |
| Cities | Grocery shoppers | University training |
| Restaurants/bars | Non-human animals | Publishing |
| Forests | Plants | Recycling |
| Cages | The Internet | |
| | Programmers | |

## Conclusion

The complex context of current discussions in Norway and around the world about the use of emerging genome editing techniques is a promising site for employing design fiction as a tool for critical inquiry. Design fiction is an approach that avoids utopian/dystopian rhetoric and visions of the future. We use it as a means for critiquing and speculating around dominant narratives, established histories and mainstream values and expectations. The scenarios, insights and perspectives that are the output of design fiction raise rather than resolve debate. In this way, it might fill a gap in more traditional social science approaches by engaging the public (for instance in public consultation processes) in both critical and creative thinking about alternative food futures. It has the potential to include perspectives and possibilities that might otherwise not rise to the surface of public debate.

In this paper we have presented the methodology and preliminary findings of a novel approach to target the ongoing debates concerning the use and release of genome editing techniques in food organisms. With this, we invite discussion on the potentials (and limitations) of creative practices compared to social science and humanities approaches used in contested and technological enterprises: How can design practices compliment traditional academic approaches to cultural issues? Do we succeed in opening up discursive spaces, to speculate alternatives, foster debate and engage publics in current debates on emerging biotechnologies and future food? How can we bring into view and make momentarily tangible a number of other futures that were otherwise missing before? At the conference, we present further results in order to initiate discussion on these issues.

## References

Dunne, A. (2009). *Hertzian Tales*. Cambridge, MA: The MIT Press.

Dunne, A. and F. Raby (2013). *Speculative Everything: Design, Fiction, and Social Dreaming*. Cambridge: The MIT Press.

Frayling, C. (1993). Research in Art and Design. *Royal College of Art Research Papers*. Volume 1, Number 1 1993/4.

Malpass, M. (2013). Between Wit and Reason: Defining Associative, Speculative, and Critical Design in Practice. *Design and Culture* 5 (3): 333-356.

Mazé, R. and J. Redström (2007). Difficult Forms: Critical Practices in Design and Research. In *Proceedings of the Conference of the International Association of Societies of Design Research*. Hong Kong: IASDR.

De Graeff N, Jongsma KR, Johnston J, Hartley S, Bredenoord AL. (2019). The ethics of genome editing in non-human animals: a systematic review of reasons reported in the academic literature. *Philosophical Transactions of the Royal Society B: Biological Sciences*, *374*(1772).

Doudna, J.A. and Charpentier, E. (2014). 'The New Frontier of Genome Engineering with CRISPR-Cas9,' *Science Magazine* 346 (6213): 1077-1088. Available at: https://science.sciencemag.org/content/sci/346/6213/1258096.full.pdf.

Forsberg, E.M., B. Hofmann, M. Kaiser, B. Myskja, R. Strand and L. Ursin (2020) 'Vurderingskriteriet *etikk*: Veileder for operasjonalisering av vurderingskriteriet etikk i genteknologiloven'.

Gene Technology Act of 2 April 1993 No. 38 Relating to the Production and Use of Genetically Modified Organisms, etc. https://www.regjeringen.no/en/dokumenter/gene-technology-act/id173031/

Helliwell R, S. Hartley, W., Pearce W. (2019). NGO perspectives on the social and ethical dimensions of plant genome-editing. *Agriculture and Human Values*, *36*(4), 779-791.

Jasanoff, S., Hurlbut, J.B. and Saha, K. (2015). CRISPR democracy: Gene editing and the need for inclusive deliberation. *Issues in Science and Technology*, 32(1), p.37.

Kaiser, M., Millar, K., Thorstensen, E. & Tomkins, S. (2007). Developing the ethical matrix as a decision support framework: GM fish as a case study. Journal of Agricultural and Environmental Ethics, 20(1), 65-80.

Myskja, B. and Myhr A.I. (2020). 'Non-safety Assessments of Genome-Edited Organisms: Should They be Included in Regulation?' Science and Engineering Ethics 26 (5):2601-2627. https://doi.org/10.1007/s11948-020-00222-4

Preston, C.J. and T. Antonsen. (In press). 'Integrity and Agency: Negotiating New Forms of Human-Nature Relations in Biotechnology' Environmental Philosophy.

Preston, C.J. and T. Antonsen (2020). 'Ethicists: We need more flexible tools for evaluating gene edited food' The Conversation, May 26 2020. https://theconversation.com/ethicists-we-need-more-flexible-tools-for-evaluating-gene-edited-food-129531

Preston, J. and Wickson, F. (2016). 'Broadening the Lens for the Governance of Emerging Technologies: Care Ethics and Agricultural Biotechnology,' *Technology in Society* 45: 48-57. https://doi.org/10.1016/j.techsoc.2016.03.001

Ricroch, A., Clairand, andW. Harwood (2017). 'Use of CRISPR Systems in Plant Genome Editing: Toward New Opportunities in Agriculture,' *Emerging Topics in Life Sciences* 1 (2): 169-82. https://doi.org/10.1042/etls20170085

Ryffel, G.U. (2012). Organic plants: gene-manipulated plants compatible with organic farming. *Biotechnology Journal*, *7*(11), pp.1328-1331.

# Section 4.

# Animal ethics, veterinary ethics, and food security

# 26. The political turn of the animal ethical discourse – the need for a virtue ethical approach

*E. van den Brandeler*
*University of Basel, Philosophy department, Steinengraben 5, 4051 Basel, Switzerland; vandenbrandeler.emnee@unibas.ch*

## Abstract

A growing body of work within the animal ethical discourse is taking a 'Political Turn'. It is primarily characterised by efforts to propose transformation of our legal and political institutions to account for a just human-animal-relationship in society. In this article, I examine the underrated potential of a virtue ethical approach, as this perspective is currently lacking in the turn's literature. For instance, we get a clearer idea of who ought to represent animals according to many of the turn's institutional reform proposals in terms of relevant political character traits. Likewise, what virtuous role modelling for politicians entails in creating 'animal friendly' moral norms, and how citizens can navigate their own correct political decision-making. Furthermore, we can better discuss the role of individuals in societal justice issues if we reframe our personal virtues as public political virtues. I conclude that including virtue ethical accounts would result in a more comprehensive political turn, not in spite of lacking a 'language of rights', but precisely because of its rich and diverse 'language of virtues'.

**Keywords:** animal ethics, political turn, interspecies justice, political virtue, collective responsibility

## The political turn in animal ethics

The animal ethics literature has long tried to include the interests of nonhuman animals (henceforth: animals) into our moral considerations. We can identify a shift from mere traditional moral philosophy – characterised by philosophers such as Regan (2004), and Singer (2016) – to political philosophy. This new body of literature is increasingly concerned with the political institutionalisation of their theories, as a response to the dissatisfaction with traditional approaches (Woodhall and Garmendia da Trindade, 2017). This is characterised in works put forward by theorists such as Nussbaum (2007), Donaldson and Kymlicka (2011) Cochrane (2012), and Garner (2013a). A few elements can be identified: (1) a broadening of the appeal to liberal political values; (2) a strong emphasis on a rights theory; (3) an emphasis on positive rights; (4) a downgrading from marginal cases; and (5) a broadly pragmatic attitude towards political engagement and compromise (Milligan, 2015: 7). Perhaps most strikingly, the political turn is made distinct by its focus on 'justice', and 'more specifically on how our political institutions, structures and process might be transformed so as to secure just human-animal-relations' (Cochrane *et al.*, 2018: 274).

Animal ethicists often argue that 'rights are the best tools to come to justice because they are designed to protect the inviolable interests of individuals, and as such they are particularly important [for vulnerable minorities] within current formations of power' (Meijer, 2016: 61). Given the set-up of our political systems, I agree this much is true, and admit that it is unsurprising and indeed appropriate that the political turn literature has focused on animal rights. Unfortunately, the introduction of actual animal rights into our legal and political institutions is still met with much resistance. Animal advocacy is largely overlooked as a social justice issue, and not yet recognised as having a legitimate political agenda (Stallwood, 2017). Even though the political turn literature itself may not be at fault for this, or bear


EurSafe 2021 – DOI 10.3920/978-90-8686-915-2_26, © Wageningen Academic Publishers 2021

responsibility for the animal rights movements' failure or success, it does play an important role in providing a comprehensive philosophical justification to support its endeavours.

In what follows, I will argue that a virtue ethical perspective is currently inexplicably missing in the turn's literature. I propose that virtue ethical insights should complement the turn's literature that currently strongly emphasises rights theories. The richness of a language of virtues, emphasis on one's moral character, as well as the long-term commitment to habituating the moral virtues, and its guidance to moral political decision-making, are crucially important. This motivates our correct treatment of animals, and our recognition that institutional reform is needed from this commitment to justice for animals. Virtue ethics should play an important role within a comprehensive political turn of animal ethics.

## Promising aspects of a virtue ethical perspective

Virtue ethics is concerned with the moral character of human beings, and instead of using the language of rights, it uses the language of virtues and vices in order to respond to moral questions (Van Wensveen, 1997). According to Aristotle, the highest human good is a state of 'eudaimonia' – often translated as flourishing – that is wanted for 'itself'. It is a goal that is shared by humans, as the type of beings they are. The moral agent should lead a life in line with the human function of rational activity in accordance with excellence or virtue, and be motivated to do so for the right reasons. Living virtuously contributes intrinsically to a person's flourishing, i.e. the best state of character a person can possess (Aristotle, 2009).

Let us first consider four elements that are characteristic for a virtue ethical perspective within animal ethics. First, the intellectual virtue of 'phronesis', or practical wisdom, takes the importance of context-specific considerations into account, and therefore provides practical guidance (Abbate, 2014). Second, the notion of acquiring virtues for character building through habituation emphasises that the language of virtues focuses on lasting change. After all, one only becomes virtuous, or flourishes, after committing consistently to a life in accordance with the moral virtues. Third, the language of virtues has a certain richness and diversity which sets it apart from other approaches. It ensures flexibility, prevents harmful extremes, and carries with it the promise of moral creativity. Furthermore, it helps us to characterise the relationship of humans with nature, which is especially helpful to animal ethical issues (Van Wensveen, 1997). And fourth, it does not need concepts such as inherent worth from which to build further, and avoids metaphysical demands by its focus on what contribution correct treatment of animals has on human flourishing (Garner, 2013b; Hursthouse, 2007). Accordingly, the virtue ethical perspective grants an uncomplicated and pragmatic approach to improving the human-animal-relationship. Admittedly, it is one thing to say virtue ethics is beneficial for animal ethics. To say it would be beneficial for the political turn as well is something else. Still, we can think of important reasons why it is.

To start, the language of virtues has important political strength. Emphasising the character traits individuals ought to possess in political decision-making can provide a different angle from which to consider the political turn of animal ethics. According to Aristotle, humans are the only 'political animal'. He regards virtue ethics not only as a moral philosophy but also as a political theory (Aristotle, 2009). With this in mind, I wish to highlight two points. First, we can consider what virtuous role modelling would look like for those who hold positions in politics. Many political turn proposals require some sort of political representation on behalf of animals, to voice their interests in our political institutions; e.g. see Smith (2012). It seems rash to assume *any* politician would be good for the job. It is also unclear whether for instance, veterinarians or animal rights activists would necessarily be best equipped for this task. Expert knowledge into the physical wellbeing of animals, or fanaticism to shut down animal agriculture at all costs, do not necessarily ensure that one goes about institutional reform in the right manner. A politician that leads a life in accordance with virtue would embody the right character traits that we praise in people with leadership roles, e.g. care, responsibility and trustworthiness (Newstead *et al.*, 2019). Second, for citizens who are politically engaged only indirectly, virtues can likewise guide them in their political decision-making. This ranges from simply perpetuating the right behavioural norms such as supporting animal advocacy through virtues of humility or empathy, to deciding what political parties and public policies one ought to endorse through virtues such as bravery or compassion. The relevant virtues thus help us determine 'what' we should do, and emphasise that we ought to have the right motivations while doing so.

Furthermore, the language of virtues creates a platform for discussing the role of individuals in social justice issues. The political turn so far largely disregards this and focuses foremost on political institutions at large and their institutional responsibilities. Even though virtue ethics reviews the moral character of the 'individual', it is nonetheless favourably applicable to matters of (interspecies) justice. Namely, virtue ethics can be useful for the political turn due to their appeal to other-regarding virtues such as e.g. compassion, empathy, and so on, since justice must not only be good for the agents themselves, but also for the other. Whilst individual action alone usually does not introduce social injustices, the roles of individuals within social justice issues are still crucial to consider.

To illustrate this further, let us reflect on how societal moral norms are shaped through virtuous role modelling by individuals. In regards to the human-animal-relationship, we can think of moral norms that range from animal friendly norms, to norms that accept and perpetuate animal cruelty. As for the former, correct treatment of animals would be highly valued. As for the latter, using animals for our own benefit across an array of handy applications (such as food, research, sport, companionship, entertainment and so forth) would be acceptable, and would not receive immediate – or perhaps any – political attention. If authors within the political turn are proposing institutional reform that will assert a more just human-animal-relationship, it is crucial we consider our current moral norms, and the virtues that enable them. Namely, our individual commitment to virtue and moral convictions 'collectively' create and maintain societal moral norms. Through virtuous role modelling, we influence each other's behaviour and shape how we as a society are inclined to regard our position towards animals. Especially virtuous role modelling of politicians remains to be considered. Arguably, those with positions in politics have more power to influence how our behaviour in relation to animal issues is – and should be – perceived. The political turn literature wishes to change our institutions, which we have grown used to, and whose existence is engrained into our daily habits. Given this, it becomes clear how valuable it is that the correct moral norms – that are in accordance with virtue – are perpetuated by political role models.

Hence, virtue ethics provides an important additional perspective for the political turn of animal ethics. It opens up discussion of what characteristics we wish for animal representatives to possess and what role virtuous role models play, in a way that has not been offered yet.

## Collective responsibility for political virtues

So far, I have claimed that individual virtues are important for the political turn. As is individual 'motivation'. Particularly, a pivotal point for right behaviour, is that the agent does not only act according to what virtue requires within a given situation, but also that the agent feels motivated to do so for the right reasons (Sandler, 2013). If we frame animal advocacy and our motivation for this only as a 'personal' virtue, the issue of justice for animals will be dependent on those who feel inclined to support justice for animals. There is a problematic tendency to think of these 'animal-friendly' motivations to act right as optional and personal choices, rather than necessary and socially engaged choices. Therefore, I suggest to regard the relevant 'animal friendly' virtues for the political turn, as 'public virtues' like Brian Treanor suggests. Their primary aim is to benefit the wellbeing of the community, and may include some benefit for the individual taking into account they are part of said community. They are character traits that bring us into virtuous relationships with our communities and environments (Treanor, 2010). Thus, community in this sense is not restricted only to humans, but includes animals and the environment we share. Similar accounts have been proposed for shared and collective virtues (Clowney, 2014). These accounts extend the aim for individual 'eudaimonia', to the effect that virtues do not only have individual value, but also social value. They highlight our collective responsibility for tackling institutionalised injustices.

As I mentioned before, our current legal and political institutions are a result of our collective action. No individual action has introduced the largescale political under-representation of animals that the political turn is trying to combat, and no individual action alone will solve it. In order to enable the required collective action and to strengthen a virtuous perspective that is politically salient, we need moral agents to possess public virtues that will benefit the community. More specifically, it would be wise to consider public 'political virtues' that would facilitate democratic decision-making and effective collective action. Candidates for political virtues could for instance be political engagement, compassion, friendliness, truthfulness, open-mindedness, justice, benevolence, and hope (Treanor, 2010). Aside from specific public political virtues, virtues such as respect for nature and the emotion of wonder would encourage people to deepen their understanding of animals and how we ought to treat them. More generally, environmental virtues emphasise the intertwined flourishing of the environment, nonhuman animals and human beings.

Establishing justice for animals requires collective action. Basing it solely on individuals would be over demanding. Here, public virtues offer the moral agent some flexibility in their efforts to be politically virtuous. First, the nature of the virtuous agent to be committed to living a life in accordance with virtue can help frame the required changes as conducive with one's own flourishing. This, in turn, reduces the demandingness of such changes. And makes it easier to incorporate virtues into one's character and act in a way that a commitment to interspecies justice requires. Second, virtue ethics recognises the importance of the context in which one finds themselves. It accepts that not everyone is in a similar position to act according to what virtue requires for the animal movement. Thus, on the one hand, virtue ethics would make achieving interspecies justice less demanding by framing it as being in accordance with one's own flourishing. And on the other hand, it would be less demanding for people in different situations. It recognises that interspecies justice is a joint effort, asking cooperation of all, though conceding this might look different for every moral agent.

## Conclusion

I have shown that it would be unwise to focus solely on pursuing the political turn through moral rights debates and to let virtue ethics remain underrepresented. I have considered the additional benefits the virtue ethical approach has to offer, but I also concede that the success of the political turn ultimately will depend on a multitude of approaches and intertwined efforts that complement each other. If we reconsider our motivations to act in accordance with individual and public (political) virtues, and likewise reframe animal advocacy as having both personal and social value, we make for a more comprehensive view of our role in the improvement of the human-animal-relationship. Therefore, I suggest reconsidering an additional language of virtues within the political turn of animal ethics.

## References

Abbate, C. (2014). Virtues and Animals: A Minimally Decent Ethic for Practical Living in a Non-ideal World. *Journal of Agricultural and Environmental Ethics*, *27*(6), 909-929. https://doi.org/10.1007/s10806-014-9505-z

Aristotle. (2009). *The Nicomachean Ethics* (Translated, p. 203). Oxford University Press.

Clowney, D. (2014). Collective environmental virtue. *Environmental Values*, *23*(3), 315-333. https://doi.org/10.3197/096327114X13947900181194

Cochrane, A. (2012). *Animal Rights Without Liberation: Applied Ethics and Human Obligations* (illustrate ed., p. 246). Columbia University Press.

Cochrane, A., Garner, R. & O'Sullivan, S. (2018). Animal ethics and the political. *Critical Review of International Social and Political Philosophy*, *21*(2), 261-277. https://doi.org/10.1080/13698230.2016.1194583

Donaldson, S. & Kymlicka, W. (2011). *Zoopolis: A Political theory of animal rights*. Oxford University Press.

Garner, R. (2013a). *A Theory of Justice for Animals: Animal Rights in a Nonideal World* (p. 197). Oxford University Press. https://doi.org/10.1093/acprof:oso/9780199936311.001.0001

Garner, R. (2013b). Indirect Duties, Virtue Ethics and Animals. In *A Theory of Justice for Animals: Animal rights in a nonideal world.* (pp. 61-76). Oxford University Press.

Hursthouse, R. (2007). Environmental virtue ethics. In R. L. Walker & P. J. Ivanhoe (eds), *Working virtue: Virtue ethics and contemporary moral problems* (pp. 155-171). Oxford University Press.

Meijer, E. (2016). Interspecies Democracies. In B. Bovenkerk & J. Keulartz (eds), *Animal Ethics in the Age of Humans* (pp. 53-72). Springer, Cham. https://doi.org/10.1007/978-3-319-44206-8_4

Milligan, T. (2015). The political turn in animal rights. *Politics and Animals*, *1*(1), 6-15.

Newstead, T., Dawkins, S., Macklin, R. & Martin, A. (2019). The virtues project: An approach to developing good leaders. *Journal of Business Ethics*, 1-18.

Nussbaum, M.C. (2007). *Frontiers of justice: Disability, nationality, species membership* (p. 487). Belknap Press of Harvard University Press.

Regan, T. (2004). *The case for animal rights* (2nd ed., p. 425). University of California Press.

Sandler, R.L. (2013). Environmental Virtue Ethics. In *International Encyclopedia of Ethics*. Blackwell Publishing Ltd. https://doi.org/10.1002/9781444367072.wbiee090

Singer, P. (2016). A utilitarian defense of animal liberation. In L. P. Pojman, P. Pojman & K. McShane (eds), *Environmental Ethics: Readings in Theory and Application.* (7th ed., pp. 96-105). Cengage Learning.

Smith, K. K. (2012). *Governing animals: Animal welfare and the liberal state*. OUP USA.

Stallwood, K. (2017). Are we smart enough to know when to take the political turn for animals? In A. Woodhall & G. Garmendia da Trindade (eds), *Ethical and Political Approaches to Nonhuman Animal Issues* (pp. 275-303). Springer International Publishing. https://doi.org/10.1007/978-3-319-54549-3_12

Treanor, B. (2010). Environmentalism and public virtue. *Journal of Agricultural and Environmental Ethics*, *23*(1-2), 9.

Van Wensveen, L. (1997). The emergence of ecological virtue language. In *Dirty Virtues: The Emergence of Ecological Virtue Ethics* (pp. 1-21). Humanity Books.

Woodhall, A. & Garmendia da Trindade, G. (2017). Introduction. In A. Woodhall & G. Garmendia da Trindade (eds), *Ethical and Political Approaches to Nonhuman Animal Issues* (pp. 1-20). Springer International Publishing. https://doi.org/10.1007/978-3-319-54549-3_1

# 27. Kantian animal ethics: it's necessity, aims and disturbance

*S. Camenzind*
*Messerli Research Institute, Unit of Ethics and Human-Animal Studies, University of Veterinary Medicine, Vienna, Veterinaerplatz 1, 1210 Vienna, Austria; samuel.camenzind@vetmeduni.ac.at*

## Abstract

In this contribution I will introduce a research field that I call 'Kantian animal ethics'. I shall make some preliminary remarks about why such a research field is needed, how it can contribute to both Kantian research and animal ethics, and the ways in which it is linked to other disciplines, such as gender studies and critical colonial studies.

**Keywords:** animal ethics, Kantian ethics, metaethics

## Kantian animal ethics

The subject referred to by the neologism 'Kantian animal ethics' is a combination of Kantian ethics and animal ethics. Kantian ethics includes research and positions in moral philosophy which have roots in Immanuel Kant's ethics, but reform and modify it in different degrees (cf. O'Neill, 1993; Wood, 2011). Animal ethics involves philosophical reflection on moral norms, values and virtues connected with the human-animal relationship (Grimm and Wild, 2016: 23). The merging of these two fields can seem strange. Kantian scholars have historically shown limited interest in Kant's conception of the human-animal relationship, and the notion that Kant saw animals as things has gone unchallenged. Equally, Kant has a very bad reputation in animal ethics. But I will argue that this dichotomy should be abandoned – that the merging of these two fields is necessary and fruitful for both disciplines.

## The potential of Kantian animal ethics

The reason why animal ethicists see little value in studying Kantian ethics is mainly that Kant denies animals moral status. For Günter Patzig, Kant's ethics seem even to be 'the peak of anthropocentric aberration' (*Gipfelpunkt der anthropozentrischen Verirrung*) (Patzig, 1986: 76). Albert Schweitzer states straightforwardly that: 'The stupidities they [Kant and other European thinkers] are guilty of in trying to maintain the traditional narrow-mindedness and raise it to a principle border on the incredible' (1923: 297). Indeed for most animal ethicists Kant's moral position on animals is an outdated relic of the eighteenth century, and one that we are glad to have left behind (cf. Camenzind, 2019: 44f.).

However, animal ethicists have largely overlooked the wide-ranging duties that Kant formulated, and his views on how we should treat animals in practice. Although Kant claims that the welfare and interests of animals do not count at all morally, and says that moral duties *to* animals are impossible, he recognizes a wide range of duties we have 'regarding' (*in Ansehung*) animals, and these go beyond contemporary animal welfare laws. An example is the requirement that owners should be grateful to their horses, or dogs, for their long service 'as if' they were members of the household. Moreover, for Kant, animals are protected by the highest category of duty (perfect duty to oneself as a moral being). This shift, from duties to others in the 'Lectures' to perfect duties to oneself as moral beings in the 'Doctrine of Virtue' is something that must be assessed with care if we are to assess Kant's position properly. But even more importantly, animal ethicists have largely overlooked the impact on the field of animal ethics of Kant's categories and terminology. This concern inter alia the division of morality into direct and indirect duties (cf. Nelson, 1932: 137ff.; Regan, 1983: 150ff.), the link between absolute moral value and dignity (of


EurSafe 2021 – DOI 10.3920/978-90-8686-915-2_27, 

animals) and Kant's concept of instrumentalization (cf. Regan, 1983: 235, 248f.), even though this, in 2005, was included in Article 3, litera a of the Swiss Animal Protection Act (for the recent debate, see Amman *et al.,* 2015, Bolliger, 2016, Camenzind, 2020).

Hence the study of Kant's ethics is important in two ways: first, it enables us to do justice to the complexity of Kant's position and avoid simply reproducing prejudices; and second, it helps animal ethicists to see how deeply Kant's moral philosophy still shapes contemporary discourse on our moral relationship with the animal world. As a research field, Kantian animal ethics will ideally point to the key concepts here, and reveal their strengths but also their weaknesses.

Criticism of Kant's views on the moral standing of animals can be dated back to Arthur Schopenhauer (1840), Albert Schweitzer (1923) and Leonard Nelson (1932), but only a few papers dealing with Kant's views on animals were published in the 1970s and early 1980s (e.g. Broadie and Pybus, 1974; Regan, 1976; Pybus and Broadie, 1978; Hoff, 1980, 1983). It is fair to say that Kant scholars have not addressed the animal question with real enthusiasm until recently. Looking at the literature, we can see that the discourse changed significantly in the mid-1990s with the work of notable scholars such as Paul Guyer (1993), Christine Korsgaard (1996), Allen W. Wood (1998) and Onora O'Neill (1998). From this point onwards, Kant's position on the human-animal relationship moved to the foreground. It was now referred to regularly in dissertations (e.g. Baranzke, 2002; Kaldewaij, 2013; Camenzind, 2020), and published work (e.g. Callanan and Allais, 2020; Franklin, 2005; Kain, 2010; Korsgaard, 2005, 2013, 2018 or Timmermann, 2005), including introductions to Kant's moral philosophy (Henning, 2016). Admittedly, some commentators discuss it as one of Kant's major mistakes (cf. Dean, 2006: 175; Franklin, 2005: 22; Kain, 2010: 210f.).

I do not wish to take position on the issue whether the animal question is the locus of a major mistake of Kant's. But I do suggest that the animal question is important for Kantians, because when we track animals in Kant's work we arrive at key questions about Kant's axiology, his concept of duty, the limitations of the so-called argument from reciprocity, and finally the Kantian foundation of ethics (cf. Camenzind, 2018, 2020, 2021). Animals are not only protected by the highest duty category. They also play a role in the development of the 'Formula of Humanity', the second formulation of the categorical imperative.

The confrontation of Kant's ethics with animal ethics encourages us to develop conceptual clarity about the nature, and limits, and the focus of, and blind spots in Kant's moral philosophy. The animal question should not be seen as a side issue in Kantian studies, since it can be used as a methodological tool to examine and challenge Kant's ethics. As regards the emerging discipline of animal theory (cf. Borgards, 2015), it would also be very valuable to examine the epistemological potential that animals have in Kant's opus as metaphors, parts of arguments, and concepts.

A further research potential should be mentioned. Contrary to the orthodox view in Western moral philosophy, which sees Kant as the founder of egalitarian human dignity (cf. Schaber, 2010, 2012; Von der Pfordten, 2016), contemporary research in gender studies and critical colonial studies has delivered another, and disturbing, picture of Kant as a racist and sexist. The structural commonalities and differences here between discrimination against animals ('*Sachen*'), women ('*passive Staatsbürger*') (cf. Camenzind, 2020: 122f.; Kleingeld, 2019; Korsgaard, 2005: 96, 100) and people of colour ('*Untermenschen*', Mills, 2017) in Kant's moral philosophy are yet to be revealed.

## References

Ammann, Ch. *et al.* (eds) (2015). Würde der Kreatur. Ethische und rechtliche Beiträge zu einem umstrittenen Konzept. Schulthess, Zürich.

Baranzke, H. (2002). Würde der Kreatur? Die Idee der Würde im Horizont der Bioethik. Königshausen & Neumann, Würzburg.

Borgards, R. (2015). Einführung. In: Borgards, R., Köhring, E., Kling, A. (eds) (2015). Texte zur Tiertheorie. Reclam, Stuttgart, pp. 7-25.

Bolliger, G. (2016). Animal Dignity Protection in Swiss Law – Status Quo and Future Perspectives. Schulthess, Zürich/Basel/Genf.

Broadie, A. and Pybus, E. M. (1974). Kant's Treatment of University of Utah Press Animals. Philosophy 49 (190): 375-383.

Callanen, J.J. and Allais, L. (eds) (2020). Kant & Animals. Oxford University Press, Oxford.

Camenzind, S. (2018). Tierversuche im Kontext der Pflicht gegen sich selbst als moralisches Wesen. In: Waibel, V. L., Ruffing, M., Wagner, D. (eds). Natur und Freiheit: Akten des XII. Internationalen Kant-Kongresses. De Gruyter, Berlin/Boston, pp. 1769-1778.

Camenzind, S. (2019). Der Paratext in Immanuel Kants Metaphysik der Sitten und seine (tier-)ethischen Implikationen. In: Hayer, B. and Schröder, K. (eds). Tierethik transdisziplinär. Literatur – Kultur – Didaktik. Transcript, Bielefeld, pp. 43-60.

Camenzind, S. (2020). Instrumentalisierung. Zur einer Grundkategorie der Ethik der Mensch-Tier-Beziehung. Mentis, Paderborn.

Camenzind, S. (2021). Kantian Ethics and the Animal Turn. On the Contemporary Defence of Kant's Indirect Duty View. In: Animals 11 (512).

Dean, R. (2006). The Value of Humanity in Kant's Moral Theory. Oxford, University Press.

Oxford.

Franklin, J. H. (2005). Animal Rights and Moral Philosophy. Columbia University Press, New York.

Grimm, H. and Wild, M. (2016). Tierethik zur Einführung. Junius, Hamburg.

Guyer, P. (1993). Kant and the Experience of Freedom. Essays on Aesthetics and Morality. Cambridge University Press, Cambridge.

Henning, T. (2016). Kants Ethik. Eine Einführung. Reclam, Stuttgart.

Hoff, Ch. (1980). Immoral and Moral Uses of Animals. The New England Journal of Medicine, 302 (2): 115-118.

Hoff, Ch. (1983) Kant's Invidious Humanism. In: Environmental Ethics, 5 (1): 63-70.

Kain, P. (2010). Duties Regarding Animals. In: Denis, L. (ed.). Kant's Metaphysics of Morals. Cambridge University Press, Cambridge, pp. 210-233.

Kaldewaij, F. (2013). The Animal in Morality. Justifying Duties to Animals in Kantian Moral Philosophy. PhD thesis, Utrecht University, Utrecht.

Kleingeld, P. (2019). On Dealing with Kant's Sexism and Racism. SGIR Review 2 (2): 3-22.

Korsgaard, Ch.M. (1996). The Source of Normativity. Cambridge University Press, Cambridge.

Korsgaard, Ch.M. (2005). Fellow Creatures: Kantian Ethics and Our Duties to Animals. In: Peterson, G. B. (ed.) (2005). The Tanner Lectures on Human Values, Vol. 25. University of Utah Press, Salt Lake City, pp. 77-110.

Korsgaard, Ch.M. (2013 [2011]). Interacting with Animals: A Kantian Account. In: Beauchamp, T. L./Frey, R. G. (eds). The Oxford Handbook of Animal Ethics. Oxford University Press, Oxford, pp. 91-118.

Korsgaard, Ch. M. (2018). Fellow Creatures. Our Obligation to Other Animals. Oxford University Press, Oxford.

Mills, Ch.W. (2017). Black Rights/White Wrongs. The Critique of Racial Liberalism. Oxford University Press, Oxford.

Nelson, L. (1956 [1932]). System of Ethics, trans. Norbert Gutermann. Yale University Press, New Haven.

O'Neill, O. (1993 [1991]). Kantian Ethics. In: Singer, P. (eds): A Companion to Ethics. Blackwell. Oxford, pp. 175-185.

O'Neill, O. (1998). Kant on Duties Regarding Nonrational Nature II. Necessary Anthropocentrism and Contingent Speciesism. Proceedings of the Aristotelian Society, Supplementary Volumes 72: 211-228.

Patzig, G. (1986). Der wissenschaftliche Tierversuch unter ethischen Aspekten. In: Hardegg, W. and Preiser, G. (eds). Tierversuche und medizinische Ethik. Beiträge zu einem Heidelberger Symposium. Olms, Hildesheim, pp. 68-84.

Pybus, E.M. & Broadie, A. (1978). Kant and the Maltreatment of Animals. Philosophy 53 (206): 560-561.

Regan, T. (1976) Broadie and Pybus on Kant. Philosophy 51 (198): 471-472.
Regan, T. (2004 [1983]). The Case for Animal Rights. University of California Press, Berkeley/Los Angeles.
Schaber, P. (2010). Instrumentalisierung und Würde. mentis, mentis.
Schaber, P. (2012). Menschenwürde. Reclam, Stuttgart.
Schopenhauer, A. (1988 [1840]). Preisschrift über die Grundlage der Moral. In: Schopenhauer, A. Werke in fünf Bänden. Ed. by L. Lütkehaus, vol. III. Haffmans, Zurich, pp. 459-632.
Schweitzer, A. (1987) [1923]. The Philosophy of Civilization. Prometheus, New York.
Timmermann, J. (2005). When the Tail Wags the Dog: Animal Welfare and Indirect Duty in Kantian Ethics. Kantian Review 10: 128-149.
Von der Pfordten, D. (2016). Menschenwürde. Beck, München.
Wood, A.W. (1998). Kant on Duties regarding Nonrational Nature I. Proceedings of the Aristotelian Society, Supplementary Volumes 72: 189-210.
Wood, A.W. (2011 [2008]). Kantian Ethics, Cambridge University Press, New York.

# 28. Relating morally to farmed salmon – fellow creatures and biomass

*H.M. Sandvik* and B.K. Myskja*
*Norwegian University of Science and Technology – NTNU, Department of Philosophy and Religious Studies, Dragvoll alle 40, 7491 Trondheim, Norway; hannah.m.sandvik@ntnu.no*

## Abstract

Cora Diamond has criticized capacity-based approaches to determining the moral status of animals, arguing instead that the morally significant fact is that we have relationships to animals as our fellow creatures. This paper explores implications of her approach to fish and the practice of fish farming. Fish differ from most other animals due to their appearances and under-water existence, and it is not obvious that fish belong to our fellow creatures, and – if so – what it means for our treatment of them. In particular: if fish are fellow creatures, can we treat them in the way done in contemporary salmon farming? Iris Murdoch points out that moral differences are conceptual differences, that is differences in how we see the world. Similarly, Diamond argues that we should not consider 'animal' or 'human' as biological classifications – they are conceptual configurations that shape the way we think and make sense of the world. In this article, we explore the implication of the fellow creature concept for the case of fish, which challenge our ordinary understandings of companionship with animals. We argue that farmed salmon should be consibdered as a special kind of fellow creatures living in water, and discuss how scientific research on biological features of fish may influence how we see them. We also sketch how Diamond's approach implies a need for reform of current salmon farming practices.

**Keywords:** fish, Cora Diamond, animal ethics, moral concepts

## Fish – an elusive being

Our attempts to understand our moral relations with animals are often blocked by a sense of unreality, Mary Midgley writes (1998: 9). These creatures are both similar and close to us, and yet utterly strange and distant. Humans interact with them, love them, kill and eat them. How do we reconcile the idea of animals as living, feeling creatures on the one side with the fact that they are also something that humans eat? All in all, they are difficult to fit into our moral picture, and when we try to make room for them, this attempt is often met with resistance.

Fish, however, make for an even more difficult case as far as our moral relations with animals go. Many have pointed out how their difference from us makes it difficult to identify with them. This includes their lack of facial and audible expressions, their physical coldness, that they have scales instead of skin or fur, their lack of distinguishing features and their habitat underneath a surface, making their lives less accessible (Børresen, 2007: 9; Hursthouse, 2000: 223). This is the often-unrecognized moral challenge of the otherness of fish. Perhaps the title of a short article by Peter Singer hits the crux of the issue: 'If fish could scream' (Singer, 2010). If they could express their pain, we would understand that our treatment of them is in fact causing them pain and that this is morally wrong.

This distance between us and the fish is reflected in the way we think about and treat them. Fish have been taken to belong to an altogether different category than other animals, and their ability to feel pain has been heavily disputed. Regan doubted that they were subjects-of-a-life (Regan, 2004: 417), indicating how they escape our regular animal categories. It is still not uncommon for people who


EurSafe 2021 – DOI 10.3920/978-90-8686-915-2_28, © Wageningen Academic Publishers 2021

regularly eat fish to call themselves vegetarians. In Norwegian sea-farming, the word biomass is an accepted term for the living fish. Fish, all in all, have not been taken to merit the same consideration as other animals.

## Animals as fellow creatures

Thinking about animals and what kind of creatures they are raises questions about how we relate to them, how we treat them, and about how we should proceed to answer questions about morally acceptable ways of having them in our lives. Another philosopher who has written about the sense of unreality Midgley refers to is Cora Diamond. Diamond writes about 'the difficulty of reality', which she describes as 'experiences in which we take something in reality to be resistant to our thinking about it, or possibly to be painful in its inexplicability, difficult in that way, or perhaps awesome and astonishing in its inexplicability.' (Diamond, 2008: 45-46). This applies to reality broadly and not just experiences with animals, but animals present a particularly clear case of this kind of struggle, and understanding what Diamond means by it is useful for understanding her own attempt to say something about what animals are to us.

Commonly, in our response to these difficulties, we tend to 'deflect' from them, Diamond writes. The term 'deflection' is one she borrows from Stanley Cavell, and she uses it to describe a manoeuvre where 'we are moved from the appreciation, or attempt at appreciation, of a difficulty in reality to a philosophical or moral problem apparently in the vicinity.' (Diamond, 2008: 57). In other words, instead of confronting the real difficulty of our relationship with animals, we ask instead a set of easier questions: How do we differ from animals, and in which ways are we similar? What rights should animals have based on these similarities or dissimilarities, and what are the morally relevant facts about them? What we have done, is avoiding the difficult problem by re-articulating it in terms that makes it more manageable (Morag, 2020: 195).

Many of the dominant theories in animal ethics start with such questions. Peter Singer and Tom Regan, for example, argue that there are morally salient capacities shared by humans and other creatures that are decisive for our obligations towards them. On this view, biological facts about the neurological basis for consciousness in a species are decisive for how we ought to treat its members. The matter of establishing the moral status of an animal, then, becomes a matter of figuring out which capacities and factors should count, and how we can establish and measure them with sufficient certitude.

For Diamond, this is pure deflection, posing the challenge in the wrong terms. To be more specific, there are three traits which make them problematic from her point of view (Monsó and Grimm, 2019). (1) They are 'rationalistic', meaning that they don't consider emotions to be relevant for discussions on animal rights; (2) they are 'naturalistic' because they argue that there are empirical facts about animals that form the basis for establishing how we should treat them; and (3) they are 'reductionist' in the sense that they place primacy on theory over the plurality of our moral lives.

Diamond suggests an altogether different approach to thinking about animals. The point is that we are never confronted with beings with empirically discoverable similarities and dissimilarities to us, based on which we apply some general principle on how we must act (Diamond, 1991: 351). Rather, they are our 'fellow creatures'.

Diamond doesn't provide a definition of fellow creatures, at least not by any regular standards of how definitions should look, and understanding what she means by it can be challenging. A standard definition would be reductionist like the approaches she criticizes. However, her elucidation of the concept provides direction. 'Fellow creatures' serves as an alternative concept to the emphasis on interests or capacities, and this tells us something about what she does not mean. She writes that it is an idea of 'a being in a certain boat, as it were, of whom it makes sense to say, among other things, that it goes off into Time's enormous Nought, and which may be sought as company' (*ibid.*). What is meant by company here should also be specified. There is an obvious sense in which animals are sought as companions, as demonstrated by the amount of people who own, for example, dogs or cats. But Diamond has a broader understanding of companionship in mind here. She means that we are share a form of live in that we are living, mortal beings (Diamond, 1978: 474). In other words, she wants to show how we share with animals a form of life in that we will all die. However, this concept of 'animals as our fellows in mortality [...] depends upon a conception of 'human' life' (*ibid.*).

To explain this further, Diamond invites us to think about how we do not find it acceptable to eat other humans, even after they have died. The reason why we don't eat humans has nothing to do with the avoidance of suffering, recognition of interests or any other thinkable morally relevant characteristics – the person in question may have died of natural and painless causes – but even so, we wouldn't prepare any of that person's limbs for dinner. Similarly, even the most enthusiastic of carnivores find the idea of eating their pets unthinkable. Humans and pets are simply not something we eat.

Thus, our moral obligations towards animals are not based on a definition of species characteristics but on a fundamental human understanding of what we are. What kind of values we have and how we value is something that is constituted through the forms of life we participate in. Just like being human is significant in moral thought, so our idea of what animals are is not given to us independently of our ways of thinking about and responding to them (Diamond, 1978: 476). This is the given framework for moral reflection on our interactions with other animals. A critique of Diamond's approach from a capacity-based perspective can be found in McMahan (2005), but due to space limitations, we will not attempt to respond to this criticism here. Our modest aim is to sketch some of the moral implications of Diamond's position for salmon farming.

## 'Fish' as a moral concept

If we return to the issue of fish, the question that arises is whether it makes sense to think of them too as our fellow creatures. There is something telling about the metaphor of the boat Diamond uses to describe our relationship with animals. Thoughts go to the story of Noah's ark, where Noah saved a pair of each animal species from the great flood. It goes without saying that he did not make fish companions on the Ark; a boat is exactly where a fish does not want to be.

It can seem, then, that if it is reasonable to consider animals as our fellow creatures, then this is something the fish is excluded from, partly due to its alien nature. Diamond says that when we talk about how we ought to treat certain animals, this is an 'extension to animals of modes of thinking characteristic of our responses to human beings' (Diamond, 1978: 474). Words like biomass prevent this kind of extension, turning fish into what Diamond calls a stage 'in the production of a meat product' (Diamond, 1978: 475) placing them outside the range of fellow creatures in a shared moral world. She points out that sometimes animals are vermin, conceptually placed as unwanted elements in our lifeworld, and hence not fellow creatures. To the extent that the concept 'fish' encompasses the alien features described above, it extends the distance between them and other animals, making it difficult regarding them as fellow creatures. Arguably, it is doing a similar conceptual job to vermin and biomass in excluding a creature from our moral world.

Which concepts we use to denote something has implications for how that thing is taken into consideration. Diamond argues that concepts such as 'animal' and 'human' are not biological concepts at all – they are 'conceptual configurations' that explain how we make sense of the world. In her development of this idea, Diamond leans on Iris Murdoch, and they are both influenced by Wittgenstein. What they all oppose is the idea that concepts correspond to a world that is given prior to moral thought and life. Diamond writes: 'What life is, what death is, what a human being is, what an animal – these things are not given to ethics by biology or metaphysics understood as external and prior to ethics but are rather 'understood through moral thought.' (Diamond, 2010: 59). In other words, the concepts of human being and animal are not simply handed down to us by biology and tradition, they are works of moral thought and imagination and derived from human practices.

## Criticizing concepts and worldviews

At this point we encounter the main challenge of Diamond's approach: how do we overcome deeply ingrained aspects of our worldview that we suspect to be morally untenable? This is also a question about how moral philosophy can be something different than a moral anthropology or psychology, empirically describing conceptual practices as historical practices. In other words, how do we go from a description of how fish are perceived and treated to any kind of assessment of how they should be treated?

Our understanding of concepts is not a stable thing, as Murdoch points out. She invites us to think about the world in which our moral concepts move about as 'cloudy and shifting'; brought about by the use of our moral capacities. We can consider, for example, how our conception of fish has gradually changed, challenging the understanding of it we presented in the introduction. There is more and more conclusive evidence that fish both feel pain and fear and that they are intelligent beings with memory, ability to learn and rich social lives, including creativity and cooperative behaviour (Børresen, 2007). If we return to Singer's statement that fish do not scream, this is not quite true. Fish communicate by sound, although we usually cannot hear it (McCauley and Cato, 2000). It has become more common to talk about fish welfare and what can be done to improve it. Additionally, if we pay attention to the practices of people who interact with fish, the conception of it as outside of the scope of any conception of meaningful relations is challenged. Hobby fishers, researchers and fish farm workers alike express concern for the well-being and even admiration for the fish they work with (Lien, 2015).

Now that science has made it likely that fish experience both fear and pain, this knowledge reinforces what are assumptions shared by many of those who regularly interact with fish. In other words, even though Diamond is critical of the moral individualism Singer defends, this doesn't mean that capacities and characteristics of individuals are unimportant. By observing and learning more about that which is unknown to us, we can come to see it in a different light. For instance, Diamond writes about how she used to think that friendship with animals was limited to a few species only, but after seeing movies about the relation between whales and their Greenpeace rescuers, she realized that she had held a narrow view of which friendships are possible (Monsó and Grimm, 2019: 12). Learning more about the whales made her see them in a new light. Likewise, few books have contributed more to promoting compassion for animals than Singer's 'Animal Liberation' and many readers report giving up meat entirely after reading it. However, as Hursthouse points out, this is not necessarily because they have been convinced by Singer's utilitarian doctrine, but because of its vivid depiction of animal suffering (Hursthouse, 2019: 119).

## Paying attention and virtues

The phenomenon of seeing things in a new light is a familiar one to most of us. But what happens is more than acquiring new facts on the basis of which we form a judgment about how something should be treated or valued. Diamond is influenced by Murdoch, who writes that morality is a matter of attention, not will. She describes this as 'a just and loving gaze directed upon an individual reality' (Murdoch, 2001: 33), putting the idea of moral vision up against the idea of morality as a choice. For the purposes here, we can take fishing as an example. Learning how to fish is more than putting bait on a hook and seeing what happens; it is a practice where the person fishing over time learns a set of skills about how to read the weather, read the ocean, and read the fish. To outsiders, it might seem as though they perceive the world a little differently (Moi, 2017: 226). It may be objected that fishing is a skill rather than an expression of moral perception. However, for Diamond, even when perform such tasks, we give expression to a moral demeanour, as Cordner and Gleeson have argued (Cordner and Gleeson, 2016: 59).

This opens up a space for exploring the role of the virtues. Diamond has little to say about any kind of positive program she may have with regards to how we may or may not treat animals, but she does give content to the idea of an answer by suggesting appropriate virtues for our interaction with animals, analogous to human-related virtues. She mentions several, including charity, fairness, respect for independence and pity (Diamond, 1978: 474-475).

These virtues do not turn animals and fish into beings with moral status, as morality on this account is not defined by certain principles telling us how to act. It is the other way around; morality is part of our moral vision of the world. Our distinctions are conceptual rather than biological, even if they have a biological basis, and saying that moral judgements can be based in the biological similarities between different creatures fails to take into account the way in which morality is immersed in these conceptual distinctions. We do not change our moral perception of fish merely by stating these scientific facts, since fish are on the border of our moral world. However, by recognizing fish as fellow creatures, we learn to see them differently and include them in a certain moral vision. This includes a realization that they are our fellows in mortality, and have lives of their own, both recognizable and alien to the lives of humans. Bringing this to our attention, includes seeing that a virtuous interaction with fish means treating them with charity and fairness, as individual beings with independent lives.

This changed moral gaze includes a realization that certain concepts are integrated aspects of untenable practices, such as calling live fish 'biomass'. This word only makes sense in a form of aquaculture that is cruel, and is an instrument enabling this particular kind of cruelty. There are, however, resources in our existing practices and stories for recognizing fish as fellow creatures, just as there are resources for seeing them as nothing more than a stage in food production. Drawing attention to our knowledge about the life of salmon, wild and farmed, enables a novel moral vision with implications for how we ought to live with them.

## Conclusion

Cora Diamond criticizes approaches to animal ethics that argue that moral status should be determined on the basis of capacities and suggests that a more adequate idea for understanding our relationships to animals is our experience of them as fellow creatures. Fish are not readily seen as fellow creatures, but we argue that there are resources in human relations and scientific knowledge for such an inclusion. Among other things, this involves discarding morally untenable words as 'biomass', which prevents us from seeing them as beings with whom we share a form of life. If we recognize fish as our fellow creatures, we have to reconsider our treatment of them in light of this and reform or end practices that are not consistent with this conception.

## References

Børresen, Bergljot. 2007. *Fiskenes Ukjente Liv*. Oslo: Solum Bokvennen.

Cordner, Christopher, and Andrew Gleeson. 2016. 'Cora Diamond and the Moral Imagination'. *Nordic Wittgenstein Review* 5(1): 55-77.

Diamond, Cora. 1978. 'Eating Meat and Eating People'. *Philosophy* 53: 465-79.

Diamond, Cora. 1991. *The Realistic Spirit. Wittgenstein, Philosophy, and the Mind.*

Diamond, Cora. 2008. 'The Difficulty of Reality and the Difficulty of Philosophy'. In *Philosophy & Animal Life*, New York; Chichester, West Sussex: Columbia University Press, 43-89.
Diamond, Cora. 2010. 'Murdoch the Explorer'. *Philosophical Topics* 38(1): 51-85.
Hursthouse, Rosalind. 2000. *Ethics, Humans and Other Animals*. London and New York: Routledge.
Diamond, Cora. 2019. 'Virtue Ethics and the Treatment of Animals Virtue Ethics and the Treatment of Animals'. (August): 1-27.
Lien, Marianne E. 2015. *Becoming Salmon: Aquaculture and the Domestication of a Fish*. Oakland, California: University of California Press.
McCauley, Robert D., and Douglas H. Cato. 2000. 'Patterns of Fish Calling in a Nearshore Environment in the Great Barrier Reef'. *Philosophical Transactions of the Royal Society B* 355: 1289-93.
McMahan, Jeff. 2005. 'Our Fellow Creatures'. *Journal of Ethics* 9(3-4): 353-80.
Midgley, Mary. 1998. *Animals and Why They Matter*. Athens, Georgia: University of Georgia Press.
Moi, Toril. 2017. *Revolution of the Orginary: Literary Studies after Wittgenstein, Austin, and Cavell*. Chicago; London: The University of Chicago Press.
Monsó, Susana, and Herwig Grimm. 2019. 'An Alternative to the Orthodoxy in Animal Ethics? Limits and Merits of the Wittgensteinian Critique of Moral Individualism'. *Animals* 9(12).
Morag, Talia. 2020. 'Comparison or Seeing-as? The Holocaust and Factory Farming'. In *Morality in a Realistic Spirit. Essays for Cora Diamond*. Andrew Gleeson and Craig Taylor (eds). New York; London: Routledge, 194-214.
Murdoch, Iris. 2001. 'The Idea of Perfection'. In *The Sovereigny of Good*, New York: Routledge, 1-44.
Regan, Tom. 2004. *The Case for Animal Rights*. Berkeley: University of California Press.
Singer, Peter. 2010. 'If Fish Could Scream'. *Project Syndicate*. Available at: https://www.project-syndicate.org/commentary/if-fish-could-scream?barrier=accesspaylog.

# 29. Respect and intrinsic value – Kantian reconstructions of key terms in Norway's Animal Welfare Act[1]

*M.F. Trøite and B.K. Myskja**
*Department of Philosophy and Religious Studies, Norwegian University of Science and Technology, – NTNU, 7491 Trondheim, Norway; bjorn.myskja@ntnu.no*

## Abstract

Animals are due respect and have intrinsic value according to the Norwegian Animal Welfare Act. However, Norwegian animal husbandry, such as salmon aquaculture, includes generally accepted practices that are contrary to an everyday understanding of 'respect' and 'intrinsic value'. A law with no connection between its letter and the actual practices it is meant to regulate, is self-undermining. In order to bridge the gap between law and ethics, we analyse the Act's use of respect and intrinsic value in light of two Kantian approaches to animal ethics to see whether they can give consistent meaning to the letter of the law and provide directions for how the law can be applied in practice regarding these concepts. We find that on the first account, the whole industry must be abolished if the letter of the Act is meant to have practical significance. On the second, fundamental reform is required.

**Keywords:** animal ethics, animal husbandry, aquaculture, moral status, ethics and law

## Introduction

The Norwegian Animal Welfare Act (AWA) states that its intention is to promote respect for animals' intrinsic value. Despite this explicit value statement, Norwegian animal husbandry, such as fish farming, involves numerous generally accepted practices that appear to be contrary to everyday, as well as philosophical, understandings of 'respect' and 'intrinsic value'. Admittedly, there are differences between legal language and that of common morality or moral philosophy, but a law with no connection between its letter and the actual practices it is meant to regulate, is self-undermining. In order to bridge the gap between law and ethics, we will analyse these moral concepts in light of recent contributions to animal ethics and draw on the political debate concerning the Act and its preparatory works.

Respect and intrinsic value are common moral terms and are used in several ethical theories, but they serve a fundamental role in Kantian ethics. We will therefore examine whether two relatively recent Kantian approaches to animal ethics can give consistent meaning to the letter of the law and provide directions for how the Act should be applied in practice. The first approach holds animal nature to be an end-in-itself, because it is a source of value and basis for obligation. Therefore, we have duties to animals as beings with intrinsic, inherent value that are due respect in their own right. The second approach holds that humans only have indirect duties with regard to animals, so when we say that they are due respect, this must be understood in a symbolic sense. However, indirect duties still imply that animals must be regarded as beings that should be valued for their own sake, i.e. as having intrinsic, final value regardless of their instrumental value for us.

[1] This paper presents a part of the as yet unpublished master's thesis 'Exploring the Potential of Clinical Ethics Support Services for a Small Animal Hospital' submitted by M. Long (formerly Stademann) at the University of Veterinary Medicine Vienna which includes a comparison of methods of clinical ethics support services.


EurSafe 2021 – DOI 10.3920/978-90-8686-915-2_29, © Wageningen Academic Publishers 2021

There is disagreement between these accounts on the source of value and the basis for obligation, as well as the practical consequences for our treatment of animals. However, regardless of which Kantian approach we follow, several well-established animal husbandry practices must change if the letter of the Act is meant to have practical significance.

## Respect and intrinsic value in the Animal Welfare Act

In the AWA's first paragraph it is stated that '[t]he intention of this Act is to promote good animal welfare and respect for animals' (AWA, 2009: §1). The preparatory works to the Act refers to 'respect' as an independent value that the law protects in addition to animal welfare. Incorporation of the term respect is intended to create attitudes and strengthen the position of animals in society, where 'ethical considerations beyond pure welfare considerations concerning how animals should/should not be treated shall be emphasized' (MAF, 2008: 93).

This clearly signifies a moral intent behind the Act's terminology, further underscored in the Act's third paragraph which states that 'Animals have an intrinsic value which is irrespective of the usable value they may have for man' (AWA, 2009: §3). According to the preparatory works, 'intrinsic value' means that animals should be treated responsibly and with respect, regardless of any economic or other useful value for humans. It is further emphasized that good animal welfare should be regarded as important first and foremost for the sake of the animals themselves (MAF, 2008). However, the practical consequences of this moral status are uncertain. Elsewhere in the Act, animals are accorded legal status as 'things' and the preparatory works says that 'intrinsic value' should not be interpreted too broadly (CBI, 2009). The Minister of Agriculture and Food at the time held that the term should not have any animal welfare consequences, but merely be of symbolic value (CBI, 2009). If 'symbolic' means that the term intrinsic value does not add anything extra to the law, it appears as meaningless and the Act would be better without it. If the term on the other hand is meant to have a practical function, this function must be clarified by identifying which moral considerations the term is meant to serve and its practical consequences.

## Kantian perspectives on respect and intrinsic value

As stated above, we will use a Kantian framework for this analysis. We assume the Act is derived from a common morality framework that is wider and less consistent, but in order to determine practical implications of the moral language of the Act, a consistent framework is necessary. This does not imply that we believe that a law should be consistent or that this Act has an implicit Kantian basis, but that we take this to be a reasonable analytic approach.

According to an influential Kantian approach, 'some fact or feature is an appropriate object of respect if inappropriate consideration or weighing of that fact or feature would result in behaviour that is morally wrong' (Darwall, 1977: 40). Entities that possess characteristics that should be taken into account in assessing how we should act, are therefore subject to respect, where respect is displayed when these characteristics place restrictions on our actions.

In Kantian theory, intrinsic value can be interpreted in two main ways: (1) as a conditional value; and (2) as an unconditional value. This is based in a distinction between what derives its value from a source outside itself, and what has value in itself. Entities that derive their value from a source outside themselves have conditional value and status as things. If the thing is valued as a means to an end, it has 'instrumental value', and if it is valued for its own sake, it has 'final value' (Korsgaard, 1983). Only entities that are the source of value are understood as possessing unconditional value and having value in themselves. Unconditional value is often called 'inherent value', understood as a categorical value – a value an entity either has or does not have. Everyone who has this value has equal moral status with equal claim to respectful treatment (Regan, 2004). In the following, we will use 'inherent value' when we refer to intrinsic value understood as an unconditional value.

### *Intrinsic value as inherent value*

Immanuel Kant restricted inherent value to humans, i.e. rational beings possessing a free (what Kant calls autonomous) will. According to him, genuine value stems from obligating oneself to universal laws, meaning that value and obligation spring together from an autonomous will. Thus, only rational beings can have inherent value and should be regarded and treated as an end in itself (Korsgaard, 2004, 2018).

According to Christine Korsgaard (2004, 2018), limiting inherent value to rational beings is the result of conflating two different conceptions of being the source of value: (1) being the source of valid, normative claims; with (2) being the source of obligation as a participant in moral legislation based on a common, universal standard of reason. Korsgaard claims that this conflation skips an important step in a two-step valuation process. The original reason why we human beings value something as an end for our actions, is not that the principle for our action can be universalized, it must be that we perceive something as good for ourselves: '"What matters to me," the human being in effect says to himself, "really matters, and is worth pursuing, because I matter"'(Korsgaard, 2004: 93). Since humans are not the only animals that perceive things as good (or bad) for themselves, this original form of valuation is not special to rational beings. However, as rational beings we must then – if the action is to be morally acceptable – universalize the principle on which the action is based. This means that our rational system of values has a basis that we share with most other animals, namely being entities that matters to ourselves from our own perspective. Since this is a condition for value and the source of normative, valid claims, animals must be ascribed inherent value.

Understanding intrinsic value as inherent value, maintains a common moral understanding of respect for intrinsic value when applied to humans. However, this leads us far beyond the understanding of these terms as expressed in the Act. Respect for a being's inherent value in Kantian theory means regarding and treating it as an end in itself and never as a mere means. To act contrary to the interests of such beings for one's own ends or to kill them for other reasons than their own good, is to treat them as mere means. If we seek to avoid a mismatch between animal husbandry practices and the moral concepts of the Act according to this account, most commercial animal husbandry must therefore be phased out – a demand supported by leading animal ethics proponents, including Regan and Singer. However, since it is stated in the preparatory works that referring to intrinsic value is not contrary to killing animals for human ends in this particular context (CBI, 2009), Korsgaard's theory cannot establish a consistent foundation for the key terms in the Act.

### *Intrinsic value as final value*

Rejecting an understanding of the Act's 'intrinsic value' as inherent value, does not mean that the moral terms in the Act must be understood as meaningless from a Kantian perspective. O'Neill (1998) holds Kant's thesis that 'rational nature exists as an end in itself' to be a necessary precondition for moral action. Since animals lack rational nature, they lack the ability to act morally as they can neither obligate nor be obligated by rational nature, and thus, cannot be accorded inherent value. However, although animals cannot be ascribed inherent value, we are not free to treat animals as we want. We have a duty to act in ways that promote and cultivate our own moral feelings, such as compassion, because these feelings help us to act morally right towards other humans. Treating an animal badly will weaken these feelings, and we do wrong towards ourselves. Hence, we have a duty towards ourselves with regard to animals, and therefore an indirect duty to treat animals well.

Jens Saugstad (2000) points out that even if animals are not the target of our moral duty, we can only promote our moral feelings if we value them for their own sake. This means that we must have compassion for the animals themselves to cultivate our sense of compassion towards other humans. We must therefore separate the reason for our duty to treat animals well, from the attitude we are obliged to have towards animals, where we are required to regard them as worthy good treatment for their own sake and not just in light of instrumental ends. Thus, although animals have the status of things, which means that they can be replaced by an equivalent, they must at the same time be regarded as beings with intrinsic, final value.

On this account, the attitude we should have towards animals have similarities to the respect we should have towards humans. Although the cultivation of our capacity for morality is the reason for our duty, it is the animal's wellbeing that place limits on the moral acceptability of our acts. Poor treatment of animals will therefore be morally wrong in itself. This also means that if one takes care of an animal's needs only because it is illegal not to do so, it gives a good reputation, it benefits production or because we become better fellow human beings, it will not qualify as an expression of respect for animals as beings with intrinsic, final value.

If we interpret the term intrinsic value in the AWA as a final value in accordance with the account above, it may be understood as a symbolic expression of the attitude we as rational, moral actors should have towards animals. This explains how animals can be accorded intrinsic value and should be treated well for their own sake, regardless of their usefulness for us – while also having legal status as things that we may use and kill for human ends.

## Respect for intrinsic value in practice

Salmon is the number one farmed animal in Norway, with more than 890 million salmon kept by the industry per year (Dyrevernalliansen, 2019). The basic principle of industrial, intensive aquaculture is cost-effective production. Increased speed of growth and increased numbers and density of animals are crucial factors for improved cost-efficiency (Fraser, 2008). This form of animal husbandry has led to significant welfare challenges such as deformation, diseases and high mortality. In 2018, approximately 53 million salmon died or had to be killed and discarded because of such welfare problems (Saue, 2018). Currently, the industry's predominant problem is salmon lice infection causing both suffering and prematurely death from the infections and from more or less experimental delousing treatments, including mechanical removal, thermal or chemical treatment (Overton *et al.*, 2018). Nevertheless, the industry continues to expand despite unsolved welfare problems. An apparently humane delousing method is using cleaner fish that feed on the lice. However, insufficient knowledge and research on the cleaner fish species-specific needs, has resulted in up to 100% mortality during the first production cycle (Bævre-Jensen, 2019). These fish are also covered by the Act and should be treated respectfully as individuals with intrinsic value, something that is neglected in the Directorate of Fisheries' description of their use as 'a positive contribution to an environmentally friendly production' (DF, 2017), and by industry representatives as 'an effective form of biological delousing' (Biomar, 2021) with significant environmental and economic gains.

These practices and statements can hardly be said to express respect and compassion for these animals as individuals whose wellbeing is of moral importance for the fish's own sake. This is an impression that is enhanced by the industry's description of farmed fish as 'biomass' and fish death as 'wastage' (Saue, 2018).

Based on both Kantian interpretations of the Act's moral concepts, intensive animal husbandry practices fail to express respect for the intrinsic value of animals regardless of whether we understand their value as inherent or final. Despite their fundamental differences, both approaches agree that animals share characteristics with humans that are due respect – if not for inherent value, so at least for final value. Either way, animals ought to be valued and treated well for their own sake, but measures for improved animal welfare within intensive animal husbandry, such as salmon aquaculture, are mainly based on instrumental considerations. This reduces the animals to mere instrumental expressions of human needs (Röcklinsberg *et al.*, 2014), contrary to both Kantian reconstructions of the right kind of attitudes demanded by respect for their intrinsic value.

Furthermore, Kantian ethics' attention to the connection between attitudes and actions, entails that if one wants to promote animal welfare in practice, it is necessary to understand motivations that prevent a respectful attitude. Injured, deformed and apathic fish is not a result of individual moral failures, but a necessary consequence of institutionalized practices. These welfare problems can therefore be understood as not only expressive of reducing animals to mere means but also preventing the development of respectful attitudes in all who partake in this industry. The systematic neglect of the intrinsic value as understood by both Kantian reconstructions, is in this way wrong not just towards the animals, but also towards those who participate, as moral agents. The suffering of and restrictions on farmed salmon lives are also well known, at least in Norway, making most consumers or citizens participants in the practices. This does not reduce the individual responsibilities of those who are directly involved but brings attention to the political responsibilities shared in democratic countries that accept morally problematic forms of animal husbandry. We become participants in moral wrongdoing, and thus, demean ourselves by accepting such practices.

## Conclusion: reform or abolition

Although the current aquaculture practices are unacceptable on both Kantian reconstructions, their implications are radically different regarding how to deal with this. Korsgaard's inherent value approach implies a radical abolition of any farming of animals for food as long as it is unnecessary for human survival, in line with animal rights abolitionism (Francione and Garner, 2010). Given that the food produced in Norwegian salmon aquaculture can be replaced by vegan alternatives, there is no justification for depriving these animals of their lives or containing and breeding them for human purposes. As these radical implications are far from the intentions underlying the AWA, the inherent value interpretation cannot be read as an interpretation of the Act. Instead it is a fundamental critique of the Act and the view on non-human animals expressed in the Act and the animal husbandry traditions it is meant to regulate.

However, animal husbandry, including aquaculture, might be acceptable under the final value interpretation of O'Neill and Saugstad, if animals are not subjected to foreseeable suffering and are regarded and treated as individuals who are allowed to engage in species-specific behaviour. It is not clear what this means in practice, as species-specific behaviour of the Atlantic salmon includes a migratory life cycle where they return to their natal river for spawning and then die. One can argue that this is an essential aspect of their nature and wellbeing that ought to be respected in practice, preventing any salmon farming. One can even argue that it is unacceptable to end this life cycle prematurely by killing them. A more pragmatic interpretation leaves room for keeping salmon in pens, based in the argument that farmed salmon already are bred to become a different species suitable for life in pens. In this case, the required reforms include reducing growth speed and number of individuals per pen in order to prevent foreseeable injuries, diseases and leave room for self-determined activities. Active use of new individual salmon monitoring technologies and genetic editing to ensure optimal life conditions and individualized handling can therefore be understood as part of the required reform.

## References

Animal Welfare Act (2009). Act of June 19, 2009 no. 97 on animal welfare. Available at: https://www.regjeringen.no/en/dokumenter/animal-welfare-act/id571188.

Biomar (2021). Hvorfor rensefisk? Available at: http://www.biomar.com/no/norway/arkiv/produkt/symbio/hvorfor-rensefisk/.

Bævre-Jensen, M. (2019). Kjenner vi rensefisken vår. Available at: https://forskning.no/fiskehelse-fiskeri--og-havbruksnaeringens-forskningsfond-partner/kjenner-vi-rensefisken-var/1358447.

Parlimentary Committee on Business and Industry [CBI] (2009). Innstilling fra næringskomiteen om lov om dyrevelferd. Available at: https://www.stortinget.no/globalassets/pdf/innstillinger/odelstinget/2008-2009/inno-200809-056.pdf.

Darwall, S.L. (1977). Two Kinds of Respect. Ethics 88(1): 36-49.

Dyrevernalliansen (2019). Fakta om fiskeoppdrett og oppdrettsfisk i Norge. Available at: https://www.dyrevern.no/resources/files/fisk-faktaark.pdf.

Directorate of Fisheries [DF] (2017). Vi må også ta vare på Leppefisken. Available at: https://www.fiskeridir.no/Yrkesfiske/Tema/Leppefisk/Kvifor-ta-vare-paa-leppefisken.

Fraser, D. (2008). Animal Welfare and the Intensification of Animal Production. In Thompson P.B. (ed.) The Ethics of Intensification. Springer, Dordrecht, the Netherlands, pp. 167-189.

Francione, G. and Garner, R. (2010). The Animal Rights Debate: Abolition or Regulation? Columbia University Press, New York, USA, 273 pp.

Korsgaard, C.M. (1983). Two Distinctions in Goodness. The Philosophical Review, 92(2): 169-195.

Korsgaard, C.M. (2004). Fellow creatures: Kantian ethics and our duties to animals. Tanner Lectures on Human Values 24: 77-110.

Korsgaard, C.M. (2018). Fellow Creatures – our obligations to the other animals. Oxford University Press, Oxford, UK, 272 pp.

Ministry on Agriculture and Food [MAF] (2008). Om lov om dyrevelferd. Available at: https://www.stortinget.no/globalassets/pdf/innstillinger/odelstinget/2008-2009/inno-200809-056.pdf.

O'Neill, O. (1998). Kant on Duties Regarding Nonrational Nature. Proceedings of the Aristotelian Society 72: 189-228.

Overton, K., Tim Dempster, T., Oppedal, F., Kristiansen, T.S., Gismervik, K. and Stien, L.H. (2018). Salmon lice treatments and salmon mortality in Norwegian aquaculture: a review. Reviews in Aquaculture 11(9): 1-20.

Regan, T (2004). The Case for Animal Rights. University of California Press, Berkeley, USA, 474 pp.

Röcklinsberg, H., Gamborg, C. and Gjerris, M. (2014). A case for integrity: gains from including more than animal welfare in animal ethics committee deliberations. Laboratory Animals 48(1): 61-71.

Saue, O.A. (2018). Heile 30 mill. matfisk er rapportert daud so langt i 2018. Available at: https://ilaks.no/i-2018-er-svinnet-dobbelt-sa-hogt-som-i-2006-heile-30-mill-matfisk-er-rapportert-daud/.

Saugstad, J. (2000). Et kantiansk syn på dyrs moralske status. In Føllesdal, A. (ed.) Dyreetikk. Fagbokforlaget, Bergen, Norway, pp. 140-162.

# 30. Don't push away your emotions – Cora Diamond and animal ethics

*A. Aigner*
*Unit of Ethics and Human-Animal Studies, Messerli Research Institute, University of Veterinary Medicine, Vienna, Medical University of Vienna, University of Vienna, 1210 Vienna, Austria; andreas.aigner@vetmeduni.ac.at*

## Abstract

Most animal ethicists hold that our moral obligations to non-human animals are based on the recognition of their individual characteristics. This view is known as 'moral individualism'. However, proponents of moral individualism such as Peter Singer and Tom Regan have been criticized for focusing too much on rational argumentation and for neglecting the role of emotions in human morality. Many moral individualists claim that ethics is primarily about creating logical consistency in the way humans think about non-human animals. But as Cora Diamond argued already in the late 1970s, emotions like pity are vital to see animals as our fellow creatures. Diamond opposes the idea that moral philosophers should leave emotions or other aspects of our complex moral lives out of discussion. Therefore, Diamond is an important critic of reductionist arguments within moral individualism. However, Diamond does not deny that an individual's particular characteristics are morally relevant, but she denies that morality should be defined only in terms of logical consistency, abstracting from our everyday practices. This paper will focus on Diamond's assumption that emotions are an important part of our moral lives and that they do not necessarily interfere with reasoning. Additionally, Diamond's idea that there are sometimes limits to our reason that nevertheless have important moral philosophical implications is discussed. Hence, the paper will examine how Diamond's approach relates to the idea that emotions can be beneficial for reasoning and that rational thinking alone is sometimes insufficient to deal with moral issues in regard to non-human animals.

**Keywords:** moral individualism, fellow creature, rationality and emotions

## Introduction

Imagine two moral philosophers who try to convince people without a philosophical background that humans should morally consider non-human animals.[1] Specifically, each of them writes a critical comment on the topic of pig farming that is published in a popular, semi-scientific journal with high distribution and impact. These comments are based on different approaches. Philosopher A claims that an individual capable of feeling pain should not be harmed for trivial reasons, no matter if this individual is a human, a dog, or a pig. The philosopher highlights the principle of equal consideration of equal interests by referring to the idea that these beings – as individuals – share an interest in not being harmed or killed. Hence, how an individual is to be treated does not depend on their species membership, but rather on their individual characteristics that are derived from their sentience. The philosopher also argues that, in case of conflicting interests between individuals, these interests must be balanced against each other. For example, a pig's vital interest in not being killed outweighs a human's trivial interest in eating the animal or earning money by selling the produced meat – especially when there are alternatives to eating meat or other ways to earn money. Therefore, he concludes that it is morally wrong to kill the pig. By contrast, philosopher B tells a fictional story about the relationship between Laura, a pig farmer's

[1] In the following, the term 'animals' is used for 'non-human animals'.


EurSafe 2021 – DOI 10.3920/978-90-8686-915-2_30, © Wageningen Academic Publishers 2021

child, and a friendly pig called Billie. Due to the reciprocal interactions between the two, Laura becomes emotionally attached to Billie – whom she now perceives as a friend – and asks her father to spare the animal's life. But Laura's father explains to her that the family has been raising pigs for generations and that he earns their living from it. Laura is unable to prevent Billie from being killed and is deeply saddened when she learns about the animal's death. For her, it is inconceivable that her friend is now food or meat.

How do these two approaches differ in terms of the effects they might have on readers? Does the second approach offer an alternative or a complement to philosopher A's principle-oriented approach that further helps raise awareness of potential moral problems in our relationships with animals? In this paper, these and other questions will be approached by showing that the examples of philosopher A and philosopher B can be associated with two different lines of thought in animal ethics. More specifically, philosopher A's arguments are nearly identical to some of Peter Singer's arguments in 'Animal Liberation' (2009), first published in 1975 and considered one of the 'classics' of animal ethics, and 'Practical Ethics' (2011). Philosopher B's approach conveys Cora Diamond's (1991) idea that philosophy is not only about formulating abstract principles but also benefits, for example, from the consideration of literary work; thereby she argues for considering all of human's moral capacities, which exceed mere reasoning/rational argumentation. In her paper 'Anything but Argument?' (1991), Diamond suggests that engaging in literature can raise people's awareness of potential moral problems in human-animal relationships and thus support their reasoning abilities. In other papers such as 'The Difficulty of Reality and the Difficulty of Philosophy' (2008), she states that certain experiences can confront humans with the limits of their reasoning abilities. Such emotional experiences may lead people to question certain practices of animal use.

In the following, Diamond's critique of Singer and other proponents of what Alice Crary (2018: 158) calls 'traditional moral individualism' will be discussed. Subsequently, Diamond's view of emotions as an important part of human morality will be examined. The aim of the paper is to discuss how Diamond's approach relates to the idea that raising awareness of our emotional responses to animals can be beneficial to reasoning. Her approach can be understood as a supplement to principle-oriented ethical theories by broadening our understanding of morality.

## Traditional moral individualism

Ideally, philosopher A's arguments make readers aware of possible inconsistencies in the ways they think about animals (e.g. treating equal interests unequally). Unlike philosopher A – who argues along the lines of Singer's theory –, philosopher B does not refer to an ethical principle from which the readers can clearly deduce what is morally wrong or right. Singer is a representative of the predominant form of animal ethics known as 'moral individualism'. Tom Regan's (2004) book 'The Case for Animal Rights', first published in 1983, is another 'classic' of animal ethics. In contrast to Singer's utilitarianism, Regan defends a rights approach that highlights the idea that individuals with inherent value have moral rights, but the line of deductive thought remains. Other examples for moral individualistic approaches are Bernard E. Rollin (2017) and Jeff McMahan (2005). Although moral individualists differ in their respective arguments, they all share one basic idea. This idea is found in James Rachels' (1999) definition of moral individualism: '[H]ow an individual may be treated is to be determined, not by considering his group memberships, but by considering his own particular characteristics' (*ibid.*: 173). Simply put: We ought not to treat an animal as we please just because the individual in question is not human.

Traditional moral individualism (Crary, 2018) aims to create logical consistency in the way humans think about and treat animals. Moreover, traditional moral individualists seem to rely on the idea that emotions interfere with the reasoning process. Hence, emotions are 'left aside in the discussion of what is owed to animals. Ethical theory is construed as a close-to-scientific endeavour, where anything that might cloud or distort rational discussion is shunned' (Monsó and Grimm, 2019: 10). Various authors criticize traditional principle-oriented approaches to animal ethics for neglecting the role of emotions like pity or contextual factors in humans' relationships to animals (cf. Diamond, 1978, 1991; Donovan, 2007; Midgley, 1998; Palmer, 2010). However, Diamond does not deny that individual characteristics are morally relevant. Diamond's approach can be described not only as an alternative approach to animal ethics (Crary, 2018) but also as a complement to traditional approaches to animal ethics (Monsó and Grimm, 2019). The point is that many philosophers and psychologists claim that both reason and emotion are important for morality. Moral emotions such as anger, disgust, shame, or guilt have a motivational function and can increase moral concern for others (Kasperbauer, 2015; Prinz, 2014). Some authors state that affects like emotions and moods also have an informative function and can even be 'crucial for good judgment' (Clore and Huntsinger, 2007: 395). Other authors even discuss whether emotions have a primary function in moral judgments (Greene and Haidt, 2002).

This reflects Diamond's assumption that moral philosophers should not merely focus on formulating abstract principles aimed at creating logical consistency as this fails to recognize the complexity of human moral life (cf. Monsó and Grimm, 2019: 13). Moreover, Diamond (1978: 474-475) problematizes the view that moral consideration of animals is primarily based on the recognition of biological facts. To give an example: According to Singer (2009), in order to know how an individual should be treated, it is important to consider whether the individual has the capacity to feel pain and thus has interests. In Diamond's (1978: 466) opinion, Singer and others are right in assuming that animals can feel pain just as humans can, but she doubts that knowing this fact would have any meaning for us if we were 'devoid of all human imagination or sympathy' (*ibid.*: 479). Her notion of animals as 'fellow creature[s]' (*ibid.*: 474) refers to the idea that they are not just something we define in terms of biological facts upon which we can base our abstract ethical arguments. 'The response to animals as our fellows in mortality, in life on this earth [...], depends upon a conception of "human" life' (*ibid.*: 474). This notion of being human involves embodied experiences, for example '[t]he awareness we each have of being a living body' (Diamond, 2008: 74). Besides these experiences, established practices, language and narratives structure our imaginations about what it means to be a human and what is morally important. In other words: Morality cannot be based on an impartial or 'neutral' perspective. Pity for animals, for example, 'depends upon a sense of human life and loss' (Diamond, 1978: 478).

## Diamond and emotions

Unlike philosopher A's arguments, philosopher B's fictional story possibly aims to elicit emotional responses in readers that subsequently cause them to rethink their views of animals. The example of Laura and Billie shows that a pig can be considered a friend. But here, several critical questions arise: From a methodological standpoint, is there a systematic way to 'use' emotions in philosophy? Is evoking emotional responses a manipulative way to mask a lack of good argumentation? Which emotional responses are desirable from a moral philosophical perspective (e.g. pity, compassion, anger)? In addition to these questions, it seems unclear to what extent people actually emotionally respond to a story like the one about Laura and Billie. And even if they do, it is not clear, for example, that this will cause them to change their attitudes toward meat consumption as well. However, the same is true for philosopher A's abstract arguments about the moral relevance of certain biological facts like sentience. Even if people understand why their behaviour can be considered inconsistent according to the argument (because they treat beings with the same morally relevant characteristics differently), this 'insight' does not necessarily lead them to become vegetarians or vegans or to stop eating industrially produced meat.

With respect to Diamond's approach, these questions cannot be clearly answered because they are based on presuppositions that she does not share. Some of these questions start from the idea that moral philosophy is only about creating logical consistency and that philosophers should provide people with reasons to morally consider animals 'which are reasons for anyone, Martian or human being' (Diamond, 1978: 477). But Diamond (1978) criticizes the assumption that the callous ways in which humans treat animals 'cannot be condemned without reasons which are reasons for anyone, no matter how devoid of all human imagination or sympathy' (*ibid.*: 479). Diamond does not attempt to provide such reasons (at least not in the form of Singer's arguments) or to give a systematic account of how exactly philosophers can elicit specific emotional responses that will lead people to adopt the desired stance (e.g. become vegetarians).

In her essays 'Eating Meat and Eating People' (1978) and 'Anything but Argument?' (1991), Diamond notes that imagination, sympathy, and affection are important factors in the moral consideration of animals. However, she does not claim that emotions are in strict opposition to rationality and reasoning. Rather, she assumes that 'thinking well involves thinking charged with appropriate feeling' (Diamond, 1991: 298). Based on this background, Diamond (1991: 307) seems to suggest that 'imaginative literature' such as novels and poems can help us to enhance our reasoning abilities 'by deepening our emotional life and our understanding of it' (*ibid.*: 299). According to her, literary work can be 'of the greatest importance in developing and strengthening our moral capacities' (*ibid.*: 307) – which are already there. As Crary (2020) states, Diamond's 'idea is that works of literature may thus, in virtue of literary features that engage us emotionally, present us with genuine, critically demanding modes of thought' (*ibid.*: 118). Hence, inviting people 'to respond emotionally' (Diamond, 1991: 303) is not the same thing as inviting them to respond in an irrational way.

In 'Eating Meat and Eating People' (1978: 473-475), Diamond refers to a poem by Walter de la Mare titled 'Titmouse'. The poem is about a nimble bird, a titmouse, and describes how the lively animal appears out from nowhere, keeps us company and then flutters away again. Diamond notes that this poem differs from arguments like Singer's and does not refer to biological facts about the bird, but still evokes the idea of a living being, a 'fellow creature' worthy of moral concern. The poem stimulates our human imagination. However, in 'The Difficulty of Reality and the Difficulty of Philosophy' (2008), Diamond suggests that emotional responses to certain experiences can also signal that human imagination and our ability to reason have reached their limits. Diamond (2008) calls such an experience a 'difficulty of reality', a term she borrows from John Updike. She states that it is an experience 'in which we take something in reality to be resistant to our thinking it, or possibly to be painful in its explicability, difficult in that way, or perhaps awesome and astonishing in its inexplicability' (Diamond, 2008: 45-46). As an example, Diamond (2008) refers to a story by J.M. Coetzee in 'The Lives of Animals' about the fictive novelist Elisabeth Costello who is 'haunted by the horror of what we do to animals' (*ibid.*: 46). Costello feels 'wounded by this knowledge, this horror, and by the knowledge of how unhaunted others are' (*ibid.*: 46). But the difficulty of reality is not only a negative experience. According to Diamond, we can be astonished and perplexed by the fact that animals are similar but also dissimilar to us; they can be companions but at the same time seem unfathomable and distant to us (*ibid.*: 61). Sometimes it is incomprehensible that animals 'should be capable of incomparable beauty and delicacy and terrible ferocity' (*ibid.*: 61). In accordance with Diamond's use of the word 'astonishment', Maria Balaska (2019) describes it as an emotionally positive or negative experience that has an unsettling effect on us because language sometimes does not provide the means to make sense of this experience.

Diamond's assumption that morality does not only depend on the recognition of biological facts can be associated with the idea that emotions have an informative and a motivational function. Emotions tell us what matters to us in a given situation, and they dispose us to act in a specific way (Schmid, 2018). For example, guilt can motivate us to right a wrong we have done. From a cognitive psychological point of view, our attitudes about other people are influenced not only by the information we have about those individuals themselves, but also by the information we have about the feelings they elicit in us (Clore and Huntsinger, 2007: 393). This seems to imply that emotions do not necessarily weaken our reasoning abilities or our moral capacities, especially if we are aware of how emotions affect us. However, emotions pose a problem in that they are often based on automatic responses that are not subject to rational control (Kasperbauer, 2018: 87). Diamond's references to literary works in her philosophical reflections can be understood as an invitation to use our imagination (sometimes in a cognitively demanding way). This can help us to take a broader view of our moral lives, including our emotions. By trying to better understand our attitudes and feelings toward animals, we may also be able to improve our moral capacities. However, Diamond's notion of the difficulty of reality refers to the idea that human imagination is sometimes limited and concepts already in place seem to fail. While our imagination helps us extend our understanding of being a living, embodied creature to animals (an extension that is based on our notion of being human and established practices), a difficulty of reality brings with it an inexplicable emotional experience that can lead us to question established practices. This perceived 'failure' is also a possible starting point for rethinking our human self-understanding and how humans should treat animals. Emotional responses – like the horror felt by Costello – are an indicator of such a perceived 'failure'. Another example is the one of Laura and Billie: Laura is unable to put together the thoughts that Billie is her friend and food/meat at the same time. However, this emotionally upsetting event not only brings about an unimaginable situation for Laura, but it also has consequences for her, for example, breaking with her father, starting to live vegan, or finding a way to deal with the situation that is satisfactory to her and allows her to continue her father's work on the pig farm.[2]

## Conclusion

The aim of this paper was to show that Diamond, in contrast to traditional moral individualists, addresses questions of morality not only on an abstract level, but also relates them to our practical lives, to relationships, imagination, and emotion. Because it is important for her to consider all of our moral capacities such as our imagination, she often refers to literary works in her essays. This leaves some questions unanswered, such as whether literature can be systematically used in philosophy to elicit certain emotional responses that simultaneously lead people to form rational moral judgments. However, Diamond's work has important implications for people concerned with animal ethical questions. Diamond invites us to focus not only on simplistic principles and leave emotion or imagination aside in philosophy. The idea to include the topic of emotions in animal ethics has to do with the debate on moral individualism. Namely, the basic idea of moral individualism that the characteristics of an individual are morally relevant cannot be overcome in animal ethics. Diamond's approach can therefore be seen as a complement to traditional approaches to animal ethics, not an alternative (Monsó and Grimm, 2019). But the use of literature, for example, is a way to 'enrich' rational moral philosophical argumentation with further perspectives by relying on our creative potential. With her notion of the difficulty of reality, however, Diamond seems to assume that our potential as rational and imaginative beings is sometimes limited by the fact that there are things that do not make sense to us (like certain practices of animal use). Emotional responses such as Costello's or Laura's imply that such a 'failure' can be seen as an alternative starting point for a potential redefinition of the human-animal relationship, apart from extending established notions of what it means to be human to animals (as our fellow creatures).

## Acknowledgments

I thank Konstantin Deininger, Herwig Grimm and two reviewers for their helpful comments.

[2] I owe this consideration to Konstantin Deininger.

## References


Balaska, M. (2019). Wittgenstein and Lacan at the limit. Meaning and astonishment. Palgrave Macmillan, Cham, Switzerland.

Clore, L.G. and Huntsinger, J.R. (2007). How emotions inform judgment and regulate thought. Trends in Cognitive Sciences, 11(9): 393-399. https://doi.org/10.1016/j.tics.2007.08.005

Crary, A. (2018). Ethics. In Gruen, L. (ed.) Critical Terms for Animal Studies; Lori, Ed.; University of Chicago Press, Chicago, USA, pp. 154-168.

Crary, A. (2020). A brilliant perspective: Diamondian Ethics. In Gleeson, A. and Taylor, C. (eds) Morality in a realistic spirit. Essays for Cora Diamond. Routledge, New York, pp. 113-131.

Diamond, C. (1978). Eating meat and eating people. Philosophy, 53(206): 465-475. https://doi.org/10.1017/S0031819100026334

Diamond, C. (1991). Anything but argument? In Diamond, C. The realistic spirit. Wittgenstein, philosophy, and the mind. The MIT Press, Cambridge, pp. 291-308.

Diamond, C. (2008). The difficulty of reality and the difficulty of philosophy. In Cavell, S., Diamond, C., McDowell, J., Hacking, I. and Wolfe, C. Philosophy & Animal Life. Columbia University Press, New York, pp. 43-90.

Donovan, J. (2007). Attention to suffering. Sympathy as a basis for ethical treatment of animals. In Donovan, J. and Adams, C.J. (eds) The feminist care tradition in animal ethics. Columbia University Press, New York, pp. 174-197.

Greene, J. and Haidt, J. (2002). How (and where) does moral judgment work? Trends in Cognitive Sciences, 6(12): 517-523.

Kasperbauer, T.J. (2015). Rejecting empathy for animal ethics. Ethical Theory and Moral Practice, 18: 817-833. https://doi.org/10.1007/s10677-014-9557-1

Kasperbauer, T.J. (2018). Subhuman. The moral psychology of human attitudes to animals. Oxford University Press, New York.

McMahan, J. (2005). Our fellow creatures. The Journal of Ethics, 9: 353-380. https://doi.org/10.1007/s10892-005-3512-2

Midgley, M. (1998). Animals and why they matter. University of Georgia Press, Athens.

Monsó, S. and Grimm, H. (2019). An Alternative to the orthodoxy in animal ethics? Limits and merits of the Wittgensteinian critique of moral individualism. Animals, 9(12): 1057. https://doi.org/10.3390/ani9121057

Palmer, C. (2010). Animal ethics in context. Columbia University Press, New York.

Prinz, J.J. (2014). Is empathy necessary for morality? In Coplan, A. and Goldie, P. (eds) Empathy. Philosophical and psychological perspectives. Oxford University Press, New York, pp. 211-229.

Rachels, J. (1999). Created from animals. The moral implications of Darwinism. Oxford University Press, Oxford.

Regan, T. (2004). The case for animal rights (2nd ed.). University of California Press, Berkeley.

Rollin, B. (2017). The unheeded cry. Animal consciousness, animal pain, and science. Revised edition. University of Missouri Press, Columbia.

Schmid, H.B. (2018). Collective emotions. In Jankovic, M. and Ludwig, K. (eds). The Routledge handbook of collective intentionality. Routledge, London: pp. 152-161.

Singer, P. (2009). Animal liberation. The definitive classic of the animal movement (4th ed.). Harper Collins, New York.

Singer, P. (2011). Practical ethics (3rd ed.). Cambridge University Press, New York.

# 31. Meat provocation – can animal ethics actually rest on rational arguments?

*M. Huth[1*] and M.E. Hodec[2,3]*
*[1]Messerli Research Institute, Vienna, Austria; [2]University of Prague, Prague, Czech Republic; [3]University of Innsbruck, Innsbruck, Austria; martin.huth@vetmeduni.ac.at*

## Abstract

The aim of this paper is to investigate in a strategy that is in our view pervasive not only, but particularly in animal ethics. As emotionally charged field of argumentation, the debates seem to be susceptible to provocative strategies. The underlying, potentially provocative, hypothesis is that provocation is an indispensable, yet often tacit vehicle of ethical debates (as inert habits are often called into question), which is concealed by the claim to provide rational arguments. This can be illustrated with moral individualists that build the mainstream in animal ethics, represented e.g. by Peter Singer, whose theory has been debated controversially, but also by Tom Regan or Jeff McMahan. We are not going to tackle the question how one should evaluate this issue, however, we seek to understand why they are often seen as 'agents provocateurs' and how they try to trigger reactions on the part of his readers, opponents or supporters. But also relationalists such as Cora Diamond or Alice Crary use provocative statements as vehicle of explanation, e.g. when complaining about the self-proclaimed 'moral radicals' among the individualists (Crary) and their 'nagging moralistic tone' (Diamond). Through our analysis, we aim to point out the function of provocation and of stirring emotions drawing from resources in philosophical analyses of rhetoric and argumentation theory from Aristotle to current debates. In conclusion, we raise the question if there should be an ethics of ethical debates that is sensitive for the benefits and the offensiveness of provocation.

**Keywords:** provocation, argumentation, moralism, rhetoric, ethics of ethics

## Introduction

In this paper, we want to highlight that provocation is a pervasive – and even virtually indispensable – strategy in ethics, and particularly in animal ethics. Provocation is understood as challenging commonsensical convictions of a specific group; crucially, this implies a potential emotional challenge of the addressed individuals of this very group. Notably, we do not assume that provocation is just the attempt to offend someone since a talk or text can also be, for instance, 'thought-provoking', and this can be regarded as sort of a 'productive' challenge (Griffiths, 2016).

As emotionally charged field of argumentation, the debates regarding animals seem to be particularly open to provocative strategies. This applies to public debates e.g. on meat consumption; involved persons frequently even regard activism in the field existential issue (Herzog and Golden, 2009). But it also applies to the academic discourse on animal ethics, not the least because leading philosophers – from the beginning of the consolidation of the discipline on – consider themselves as activists, too (Regan, 2004; Singer, 1975).

At the same time, philosophers usually insist that their theories consist only of rational argumentation and that they would particularly not use methods that (should) stir up emotions or try to influence listeners on a non-rational level (De Lazari-Radek and Singer, 2014). However, we want to argue that provocation is pervasive in ethical argumentation because this is a sine qua non to uncover pitfalls of


EurSafe 2021 – DOI 10.3920/978-90-8686-915-2_31, © Wageningen Academic Publishers 2021

the lived morality – and, to put it a bit provocatively, in order to attract attention as a sine qua non for being invited to give talks and to publish papers. In academic discourses, Berkeley's assertion that *esse est percipi* (Berkeley, 2004), that existence means being perceived or recognized, is particularly significant.

Basically, our analysis draws from Aristotle's influential theory of rhetoric (2010), in which he suggests that statements never rely solely on rationality (the *logos*), but also on the character (and the appearance) of the speaker (the *ethos*) and not the least on affecting the audience (the *pathos*). Without provoking (which also means shocking, disturbing, annoying, surprising), argumentation will hardly find a resonance on the part of readers or listeners.

Bernard Rollin rightly stresses that the violation of the common sense can be regarded as part and parcel of the establishment of new perspectives in different discourses (Rollin, 1998). Copernicus, Darwin, Freud developed theories that have been conceived of as veritable insults for the (narcissistic) human self-understanding; we are not the centre of the universe (Copernicus), we are not the pinnacle of creation (Darwin), and we are not even masters in our own house as the It and the Super-Ego thwart our plans (Freud). These were substantial provocations. But actually also less spectacular scientific revolutions can be considered provocations of our alleged knowledge and self-understanding as has been pointed out in Thomas Kuhn's analysis of paradigm shifts (Kuhn, 1970); we want to contend that this applies to philosophical theories, too. Descartes understood his 'Meditations' as entirely new and radical approach to philosophy (and it had a significant impact on the understanding of bodies and mind including the infamous conceptualization of animals as mere machines); Kant conceived of his 'Critique of Pure Reason as Copernican Revolution'. Similarly, many animal ethicists were and are convinced that they open new paths of understanding animals, humans, and human moral obligations towards animals (McMahan, 2005; Regan, 2004 [1983]; Singer, 1975). But we want to stress that philosophical argumentation that seeks to violate a more or less tacit common sense regarding the legitimacy of behaviours that relate to animals (including meat consumption, wearing leather, using animals in experiments or using animals as guide dogs for blind people or therapy animals, to name only a few) will not do without provocations that are not only purely rational reconsiderations of allegedly all-too clear convictions.

## The 'moral radicals'

In a similar vein as Wittgensteinian philosopher Cora Diamond, Alice Crary conceives of some authors committed to the paradigm of moral individualism, such as Peter Singer, Tom Regan or Jeff McMahan, as 'moral radicals' (2016). What these authors have in common is their attempt to radically challenge an alleged common human self-conception that seems to be crystallized in the metaphysical *differentia specifica*: Starting from the ancient foundations of Western philosophy, many philosophers insinuated that we are animals plus X, i.e. animals that have a particular characteristic that makes us unique (e.g. rationality, autonomy, tool use, art, etc.). This corresponds to a common self-understanding as humans – at least in the Western world. Moreover, we find ourselves in common practices and norms, which specifically or even exclusively protect human beings and their flourishing irrespectively of their actual individual characteristics (think, for instance, of human rights and further legal regulations).

In contrast, Singer, Regan or McMahan (among many other animal ethicists) highlight that we are, as members of a particular biological species, 'nothing special' since evolution theory has shown clearly that we are close relatives of animals (see particularly McMahan, 2002: 11f.; Singer, 2011, particularly the chapter on 'Differences between Humans and Animals'); Correspondingly, the language of 'human and nonhuman animals' displays their strategy of a philosophical alignment of 'us and them' and of questioning and thwarting an alleged human chauvinism. This is mirrored in the fact that relevant moral individualists do not only attack philosophical theories but mundane practices (such as meat consumption) and convictions. Broad populations are found guilty of neglecting the moral considerability of animals, which is traced back particularly not to being human, but to specific individual characteristics (Huth, 2020). These characteristics are usually 'cognitive capacities' (Delon, 2014), for instance, the ability to suffer consciously and to have future preferences (Singer, 2011), being a subject-of-a-life (Regan, 2004) or having self-awareness (McMahan, 2002).

Clearly, this is not only a provocation of traditional philosophers in the proverbial ivory tower, it is also a questioning of a human self-understanding that is, in our view, 'not constituted' but 'reflected and expressed' by philosophers that adhere to a differentialism between humans and animals. The actual provocation consists of the assertion that we as humans are nothing but animals among animals and that the mere fact of being a *homo sapiens* has no moral significance; any proliferation of normativity would be naturalistic fallacy. Regarding morality, you are nothing special or if you are, then only because you have specific abilities that not all beings, even not all human beings, share. Consequently, McMahan speaks of a threshold of respect (McMahan, 2002: 248), and it is a matter of fact that many humans are in some phases of their lives or generally below this threshold.

This provocation is, in our view, ambivalent. On the one hand, it is attractive because many people actually already feel that the use or abuse of animals in human practices is morally problematic so that a manifest ambivalence with regards to phenomena such as factory farming can be detected empirically (Boogard *et al.,* 2011). Given the fact that many people feel powerless in the face of agricultural structures, they might experience that the provocation just reflects what they already would like to find publicly expressed.

On the other hand, the provocation is frequently regarded as repugnant. First, it addresses almost everyone. It, thus, implies the construction of a more or less homogeneous group that is described as fundamentally chauvinist and ignorant regarding the moral consideration of animals. A huge immoral enemy is created, and a moral minority is downright obliged to argue and mobilize (the outsiders) against this opponent. Therefore, it is insinuated that one could either stand on the right side (adopting the criteria proposed by moral individualists) or on the wrong side (foregoing these criteria). Second, these authors seem to invite us to dismiss any understanding of human existence that exceeds the mere membership to a particular species. This can be regarded as an in some respect reductionist understanding of being human. Let us emphasize that by this assertion we precisely do not want to revive the metaphysical *differentia specifica* or intend to deny that the 20th century scientific development in the aftermath of the Darwinian revolution compels us to find a new definition of 'the human'. We do not intend to throw out the baby with the bathwater when we highlight the provocative dimension, which is implied in such an argumentation. However, it simply does not take seriously that we 'distinguish ourselves' from animals (and that there is hardly any definition of the human that does not use the animal as contrast foil; see, for instance, Agamben, 2003). It seems as if authors such as McMahan or Singer want to denounce that we speak of animals and humans 'from a human perspective', which is, in addition, related to a specific socio-cultural perspective. Within the framework of moral individualism, phenomena such as piety and related practices might appear irrational or are portrayed as squeamish or ridiculous (Huth, 2020). Cora Diamond makes this visible referring to the consumption of human flesh of victims of accidents or to the desecration of corpses during warfare; in Singer, such acts can hardly be assessed as genuinely morally inacceptable because corpses lack an ability to suffer and self-awareness (Diamond, 1978: 467, 476). Third, some moral individualists are accused of denouncing speciesism at the expense of 'ableism' (Crary, 2016), i.e. a discrimination or devaluation of individuals with disabilities. This becomes particularly visible in the 'argument from marginal cases' (McMahan, 2002: e.g. 248). McMahan argues that human beings lacking relevant cognitive capacities are at the margins or beyond moral considerability while animals having these capacities are not. Singer, for instance, contends that the readiness to perform research by using animals should be equated with the readiness to perform the same experiments using humans with disabilities being at the same cognitive level as the animals (2011). Similarly, Regan's thought experiment of the lifeboat case can be interpreted as displaying a sort of ableism. This thought experiment can be outlined as follows: We shall imagine five persons (four humans and one dog) in a lifeboat that is suitable for four individuals only, and – all passengers will drown if none of them is sacrificed for the other four persons. Regan claims that the dog should be sacrificed for the others because for an animal death has a minor significance – 'if' none of the human persons is cognitively impaired. In such a case the impaired person would be the one to leave the lifeboat because now for the dog the drowning would be the greater harm (Regan, 2004: 324). However, while these authors usually allege that being a marginal case is a normative fact, we want to contend that marginality is reliant on a process of marginalization (Campbell, 2009) that operates, for instance, through normalization, but also tacit or explicit criteria that distinguish the full-blown person from lesser-than-persons such as embryos, severely cognitively impaired individuals or people suffering from dementia (Agamben, 2003; Crary, 2016). Such a process is efficacious against the backdrop of a basic understanding of being human that exceeds the mere species affiliation and includes individuals using a broader, but vague concept of humanity as basis. This basis is being attacked by 'moral radicals'.

## The moderates

Cora Diamond and Alice Cary are but two authors who call into question whether the strategy of moral individualism is adequate. This is mirrored in the introduction to Cora Diamond's famous paper 'Eating Meat and Eating People':

> This paper is a response to a certain sort of argument defending the rights of animals. Part I is a brief explanation of the background and of the sort of argument I want to reject. Part II is an attempt to characterize those arguments: they contain fundamental confusions about moral relations between people and people and between people and animals. And Part III is an indication of what I think can still be said on – as it were – the animal's side. (Diamond, 1978: 465)

Like Crary (2010, 2016), Diamond takes shared intuitions, practices and concepts as staring point. This includes a shared human self-understanding within a socio-cultural infrastructure and the differential relations we have to animals in contrast to humans; we 'simply' do not consider ourselves as or like animals, and this is not dismissed as irrational but understood as part and parcel of our human rationality. This marks a stark contrast to authors such as McMahan, Regan or Singer. These 'relationalists' consider themselves as the moderate, non-radical protagonists of the discourse of animal ethics. While authors such as Singer, Regan or McMahan contest the understanding of the human as being genuinely morally different from animals, Diamond rejects this contestation: '[T]here is a difference between human beings and animals which is being ignored' (Diamond, 1978: 468).

However, there we detect a provocative strategy in these authors, too. First, even though this supposedly less radical strategy is indulgent with shared, traditional practices that treat animals differently than humans, they contend that the current treatment of animals is morally flawed: '[O]ur relationship to animals show confusion or hypocrisy.' (Diamond, 1978: 473) Of course, this assertion oscillates between being a matter of course and being a provocation. Those who can identify with this diagnosis will be encouraged to applaud. Those who cannot immediately identify with this claim could ask why a differential treatment of animals is to be conceived of as confused or hypocritical. Second, it seems to us as if the key provocation is the manner in which Diamond or Crary challenge moral individualism. We can detect provocations, for instance, in the assertion that relevant authors – and even vegetarians in general – display a 'nagging moralistic tone' (Diamond, 1978: 469); but also in the statement that, within the framework of moral individualism, it would be squeamish not to eat victims of car accidents (consider the title of Diamond's paper 'Eating Meat and Eating People') or that „there is absolutely nothing queer, nothing at all odd, in the vegetarian eating the cow that has obligingly been struck by a lightning' (1978: 468). The ultimate ironical provocation seems to us Diamond's claim that Singer would be unable to explain why it is not speciesism to decline sex with a consenting adult gorilla. Even though one could see a problematic tendency in Singer's utilitarianism that pertains to ableism or to the ignorance of a human self-understanding, this concept would not necessarily imply that this would be a morally problematic discrimination (simply because it is in hardly anyone's interest to have sex with gorillas and vice versa; exceptions might prove the rule).

What could also be regarded as provocative, but in a different respect, is Crary's and Diamond's styles of argumentation (which show significant similarities). Neither Diamond nor Crary operate with logical deductions or syllogisms to support their view; philosophers who expect a strictly logical methodology might be disappointed. Instead, they make use of a more Aristotelian strategy that can be understood as protreptic and as demonstrating pre-existing shared perspectives and values (Huth, 2016). Both highlight that animals are fellows (fellow creatures, fellow travellers in life, fellows in mortality, etc.) and that there is a moral obligation that would be already immanent in, i.e. explicable referring to, our perceptions. This strategy implies the assumption of another 'we' than in the moral radicals. The potential provocation lies in the allegation that we share a specific perceptual relation to the world and that, in a pluralistic society, the meaning and significance of being human and of animals can be taken for granted. We are not as near as critical to insinuate that a significant group of people would 'see' animals as Cartesian machines. But it is a strong claim that ethics' task (merely) consists of making explicit what is already there in 'anyone's' preconscious perceptions.

## Addendum: the denial

Finally, we want to mention briefly Timothy Hsiao' attempt to provoke at least thoughts through his defence of meat consumption. According to Hsiao, the conceptual connection between the sentience and a moral status of animals is inexistent because animal interests (in welfare) are not moral interests:

> Since animals lack moral status and can be used for the purpose of satisfying a legitimate welfare interest (i.e. nutrition), then their use for this end is morally permissible in principle. Acts of harming animals that are disproportionate to the end of nutrition, would, under this analysis, rightly be immoral. Perhaps this requires drastic changes to current practices, perhaps it doesn't. My goal here is not to defend current practices involving the use of animals for food, but to critique the more general idea that we ought to be vegetarians because nourishment is not a morally good reason to cause pain to animals or to support practices that cause pain to animals. (Hsiao, 2015: 290)

Hsiao assures us that his only aim is the objection of the allegedly all-too unquestioned conviction that animals do have moral status. This should not be more than a contribution to a reasonable debate that is first of all centred in arguments. However, even though it is doubtless that he argues rationally and logically sound, the question emerges whether he really does not consider that his text flies in the face not only of a vast majority of animal ethicists, be they individualists or relationalists, but also of any vegan and vegetarian. The suspicion emerges that Hsiao's approach relies heavily on the provocative dimension of the denial of a correlation between sentience and moral status, which plays a significant role in the vast majority of ethical theories as well as in common convictions.

## The restrained conclusion

We tried to make visible that there is a provocative strategy in ethics, particularly in animal ethics, which can be made visible in referring to some of the most received texts in the discourse. However, we did not want to defend the view that provocation in ethics is per se problematic, let alone illegitimate. Quite on the contrary, we want to contend that provocation is an inevitable part of theories that are meant to be more than a mere reproduction of already established position and opinions. Rollin (1998) rightly stresses the significance of the violation of common sense. Moreover, academic and public discourses might be reliant on more than cascades of syllogisms and deductions, but also on 'putting the audience into a certain frame of mind' (Aristotle, 2010: 1356 a 3; this is a very cautious translation of the passage on the *pathos*), not the least through provocation. But this begs the question whether there are ethical limits of this strategy, for instance, if the defence of animals shifts into ableism or if philosophers, or opinions, are being ridiculed or assessed in a manner that does not justice to their arguments. While we want to highlight that provocation might be indispensable for discourse, we want to contend that an ethics of ethical argumentation would be definitely worth considering.

## References


Agamben, G. (2003). The Open *Man and Animal*. Stanford: Stanford University Press, USA, 120 pp.

Aristotle (2010). Rhetoric. Translated by W.R. Roberts. Edited by W.D. Ross. Cosimo: New York, USA, 174 pp.

Berkeley, G. (2004). A Treatise Concerning the Principles of Human Knowledge. Kessinger Publishing: Whitefish, Montana, USA, 80 pp.

Boogaard, B.K., Bock, B.B., Oosting, S.J., Wiskerke, J.S.C., Van der Zijpp, A. (2011). Social Acceptance of Dairy Farming: The Ambivalence Between the Two Faces of Modernity. In: Journal of Agricultural and Environmental Ethics (2011) 24: pp. 259-282. https://doi.org/10.1007/s10806-010-9256-4

Campbell, F.K. (2009). Contours of Ableism. The Production of Disability and Abledness. Palgrave MacMillan: London, UK, 231 pp.

Crary, A. (2010). 'Minding What Already Matters. A Critique of Moral Individualism.' Philosophical Topics Vol. 38, Nr. 1 (2010): 17-49.

Crary, A. (2016). Inside Ethics Inside Ethics On the Demands of Moral Thought. Harvard University Press: Harvard, USA, 2016.

De Lazari-Radek, K., and Singer, P. (2014). The Point of View of the Universe. Sidgwick and Contemporary Ethics. Oxford University Press: Oxford, UK.

Delon, N. (2014). Against Moral Intrinsicalism. In: Animal Ethics and Philosophy. Questioning the Orthodoxy, Hadley, J., Aaltola, E., Eds, 2014, Rowman and Littlefield: New York, USA, London, UK, pp. 31-45.

Diamond, C. (1978). Eating Meat and Eating People.' Philosophy 53 (206), pp. 465-479.

Griffiths, M. (2016). On provocation, fascination, and writing in philosophy of education. In: A. Fulford and Hodgson, N. (eds): On provocation, fascination, and writing in philosophy of education

Herzog, H.A., Golden L.L. (2009). Moral Emotions and Social Activism. The Case of Animal Rights. In: Journal of Social Issues 65 (3), pp. 485-498.

Huth, M. (2016). Humans, Animals, and Aristotle. Aristotelian Traces in the Current Critique of Moral Individualism. Labyrinth: An International Journal for Philosophy, Value Theory and Sociocultural Hermeneutics, 18, pp. 117-136.

Huth, M. (2020). How to Recognize Animals' Vulnerability: Questioning the Orthodoxies of Moral Individualism and Relationalism in Animal Ethics. Animals 10 (2), 235.

Kuhn, Th. (1996). The Structure of Scientific Revolutions. University of Chicago Press: Chicago, USA, 240 pp.

Regan, T. (2004). The Case for Animal Rights. Berkeley: The University of California Press.

Rollin, B. (1998). The Unheeded Cry. Animal Consciousness, Animal Pain and Science. Iowa State University Press, Ames, USA, 330 pp.

Singer, P. (1975). Animal Liberation: A new Ethics for our Treatment of Animals. Duckworth, London, UK.

Singer, P. (2011). Practical Ethics. Cambridge University Press: Cambridge, USA.

# 32. The ethics and mindedness of insects

*M.S. van Loon* and B. Bovenkerk*
*Philosophy Group, Wageningen University & Research, Wageningen, the Netherlands;*
*martijn.vanloon@wur.nl*

## Abstract

Several insect species, including flies, have been proposed as feed for pigs and poultry. The larvae of these flies can convert organic left-over streams into high quality animal protein, while requiring a low-value input: the substrate for mass-rearing these insects could include pig and poultry manure, resulting in a low environmental impact due to its circular use of resources. This proposed innovation has allure, because it seems to reconcile conflicting economic and environmental interests. However, this proposal raises several questions, for example, about insect sentience. Sentience is often defined as feeling, and being equated with phenomenal consciousness, which has been seriously criticized within philosophy of mind. It is disputed whether phenomenal consciousness is necessary for explaining and predicting behaviour, and thus deemed irrelevant for understanding a subject's mental life altogether, subsequently undermining its moral significance. Peter Carruthers provides an alternative approach to animal suffering and moral status, without invoking phenomenal consciousness. In this paper, we explore his alternative approach to animal mentality and moral status: 'suffering without subjectivity'. Finally, we discuss the possible implications for ethics and a potential insect industry.

**Keywords:** phenomenal consciousness, sentience, moral status

## Introduction

According to the Food and Agricultural Organization, the predicted growth of the human population demands an increase in production of high quality protein, while agriculture, specifically animal husbandry, is under pressure because of its environmental impact. The use of insects as food and feed is proposed as a sustainable development (Van Huis and Oonincx, 2017), that could reconcile these conflicting ecological and economic interests. Hence, FAO aims for the development of large scale insect rearing (FAO, 2012). Two fly species, the black soldier fly (*Hermetia illucens*) and house fly (*Musca domestica*), receive increased attention as animal feed, since the growth of larvae requires low value input, i.e. agricultural sidestreams such as manure, thus converting 'waste' into high value animal protein. The circular use of resources results in a lower environmental impact compared to conventional feed with high environmental footprints, i.e. soy and fish meal (Parodi *et al.*, 2020).

Although this innovation has allure, Gjerris *et al.* (2016) have pointed out several ethical questions that are in desperate need of an answer, including uncertainties about insect sentience, welfare conditions, as well as broader animal ethical considerations. The first point is of particular importance, since the capacity for sentience is a common (minimum) criterion for the attribution of moral status in animal ethical theories. Singer (2016) has explicitly stated, following assertions that honeybees might have the capacity for subjective experience (Barron and Klein, 2016), that if insects have feelings, they should be attributed moral status. It is disputed, however, whether insects meet the possession-conditions to be attributed moral status. Moreover, the exact definition of sentience differs between authors, and varies from being synonymous with consciousness, to the capacity for pain and suffering. Broom's (2019) definition of sentience, i.e. the capacity to have feelings, is probably the most common.



Hanna Schübel and Ivo Wallimann-Helmer (eds.) ***Justice and food security in a changing climate***
EurSafe 2021 – DOI 10.3920/978-90-8686-915-2_32, 

Yet, being able to 'feel' does not clarify much, since it does not tell us what mental capacities, say, an insect, needs to possess in order to qualify as a sentient creature, i.e. its exact cognitive or emotional abilities, and perhaps more importantly: the underlying physical structures thereof. Carruthers (2007, 2020) has argued that animal ethicists envisage 'phenomenal' consciousness when they refer to sentience or feelings, that is, consciousness conceptualized from the organism's own experiential point of view. According to Carruthers, however, the vast majority of animal species do not possess the complex cognitive architecture that phenomenal consciousness demands. Interestingly, however, Carruthers argues that phenomenal consciousness should not be regarded as the 'magic bullet' for moral status, because animals could also be 'suffering without subjectivity' (Carruthers, 2004). As a result, insects and other invertebrates – although not being sentient creatures on Carruthers' account – could be included into our ethical deliberations, because they are minded creatures that possess certain mental capacities that could warrant sympathy and moral concern.

In this article, we reflect on Carruthers' argumentation, which consists of three main points: (1) animal ethicists are concerned with capacity for phenomenal consciousness when they talk about sentience; and while (2) phenomenal consciousness is largely absent amongst the vast majority of species in the animal kingdom; yet (3) most animals do possess 'access' conscious mental states, which, as he argues, warrant sympathy and moral concern. Given that Carruthers ultimately discards sentience as a morally relevant mental capacity altogether, we will not engage in any detailed theoretical discussions on consciousness, and, furthermore, we remain agnostic about the definiteness of his theory of consciousness. Rather, we explore whether his notion of 'suffering without subjectivity' is a realistic alternative approach to moral status and the minds of insects, and its implication regarding a potential insect industry. First, we briefly sketch how recent developments and discussions on invertebrate sentience have become of practical ethical importance, given the intended agricultural innovation as described above. Second, we discuss Carruthers contributions to consciousness studies and invertebrate minds. Finally, after analysing Carruthers' alternative approach to attributing moral status, we address several shortcomings, and address its application to a new insect industry.

## Invertebrate sentience

As mentioned, Barron and Klein (2016), have argued that honeybees could have the capacity for subjective experience. Others have shown that various invertebrate species, such as squid and octopus, show extensive cognitive, emotional and behavioural complexity (Mendle, 2011), and some maintain that fruit flies have free will (Brembs, 2011). This has caused increased moral awareness (Drinkwater *et al.*, 2019), attention for welfare considerations (Carere and Mather, 2019), and an overall interest in ethics and invertebrate sentience (Mikhalevich and Powell, 2020).

The emergence of a new insect industry poses the question whether insects, such as flies, possess the capacity to phenomenally experience pain and suffering. Methods such as freezing and blanching are allowed in order to pre-process and transform larvae into feed, because The European animal welfare Directive 98/58/EC for animal farming does not apply to invertebrates. Yet, these animals, even in the larval stage, sense and avoid stimuli that we, as humans, would perceive and describe as painful (Tracey *et al.*, 2003). Thus, arguably, invertebrates should be given the benefit of the doubt, and the precautionary principle should be applied (Dutch Council on Animal Affairs, 2018).

However, there are several caveats: first, 'invertebrates' are not part of official taxonomy, and, moreover, the 'shared lack' of a backbone does not signify much; invertebrates cover roughly 95% of all animal species, and thus there is tremendous diversity, variation, and phylogenetic distance, including significant differences in brain organization between invertebrate species. Hence, we believe it is unwarranted to generalize or extrapolate species-specific findings. Second, although nociception, i.e. the presence of a sensory nervous system that enables a particular organism to 'perceive' noxious stimuli, should be regarded as a necessary condition in order to 'feel' pain, it is generally disregarded as a sufficient condition. The latter would require an additional capacity, that is, the capacity for subjective experience. That is, being able to phenomenally experience a particular mental state. Third, arguably, putting the precautionary principle in place demands indicators of sentience on the level of order (Birch, 2017), in our case *Diptera*. Hence, appealing to indicators of subjective experience in (adult) honeybees (i.e. *Hymenoptera*) will probably not suffice.

Thus, despite several findings on relevant and related mental capacities in different invertebrates, it certainly does not follow that flies, and, more specifically fly larvae, consciously experience the 'hurtfulness' of pain. This would require the capacity to combine or integrate information of (potential) tissue damage with other information and mechanisms. In other words, while nociception is a reasonable indicator (or perhaps: prerequisite) for pain experience, in itself it does not provide the possession-conditions for sentience. Only if one would equate perception with phenomenal consciousness, as proposed by Tye's (2017) 'Simple View', then it follows that fly larvae feel pain when they are being cooked alive. Carruthers (2020), however, argues that Tye's 'theory free' approach is misguided, since it cannot explain all kinds of mental phenomena, granted that all kinds of mental states and subsequent behaviour can be guided by wholly (phenomenally) unconscious perception and processes. Although we do not wish to commit to any theory of consciousness, we concur with Carruthers that the question of animal sentience cannot be properly addressed in the absence of a coherent theory of consciousness, i.e. a theory that can systematically explain the functional and/or neural correlates of phenomenal consciousness, and provides us with the objective physical structures that are constitutive of subjective experience.

## Sympathy for insect mentality

As outlined in the introduction, Carruthers' argument consists of three main points, and starts out with a conceptual critique of sentience, and states that animal ethicists must be referring to phenomenal consciousness. We assume that his first point is correct, and does not require much explanation. It is common to allude to the 'what it is like'-character of experience in discussing or describing sentience, echoing Nagel's (1974) influential paper. In other words, sentience or the capacity to feel indicates consciousness conceptualized from the organism's own experiential point of view; it describes the first-person or subjective experience of mental states. Thus, sentience appears 'private', and, as a result, inaccessible and unknowable from a third-person perspective. As a result, phenomenal consciousness has led to various famous thought-experiments (e.g. Jackson, 1986), and, disputably, profound metaphysical difficulties. Indeed, some philosophers have maintained that we will never be able to fully understand it scientifically (McGinn, 1991), or at least not without being non-reductionist (Nagel, 2012).

Carruthers dismisses the mystification of phenomenal consciousness, and argues that phenomenal consciousness can be reductively explained in physicalistic terms. According to Carruthers (2020), the 'explanatory gap' (Levine, 1983) only arises because we, as humans, employ distinctive acquaintance-based indexical concepts to describe our experiences, i.e. humans can talk about their, for example, perceptual experience, in terms that have no descriptive content (e.g. 'this' red is darker than 'that' red) (Carruthers, 2020). The notorious 'hard problem' (Chalmers, 1995) is merely a result of conceptual confusion. According to Carruthers, global-workspace theory (Baars, 1997) can provide satisfactory physicalistic explanations on how such experience arises from objective processes, i.e. as nonconceptual representations in working memory.

Although Carruthers does not deny sentience in animals categorically, he maintains it is unlikely that most animals possess the cognitive complexity that global-workspace theory demands. But we need not dwell on this point, because, ultimately, Carruthers discards the moral significance of sentience altogether, and argues other, less cognitively demanding, mental capacities could warrant moral concern, because animals are subject to other kinds of consciousness: animals can consciously perceive their environment, can be awake, asleep, or comatose, and thus possess (varying degrees) of access conscious states.

Access consciousness states are described by Block (1995) as the mental states (e.g. sensations, perceptions, beliefs, desires) that are available to a wide variety of cognitive systems, and are involved in intra-mental relations, e.g. visual information that is available for rational control, speech and behaviour, without any distinctive experiential character. Thus, phenomenal consciousness presupposes the possession and availability of access conscious mental states, but not the other way around. That is, the sensorimotor capacity to perceive and respond to pain (i.e. nociceptive withdrawal reflex), can occur without phenomenal experience. One need not be phenomenally conscious of the 'hurtfulness' of pain sensations.

These kind of access conscious states could be sufficient for moral status, as Carruthers sustains, since the relevant harm is not the 'experience of' pain, but rather, the fact that pain can frustrate desires. In other words, even 'unfelt' (i.e. phenomenally unconscious) pain has an impact on the animal's ongoing mental life, by, for example, frustrating its desire to eat (while, similarly, the animal may not 'feel' hungry). Thus, according to Carruthers, the harm of unpleasant sensations, like pain, are not morally bad because they 'feel' unpleasant, but rather because it impedes an organism to fulfil particular desires which appear valuable for the organism. Consequently, the organism 'wants' to get rid of the sensation, not because it has a qualitative sensory experience of, say, a 'sharp' or 'burning' pain, but because it frustrates its ability to satisfy its desires or to act upon its preferences.

Thus, Carruthers' account relies on the presence of access conscious states, and takes the capacity for 'desire-frustration' to have moral weight, because it is an appropriate mental state to have sympathy for. Finally, Carruthers conveys that '(..) sympathy can be independent of questions of consciousness', and 'To be sympathetic towards the plight of a creature is to be motivated to help it, grounded in the third-personal appreciation of its needs and mental states' (Carruthers, 2020: 176). This supposedly leads to 'objective' harms, which can be properly understood from a third-person perspective, rather than a 'subjective' harm, which can only be understood from a first-person perspective. Hence the notion 'suffering without subjectivity' (Carruthers, 2004).

The question remains, then, whether insects can be attributed mental states such as beliefs and desires. Following Dennett's (1987) intentional stance, Carruthers (2007) argues that (at least) jumping spiders and honeybees could be considered 'genuine believer-desirers', because they display sophisticated navigation abilities. He concludes that 'Invertebrates believe things, want things, and make simple plans, and they are capable of having their plans thwarted and their desires frustrated' (Carruthers, 2007: 296). However, he argues that, despite the fact that invertebrates possess minds, which could warrant sympathy and moral concern, we are not required to sympathize, because invertebrates make no direct claim on us, or so he states. We believe the latter is untenable in the context of industrial insect production, in which many hundreds of billions of insects would be kept and killed as mini-livestock. In this case, arguably, one is obligated to feel sympathy towards insects, or at least responsibility.

## Analysis and some critical remarks

The question whether invertebrates possess the capacity for phenomenal consciousness, and thus whether they can feel pain (or pleasure, for that matter) proves of no additional value, according to Carruthers. Hence, he finds himself in practical agreement with Dawkins (2017), and propounds that we should leave the question of sentience, and we should focus on studying health, welfare, and emotional wellbeing in animals instead. What we need is a better understanding of (the range) of mental states a specific animal possesses. That is, given that insects possess a mind, and are subject to access conscious states, we can 'third-personally' study the behaviour and underlying cognitive processes and neural correlates thereof, in an attempt to find what is good for insects, without having to identify ourselves with their (according to Carruthers, alleged) feelings. However, although Carruthers (2019) is a staunch reductionist, and claims that beliefs are (paradigmatically phenomenally unconscious) properties of the brain, he does not provide a physical explanation for the existence of these beliefs (or 'belief-like') states in insects. Thus, it remains unclear how beliefs and desires in insects are realistically construed and to be studied objectively.

If we assume insects possess these kinds of mental states, however, it does warrant sympathy and moral concern with respect to insects. The absence of sentience may indeed not matter very much for the inclusion into animal ethical considerations. That is, if one can concur with Carruthers' explanations. If that is the case, it seems that objective theories of well-being (e.g. 'satisfaction of desires'-approaches) appear compatible, and, could therefore be properly applied to insects. However, the latter would require much more empirical research, studying the behaviour and underlying beliefs and desires of these flies, e.g. regarding food preferences (substrate), (population) density, abiotic conditions, and so forth.

Given our agnosticism regarding theories of consciousness, it is also worth to remark that Carruthers' theory employs the traditional conception of cognition, that is, representationalism. These kinds of representationalist theories have been subject to exceeding criticism: theories of embedded embodied cognition illustrate the limits of traditional theories. Given the fact that the species we are concerned with have evolved in microbially rich environments, and have complex couplings between their central nervous system, immune system and environment, other theories could entail different conclusions about consciousness in insects.

## Conclusions

The account provided by Carruthers provides an alternative approach to include invertebrates, including insects, into our ethical deliberations. The frustration of desires can be considered as a sufficient condition for moral standing, that is, even though insect's mental states may be phenomenally unconscious, i.e. not sentient. Note, however, that the (supposed) absence of feelings, as maintained by Carruthers, entails that we cannot not realistically identify ourselves with the feelings of most animals, hence empathy as a means to morality is problematic, or so he states. To discard empathy is likely to face serious criticism by ethicists. Furthermore, if one agrees with this account, much more empirical research is necessary to get an objective and species-specific account of what is desired by and valuable for these insects, before they are going to be adopted as 'mini-livestock'. The mistakes made in the past in designing animal husbandry systems will be repeated otherwise.

## References

Baars, B. (1997). *In the Theatre of Consciousness*: Oxford University Press.

Barron, A.B. & Klein, C. (2016). What insects can tell us about the origins of consciousness. *Proceedings of the National Academy of Sciences of the United States of America, 113*(18), 4900-4908. https://doi.org/10.1073/pnas.1520084113

Birch, J. (2017). Animal sentience and the precautionary principle. *Animal Sentience, 17.*

Block, N. (1995). A confusion about a function of consciousness. *Behavioral and Brain Sciences, 18*, 227-247.

Brembs, B. (2011). Towards a scientific concept of free will as a biological trait: spontaneous actions and decision-making in invertebrates. *Proceedings of the Royal Society B: Biological Sciences, 278*(1707), 930-939. https://doi.org/10.1098/rspb.2010.2325

Broom, D. (2019). Sentience. In J. C. Choe (ed.), *Encylopedia of Animal Behaviour* (2nd ed., Vol. 1, pp. 13-133): Elsevier, Academic Press.

Carere, C. & Mather, J.A. (2019). *The welfare of invertebrate animals* Animal Welfare; v18.

Carruthers, P. (2004). Suffering without Subjectivity. *Philosophical Studies: An International Journal for Philosophy in the Analytic Tradition, 121*(2), 99-125.

Carruthers, P. (2007). Invertebrate Minds: A Challenge for Ethical Theory. *Journal of Ethics, 11*(3), 275-297.

Carruthers, P. (2020). *Human and Animal Minds: The Consciousness Questions Laid to Rest.* New York: Oxford University Press.

Dawkins, M. S. (2017). Animal welfare with and without consciousness. *Journal of Zoology, 301*(1), 1-10. https://doi.org/10.1111/jzo.12434

Dennett, D. C. (1987). *The intentional stance.* Cambridge: MIT Press.

Drinkwater, E., Robinson, E.J.H. & Hart, A.G. (2019). Keeping invertebrate research ethical in a landscape of shifting public opinion. *Methods in Ecology and Evolution, 10*(8), 1265-1273. https://doi.org/10.1111/2041-210x.13208

Dutch Council on Animal Affairs (2018). *The Emerging Insect Industry; Invertebrates as production animals.* The Hague, the Netherlands: Raad voor Dieraangelegenheden.

FAO. (2012). *Assessing the potential of insects as food and feed in assuring food security, summary report.* Paper presented at the FAO Technical Consultation Meeting, Rome, Italy.

Gjerris, M., Gamborg, C. & Röcklinsberg, H. (2016). Ethical aspects of insect production for food and feed. *Journal of Insects as Food and Feed, 2*(2), 101-110. https://doi.org/10.3920/jiff2015.0097

Jackson, F. (1986). What Mary Didn't Know. *Journal of Philosophy, 83*, 291-295.

Levine, J. (1983). Materialism and qualia: the explanatory gap'. *Pacific Philosophical Quarterly, 64*, 354-361.

McGinn, C. (1991). *The Problem of Consciousness: Essays Toward a Resolution* (Vol. 102): Blackwell.

Mendle, M., Pal, E.S. & Cittka, L. (2011). Animal behaviour: Emotion in invertebrates? *Current Biology, 21*(12), 463-465.

Mikhalevich, I. & Powell, R. (2020). Minds without spines: Evolutionary inclusive animal ethics. *Animal Sentience, 29*(1).

Nagel, T. (1974). What Is It Like to Be a Bat? *The Philosophical Review, 83*(4), 435-450.

Nagel, T. (2012). *Mind and Cosmos: Why the Materialist Neo-Darwinian Conception of Nature is Almost Certainly False.* Oxford: Oxford University Press.

Parodi, A., De Boer, I.J.M., Gerrits, W.J.J., Van Loon, J.J.A., Heetkamp, M.J.W., Van Schelt, J., ... Van Zanten, H. H. E. (2020). Bioconversion efficiencies, greenhouse gas and ammonia emissions during black soldier fly rearing – a mass balance approach. *Journal of Cleaner Production, 271*, 122488. https://doi.org/10.1016/j.jclepro.2020.122488

Singer, P. (2016). Sting in the tail. *New Scientist, 232*(3094), 20-21. https://doi.org/10.1016/S0262-4079(16)31838-3

Tracey, W.D., Wilson, R.I., Laurent, G. & Benzer, S. (2003). painless, a Drosophila Gene Essential for Nociception. *Cell, 113*(2), 261-273. https://doi.org/10.1016/S0092-8674(03)00272-1

Tye, M. (2017). *Tense Bees and Shell-Shocked Crabs: Are Animals Conscious?* Oxford University Press.

Van Huis, A. and Oonincx, D.G.A.B. (2017). The environmental sustainability of insects as food and feed. A review. *Agronomy for Sustainable Development: Official journal of the Institut National de la Recherche Agronomique (INRA), 37*(5), 1-14. https://doi.org/10.1007/s13593-017-0452-8

# 33. What if we lack a licence to kill – thinking out-of-the-box in our relationship with liminal rodents

*M.A.A.M. van Gerwen[1*], J. Nieuwland[1,2] and F.L.B. Meijboom[1,2,3]*
*[1]Centre for Sustainable Animal Stewardship (CenSAS), Department Population Health Sciences, Faculty of Veterinary Medicine, Utrecht University, Yalelaan 2, 3584 CM Utrecht, the Netherlands; [2]Adaptation Physiology Group, Department of Animal Sciences, Wageningen University & Research, De Elst 1, 6708 WD Wageningen, the Netherlands; [3]Ethics Institute, Faculty of Humanities, Utrecht University, Janskerkhof 13, 3512 BL Utrecht, the Netherlands; m.a.a.m.vangerwen@uu.nl*

## Abstract

During previous EurSafe conferences we presented our work about the treatment of liminal rodents in pest management. When it comes to moral status and animal welfare, these animals are generally overlooked. We found that stakeholders involved with pest management feel the need to take the moral position and welfare of liminal rodents more seriously. The outcomes of this study were the start of a multi-stakeholder project to develop an assessment frame for a more responsible rodent management. In order to facilitate ethical decision-making in pest management, various authors have indicated animal research ethics as a valuable source. In this paper we question the relevance of animal research ethics for dealing with liminal animals. Our main concern is that animal research ethics seem to start with the assumption that animal interests can be infringed upon when good reasons are given as a justification. Anectodical information from professionals in the field of rodent management indicates that it is possible to leave the current default position and minimize or even abolished such infringement of rodent interests, while still addressing the nuisance experienced by humans. We aim to explore the potential of non-killing methods as a means to mediate liminal rodent – human conflicts by employing a thought experiment in which we retract the 'licence to kill'. We elaborate on this informed by the concept of liminal rodents as denizens, looking into ecological and socio-cultural carrying capacity and ways to overcome stigmatisation due to feelings of fear and disgust. With this paper we hope to inspire other scientists and professionals in the field to think out-of-the-box when it comes to the relation between humans and liminal rodents.

**Keywords:** animal ethics, animal welfare, liminal animals, commensal rodents, human-animal relationship, rodent management

## Introduction

During earlier EurSafe conferences in 2018 and 2019 we presented our work about the treatment of commensal rodents (brown and black rats and house mice) in pest management. Discussions about moral status and welfare of animals usually focus on livestock, pets or lab animals. Commensal rodents living at the fringes of human societies are generally overlooked. These animals are neither fully wild, nor domesticated. In between nature and culture, these animals can be considered liminal (Donaldson and Kymlicka, 2010). In most countries people kill liminal rodents, often with methods inflicting significant levels of suffering (Broom, 1999; Mason and Littin, 2003). We found that stakeholders involved in rodent management feel the need to take the moral position and welfare of liminal rodents more seriously (Van Gerwen and Meijboom, 2018). They urge for a better application of preventive measures by all people involved in rodent control. A survey among 129 Dutch pest controllers (Van Gerwen *et al.*, 2020) showed that they consider animal welfare during their work and see more room to improve the welfare of liminal rodents, mainly through prevention. They see differences in the welfare


EurSafe 2021 – DOI 10.3920/978-90-8686-915-2_33, © Wageningen Academic Publishers 2021

impact of different control methods used. Furthermore, they indicate that in different real-life scenarios a different weight may be attributed to animal interests in comparison to the human interests at stake. This weighing may have consequences for the methods to use. Together with stakeholders we work to develop an assessment frame for a more responsible rodent management, in which the moral position and welfare of commensal rodents is included.

Various authors have previously indicated animal research ethics as a valuable source in order to facilitate such ethical decision-making. This includes the specific way of justifying research by means of a harm-benefits analysis and application of the '3Rs principle' (replacement, reduction and refinement) in order to prevent unnecessary suffering (Meerburg *et al.*, 2008; Yeates, 2010). Replacement means the application of non-lethal or preventive methods or solutions that do not have a direct impact on the rodents itself. Reduction can be applied by designing rodent management in such a way that as few animals as possible become victim of the intervention. Refinement is applied by choosing methods that inflict the lowest possible welfare impact.

Although the ethical principles from animal research seem promising for implementing animal ethics and welfare in rodent control, in this paper we question the use of animal research ethics for the treatment of liminal rodents. Our main concern is that animal research ethics seem to start with the assumption that animal interests can be infringed upon when good reasons are given as a justification (Nieuwland and Meijboom, 2021). In daily practice it seems that, when it comes to rodent management, people will quickly allow for any form of management when this resolves the nuisance experienced. In that sense, one rat or mouse is already one too many. The harm-benefit analysis then plays out to the disadvantage of rodents in most cases. As a result, infringing animal interests including killing is still part of the default intervention. The presumed license to kill is always there to be used. Besides the ethical problems that rise from this, we believe the approach with the license to kill as default prevents people from creative thinking. It is focussed on the conflict only and does not leave much room to work on creative ways to establish a sustainable co-existence between liminal rodents and humans.

Anectodical information from professionals (personal communications) in the field of rodent management indicates that it is possible to leave the current default position and minimize or even abolished such infringement of rodent interests, while still addressing the nuisance experienced by humans. Also, the respondents of our survey (Van Gerwen *et al.*, 2020) have faith in preventive methods. They indicate that on average 62% of the nuisance could possibly be solved by preventive methods solely. It seems that there is a lot to gain in a proper application of preventive methods, both for maximizing human benefits as for minimizing harm to liminal rodents.

In this paper we aim to explore the potential of preventive – non-killing – methods as a means to mediate liminal rodent – human conflicts by employing a thought experiment in which we retract the 'licence to kill' from the list with control options. What would the relationship between humans and liminal rodents look like if killing was not allowed? Are new more creative forms of this relationship possible? We will elaborate on this approach inspired by the concept of liminal animals by Donaldson and Kymlicka (2011), anecdotical information from professionals (personal communications) in the field and literature on the psychology of moral decision-making.

## Liminal rodents as denizens

The term 'liminal animals' (Donaldson and Kymlicka, 2011) refers to animals that are not fully wild nor domesticated. Rats and mice are considered liminal animals. They live in-between nature and culture amidst people. While they lack close interactions with humans, liminal animals greatly rely on human societies for their survival. Not on us directly, but on the food, shelter and nesting possibilities we offer them. They are experts in using the opportunities we, for a large part unintended, create for them. As they make use of these opportunities, liminal animals are subsequently stigmatised as aliens or invaders who do not belong and do not have the right to be there (Donaldson and Kymlicka, 2011).

Donaldson and Kymlicka start with domesticated animals, which they believe should be endowed full membership of society in terms of citizenship. Liminal animals resemble the status of denizens, not fully part of society but also not completely detached. As such, liminal animals are eligible to a more limited range of entitlements in comparison to domesticated animals, including the right to live their lives. Liminal rodents are to a great extent protected against encroachment of the mixed human-animal society at which fringes they live their lives. Donaldson and Kymlicka describe how liminal rodents navigate these often-urban territories. As these animals are opportunists, they are able to adapt to many different circumstances and move into new niches. Denizenship protects liminal animals against most control methods that are often used in current pest management. These animals not only have a right to life, they also have political entitlements that call for mediation whenever human-animal conflicts arise. Humans have to make every reasonable effort to honour these political entitlements, and to adjust their way of life to accommodate co-existence with liminal animals. We realize that for many people this may be a bit far out-of-the-box. However, we believe that we need to think further than we do normally in order to find creative future proof solutions for the 'rodent problems' many people experience. Liminal rodents will always be around us and belong to the city whether we like it or not. We thus rather invest in a relationship with them instead of being in a conflict situation forever.

One way of averting conflict is to allocate space. As one can imagine, there are locations where rats and mice cannot be allowed because of certain hygiene and health risks. On these locations neither other animals are allowed for that reason. Examples are hospitals, food storage locations and supermarkets. In a society in which liminal rodents and humans co-exist these areas could be marked as human areas that are closed off for rodents and other animals. Next to these types of locations, there should also be areas where liminal rodents are welcome and may live their lives. These areas are free from pest control activities, liminal rodents are left alone by humans and may fulfil certain responsibilities for the whole society. Rats in particular are very good at something we humans find often challenging, namely: getting rid of organic waste. By eating our left-overs and waste, they are recycling champions. This seems like a win-win situation doesn't it? Why would we then continue our rodent control activities at waste disposal companies?

Drawing the line between different areas as described above, is something to deliberate on together as a society. Thereby taking both interests of animals and humans into account. Obviously, the eventual decision about how to share the city habitat should be clear for all inhabitants, including liminal rodents. We should thus communicate the eventual decision in a language they understand. We believe communicating by altering the ecological carrying capacity of certain location by means of preventive methods can be useful.

## Carrying capacity

Preventive methods to manage liminal rodent populations can reduce ecological carrying capacity, improve rodent welfare, protect both the environment and other animals and benefit the effectiveness of the intervention. Preventive methods may consist of all forms of interventions that lower rodent carrying capacity of an area and therefore prevent rodent populations to establish or flourish. In order to influence the ecological carrying capacity (ECC) we need location specific knowledge about the ecology and behaviour of liminal rodents and the environmental characteristics. Furthermore, we need to know how many individuals live in a certain area. At the moment, much remains unknown about the lives and ecology of liminal rats (Parsons *et al.*, 2015). We simply do not know how many rats and mice there are.

The ECC is one way to look at carrying capacity. There is another type of carrying capacity we, and professionals in the field, believe to be very important to consider in the relation between humans and liminal rodents. We can call this the socio-cultural carrying capacity (SCCC). The term socio-cultural carrying capacity is also used in the context of (sustainable) tourism. It there refers to the psychological threshold beyond which locals will negatively perceive and act against tourism (Mansveld and Jonas, 2006). Using this definition for the context of liminal animals we here define the socio-cultural carrying capacity as:

> the threshold beyond which humans will negatively perceive and act against the presence and activities of liminal rodents and human quality of life can no longer be safeguarded.

How to assess the SCCC? This, we believe, is a challenge and only at its very beginning. SCCC appears to be quite subjective since for some people one rat or mouse is already one too many, as for others these animals clearly belong in the city. Liminal animals are frequently stigmatised (Donaldson and Kymlicka, 2011) and fear and antipathy of liminal rodents has been embedded in European culture for centuries (Meerburg *et al.*, 2008). These rodents are associated with filthy environments and illness. Besides these subjective or psychological components, SCCC depends on risks of the presence of liminal rodents for public safety and health, food losses and damage to human property. In order to identify risk in a quantifiable manner, we propose to perform location specific risk assessments that take into account the number of liminal rodents, the pathogens they carry, the risk for pathogens to transfer to and infect humans or other animals, etc. In Europe these types of assessments are already performed for invasive alien species following Regulation (EU) No 1143/2014.

The combination of ecological and socio-cultural carrying capacity can inform the ways in which we shape urban landscapes, drawing attention to the conditions required for both rodents and humans. The goal is to search for a balance between ecological and SCCC. At which number of liminal rodents this balance is found is location specific. This may mean that at some locations ECC needs to be lowered (less opportunities to nest and feed for liminal rodents), whereas SCCC needs to be increased (more acceptance of the presence and activity of liminal rodents by humans).

## How to deal with stigmatisation and disgust?

Increasing the SCCC, as explained before, partly depends on the subjective experiences of individuals. It appears that in general, liminal rodents are stigmatised (Donaldson and Kymlicka, 2011) and trigger feelings of disgust (Haidt, 2001; Meerburg *et al.*, 2008). These aspects make it challenging to ensure more acceptance of liminal rodents by humans. We propose two ways to deal with stigmatisation and disgust. One is to build in anti-stigma safeguards, as Donaldson and Kymlicka (2011) propose. This could be done by protecting the interests of liminal rodents through legislation. Setting up legal protection of liminal animals and their basic interests helps to prevent bias (including disgust and stigma) to result in unjustified disadvantage.

The other is trying to alleviate the bias of disgust and fear more directly in moral decision-making. Basic ethical principles that call upon reason, such 'do not harm unnecessarily', may have difficulty addressing these biases if they insufficiently engage individuals at a more emotional level. In other words, even if one is rationally open to the idea of not harming unnecessarily, such a principle could falter in the midst of aversion, repulsion and disgust. A fruitful way to overcome such psychological currents running underneath one's moral agency is to foster compassion. In interspecies conflict, compassion can function as a key moral faculty, as it requires one to engage with the experience of all sentient beings involved. Moreover, this engagement has to be wholeheartedly, not merely rational (Nieuwland and Meijboom, 2021). Of course, emotional involvement has its downsides in moral deliberation, for example, as empathy can increase moral concern for those with whom you share affinity in some way (Bloom, 2017). Then, moral consideration is channelled through the already existing emotional attachment between you and the other. You feel the others pain, like when your child or beloved spouse hurts her or himself. Compassion, however, can overcome these forms of partiality and many other forms of bias. It calls for a more encompassing form of engagement, 'characterized by feelings of warmth, concern and care for the other, as well as a strong motivation to improve the other's wellbeing. Compassion is feeling for and not feeling with the other' (Singer and Klimecki, 2014: r875). Also, if the other is a rat, associated with aversion, through a compassionate outlook we can recognize their suffering as well.

While it is difficult to foster compassion on the spot, caught up in interspecies conflict, there are ways of communicating that push in this direction. For example, perhaps assisted by storytelling, one can facilitate others to open up to the lives of liminal rodents. Not only in terms of ecological facts, but in terms of how rodents experience the space that they share with humans, including their fears and joys (Nieuwland and Meijboom, 2021). Also, one has to make the consequences of current rodent control in terms of animal welfare more transparent. In current rodent control there is a tendency for humans to keep control measures at a distance (Jackson, 1980 in: Meerburg *et al.*, 2008; Van Gerwen and Meijboom, 2018). Rodents die often out of sight and people see little of the suffering they go through. While animal protection NGO's show video footage of animals in slaughterhouses or pet animal abuse and neglect, they hardly show such footage of animals involved in pest control. Would it change attitudes when people would see more of how rodent control is performed and what is means for animal welfare? We know that animals can actually find rats and mice cute and have feelings of care for them. The fact that these animals are kept as pets with their own social media accounts shows that this is possible too. Furthermore, we sometimes see situations in which humans help liminal mice or rats in trouble. In early 2019, for example, a video made in Germany was picked up by various international media (BBC, 2019). Local individuals together with animal rescue persons and firefighters freed a brown rat that was trapped in a manhole cover. In the same year, in the Netherlands a man called the animal ambulance to free a mouse after he found the mouse trapped on a glue trap he fabricated and placed himself (AD, 2019). Efforts to stimulate an attitude shift among pest controllers could also be integrated in pest management education. However, we did not see any effect of an ethics course in the attitudes of Dutch pest controllers participating in a survey (Van Gerwen *et al.*, 2020). We believe there is still room to improve this type of education and its content.

## Concluding remarks

Although animal research ethics form a good start to embed ethical decision-making in rodent management, they still allow a license to kill as part of default. We explored the potential of preventive – non-killing – methods as a means to mediate liminal rodent – human conflicts by employing a thought experiment in which we retract the 'licence to kill' from the list with control options. We did so by: (1) giving denizenship to liminal rodents; (2) looking into ecological and socio-cultural carrying capacity; and (3) ways to overcome stigmatisation due to feelings of fear and disgust. With this paper we hope to inspire other scientists and professionals in the field to think out-of-the-box when it comes to the relation between humans and liminal rodents.

## Acknowledgements

We would like to thank all our students – Mark, Simcha, Bas, Lucca and Tamar – who enthusiastically contributed to this paper for their input and creative thinking about the ethical aspects of rodent management.

## References

AD (2019). Man zet val voor muisje, krijgt medelijden en belt dierenambulance. Available at: https://www.ad.nl/den-haag/man-zet-val-voor-muisje-krijgt-medelijden-en-belt-dierenambulance~a7ace9f2/?referrer=https%3A%2F%2Fwww.google.com%2F.

BBC (2019). Fat rat saved from manhole by German animal rescue. Available at: https://www.bbc.com/news/world-europe-47370291.

Bloom, P. (2017). Against Empathy: The Case for Rational Compassion. HarperCollins Publishers, New York, NY, USA.

Broom, D.M. (1999). The welfare of vertebrate pests in relation to their management. In: Advances in Vertebrate Pest Management, ed. P.D. Cowan and C.J. Feare, 309-329.

Donaldson, S.; Kymlicka, W. (2011). Zoopolis: A Political Theory of Animal Rights. Oxford University Press, New York, NY, USA.

Haidt, J. (2001). The Emotional Dog and Its Rational Tail: A Social Intuitionist Approach to Moral Judgment. Psychological Reviews 108 (4): 814-834.

Mansveld, Y. and A. Jonas. (2006). Evaluating the socio-cultural carrying capacity of rural tourism communities: a 'value stretch' approach. Journal of Economic and Social Geography 97: 583-601.

Mason, G. and K.E. Littin. (2003). The humaneness of rodent pest control. Animal Welfare 12: 1-37.

Meerburg, B.G., F.W.A. Brom and A. Kijlstra (2008). Perspective, The ethics of rodent control. Pest Management Science 64: 1205-1211.

Nieuwland and Meijboom (2021). 'Eek! A Rat!'. In: Bovenkerk B. and J. Keulartz (eds) Animals in our Midst: The Challenges of Co-existing with Animals in the Anthropocene. Springer International Publishing, New York, NY, USA.

Parsons, M.H., Sarno, R.J. & Deutsch, M.A. (2015). Jump-Starting Urban Rat Research: Conspecific Pheromones Recruit Wild Rats into a Behavioral and Pathogen-Monitoring Assay. Frontiers in Ecology and Evolution 3.

Singer, T. & Klimecki, O.M. (2014). Empathy and compassion. Current Biology 24: 875-878.

Yeates, J. (2010). What can pest management learn from laboratory animal ethics? Pest Management Science 66: 231-237.

# 34. Viennese hamsters and the interspecies politics of urban space

*B. Kristensen*
*University of Oregon, Environmental Studies Program & Department of Philosophy, 5223 University of Oregon, Eugene, OR 97403, USA; brk@uoregon.edu*

## Abstract

I consider the interspecies political implications of Vienna, Austria's urban-dwelling European hamsters (Cricetus cricetus). In July 2020, the International Union for Conservation of Nature updated the status of the European hamster on its Red List of Threatened Species to 'critically endangered,' warning that the species is likely to go extinct in the next 30 years. While anthropogenic factors are largely responsible for the current state of the species, the adaptability of these synanthropic hamsters has also led to a unique trajectory in their hope for survival; they have been successful as a liminal species inhabiting urban environments where they have also received prominent attention and protection. I look to the former disregard for their presence as laying the groundwork for closer contemporary ties between humans and hamsters in the city of Vienna. Through Val Plumwood's intentional recognition stance, I propose that the interactions between humans and hamsters in Vienna exemplify the recognition of hamster agency and their political claim on space. The story of the Viennese hamsters also pushes against conventional narratives that often fail to account for the dependence of liminal animals on humans, and it emphasises a call toward not just protecting the abstract category of species, but also the unique relationships between hamsters, humans, and the urban environment of the city.

**Keywords:** animal studies, European hamster, intentionality, urban wildlife, endangered species

Once ubiquitous to steppes and grasslands from Siberia to Eastern France, the European hamster (*Cricetus cricetus*), also referred to as the common hamster, has inhabited the European continent as far back as the Pleistocene. From the 20$^{th}$ century on, many subpopulations have gone extinct or have seen their numbers drop significantly. In fact, the European hamster is more than likely the fastest declining species of mammal in Europe and is at risk of going extinct within 30 years (IUCN, 2020). In certain countries including France, Germany, Austria, Switzerland, the Netherlands, and Belgium they have been given special protections for some time, however it was not until July 2020, that the International Union for Conservation of Nature announced the addition of the European hamster to its Red List of Threatened Species. Overnight, they went from being deemed 'Least Concern' to 'Critically Endangered' (IUCN, 2020).

The struggles faced by European hamsters illuminate how speciesism operates spatially through the environment. Common discussions of speciesism often fail to account for what has been termed 'environmental speciesism' (Gaard, 2017; Springer, 2021). This discourse highlights how environmental burdens are distributed and experienced according to species. Speciesism can be defined as 'the unjustified disadvantageous consideration or treatment of those who are not classified as belonging to one or more particular species' (Horta, 2010: 245). In the context of the socio-political construction of the environment, one way environmental speciesism manifests is when humans fail to account for the presence and perspectives of animals living alongside them. This is often not intentionally malicious, but rather can be understood as unintentionally perpetuated due to what Erika Cudworth (2008) refers to as anthroparchy, 'a social system, a complex and relatively stable set of relationships in which the


EurSafe 2021 – DOI 10.3920/978-90-8686-915-2_34, © Wageningen Academic Publishers 2021

'environment' is dominated through formations of social organization which privilege the human' (33-34). This representational and relational injustice leads to unintended compounding impacts on animals.

Much of the preferred habitat of the European hamster is ideal for agricultural development, and this is one way in which their lives intersect with humans. Hamsters are adaptable and have learned to live on agricultural lands. They do well where grains or legumes are being grown, due to the shade and shelter from predators that these crops offer. However, when large tracts are harvested or ploughed all at once, they are left without a habitat, shelter from predators, or food. The impact of monocropping of maize in particular has had a profound impact (Tissier *et al.*, 2017). Biologists researching European hamsters in France discovered that hamsters consuming diets primarily consisting of maize were deficient in vitamin B3. This leads to a condition known as pellagra (Tissier *et al.*, 2017). Hamsters' simplified diets within such landscapes have horrifying side effects including neurological impacts leading to frequent cannibalization of their young. Only around five percent of pups from hamsters consuming maize-heavy diets live to the point of being weaned (Hood, 2017). More sustainable agricultural practices that diversify crops grown and stagger harvesting would eliminate some of these issues of malnutrition.

However, European hamster populations began declining as early as the 1920s, prior to contemporary intensive agriculture. Drops in reproduction rates have been observed since at least the 1950s (Surov *et al.*, 2016). During the 20th century, female hamsters birthed an average of 20 pups a year, whereas they now often have as few as five (IUCN, 2020). Hamsters themselves were once harvested for fur, which caused untold suffering and death for millions, and also likely had an intergenerational impact. Long considered pests, they were poisoned, trapped, drowned, and gassed (Surov *et al.*, 2016). The impact of past hunting and trapping of hamsters may extend to the genetic level. With life spans of one to two years, they must reproduce two to three times per year to maintain population levels (Surov *et al.*, 2016). Although hunting and trapping has largely ended, it may still be responsible for current drops in reproductive levels. Important to consider in this context is the time of year that hamsters were trapped, not necessarily the number killed. The quality of their fur is higher if taken right after hibernation, so hamsters that came out of hibernation earliest were most frequently killed. Those that come out of hibernation earliest are also more likely to have three rather than two litters per year. This means that over time, more reproductively active hamsters were unknowingly selected against (Surov *et al.*, 2016).

These examples illustrate some largely unintentional and unpredictable negative implications of environmental speciesism whereby humans did not account for the presence and perspectives of other animals in their interactions with the land. It is crucial to consider these negative impacts; however, I want to explore what can be interpreted in a positive light, a deepening of dependency and connection resulting from the past disregard of hamster agency. This is the focus of my project as I maintain that beneficial relationships and agency born out of past relational injustices often goes unnoticed, especially within liminal species, those animals existing somewhere between human understandings of wild and domestic (Donaldson and Kymlicka, 2011).

A demand and acknowledgment of entangled relations is displayed in the context of European hamsters occupying urban space. As discussed, the species' historic ways of living, even at the genetic level, have been transformed. Ironically, the encroachment of humans into the historic territory of the European hamster, and the disregard of their presence and perspectives, have also led to a chain of events in which one of the largest and most successful populations has emerged in the middle of Vienna, Austria where more than 3,000 individuals occupy parks, gardens, school grounds, and graveyards (Feoktistova *et al.*, 2013). One of the city's major populations lives at the central cemetery, where hamsters can be seen scuttling between graves, often consuming flowers left at the sides of gravestones, and eating the wax from candles.

Industrialization and the growth of intensive agriculture took place alongside the emergence of radically different yet well-suited habitats for hamsters in urban green areas within cities such as Vienna. As a synanthropic species adaptable to living in association with or near humans, European hamsters have a proclivity toward liminality. Thus cities provide an alternative hope for the continuation of their species. It is common for animals to take advantage of life in urban environments. However, such liminal animals are often viewed as pests and offered little legal protection or recognition (Meijer, 2019). The Viennese hamsters are an exception to the persecution many liminal animals face. Thanks to multinational and physical borders, subpopulations in Vienna and other cities in Central and Western Europe have received far more legal protection than other hamsters. They are walled-off from leaving many of the habitats they occupy in Vienna, and have significantly different political standing than their wild counterparts.

European hamsters have received strict protections under the Bern Convention and the European Commission's Habitats Directive (Nechay, 2000). Millions of euros are spent protecting the species. If buildings are constructed or the landscape is altered in an area occupied by hamsters in Austria, there are strict measures that must be followed. Austria has a well-known 'hamster commissioner,' a landscape consultant who specialises in hamster-friendly construction (Kott, 2018). For example, in 2018, a €50 million renovation to university grounds in Vienna could only move forward with his involvement since the campus was home to 50 hamsters. The hamster commissioner tracked the hamsters and mapped out the location of the burrows on site, fencing off those that were active. Once construction in one area was finished, he would coax the hamsters to another location.

The Viennese hamsters have developed positive notoriety. The city's tourism website even provides a map of the city's major hamster habitats to interested visitors. There are also wildlife-based tours of the city. In Vienna it is clear that European hamsters are not viewed as pests. Importantly, they are also not invisibilised. As Donaldson and Kymlicka (2011) maintain, liminal animals are often rendered invisible by our structuring of the world without them in mind. One way this is exemplified is by not noticing their presence among us. This also manifests through what is often a failure to recognise how deeply many animals' lives are wrapped up in human presence. The media often speaks of how dependent humans are on pollinators, for example, but animals dependent on human presence are less likely to receive attention. One notable recent example of this played out during the Covid-19 pandemic. Many news stories shared that animals were thriving in the absence of humans. This was the case in some circumstances, but was by no means the norm. In parts of India, for example, many liminal animals including free-roaming dogs and cows, as well as rhesus macaques depend on human presence to the degree that lockdowns led to risks of starvation for them (Kristensen, 2020). Such examples flip the usual conception of backgrounding, which Plumwood (1993) describes as a denial of human dependencies on the more-than-human world. Here instead we find that humans often background themselves in failing to recognise those animals dependent on them.

The protective measures taken in Vienna exemplify recognition of the hamsters' dependence on and connection to humans, and the acknowledgement that they have laid legitimate claims within the city. The Viennese hamsters therefore occupy human understanding in a way that disrupts conventional invisibilised viewpoints of liminal animals. Moral approaches that include animals tend to focus heavily on negative duties (Donaldson and Kymlicka, 2018). Leaving animals alone or 'doing no harm' are often prioritised over considering positive interactions that we might have in benefiting animals. This is problematic in cases where an animal's inseparable connections to humans are denied. Consider Kyle Powys Whyte's (2017) discussion of Indigenous North American perspectives on species conservation. He states, 'while we may embrace the value of a species such as the polar bear – even when we may have never interacted with one – it is also true that we are unlikely to invoke the polar bear in the absence of also invoking the species' significance to particular human and nonhuman communities for whom it has long, local, complex, and unique relationships' (207). Importantly, this is not to deny the individual moral standing of animals, but rather to recognise that we deny a huge part of who an animal is and their unique place in the world by failing to account for their relationships (with both humans and other animals) as necessarily constituting who they are.

The interactions between humans and hamsters in Vienna exemplify an acknowledgement of positive duties on the part of humans. This emphasis on the active engagement of humans with other animals, and the push against invisibilisation can be elucidated further by looking to Val Plumwood's concept of the intentional recognition stance. Plumwood (1993) grounds relational recognition of other animals in intentionality. She states: 'To all living creatures we may clearly ascribe a teleology or overall life-goal, for the sake of which its parts are organised, as well as many other teleological concepts' (Plumwood, 1993: 135). She maintains that we often already operate in ways that deny the neutrality or invisibility of animals in how humans tend to characterise the actions of animals as agental. Yet, such an understanding of animals is often seen as emotional and irrational, and not in line with dominant reason. Impartiality is encouraged and it is understood as preferable to remain detached from the lives of other animals. This is commonly the approach when it comes to threatened animals in the wild. In such cases, interference is often to resolve anthropogenic harms with the goal of re-establishing an existence that is largely independent of humans. Therefore, the example of the conservation of species of animals like the European hamster within urban environments and human entanglements breaks from this typical form of detachment, exclusion, and human-animal dualism.

By conceptualising human-hamster entanglements through Plumwood's intentional recognition stance, the Viennese hamsters are understood as dialogical individuals capable of shaping our shared world(s). This is framed through the recognition of their intentionality, but also through an emphasis on humans intentionally fostering interspecies relations. European hamsters have been inextricably altered through interactions with humans, both in the shifts of their geographical locations and physical presence in human-altered landscapes. They have likely even been altered at the genetic level. In Vienna we find the beginnings of an acknowledgement of the potential for a mutually constructed interspecies existence in the context of an urban environment. Intention is at the core of this acknowledgement, as the hamsters themselves are taken seriously as laying a justified claim to this space. The Viennese hamsters thus put forward a challenge to conventional conceptions of urban space, and push us toward what Jennifer Wolch (1998) refers to as a '*zoöpolis*', a 're-enchanted city' in which the urban environment is re-naturalised and the city becomes a space of reintegration for humans and animals (123-124).

Understanding the Viennese hamsters through the intentional recognition stance is not focused on moral hierarchy or an ethics of comparison to humans. Rather, I emphasise moral relationships grounded in an active recognition of hamster agency. As Plumwood (1998) states, 'Intentional description is essential to being able to represent agency, the view of the other as an originator of projects that demand our respect' (399). Through this lens, the Viennese hamsters invite us to expand our conception of agency beyond the limited and reductive Cartesian notion of the rational mind, and to remain open to thinking beyond the human as the orator of intentionality. When faced with threats and the eradication of their habitats, it is likely that animals do not have the opportunity to transform human social and political systems through verbalising their concerns (Meijer, 2019). One option they have is to relocate to a new habitat. Eva Meijer (2019) asks us to imagine scenarios like this with a comparison to human movements of relocation such a squatting, which is often done out of necessity but also takes on political influence. While we may assume that the Viennese hamsters are not consciously choosing to make a political statement when they occupy lands within the city, their presence still challenges prevailing notions of the human domination of space. In the human context, squatting arises out of an actual need for impoverished and houseless people, but it also exists as a political act. As Meijer (2019) states, 'Squatters challenge the existing power hierarchy as expressed in legal and financial constructions, even if they do not explicate this. Their physical act is the protest, not the words used to defend it' (175). The Viennese hamsters undoubtedly chose to occupy areas of the city out of the need to survive, and populations have adapted well in doing so, but they have also had a transcending political impact.

The Viennese hamsters have been empowered through a course of events tied to past acts of environmental speciesism. The former disregard for their presence laid the groundwork for closer contemporary ties between humans and hamsters in Vienna, and toward the recognition of hamsters as intentional participants in political considerations within the city. I propose that the case of the Viennese hamsters is revelatory of possibilities not just for hamsters but for a variety of other animals occupying what are generally understood as human spaces in an ever more urbanised world. The response to hamsters in Vienna shows that humans are capable of recognising more-than-human agency in the continued co-construction and reimagining of urban space. Further, the Viennese hamsters challenge misconceptions that the city is merely a human habitat. As the City of Vienna's website states, we should be aware that the city is a home to humans and animals alike.'[1] Crucially, this case also exemplifies a desire to protect not just the abstract category of species, but the specific unique relationships between hamsters, humans, and the urban environment in Vienna.

## References

Cudworth, E. (2008). 'Most Farmers Prefer Blondes': The Dynamics of Anthroparchy in Animals' Becoming Meat. Journal for Critical Animal Studies 6, Issue 1: 32-45. Available at: http://www.criticalanimalstudies.org/volume-vi-issue-i-2008/.

Donaldson, S. and Kymlicka, W. (2018). 'Rights' in Critical Terms for Animal Studies. Gruen, L. (ed.) The University of Chicago Press, Chicago, 448 pp. 320-336.

Donaldson, S. and Kymlicka, W. (2011). Zoopolis: A Political Theory of Animal Rights, Oxford University Press, Oxford, 329 pp.

Feoktistova, N.Y., Surov, A.V., Toppinetz, N.N., Kropotkina, M.V., Bogomolov, P.L., Siutz, C., Haberl, W., and Hoffman, I.E. (2013). The Common Hamster as a Synurbist: A History of Settlement in European Cities. Zoologica Poloniae 58 3-4: 113-126.

Gaard, G. (2018). 'Feminism and environmental justice' in The Routledge Handbook of Environmental Justice. Holifield, R., Chakraborty, J. and Walker, G. Routledge, New York, 696 pp. 74-88.

Hood, M. (2017). Corn turning French hamsters into deranged cannibals: research. PHYS.ORG. Available at: https://phys.org/news/2017-01-corn-french-hamsters-deranged-cannibals.html.

Horta, O. (2010). What is Speciesism?. Journal of Agricultural and Environmental Ethics 23: 243-266.

IUCN, International Union for Conservation of Nature (2020). Almost a third of lemurs and North Atlantic Right Whale now Critically Endangered – IUCN Red List. Available at: https://www.iucn.org/news/species/202007/almost-a-third-lemurs-and-north-atlantic-right-whale-now-critically-endangered-iucn-red-list.

Kott, L.J. (2018). With Europe's Hamsters At Risk, Better Call The 'Hamster Commish'. NPR. Available at: https://www.npr.org/2018/10/14/650605114/with-austrias-hamsters-at-risk-better-call-the-hamster-commish.

Kristensen, B. (2020). Welcome to the Viralocene: Transcorporeality and Peripheral Justice in an Age of Pandemics. Medium. Available at: https://medium.com/@bjornkristensen/viralocene-66a954260487.

Meijer, E. (2019). When Animals Speak: Toward an Interspecies Democracy. New York University Press, New York: 291 pp.

Nechay, G. (2000). Status of Hamsters *Cricetus cricetus*, *Cricetus migratorius*, *Mesocricetus newtoni*, and Other Hamster Species in Europe. Convention on the Conservation of European Wildlife and Habitats (Bern Convention). Council of Europe Publishing, Strasbourg: Issues 106-110.

Plumwood, V. (1993). Feminism and the Mastery of Nature. Routledge, New York: 239 pp.

[1] City of Vienna, 'European hamsters as city dwellers.' Available at: https://www.wien.gv.at/umweltschutz/naturschutz/biotop/feldhamster.html.

Plumwood, V. (1998). Intentional Recognition and Reductive Rationality: A Response to John Andrews. Environmental Values 7, 4: 397-421.

Springer, S. (2021). 'Check Your Anthroprivilege! Situated Knowledge and Geographical Imagination as an Antidote to Environmental Speciesism, Anthroparchy and Human Fragility' in Vegan Geographies: Spaces Beyond Violence, Ethics Beyond Speciesism. Hodge, P., McGregor, A., Springer, S., Veron, O. and White, R.J. (eds) (in print)

Surov, A., Banaszek, A., Bogomolov, P., Feoktistova, N., Monecke, S. (2016). Dramatic global decrease in the range and reproduction rate of the European hamster *Cricetus cricetus*. Endangered Species Research 31: 119-145.

Tissier, M.L., Handrich, Y., Dallongeville, O., Robin, J.P. and Habold, C. (2017). Diets derived from maize monoculture cause maternal infanticides in the endangered European hamster due to a vitamin B3 deficiency. Proceedings of the Royal Society B: Biological Sciences 284: 20162168.

Whyte, K.P. (2017). 'Our Ancestors' Dystopia Now: Indigenous conservation and the Anthropocene' in The Routledge Companion to the Environmental Humanities. Christensen, J., Niemann, M. and Heise, U.K. (eds) Routledge, New York: 506 pp. 206-215.

Wolch, J. (1998). '*Zoöpolis*' in Animal Geographies: Place, Politics, and Identity in the Nature-culture Borderlands. Wolch, J. and Emel, J. (eds). Verso, New York: 310 pp. 119-138.

# 35. Four models of the veterinarian – owner relationship

*M. Huth*
*Messerli Research Institute, Unit Ethics and Human-Animal-Studies, University of Veterinary Medicine Vienna/University of Vienna/Meduni Vienna, Austria; martin.huth@vetmeduni.ac.at*

## Abstract

The aim of this paper is to modify Linda and Ezekiel Emanuel's broadly debated four models of the physician – patient – relationship in order to make it applicable to veterinary ethics. This should bring the relevance of the relationship between veterinarian and owner to the fore. Though the triad of veterinarian, animal and owner is often used as a starting point for ethical analysis, the owner often plays a tangential role in the considerations. While, for instance, Bernard Rollin stresses that veterinarians shall dedicate primary allegiance to the animal, this paper takes the interdependence of the veterinarian's and the owner's responsibility seriously. Instead of comprehending the owner either as an obstacle or as compliant, s/he is regarded as crucial factor within veterinary practice. With this change of perspectives, the question emerges how the relationship between veterinarian and owner can be shaped. Four different models will be outlined, which are neither mere descriptions nor normative guidelines. Rather, they are conceivable as basic orientation to better understand roles and responsibilities of veterinarians. They comprise: (1) the advocatory model, in which the decision-making is tacitly or overtly conferred to the veterinarian; (2) the informative model, in which the veterinarian provides relevant information about diagnosis, prognosis and therapy options, and the owner is to decide on their own; (3) the psychological model, in which the veterinarian is primarily focused on the owner's psychic struggles to facilitate decision-making; and (4) the empowerment model, in which veterinarian and owner participate in a shared process of decision-making. An analysis of the strengths and pitfalls of each model can, finally, contribute to the awareness on the part of veterinarians for how to orient and arrange these relationships.

**Keywords:** veterinary responsibility, decision-making, moral distress, empowerment

## Introduction: Is the triad of veterinarian-owner-animal taken seriously?

In many theories of veterinary ethics, we find the implicit or explicit assumption that ethical dilemmas among professionals frequently result from conflicts between the animal's health-related interests and the owner's perspective or decision-making (Grimm and Huth, 2017). As a paradigmatic example, we can take Bernard Rollin's broadly received distinction between the model of the mechanic and the model of the paediatrician (2006). While the 'mechanic' symbolizes the veterinarian who reifies bodies in order to gain money through fixing damages or scrapping these bodies by euthanasia, the 'paediatrician' has primary allegiance to the animal, and particularly not to the owner (2006: 27). Rollin alleges that there is a particular attitude among some veterinarians, which makes them understanding themselves as being committed primarily to owners; at the same time, a vast majority would consider themselves as focusing primarily on the animal.

There is empirical evidence that ethics is expected not only to support the finding of morally justified ways to deal with moral dilemmas, but also, even more basically, to support the understanding of moral conflicts (Dürnberger, 2019) in order to support the coping with the moral distress (which is pervasive among veterinarians; Fawcett and Mullan, 2018). On the This reflects – on the level of shared moral intuitions – that, if moral obligations can be fulfilled better if their implications are understood more thoroughly, one ethicists' task is to contribute to this understanding. Thus, the question emerges whether we gain such a better understanding of morally difficult situations if we conceive of the triad


EurSafe 2021 – DOI 10.3920/978-90-8686-915-2_35, © Wageningen Academic Publishers 2021

of veterinarian, owner and animal as split into different relations, first, between animal and owner and, secondly, the one between veterinarian and animal (the relationship between animal and owner is often seen as basis for, but basically as somewhat outside of the triad). If we do this (and it seems as if this has been and still is a standard approach in veterinary ethics), we produce a potentially insuperable gap between the owner's decision-making and the veterinarian's expertise.

## A shift in perspectives

If we follow a pervasive line of thinking in veterinary ethics, the owner is understood as either an accomplice or as obstacle in the clinical encounter between veterinarian and animal; or the owner is reduced to an extra. It can hardly be denied that this perspective has its root in particular experiences of veterinarians who often face a tension, if not contradiction, between their own expertise and the owner's will or decisions (Fawcett and Mullan, 2018; Rollin, 2006). However, such thinking runs the risk of underestimating the practical significance of the much-vaunted triad of the veterinarian, the owner, and the animal. If we take this triad seriously, the role of the owner has to be recognized properly. The owner is not only a significant player in this triad as s/he might have their own perspective and moral convictions, which lead them to decide in particular ways within clinical encounters. Particular decision-making and, a fortiori, general habits and whole lifestyles have a significant impact on animal health. This has been illustrated with the examples of obesity (Sandøe *et al.,* 2014) and anorexia (Rollin, 2006). In both cases, the owner's dietary habits caused severe health problems on the part of the animal. While obese persons tend to underestimate or not to recognize adiposity in their animal, the example of the anorectic person in Rollin's description shows an inability to recognize poor nutrition. It is, thus, clear that the relationship between owner and animal is an immanent part of health and of etiologies. Therefore, a sustainable veterinary practice should also regard the owner's perspective and their disposition for treating the animal.

Since the relationship between veterinarian and owner is a crucial resource in veterinary practice, it is worth having a closer look on this relationship in order to reflect its possible structure and the way 'how' a veterinarian could relate to the client. In the upcoming section, I will outline four different models that should contribute to a better understanding of relationship. Certainly, it need not be a close one and could also be fragmentary or rudimentary. However, generally, it could be regarded as a valuable prerequisite of salutogenesis (the constitution of health), prevention, treatment and rehabilitation.

The elaboration of these models draws from Linda and Ezekiel Emanuel's famous 'Four Models of the Physician-Patient Relationship' (1992). The authors propose the distinction between the paternalistic, the informative, the interpretative, and the deliberative model. The list in the upcoming section is clearly oriented by these models, however, for several reasons, they cannot be directly transferred to veterinary medicine and must be modified. First, they are oriented towards a dyadic relation between a human, potentially self-determining patient and the physician while veterinary encounters are usually triadic and the decision-making is advocatory. Second, the tension between autonomy (*voluntas aegroti suprema lex*) and health (*salus aegroti suprema lex*) is less important than in human medicine since animals would hardly ever decide to visit the veterinarian voluntarily at all (Huth, 2020). Third, clinical encounters are embedded in different structures with diverse economic and practical preconditions; a companion animal is another 'kind' of animal than a livestock animal with regards to practices, institutions and the available knowledge stock (Huth, 2020; Huth *et al.,* 2019). Fourth, decision-making is often constrained by individual economic situations. Fifth, veterinarians are dependent on clients who are willing to consult and to pay them. Therefore, s/he might be – at least in some situations – inclined to fulfil wishes that go against their own convictions.

It is important to note that the upcoming outline of four different models of the relationship between veterinarian and animal owner is neither an attempt to describe real situations or relationships nor to provide ready-made ethical guidelines for veterinarians in order to shape such relationships. Rather, they shall describe specific structures that underlie different particular relationships. These will hardly ever comply with one of these so to speak idealized structures. However, what we can gain from the identification of these structural models is a better understanding of the relevance and the differential nature of relationships between veterinarian and owner and how these relationships shape the triad of veterinarian, animal and owner, i.e. also how this shapes the relationship between veterinarian and animal and how it pre-determines clinical encounters.

## Four models of the veterinarian – owner – relationship

In what follows, I propose four different models that describe different possible structures of the relationship between veterinarians and owners.

### *1. The advocatory model*

If owners struggle with decision-making, the decision could be tacitly or overtly conferred to the veterinarian. The advocatory model represents a modification of the paternalistic model in Emanuel and Emanuel who point out that within this frame the physician's role is understood as expert and who pursues not only first and foremost, but exclusively patient's health while neglecting autonomy (1992: 2221-2222). In contrast, the veterinarian as advocatory decision-maker emerges from a situation in which the owner appears to be unwilling or unable to make a decision – but willing to take advice. Conceptualizations of the relationship between veterinarian and owner often suggest that it represents a relationship between a service provider and a client or customer who decides what sort of service they will demand. But within the advocatory model the role of the veterinarian changes. S/he is not only regarded as expert in knowledge and medical practice, but also as expert in (moral) decision-making. The tacit insinuation is that there is a purely objective knowledge about the most efficient, but also the 'morally' right treatment.

This can be conceived of as a situation between boon and bane. Veterinarians might find themselves in the situation to decide in all conscience for or instead of the owner. This could not only be the opportunity to treat animals according to the best knowledge but also a psychological and moral burden and, thus, a contribution to the emergence of moral distress. Moreover, veterinarians should be aware that this could lead to tensions between the owner's implicit moral beliefs and the veterinarian's decision. Retrospectively, the owner might be discontent with the decision for treatment, omission or euthanasia. What is more, considering long time veterinary care, if the owner's implicit beliefs are at odds with the veterinarian's advice, the client's compliance to treat the animal accordingly might remain superficial and inefficient.

### *2. The informative model*

Similarly to the informative model in Emanuel and Emanuel, in the logic of this model, the veterinarian is expected to provide the relevant information about prophylaxis, diagnosis, prognosis and therapy options; the owner is to decide on her/his own. It can be assumed that relationships that more or less conform to this model occur in the sphere of livestock farming (particularly when we face a focus on mere stock conservation in contrast to an individualized curative practice). Here, regarding the owner's perspective, the economic constraints and typical practices are more likely to conceal the animals' vulnerabilities than in companion animal keeping (Huth, 2020). But it can be assumed that similar relationships can also emerge between companion animal keepers and veterinarians.

The probably most obvious advantage of such a relationship between veterinarian and owner is the fact that the owner will never feel pushed or manipulated and will, thus, act efficiently according to a made decision. Moreover, the fact that the owner might know their animal best can support appropriate prophylaxis, diagnosis and decision-making. However, there are also of possible pitfalls. For example, the owner might be overstrained with the information (Springer *et al.*, 2019) and/or overburdened with deciding on her/his own (it is remarkable that moral distress on the part of owners is a rarely addressed phenomenon). In addition, the veterinarian might feel quite powerless. While being the one to perform interventions in the bodily integrity of the animal, the veterinarian is virtually excluded from a process of deliberation as basis for decision-making and treatment.

### 3. The psychological model

This model is a modification of Emanuel's and Emanuel's 'interpretive model', in which the patient is introduced as individual whose values are, in the first instance, inchoate (1992: 2222). According to Emanuel and Emanuel, the physician is in charge of elucidating and making coherent these values. In the psychological model, the veterinarian is not the least focused on the owner's psychic struggles, be it overwhelming compassion with the animal, grief, economic limits, etc., but also previously more or less implicit moral beliefs as prerequisites of decision-making.

The advantage is that the owner can decide on her or his own and receive help if s/he is overburdened with an orientation in her/his own values. As a result, the owner might come to a decision and would act, as far as s/he can contribute to the animal's treatment, accordingly in an efficient way. However, first, the veterinarian is, again, not really part of the decision-making. S/he might feel like a mere hench (wo)man of the owner. Second, this strategy of making allegedly pre-existing values explicit can imbued with the veterinarian's own convictions. Emanuel and Emanuel obviously think of a kind of Socratic strategy that helps the patient to gain self-understanding, however, it can be doubted whether the physician's or, in our considerations, the veterinarian's point of view does not influence the decision-making at all (as if the expert would be an impartial observer of the client's psyche). Third, the veterinarian is actually not a psychologist, though psychological skills are significant for this profession. Fourth, from the client's perspective, this could evoke a feeling of being lectured or, even worse, it could be experienced as intrusive.

### 4. The empowerment model

Derived from and modifying the 'deliberative model' in Emanuel and Emanuel (1992), the empowerment model signifies a relationship in which the veterinarian operates as a kind of expert 'and' counsellor. Veterinarian and owner participate in a genuinely shared process of decision-making, albeit the owner is understood as in need of specific support. This support is neither reducible to mere information about physiological facts nor to the assistance in making values explicit and coherent. The veterinarian has the task to inform and to empower the owner (Huth, 2019), but also to reveal the veterinarian's own view. Only by making explicit her/his own conviction, the relationship between veterinarian and owner will become a relationship of trust.

The advantages are pretty obvious: The owner is part of the decision-making and will probably be compliant with its practical implications regarding the treatment of the animal. At the same time, s/he will not be overburdened by facts and responsibilities due to a process that is intended to empower her/him. The veterinarian can expect that her/his recommendations will be acknowledged, yet not uncritically accepted. However, this procedure is time consuming and dependent on a successfully respectful relationship. The owner is required to be able to understand information and to be willing to take advice. The veterinarian is required to accept that her/his expertise is only one part of the shared process of decision-making. And not the last, the owner has to be open to be committed to an ethical orientation, which is acceptable from the veterinarian's perspective. Finally, the asymmetry in expertise and experience might lead to unintended influences on the owner's perspective and decision-making.

As noted, these models are not meant to describe really existing situations, but to make general structures explicit. It is likely that different structures are appropriate in different situations. Therefore, an awareness of the various configurations of the veterinarian-owner relationship can contribute to an increased sensitivity for the veterinarian's role in the triad of veterinarian, owner, and animal.

## Conclusion

This paper pursued a modification of Linda and Ezekiel Emanuel's four models of the physician-patient relationship in order to make them suitable for veterinary ethics. By the same token, the relevance of the relationship between veterinarian and owner for ethical considerations should be made visible. Albeit the triad of veterinarian, animal and owner is often used as a starting point for ethical analysis, the owner is often underrepresented in the relevant discourse. Instead of focusing primarily on the relationship between veterinarian and animal, this paper emphasized the interdependence of the veterinarian's and the owner's responsibility. The owner became more recognizable as crucial factor within veterinary practice. However, this begged the question how the relationship between veterinarian and owner should be shaped. Four different models have been outlined, which are neither mere descriptions nor normative guidelines. Rather, they were conceived as models in order to better understand roles and responsibilities of veterinarians. They comprise: (1) the advocatory model, in which the decision-making is tacitly or overtly conferred to the veterinarian; (2) the informative model, in which the veterinarian provides relevant information about diagnosis, prognosis and therapy options, and the owner is to decide on their own; (3) the psychological model, in which the veterinarian is primarily focused on the owner's psychic struggles to facilitate decision-making; and (4) the empowerment model, in which veterinarian and owner participate in a shared process of decision-making.

An analysis of the strengths and pitfalls of each model can contribute to the veterinarians' awareness for their own role in different situations and how to orient and arrange these relationships. The singular owner and her/his point of view and behaviour, the particular situation and its pragmatic requirements determine the perspective from which these models can be assessed and taken as a basis to reflect the acting of the veterinarian. This presupposes a primary responsiveness and sensitivity for the particular triad, which cannot be anticipated or implied in the four models hat have been outlined in this paper.

## References

Emanuel, E.J and Emanuel, L.L. (1992): Four Models of the Physician-Patient Relationship. JAMA 267 (16): 2221-2226.

Dürnberger, Ch. (2019): It's Not About Ethical Dilemmas: A Survey of Bavarian Veterinary Officers' Opinions on Moral Challenges and an e-Learning Ethics Course. Journal of Agricultural and Environmental Ethics 32: 891-903. https://doi.org/10.1007/s10806-019-09805-z

Fawcett A, Mullan S (2018): Managing moral distress in practice. In Practice 40 (1): 34-36.

Grimm, H. and Huth, M. (2017): One Health: Many Patients? A Short Theory on What Makes an Animal a Patient. In: Jensen-Jarolim, E. (ed.): Comparative medicine: disorders linking humans with their animals. Cham: Springer: pp. 219-230.

Huth, M (2019): Empowerment – A Veterinarian's Main Task? In: Vinnari, E, Vinnari M (eds): Sustainable governance and management of food systems. Wageningen: Wageningen Academic Publishers, pp. 86-91.

Huth, M., Weich, K., Grimm, H. (2019): Veterinarians between the Frontlines!? The Concept of One Health and Three Frames of Health in Veterinary Medicine. Food Ethics. A Journal of the Societies for Agricultural and Food Ethics 3 (1-2): 91-108.

Huth, M. (2020): On facing different kinds of animals – reflecting veterinary responsibility. Berliner und Münchener Tierärztliche Wochenschrift 133(3-4): 238-247.

Rollin B (2006): An Introduction to Veterinary Medical Ethics: Theory and Cases. Ames: Blackwell.

Sandøe, P., Palmer, C., Corr, S.A., Astrup, A., Bjørnvad, C.R. (2014): Canine and feline obesity: a One Health perspective. Vet Rec 175(24): 610-616.

Springer, S., Sandøe P., Lund, T.B. Grimm, H. (2019): 'Patients' interests first, but...'–Austrian Veterinarians' Attitudes to Moral Challenges in Modern Small Animal Practice. Animals 9: 241. https://doi.org/10.3390/ani9050241

# 36. Shifting from 'cure' to 'care' – theoretical considerations of small animal hospice and palliative care

*S. Springer*[1*] *and S. Axiak Flammer*[2]
[1] *Unit of Ethics and Human-Animal Studies, Messerli Research Institute, Veterinaerplatz 1, 1210 Vienna, Austria;* [2] *Department of Clinical Veterinary Medicine, Vetsuisse Faculty, University of Bern, 3001 Bern, Switzerland; svenja.springer@vetmeduni.ac.at*

## Abstract

In recent years, a demand for veterinary care focussing on terminally ill patients has been recognized. Rather than focussing on curing the animal patient, palliative care aims to achieve an animal's best quality of life regardless of disease outcome and to provide guidance to the animal's owners. Animal hospice can be described as a specialized form of a palliative care that cares for patients in the end stages of an illness, near death. In this paper we will first illustrate to what extent the status of the animal patient as well as prevailing veterinary aims and norms change in this specific working environment, which ultimately results in a shift of a veterinarian's self-understanding from the 'curing' to the 'caring' professional. In a second step, we will present and discuss four aspects within a veterinarian's working life – relationships, time, communication, and infrastructure – that evolve during animal hospice and palliative care work. In a final section, we will provide a brief outlook on how these theoretical considerations and developed hypotheses will be empirically investigated in a future research project.

**Keywords:** small animal hospice, small animal palliative care, professional ethics

## Introduction

There is no doubt that modern small animal practice enables high-level animal patient care that is nearly as advanced as in human medicine (Springer *et al.*, 2019). The increasing implementation of innovative technologies and methods in veterinary practice allow not only detection of diseases at an earlier stage, but also for therapeutic interventions that can cure patients that could not be treated a few years ago. However, there are situations where veterinarians might not able to heal an animal because reliable treatment, material or financial resources are not available, or the owner does not wish to submit their animal to invasive or risky procedures. The option of euthanasia comes quickly to the foreground in cases of terminal illness and perceived or anticipated suffering. There is no doubt that professionally performed termination of life will prevent pain, suffering, and stress for the animal (Rollin, 2009; Springer, 2018). However, a relatively new organized veterinary discipline that cares specifically for animals with incurable disease is increasingly being recognized: animal hospice and palliative care (Shearer, 2011, 2019).

Rather than focussing on curing the animal patient, '[p]alliative care addresse[s] the treatment of pain and other clinical signs to achieve the best quality of life regardless of disease outcome' (Shearer, 2019). Further, animal palliative care aims to provide guidance to the animal's caregivers and support them 'in making plans for living well based on the animals' needs and concerns and on the caregivers' goals for care' (Shanan and Shearer, 2017: 6). On the contrary, hospice care is a specialized form of a palliative care that focuses on caring for the animal near death (Shearer, 2019).

In this paper we aim to illustrate first, to what extent prevailing norms and veterinary medical aims shift in the context of animal hospice and palliative care compared to general veterinary practice. Second, we will discuss four relevant aspects within this specific working environment and, how these

 Hanna Schübel and Ivo Wallimann-Helmer (eds.) ***Justice and food security in a changing climate***
EurSafe 2021 – DOI 10.3920/978-90-8686-915-2_36, 

aspects will evolve in the trialogue between the animal, the owner, and the veterinarian. Finally, we will introduce a project that is an empirical investigation of the developed theoretical considerations and developed hypotheses. This project aims to gain a better understanding of the shift that occurs in a veterinary professional when the treatment focus becomes optimizing an animal's quality of life rather than attempting to cure.

## Occurring shifts in animal hospice and palliative care

In comparison to traditional disciplines in small animal practice, hospice and palliative care focusses on the care of patients with chronic and incurable diseases. Quality-of-life maintenance and symptom management become the central medical aims. Imagine the following situation: A veterinarian has a longstanding client with an eleven-year-old cat. For eleven years, all veterinary decisions and actions focused on keeping the cat healthy or to restore its health in the event of a curable illness. This situation changed when the veterinarian diagnosed a tumour near the shoulder. The tumour was found to be inoperable and the situation was expected to worsen in the following weeks and months – without any prospect for a cure. It was anticipated that the cat would lose weight, have problems walking, and that slowly the physiological parameters would worsen. This example shows that the status of the animal changes from healthy to terminally ill and that physiological norms will no longer be the sole basis for medical decisions. Treatments for the cat will primarily focus on the cat's welfare and the veterinarian's responsibility is the provision of palliative care until death. These changes ultimately lead to a shift in a veterinarian's self-understanding from a 'curing' to a 'caring' professional (Table 1). In doing so, the term curing does not exclude the fact that a veterinarian's responsibility is to care for their animal patients during treatment that may lead to recovery. Rather this differentiation between curing and caring emphasizes that a veterinarian's decision-making process during hospice and palliative care primarily focuses on the animal's comfort and quality-of-life. It can also be assumed that this care context is characterized by specific requirements, some of which will be presented as four key relevant aspects in the subsequent section.

## Relevant aspects during animal hospice and palliative care work

We assume that four key aspects change and/or come to a head during animal hospice and palliative care work. The first aspect, 'relationships', focusses on the specific trialogue between the veterinarian, owner, and patient that impacts decision-making in veterinary medicine. The second aspect, 'time', addresses not only the time required for caring, but also how the remaining time at the end of an animal's life is perceived by involved parties. The third aspect, 'communication', considers strategies and

*Table 1. Shifts related to the animal's health status, prevailing norms, medical aims, and the veterinarian's self-understanding that occur in small animal hospice and palliative care.*

| | Working field | | |
|---|---|---|---|
| | Traditional small animal practice disciplines | Palliative care | Hospice |
| Patient | | | |
| Status quo | Healthy or ill | Incurable illness | Terminal illness/dying |
| Norm | Health and physiological parameters | Patient itself and its quality of life with the disease | |
| Veterinarian | | | |
| Aim | Prevent illness or restore health | Promote individual welfare | |
| Self-understanding | Curing | Caring | |

care conversations. The fourth aspect, 'infrastructure', considers veterinarians' different work settings, interdisciplinary collaborations, and educational opportunities, which are of importance to successfully institutionalize hospice and palliative care in small animal practice.

### Aspect of relationships

In general, veterinary practice is characterized by a threefold relationship between the patient, the veterinarian and the client (May, 2013; Morgan and McDonald, 2007; Yeates, 2013). These relationships are characterized by the individual needs of all involved parties and hence, can vary dramatically in veterinary practice. For instance, veterinarians can be confronted with owners having tight emotional bonds to their animals. Undoubtedly, these tight emotional bonds can have a positive impact in that owners might be more willing to care for their animals or pay for cost-intensive veterinary treatment including animal hospice and palliative care services (Brockman *et al.*, 2008). However, it can also develop into the problem of overtreatment (Sandøe *et al.*, 2016; Yeates, 2013), which may be especially true in small animal practices offering advanced technologies and methods (Springer *et al.*, 2019). This may lead to situations where: (1) an owner seeks therapy that would – from the veterinarian's perspective – merely prolong the animal's suffering (Batchelor and McKeegan, 2012); or (2) situations where a veterinarian offers advanced treatment even though the prognosis is questionable and creates – from the owner's perspective – an 'all or nothing' decision.

In the working field of animal hospice and palliative care, relationship-oriented care is an important demand (Shearer, 2019; Yaxley, 2019) since end-of-life decisions are not only shaped by relationships but can also terminated by them. It is likely to assume, that relationships between the owner and the animal as well as the owner and the veterinarian might intensify during the care of terminally ill patients and that tight emotional bonds between the owner and animal constitute a premise to successfully care for terminally ill patients in this field. Additionally, veterinarians' actions might be influenced by the relationship to and expectations of owners when caring for an animal until death that consequently leads to more intensive and personal relationship compared to other working fields. Even though Shanan (2011) argues that veterinarians should know about the limits between treating the animal and treating the owner, it becomes clear that trust-filled owner-veterinarian relationships are important in animal hospice and palliative care settings and consider the owners' emotions and concerns. Hence, we hypothesize that relationships between the patient and the owner as well as between the owner and the veterinarian are of major significance and change in regard to their intensity and emotionality.

### The aspect of time

Time plays an important factor when caring for animals with life-limiting illness or disabilities and needs to be considered at several levels. A clear distinction can be made between the provision of hospice, on the one side, and palliative care on the other. The differentiation on a temporal level is already evident from the definitions of both services. Shanan and Shearer (2017) highlight that '[a]nimal hospice care is provided to the patient from the 'time' of a 'terminal' diagnosis through the death of the animal, inclusive of death by euthanasia or by hospice-assisted natural death.' (Shearer and Shannon, 2017: 6) On the contrary, palliative care can extend over a longer period by aiming to improve a patient's quality of life without knowing the disease outcome (Shearer, 2019). Hence, in comparison to general veterinary practice – in which usually the recovery of health determined the end of veterinary medical care in case of a curable disease – the timeframe in animal hospice and palliative care is determined by different reference points: it starts with the 'terminal' diagnosis, and 'ends' with the animal's death. In addition to the duration of treatment, the frequency and time needed for hospice and palliative care consultation also needs consideration. Different appointment types – from initial consultation, quality of life evaluation, to ongoing care consultations (Bennett and Cook, 2019) – require different time resources from the veterinarian as well as the owner.

Keeping this in mind, it is likely to assume that not only the quantity of time, but also the quality of remaining time comes into focus. We expect that duration and frequency of time affect relationships between the veterinarian, client, and animal and that the time span – determined by the terminal diagnosis and the death of the animal (whether euthanasia or natural death) – impacts the perceived quality of remaining time by veterinarians and owners. In addition to that, veterinarians need to adjust their business concept to account for the increased time hospice and palliative care consultations require so that they can effectively integrate it into their working life.

### *The aspect of communication*

The practice of animal hospice and palliative care is emotionally challenging; empathy and good communication strategies constitute an important foundation to successfully care for the animal and the owner (Moses, 2019). As animals cannot verbalize their perspective regarding future life or possible treatments, their interests can only be advocated for by their family, veterinarian, or both. In this respect, the communication between the veterinarian, the owner and further involved parties (e.g. social worker) is of major importance and often a source of support to the owners. Moses (2019) suggests a clear communication strategy that avoids euphemisms and clinical jargon, is empathetical and non-judgmental, and gives special attention to the feelings of the owner. Goldberg (2019) proposes a serious illness conversation guide that considers and integrates issues that emerge during complex end-of-life decisions. She declares that such conversation guidelines build an important 'keystone step for quality advancement of palliative care within [the] veterinary profession' (Goldberg, 2019: 399). Further, not only the method of communication but also the words and language used are relevant in both human and veterinary hospice care. For example, in human hospice care, the patients are not referred to as patients but as guests (Nemeth *et al.*, 2012). In animal hospice and palliative care owners may be referred to as 'pet parents' instead of 'pet owners' and the treatment phase as a 'stage of life' instead of the 'end of life' that is a 'journey' for the involved parties (IAAHPC, 2020). The shift in language already indicates that communication takes place on a different level compared to general veterinary practice. Keeping this in mind, we hypothesize that veterinarians working in the field of animal hospice and palliative care specifically need to pay attention to communication strategies and that word and language usage changes in this setting.

### *Infrastructural aspects*

Even though animal hospice and palliative care is still in its infancy, there is no doubt that there exists an increasing desire for these services in veterinary practice and that veterinarians and students are more interested in this field (Gore *et al.*, 2019). Important infrastructural aspects such as the style of practice, the interdisciplinarity of teams (e.g. including social workers), and the types of educational institutions (e.g. university-based or association-based) have to be reconsidered, as it has in human hospice over the past decades (Bui, 2012; Chor *et al.*, 2019).

Several different hospice and palliative care practice models have been developed. They vary from general practices who provide at-home euthanasia to mobile practices who focus on euthanasia and after-care services, practices (both mobile and stationary) that have a dedicated animal hospice team and consult on end-of-life care and pain management and support natural death (when appropriate), and specialist referral centres (internists, surgeons, emergency and critical care specialists, anaesthesiologists) that provide services to geriatric animals and/or treat chronic conditions.

Authors such as Gore *et al.* (2019) indicate that providing medical and supportive care for animals (in the home setting), offering owners emotional support and grief education, assisting referring veterinarians by providing hospice and palliative care services, and educating students are core objectives for their Pet Hospice Program at Colorado State University. We hypothesize that the consideration of specific required infrastructural aspects is of major importance to successfully institutionalize and implement animal hospice and palliative care in veterinary medicine.

## From theoretical considerations to empirical investigations

The assumption that animal hospice and palliative care experience shifts regarding a patient's health status, veterinary norms, medical aims and consequently veterinarians' normative self-understanding raises several questions: How do veterinarians handle the integration of the owners' emotional status into this context and, how does this affect relationships between them? What specific conflicts occur during animal hospice and palliative care work? On what basis will decisions be made and communicated between veterinarians, animal owners, and professional colleagues? And finally, how do veterinarians experience the shift from the curing to the caring professional and how does this shift impact veterinary decision-making during the care of chronically and terminally ill patients? Several publications explicitly argue in favour of the scientific importance of these emerging questions (Goldberg, 2016; Gore *et al.*, 2019; Moses, 2019; Shearer, 2019; Yaxley, 2019). However, most of these discussions rely on theoretical assumptions and practical experience and are not informed by systematically conducted studies. By means of an interdisciplinary research project, we aim to close this research gap within the field of small animal practice by testing formulated hypotheses related to the above-mentioned aspects. Due to the explorative character of this envisaged project, a qualitative approach will initially be taken. The qualitative data gained will build a solid basis for a subsequent systematic development of hypotheses. These hypotheses will then be examined in a second step using a questionnaire-based survey. Hence, we aim to provide some of the first empirical insights into these addressed issues by focusing on the described theoretical considerations.

## References


Batchelor, C.E., McKeegan, D.E. (2012). Survey of the frequency and perceived stressfulness of ethical dilemmas encountered in UK veterinary practice. Veterinary Record,170:19.

Belshaw, Z., Robinson, N.J., Dean, R.S., Brennan, M.L. (2018). 'I always feel like I have to rush...'Pet owner and small animal veterinary surgeons' reflections on time during preventative healthcare consultations in the United States. Veterinary Sciences 5(1):20.

Bennett, C., Cook, N. (2019). Palliative Care Services at Home: Viewpoint from Multidoctor Practice. Veterinary Clinics of North America: Small Animal Practice 49:235-338.

Brockman, B.K., Taylor, V.A., Brockman, C.M. (2008). The price of unconditional love: Consumer decision making for high-dollar veterinary care. Journal of Business Research 61:397-405.

Bui, T. (2012). Effectively Training the Hospice and Palliative Medicine Physician Workforce for Improved End-of-Life Health Care in the United States. American Journal of Hospice and Palliative Medicine 29(6):417-420.

Chor, W.P.D., Wong, S.Y.P., Mohamad Ikbal, M.F.Bin M., Kuan, W.S., Chua, M.T., Yash Pal., R. (2019). Initiating End-of-Life Care at the Emergency Department: An Observational Study. American Journal of Hospice and Palliative Medicine 36: (11)941-946.

Goldberg, K. (2019). Goals of Care: Development and Use of the Serious Veterinary Illness Conversation Guide. Veterinary Clinics of North America: Small Animal Practice 49: 399-415.

Gore, M., Lana, S.E., Bishop, G.A. (2019). Colorado State University, Pet Hospice Program. Veterinary Clinics: Small Animal Practice 49:339-349.

Hartnack, S., Springer, S., Pittavino, M. and Grimm, H. (2016). Attitudes of Austrian veterinarians towards euthanasia in small animal practice: impacts of age and gender on views on euthanasia. BMC Veterinary Research 12:26.

IAAHPC – International Organization for Animal Hospice and Palliative Care. https://iaahpc.org/. Accessed 30 December 2020.

May, S.A. (2013). Veterinary Ethics, Professionalism and Society. In: Wathes, C.M., Corr, S.A., May, S.A., McCulloch, S.P., Whiting, M.C. (eds): Veterinary and Animal Ethics: Proceedings of the First International Conference on Veterinary and Animal Ethics, September, 2011. Wiley-Blackwell, p. 50.

Morgan, C.A. and McDonald, M. (2007). Ethical dilemmas in veterinary medicine. Veterinary Clinics of North America: Small Animal Practice, 37, 165-179.

Moses, L. (2019). Pain and Palliative Care Service, Massachusetts Society for the Prevention of Cruelty to Animals-Angell Animals Medical Centre, Veterinary Clinics of North America: Small Animal Practice 49:363-371.

Nemeth, C., Rottenhofer, I., Schleicher, B., Bitschnu, K.W., Brovnyak, E., Eisl, C.S., Heck, U., Prohaska, R., Rupacher, E., Werni-Kourik, M. (2012). Prozesshandbuch Hospiz- und Palliativeinrichtung. http://www.hospiz.at/wordpress/wp-content/uploads/2017/03/prozesshandbuch_hospiz-_und_palliativeinrichtungen_02-11-2012.pdf.

Rollin, B.E. (2009). Ethics and euthanasia. Canadian Veterinary Journal 50:1081-1086.

Sandøe, Corr, S., P.; Palmer, C. (2016). Treating Sick Animals and end-of-Life Issues. In Companion Animal Ethics; Sandøe, P., Corr, S., Palmer, C. (eds) Wiley Blackwell: New York, NY, USA, pp. 196-199.

Shanan, A. (2011). A Veterinarian's Role in Helping Pet Owners with Decision Making. Veterinary Clinics of North America: Small Animal Practice 41:635-646.

Shanan, A., Shearer, T. (2017). Introduction. In: Shanan, A., Pierce, J., Shearer, T. (eds): Hospice and Palliative Care for Companion Animals: Principles and Practice. Wiley&Sons publishers, p. 6.

Shaw, J.R.; Adams, C.L.; Bonnett, B.N.; Larson, S.; Roter, D.L. (2012). Veterinarian satisfaction with companion animal visits. Journal of the American Veterinary Medical Association 240:832-841.

Shearer, T. (2011). The Role of the Veterinarian in Hospice and Palliative Care. Veterinary Clinics of North America: Small Animal Practice 41.

Shearer, T. (2019). Where Have We Been, Where We Going – Continuity from 2011. Veterinary Clinics of North America: Small Animal Practice 49:235-338.

Springer, S., Sandøe, P., Bøker Lund, T., Grimm, H (2019). Patients' interests first, but...'– Austrian Veterinarians' Attitudes to Moral Challenges in modern Small Animal Practice. Animals 9:241.

Springer, S. (2018). Moral challenges of euthanasia in small animal practice: An empirically informed analysis in veterinary medical ethics. Dissertation, Vetmeduni, Vienna.

Yaxley, P.E. (2019). Michigan State University Veterinary Hospice Care – An Academic Hospice Care, Veterinary Clinics of North America: Small Animal Practice 49:351-362.

Yeates, J.W. (2013). Animal Welfare in Veterinary Practice. Wiley Blackwell, p. 3.

# 37. Veterinarians as protectors of animals – caring and killing in the narratives of vet students

*M.J. Bubeck*
*Department of Sociology, LMU Munich, Konradstraße 6, 80801 Munich, Germany; m.bubeck@soziologie.uni-muenchen.de*

## Abstract

In veterinary medicine, animal life is 'taken care of', as becomes clear from veterinary medicine's self-description as a 'protective' profession. Unlike in human medicine, killing is a routine part of the (health) profession. Killing as a situative act is a combination of practices, techniques, and actors in different contexts, for example the situations to 'put to sleep' or to 'butcher'. Veterinary students are socialised in this ambivalence between care and killing, described in the literature as the 'caring-killing paradox', and represent it in their self-understanding as veterinary professionals. The interpretation of ten narrative-biographical interviews with veterinary students followed the coding paradigm of the Grounded Theory methodology, practices of killing are constructed as care practices in the narratives of veterinary professionalism. Therefore, this article uses a theoretical framework of care theory to analyse two situations, in a curative practice and in a slaughterhouse, to broaden the understanding of care by acknowledging killing as part of care in the students' narratives.

**Keywords:** care theory, human-animal relations, professionalism, qualitative research, veterinary humanities

## Introduction

Why do people decide to become veterinarians? It is a well-established fact from a variety of studies that the care for animals is a major motivator to become a veterinarian. A quantitative inquiry of 1,498 veterinary students in Germany on the motivation to choose a career showed 'pet-ownership' and 'a love for animals' as well as a 'scientific interest' to be the most common motives for veterinarians (Kostelnik *et al.*, 2010). Ethnographic and qualitative studies by Patricia Morris (2012) and Jenny Vermilya (2015) come to the same conclusion, with the latter specifying that the motivation of 'caring' for animals goes 'beyond the individual level to characterize the profession' (*ibid.*: 61). A 'segmented collective identity' concerning all veterinarians is necessary because the members of the profession are engaged in a wide spectrum of activities (*ibid.*: 147).

In vet school however, students learn that despite their intention – to care for animals – a major field of duty of veterinarian practice is to kill animals. Analysing qualitative empirical findings, this article demonstrates how vet students encounter two forms of killing (slaughter and euthanasia) in their training and how they are integrated into their professional self-understanding as a 'protector of animals'. Therefore, this paper gives a short overview of veterinary medicine as a sociological research field and of veterinary education in Germany. Before illustrating the empirical findings, the paper lays out the theoretical framework of care theories and the research design. Analysing two different situations from narrative interviews with veterinary students, an argument will be made on how these narratives reframe 'killing' as 'care'.


EurSafe 2021 – DOI 10.3920/978-90-8686-915-2_37, 

## Veterinary medicine as a sociological research field – caring and killing

Veterinary medicine is a health profession with a responsibility for the life and well-being of animals as well as for humans and the environment. Unlike their human counterpart however, the actions of veterinarians include not only restorative procedures but also, in relevant cases, the killing of those they are responsible for (Gardiner, 2020, Harrmann and Weich, 2016). Nevertheless, there has neither been much sociological research on veterinary medicine (Hobson-West and Timmons, 2016, for a literature review, see Bonnaud and Fortané, 2020) nor on human-animal relations in general (Hobson-West, 2007).

Previous research has focused primarily on what is called 'death work' (Morris, 2012) or the framework conditions of professional killing. Killing is an institutionalised, routinised, and regular practice in this profession (Arkow, 1985; Chang and Hart, 2002; Rollin, 1987). In dependency of veterinarians' working context, different techniques ('captive bolt pistol' or pharmaceuticals such 'barbiturate') are used and different stakeholders ('veterinarians', 'clients', 'patients') are involved. This may explain why there are multiple names for killing in this profession: putting to sleep, culling, slaughter, and terminal animal testing (Fahrion *et al.*, 2011). The differentiation of killing practices highlights the multi-contextuality of the veterinary profession (Tannenbaum, 1995) as well as the multi-spatial and multi-species ways of care and killing (Donald, 2019).

Another specific aspect of killing in a veterinarian context is its embeddedness in different (ambivalent) human-animal relations. Over the past decades, there has been a shift in the professional focus from rural work with large animals to predominantly urban companion work (Hobson-West and Timmons, 2016), as nowadays some animals are becoming 'substitute love objects and companions precisely because they can be involved in enduring relations of mutual dependency' (Franklin, 1999: 57). This increase of the number of companion animals as well as the intensification of farming systems can be viewed as major drivers of the transformation of the profession (Bonnaud and Fortané, 2020).

To depict veterinary medicine's multi-contextuality and its embeddedness in manifold human-animal relations, this study interviewed veterinary medicine students. Students encounter a variety of professional contexts in their training: food production, small and large animal medicine as well as its specific equine equivalent. The main aim of this paper is to describe how veterinary students integrate the ambivalence between 'caring' and 'killing' – sometimes referred to as the 'caring-killing paradox' (Arluke and Sanders, 1996) – into their self-understanding as veterinary professionals.

## Theoretical framework: care theory

In order to analyse the caring-killing-paradox in vet-students' narratives, a framework for defining the broad field of care is needed. Since the 1970/80's there have been feminist debates about care, e.g. reproduction and production, paid and unpaid work. Political scientist Joan Tronto defines care comparatively broad as follows: 'On the most general level, we suggest that caring be viewed as a species activity that includes everything that we do to maintain, continue and repair our 'world' so that we can live in it as well as possible. That world includes our bodies, ourselves, and our environment, all that we seek to interweave in a complex, life sustaining web' (Tronto, 1993: 103). This definition enables care to be considered a practice ('maintain, continue and repair'), while simultaneously acknowledging the complexity of the material and somatic dimensions in form of a 'life sustaining web' (*ibid.*). Care, as philosopher Maria Puig de la Bellacasa (2015, 2017) points out, is always to be understood as 'inseparably a vital affective state, an ethical obligation, and a practical labour' (*ibid.*, 2017: 140), a triad of practical, affective and politico-ethical dimensions which are always intertwined. The theoretical framework functions as an analytical lens for the empirical coding in order to see if and how exactly students frame killing as care. Therefore, the narratives were analysed to conceptualise the practical, affective and ethical dimensions of care offered by Puig de la Bellacasa in the views of veterinary students.

## Veterinary education in Germany

The curriculum of veterinary medicine is regulated by the *Verordnung zur Approbation von Tierärztinnen und Tierärzten* (TAppV), spans 5 ½ years on average and is completed with the approval as a veterinarian, which is independent from other medical specialisations. Furthermore, a total of 1,170 hours of practice are required, which includes a min. three-week internship at a slaughterhouse. The TAppV states that aside from knowledge and skills, the training is aimed at teaching 'intellectual and ethical foundations' and 'a professional attitude toward the well-being of humans, animals and the environment'. Parallels can be drawn here from Tronto's aforementioned definition, in which she emphasises the maintenance of 'our bodies, ourselves, and our environment' (1993: 103) as care.

## Study design & sample: a grounded theory approach

The findings are based on analysing narrative-biographical interviews with ten veterinary students. The interviews were collected over the span of three years (2014-2017) in three German-speaking cities (Berlin, Munich, Vienna) with a similarly sized veterinary faculty. The interviewer himself studied veterinary medicine and sociology. The interviews lasted between 45 minutes and two and a half hours. All but one interviewee were advanced students, which means they were already in the practical phase of their training or right before graduation. All of them had already concluded the slaughterhouse internship. The sample consisted of eight women and two men.

The analysis followed the coding paradigm of the Grounded Theory methodology (Strauss and Corbin, 1998) using the software MAXQDA. This approach was used because of its iterative-circular research process and ability to build an empirical-based theory. It is well established in health research and has already been used to study veterinary medicine (Morris, 2012, Vermilya, 2015).

## Empirical result 1: motivation 'to care' for animals

When asked why they wanted to become veterinarians, all ten interviewees used the narrative of 'caring' for animals. For eight of them, the profession has been something they strived for since childhood, e.g. in Petra's words:

> yes I'm probably like most veterinary students, there was actually never another professional wish (.) since I can think veterinary medicine was always (.) present for me and I have already (.) already (.) during the school time already selected my subjects there mostly targeted //mhm// for the training. (interview Petra)

Empathy towards animals – sometimes felt as a special connection with them – was the foundation of their childhood ambition for some of the interviewees like Vera:

> As a child I always empathized with animals (.) and (.) for me it was always quite bad when (.) animals are tormented or beaten (.) or have no idea (.) erm (1) and that was just (.) also often so (.) the (.) that I always had somehow a very good connection with the animals (.) so most animals (.) at least that I met (.) liked me (.) and (.) with others (.) they have (.) growled or had no idea (.) and then I thought (.) well (1) somehow I can handle animals well (.) or put myself into their shoes (.) and (.) I can bring that (.) maybe with me in this veterinary profession. (interview Vera)

Experiences with their own 'pets' and the witnessing of veterinary treatments – for example, the castration of a horse or the unsuccessful treatment of a beloved dog – often sparked a fascination for veterinary medicine. The interest in medicine manifests itself independent from species: 'I also find it interesting in humans' (interview Petra).

## Empirical result 2: killing as care

For the participants, the different embedding of 'death work' in the profession was a central issue in the interviews. Their experiences were situated in different working places such as curative practices, stables, or slaughterhouses as well as in anatomy or pathology halls. The following two exemplary contexts, which were evident in all interviews (e.g. killing of laboratory animals or culling were not), show how the act of killing is linked to a care for animals, despite the violence which is associated with the killing. In these narrations, an inherent ambivalence between killing and caring can be reconstructed from the interviews. For the interviewed students, killing them can also mean to care for the animals. This can be reconstructed in the three dimensions described by Puig de la Bellacasa: the practical, the affective and the ethical. The following two situations were typical in their narrative framing of killing animals in an avian clinic and a slaughterhouse, respectively.

## Situation a: to euthanize a woodpecker

In the interviews, euthanasia is described as one of many everyday activities in the clinic. In this example, Sarah is doing an internship at a bird clinic, which takes care of injured wildlife. In these cases, students are allowed to perform most of the treatments themselves. This situation took place during the morning visit when dead animals have to be packed up and disposed of. Sarah describes how she went to the box of a green woodpecker, which she thought was dead. Looking into its box however, she realised that the bird was still alive and probably in pain: 'so that was really (1) (deep exhalation) really bad' (interview Sarah). The vet in charge explained that the woodpecker should therefore be euthanised that day: 'we'll make him dead now' (*ibid.*). Sarah goes on to describe the process of the euthanasia, which she was allowed to carry out by herself:

> we were allowed to put him to sleep (.) so we have narcotized him (1) um (1) um (.) with=with isoflurane (.) and then we were allowed to try it ourselves (.) we tried to find the heart (.) but it actually made me very sorry because I always have so (.) negative=especially I have hit and thought to myself (.) hopefully (.) he notices nothing (.) so I mean he slept (1) //mhm// hopefully. (*ibid.*)

The woodpecker was given the anaesthetic 'isoflurane' (*ibid.*), an anaesthetic gas. Then, another anaesthetic was injected into its heart through a cannula. This practical aspect concerning the technique was important to Sarah, as she worried 'whether we can find the heart' (*ibid.*) in order to make the procedure less painful for the bird. Sarah goes on to describe how she 'suffered' with the bird. This reflects the affective dimension of the care for animals. She explains how she felt empathy for the bird during the act. The professional knowledge about narcotisation – 'so I mean he slept' (*ibid.*) – helps to link her affect to an ethical-political stand. Her profession aims to reduce suffering and to care for the animal. This is depicted in a professional identity as the 'protector of animals', even if it means to kill – or here: to euthanise (from 'euthanasia' meaning 'good death') – a 'suffering' animal.

## Situation b: killing in the slaughterhouse

The second situation focuses on the context of killing in slaughterhouses, which all interviewees had experienced during their internship. Maria, one of the interviewees, describes: 'I don't find the slaughter process itself so bad anymore (.) I always thought that everything is so tragic (.) but I find the (.) killing (.) in itself not so tragic' (interview Maria). The practical dimension of care in this example consists of the knowledge and techniques that are used to kill animals in the slaughter process ('captive bolt pistol' and a 'butcher's knife'). The killing in slaughterhouses however is not normally carried out by veterinarians. One of their tasks is to look after the animals before they are killed. Karsten, another interviewee, describes what he experienced in the cattle pens and killing floors in his internship:

> you really have a responsibility (.) even if it's only for a short time for the animal (.) but you want to be (.) yes (.) so I tried it at least (.) as respectfully and nicely in regard to the animal (.) as long as it is still (.) alive //mhm// I did (.) I mean that (.) uh (.) the ones who work there daily certainly won't do it anymore (.) because there was already some cattle (.) for which it was already over=where you really saw that (.) for the animal it's definitely better if it's gone (.) then (.) sometimes I went over to pet it and I said (.) so well you're almost there //mhm// and erm (.) maybe a little bit=sentimental (.) but it was=some were already in a state that I thought oh the poor pigs //mhm// [...] there I already have realised (.) okay (1) it is really good that there is a profession which controls if the animals are being treated respectfully (.) that you (.) also decide (.) that an animal (.) if it can no longer run (.) that it is (.) killed. (interview Karsten)

In this short extract, Karsten demonstrates that the killing of animals in the slaughterhouse is a normalised practice, which nevertheless has a different effect on the people who are new to this context, and those who work there day by day. The procedures were therefore experienced differently by him as an 'outsider', especially on their affective dimension. Through petting and talking to the cattle, he is trying to calm the animals, and he feels bad for them, when they aren't well. The third dimension, the ethical-political one, manifests itself when Karsten describes the 'responsibility' to care for the animal, ‚even if it's only for a short time' (*ibid.*). He emphasises the need for a profession which 'cares for' the animals: 'it is really good that there is a profession which controls if the animals are being treated respectfully'. Despite the fact that animals have to be killed for food production, Karsten holds on to the construction of the professional identity as the 'protector of animals'. Judging the fact that there are some animals for which 'it's definitely better if [they're] gone', the act of killing in this example can also be reconstructed as an act of care.

## Conclusion

Previous research on the caring-killing paradox has investigated the psychological cost of the fact that in some occupations, people have to kill animals despite their moral standpoint and/or motivation. This aspect has been linked to 'moral stressors' (Rollin, 1987), or high levels of stress and anxiety (Arluke, 1994), and to 'multiple emotional and physical ailments including unresolved anger and grief, depression, sleeplessness, elevated blood pressure, and destructive behaviours such as substance abuse and suicide' (Shearer, 2018: 11). Accordingly, the participants of this study continuously express discomfort in describing the killing of animals and emphasise their compassion. However, they view it as a possibility to offer the animals relief of pain and suffering. The act of killing itself can be reconstructed as an act of care in their narratives. This is very obvious in the first context of the avian clinic, where euthanasia is without question being considered as the best option for a sick bird in pain. But even in the context of a slaughterhouse, where the killing of animals happens not primarily for their own relief but rather for food production, the aspect of care is still emphasised when trying to treat the animals according to professional standards in a 'humane' way (Wilbur and Gibbs, 2020). Despite the widely differing contexts of a clinic and a slaughterhouse, all three dimensions of care according to Puig de la Bellacasa (2017) can be found in the interviewees' descriptions of the procedures. Contrasting different nuances in the narratives aside from the exemplary will be a fruitful yet vast endeavour, destined for another project.

Most research literature focuses on only one isolated context: e.g. the clinic, the slaughterhouse, or the animal laboratory. This paper aims to remain sensible to its multi-contextuality and analyses the similarities in the identity construction in the interviewees' narratives, significantly forged by the unique aspect of killing as care. As the veterinary profession is highly multi-contextual, the research must follow along in order to understand the professional identity adequately. This study shows the importance of grounding care theory in empirical investigations. Through looking at the veterinary profession, there is a unique possibility to view care in a broader context, in a more-than-human world.

## References


Arkow, P. (1985). The humane society and the human companion animal bond: Reflections on the broken bond. The Veterinary Clinics of North America. Small Animal Practice 15 (2): 455-466.

Arluke, A. (1994). Managing emotions in an animal shelter. In: Manning, A., Serpell, J. (eds) Animals and human society. Routledge, New York, USA, pp. 145-165.

Arluke, A. and Sanders, C.R. (1996). Regarding animals. Temple University Press, Philadelphia, USA, 218 pp.

Bonnaud, L. and Fortané N. (2020). Being a vet: The veterinary profession in social science research. Review of Agricultural, Food and Environmental Studies: 1-22. https://doi.org/10.1007/s41130-020-00103-1

Chang, F.T. and Hart, L.A. (2002). Implications of human-animal interactions and bonds in the laboratory. Institute for Laboratory Animal Research Journal Online 43 (1): 1-3.

Donald, M.M. (2019). When care is defined by science: Exploring veterinary medicine through a more-than-human geography of empathy. Area 51 (3): 470-478.

Fahrion, A. Dürr, S. Doherr, M.G. Hartnack, S. and Kunzmann, P. (2011). Das Töten und die Würde von Tieren: ein Problem für Tierärzte?. Schweizer Archiv für Tierheilkunde 153: 209-214.

Haarmann, D. and Weich, K.L. (2016). Geschichte der Tiermedizin. In: Borgards, R. (ed.), Tiere. Kulturwissenschaftliches Handbuch, Metzler, Stuttgart, Germany, pp. 149-158.

Franklin, A. (1999). Animals and modern cultures: A sociology of human-animal relations in modernity. SAGE Publications, London, UK, 213 pp.

Gardiner A. (2020). Killing and caring: A historical perspective in veterinary medicine. Veterinary History 20 (1): 6-26.

Hobson-West, P. (2007). Beasts and boundaries: an introduction to animals in sociology, science and society. Qualitative Sociology Review 3: 23-40.

Hobson-West, P. and Timmons, S. (2016). Animals and anomalies: An analysis of the UK veterinary profession and the relative lack of state reform. The Sociological Review 64 (1): 47-63.

Kostelnik, K. Lotz, F. Sötje, L. and Heuwieser, W. (2010). Vorstellungen von Studierenden der Tiermedizin und die berufliche Realität. Tierärztliche Praxis 38 (4): 209-218.

Morris, P. (2012). Blue juice: Euthanasia in veterinary medicine. Temple University Press, Philadelphia, USA, 230 pp.

Puig de la Bellacasa, M. (2012). 'Nothing comes without its world': Thinking with care. Sociological Review 60 (2): 197-216.

Puig de la Bellacase, M. (2017). Matters of care: Speculative ethics in more than human worlds. University of Minnesota Press, Minneapolis, USA, 265 pp.

Rollin, B.E. (1987). Euthanasia and Moral Stress. In: Loss, Grief & Care 1 (1-2): 115-126.

Shearer, J.K. (2018). Euthanasia of cattle: Practical considerations and application. Animals 8 (57): 1-17.

Strauss, A.L. and Corbin, J. (1998). Basics of qualitative research: grounded theory procedures and techniques. Sage Publications, Thousand Oaks, USA, 312 pp.

Tannenbaum, J. (1995). Veterinary Ethics: Animal welfare, client relations, competition and collegiality. Mosby, St. Louis, USA, 615 pp.

TAppV. Verordnung zur Approbation von Tierärztinnen und Tierärzten. Available at: https://www.gesetze-im-internet.de/tappv/.

Tronto, J.C. (1993). Moral boundaries: A political argument for an ethic of care. Routledge New York, USA, 226 pp.

Vermilya, J.R. (2015). Tracking 'large' or 'small': Boundaries and their consequences for veterinary students within the tracking system. University of Colorado, Boulder, USA, 173 pp. Available at: https://scholar.colorado.edu/concern/graduate_thesis_or_dissertations/08612n63v.

Wilbur, A. and Gibbs, L. (2020). 'Try it, it's like chocolate': embodied methods reveal food politics. *Social & Cultural Geography* 21 (2): 265-284.

# 38. Killing in plural? On animal life assessments by veterinarians and the role of euthanasia

*E. Deelen* and F.L.B. Meijboom*
*Faculty of Veterinary Medicine, Unit of Animals in Science and Society, Sustainable Animal Stewardship, Utrecht University, Yalelaan 2, 3584 CM Utrecht, the Netherlands; e.deelen@uu.nl*

## Abstract

In society and within the veterinary profession, a plurality of views is seen on killing of animals. This plurality of views leads to diversity in the characteristics of end of animal life (EOL) discussions. It is in that perspective remarkable that euthanasia as a concept is used consistently throughout the various disciplines of the veterinary practice. In this paper we will explore how the concept of euthanasia can function as a bridging concept in ethics for EOL decisions and how it enables comparison of veterinary actions in different disciplines of the veterinary practice. The former requires conceptual clarity. Using existing literature, we define four core elements of the concept of euthanasia in the veterinary practice, namely: (1) a deliberate decision to kill an animal; (2) the act of killing performed in such a way that it does not come with any avoidable pain, distress or suffering; (3) the intention to act in the best interest of the animal by preventing (further) suffering; and (4) the determination of the agents to act in the interest of the animal as much as possible when a conflict of interest arises. Building on this, we argue that euthanasia may not function as a bridging concept, but it has the ability to compare various actions in different disciplines of the veterinary practice. Furthermore, euthanasia can enable discussion between veterinarians and the general public on EOL related questions including the veterinary role and responsibilities at the EOL. In conclusion, we will show how this desk study functions as building block for the design of a planned empirical study on (moral) attitudes of veterinarians on EOL discussions that allows for comparing the role of veterinarians in various contexts.

**Keywords:** veterinary ethics, conceptual analysis, intentions

## Introduction

In all contexts in which animals are kept or receive veterinary care, questions about the end of animal lives (EOL) occur (Christianen *et al.*, 2016; Hartnack *et al.*, 2016; Meijboom and Stassen, 2016; Springer *et al.*, 2019). Nonetheless, there are serious differences in what kind of questions arise depending on the animal practice. For example, in a small animal practice, EOL discussions are often about quality of life or expected life span. Whereas in a farm animal practice discussions more often focus on the economic value of an animal in relation to further treatment. These differences have a multi-layered background including tradition, views on the moral position of animals and human-animal bonds (Meijboom and Stassen, 2016). Moreover, these differences are rooted in a plurality of views on killing of animals in society and within the veterinary profession. This plurality is reflected in discussions regarding when, why and who is allowed to kill an animal or even should kill an animal. Additionally, discussions focus on when one is allowed to refuse killing an animal (Meijboom and Stassen, 2016; Schukken *et al.*, 2019; Yeates and Main, 2011). Consequently, questions about consistency of veterinary ethics arise. What elements are needed in veterinary ethics to deal with the plurality of views on killing of animals? Additionally, we wonder what could support the veterinary profession in dealing with the impact of this plurality in their daily work. Several publications reported that veterinarians experience ethical dilemmas when confronted with the killing of animals. In some cases, this affects their professional and personal wellbeing due to signs of burnout and moral distress (e.g. Rollin, 2011). Although a plurality


EurSafe 2021 – DOI 10.3920/978-90-8686-915-2_38, © Wageningen Academic Publishers 2021

of views within the profession leads to differences in EOL discussions, it is noteworthy that euthanasia as a concept is used consistently throughout the various disciplines of the veterinary practice (Edwards-Callaway *et al.*, 2020; Hartnack *et al.*, 2016; Springer *et al.*, 2019). In this paper we will therefore explore whether, and if so, how the concept of euthanasia can function as a bridging concept in ethics for EOL decisions and how the concept enables the comparison of veterinary actions within different disciplines of the veterinary practice. The former requires conceptual clarity. Using existing literature, we define four core elements of the concept of euthanasia in the veterinary practice to create the needed conceptual clarity.

## The concept of euthanasia in veterinary practice

When focussing on euthanasia in veterinary practice several definitions can be found in the literature (Bekoff and Meaney, 1998; Cholbi, 2017; Kasperbauer and Sandøe, 2016; McMahan, 2002; Yeates, 2010). The Greek ευθανατος, from which euthanasia originates, literally translates to 'good death'. A part of the content is revealed by the literal meaning as it implies that we can differentiate between forms of death considered as 'good' and forms which we consider not as 'good'. This evaluative stance supposes that euthanasia is about something active that can be valued rather than a situation that happens to you. Therefore, we consider euthanasia as a concept that describes killing rather than the decisions of letting an animal die. Furthermore, euthanasia is about a decision to act rather than to omit. This leads us to a first core element of the concept of euthanasia, namely euthanasia refers to a deliberate decision to kill an animal. This first element of euthanasia is, however, still rather broad and spans over many cases in which animal lives are ended, such as procedures of slaughter and depopulation. With the first core element in mind, we compare two examples to discover two additional core elements.

Suppose one drives a car and suddenly a dog crosses the road. Imagine the car hits the dog and as a result the animal dies. This example is considered to be an accident rather than euthanasia. It lacks a deliberate decision to kill an animal as there is no intention to kill. The importance of intention can be illustrated in the following example. Two sisters decided to kill their hen with help of their car[1]. Their intention was to kill the animal and consequently to stop the suffering of the hen. This case raised a lot of public commotion including suspicions of animal cruelty. In the end, the authorities concluded that this was not a case of animal cruelty as the animal seemed to be killed instantaneously and the intentions of the sisters were not to harm the chicken.

The comparison of these two examples first illustrates that euthanasia is not only about active killing and the decision to end an animal's life but is also about the performance of an act that results in the death of an animal and to prevent (further) suffering. When we focus on the method, we argue that the 'good' in euthanasia refers to the quality of the performance of the act. Euthanasia links to conditions of an appropriate level of knowledge, competence and experience to kill without causing the animal any avoidable discomfort. This leads us to the second core element, namely the act of killing the animal is performed such that it does not come with any avoidable pain, distress or suffering. Furthermore, the comparison reveals also a third element: the intention. Intentions play a vital role to differentiate euthanasia from other forms of killing animals. In the example of the sisters it becomes clear that they had the intention to do the right thing for the animal. When we focus on the motivation, we argue that the 'good' in euthanasia refers to the intention to act in the best interest of the animal. In other forms of killing such as slaughter or depopulation, the interest of the individual animal seems to play a less central role. This conclusion brings us to the following questions: how can 'the right thing to do' be defined and, building on that, 'the best interest of the animal'? Before we address these questions, let us consider the above discussions. Based on these we can conclude that the core elements that define the

[1] https://www.rtlnieuws.nl/nieuws/artikel/4606751/vrouwen-kip-slachten-auto.

concept of euthanasia are: (1) a deliberate decision to kill an animal; (2) the act of killing performed in such a way that it does not come with any avoidable pain, distress or suffering; and (3) the intention to act in the best interest of the animal by preventing (further) suffering. From these three elements, element (3) seems to be the most specific for euthanasia as (1) and (2) are also applicable in other cases in which animals are killed. Therefore, we elaborate on (3) in more detail. We first focus on the interest of the animal in cases of end of life decisions followed by an analysis of the intention. This leads us to the fourth core element of euthanasia as a concept.

## The interest of the animal

Focussing on the interest of the animal raises the question whether an animal can have an interest in the continuation or ending of its life. Although many animals have a strong biological instinct to survive this is not the same as to having an interest in staying alive, i.e. that staying alive is good for the animal in itself. From the recognition of animals as sentient beings, we can conclude that animals can be harmed rather than damaged only. Animals have interests, at least to prevent harm, but also to achieve certain goods. However, this is not yet an answer to the question whether they have interests in the context of end of life. A sentience approach combined with a utilitarian outlook often results in a view that the death of an animal is morally neutral. This is based on the assumption that animals cannot be deprived of future welfare as long as they lack the capacity to anticipate or reflect on the future. Therefore, if an animal is killed in a way that does not lead to any avoidable discomfort, it can be in the interest of the animal. Nonetheless, there is more to say about interest of animals in end of life situations. There are even indications that some animals are capable of episodic future thinking (Broom, 2010; Clayton *et al.*, 2003) and thus have an idea of future or are aware of their life stage. Although this episodic future thinking may not be completely comparable to the capability of humans, these indications complicate the idea that animals have no, or very limited, interests related to the end of life. This observation questions the statement that killing is morally neutral. Furthermore, for an animal to have interests, an animal does not necessarily need to be aware of its interests. This entails that animals can have an interest in the continuation or ending of their life referring to what is good for themselves. The reference to the 'good' of the animal may sound odd in combination with the ending of a life, though this can be explained via the Life Comparative account theory and time-relative interest account theory. In the Life Comparative account theory, death is measured in terms of its effect on the overall value of the animal's life as a whole (Cholbi, 2017; McMahan, 2002). This theory suggests that the death of an animal is *prima facie* harmful to the animal and should be compared with the harm of suffering when life is prolonged. Depending on the prospective quality of life, death can be harmful or beneficial for the animal. In practice, one has to assess how to prevent greater harm to the animal by comparing the actual life with a death at time $t$ with a prolonged life and a death at time $x$. From this perspective, it can be in the interest of an animal to kill it if the continuation of its life leads to greater harm. However, this may lead to counterintuitive implications. Suppose one is confronted with a dog that is a companion to a family and a fish in an aquarium in the same family. The Life Comparative account theory would lead to a much lower threshold to kill the fish than the dog, because the dog's quality of life is more complex than that of a fish. Therefore, the dog can be more harmed by the killing than the fish (McMahan, 2002). The time-relative interest account (TRIA), proposed by McMahan (2002; 2016), adds a level of complexity. TRIA also considers the amount of good life of which the individual is deprived by death. However, it expands on this by including the extent to which the individual is psychologically connected to their future self and their self at the time of death. Hence, the badness of death is proportional to the strength of the victim's time-relative interest in continuing to live (McMahan, 2002). As a result, the death of a 3-year-old human child, a 3-year-old dog or a one-day old chicken depends on the time-relative interest of each of these three in continuing to live. This interest is based on: (1) the amount of good that a living being loses through death; and (2) the psychological unity over time (DeGrazia, 2016). Regarding the former, the young child loses more 'good' than the young chicken due to both the life expectancy and the complex cognitive abilities that a child already has and the chicken does not (seem to) have. Concerning the latter, the dog has a stronger internal reference between earlier and later mental states, such as past experiences or anticipations than the one-day old chicken (seems to) have. Therefore, the interest of these animals in cases of end of life situations differ. Although TRIA has its own disadvantages (Heeger, 2016), it shows that animals can have interests that have a direct impact on the evaluation of the death of the animal.

## Acting in the animal's interest

Even though animals have interests related to their end of life, there is a complicating matter: humans are making the decision when it comes to ending the animal's life. This leads to epistemic problems and questions of intention.

First, in most cases we cannot communicate with animals about their interests. Additionally, despite innovation in ethology and animal welfare science, we still lack sufficient knowledge about the interest of the animal. Cholbi (2017) proposes to let the animal keeper play an important role as a guardian in questions of euthanasia of companion animals due to their long-term relationship. This is an interesting start, although we see two problems. On the one hand, despite the daily interaction that animal owners have with their animals, animal owners might have limited knowledge in interpreting the health and/or welfare status of their animal. Furthermore, their vision can be clouded by human-centred interests. On the other hand, the proposal to consider the animal keeper as guardian could make the animal quite vulnerable. For example, in cases in which a long-term relationship is missing. The lack of this long-term relationship could be due to the intentions of the owner or the specific system in which the animal is kept. Think of a broiler farm in which chicken by definition of the practice will not get old enough to build a long-term relation. In both cases, the expertise of a veterinarian can be of essential added value to decide whether, considering the health and welfare status of the animal, it is in the best interest of the animal to resort to end the animal's life.

Second, one can wonder if killing an animal is still euthanasia when the intention of those involved is not only focused on the interest of the animal. Do we exclusively consider the killing of an animal as euthanasia when discomfort is avoided, and the act of killing is performed by a person who is only guided by the animal's interest? Or can we still speak of euthanasia when other considerations such as context are included? This is important for our aim to explore whether euthanasia can play a role in bridging the diversity in EOL decisions. An answer can be formulated with the help of the notion of 'contextually-justified euthanasia' as proposed by Yeates (2010). This viewpoint stresses the importance of context in EOL situations and argues that quality of life is also depending on the context in which the animal lives. Therefore, the situation can exist in which an animal under ideal circumstances could have a life worth living. However, because of the context in which the animal lives there are circumstances due to which one decides to kill the animal. These circumstances may be the result of various reasons such as 'an owner's unreasonableness or the fault of society' (Yeates, 2010). From this perspective, the interest of the animal is still included in the decision whether to euthanise the animal. Other contextual considerations can be included as well such that ending the animal's life can be still justified. To further elaborate on the inclusion of other considerations, we use an example from practice. An owner comes to a veterinary practice with his dog which repeatedly vomited over the past two weeks and has significant weight loss. Despite efforts by the veterinarian to convince the owner, the owner does not allow the veterinarian to perform any diagnostics. As second-best option, the dog received medication which did not stop the vomiting. In the end, the veterinarian decided to kill the animal considering that the continuation of the animal's life was not acceptable. In line with Yeates' account, we could speak of contextually justified euthanasia. It is obvious that the decision to kill the animal is not only based on animal interest but influenced by the interest of the owner. This example raises the question whether the inclusion of human-centred interest in the definition of euthanasia will not lead to arbitrary situations.

Is there a difference between acting according to the wishes of the owner and what the veterinarian did in this example? We argue that intentions are essential to make a difference. This leads to a fourth core element of euthanasia. When confronted with a conflict of interest in EOL discussions, the intention of the agents involved should still be to act in the best interest of the animal as much as possible. This allows that context and the interests of others to be included in EOL discussions, though it does not form a carte blanche to escape from professional responsibilities. One should strive to remedy the conflict of interest in such a way that one still acts in the interest of the animal.

In summary, we propose the concept of euthanasia is formed by four core elements: (1) a deliberate decision to kill an animal; (2) the act of killing performed in such a way that it does not come with any avoidable pain, distress or suffering; (3) the intention to act in the best interest of the animal by preventing (further) suffering; and (4) the determination of the agents to act in the interest of the animal as much as possible when a conflict of interest arises.

## The perspective in practice

We defined four core elements of the concept euthanasia in the veterinary practice to create conceptual and normative clarity. Now we can return to the question whether euthanasia can function as a bridging concept. The four elements show that euthanasia is not restricted to a particular context or part of veterinary practice. Furthermore, euthanasia can incorporate the characteristics of the specific disciplines of the veterinary practice. In this way the concept can bring different practices of EOL together and can deal with the characteristics of a (clinical) case. However, we would overload the concept if we aim to build a veterinary ethics for EOL decisions only on this concept. The plurality at stake in cases of animal death is more profound and includes fundamental assumptions such as about what we owe to animals, the value of live and human freedom. Differences at this level cannot be mitigated on a conceptual level only. Nonetheless, euthanasia in the way we presented in this paper enables to open and to stimulate the debate between veterinarians and the general public on EOL related questions including the role of the veterinarian and her responsibilities at the end of animal life. As first step to support this claim an empirical study among veterinarians of various disciplines of the veterinary practice seems a logical step to explore what our proposed concept implies in practice. Therefore, we have planned a study to analyse whether, and if so, how the concept enables to compare contexts and enables discussion. Additionally, we plan to explore whether the concept can provide guidance for veterinarians to address EOL related problems and so to cope with these problems more effectively. In this way our empirical study aims to focus on the (moral) attitude of veterinarians on EOL discussions that allows us to compare the role of veterinarians in various contexts.

## References

Bekoff, M. and Meaney, C.A. (1998). Encyclopedia of Animal Rights and Animal Welfare; Greenwood Press: Westport, CT, USA.

Broom D.M. (2010). Cognitive ability and awareness in domestic animals and decisions about obligations to animals, Applied Animal Behaviour Science, 126 (1-2), pp. 1-11.

Stine Billeschou Christiansen, Annemarie Thuri Kristensen, Jesper Lassen and Peter Sandøe (2016). Acta Vet Scand (2016) 58:30. https://doi.org/10.1186/s13028-016-0211-x

Cholbi, M. (2017) The Euthanasia of companion animals. In Pets and People: The Ethics of Our Relationships with Companion Animals; Overall, C., Ed.; Oxford University Press: New York, NY, USA, 2017; pp. 264-278.

Clayton, N. S., Bussey, T. J. & Dickinson, A. (2003) Can animals recall the past and plan for the future? Nature Reviews Neuroscience 4(8):685-91.

DeGrazia, D. (2016). Sentient non-persons and the disvalue of death. Bioethics. 30/7. 511-519. https://doi.org/10.1111/bioe.12250

Edwards-Callaway, L.N., M.C. Cramer, I. N Roman-Muniz, L. Stallones, S. Thompson, S. Ennis, J. Marsh, H. Simpson, E. Kim, E. Calaba and M. Pairis-Garcia (2020). Preliminary exploration of swine veterinarian perspectives of on-farm euthanasia. Animals 2020, 10, 1919.

Hartnack, S., S. Springer, M. Pittavino and H. Grimm (2016). Attitudes of Austrian veterinarians towards euthanasia in small animal practice: impacts of age and gender on views on euthanasia. BMC Veterinary Research (2016) 12:26.

Heeger, F.R. (2016). Killing animals and the value of life. In: Meijboom F.L.B. and Stassen E.N. The end of animal life: a start for ethical debate Ethical and societal considerations on killing animals, Wageningen Academic Press: Wageningen, The Netherlands: pp. 25-40.

Kasperbauer, T.J. and Sandøe, P. (2016). Killing as a welfare issue. In: The Ethics of Killing Animals; Višak, T., Garner, R. (eds); Oxford University Press: New York, NY, USA, pp. 17-31.

McMahan, J. (2002). The Ethics of Killing: Problems at the Margins of Life, Oxford UP.

McMahan, J. (2016) The comparative badness for animals of suffering and death. In: Višak, T., Garner, R. (eds) The Ethics of Killing Animals; Oxford University Press: New York, NY, USA, 2016; pp. 65-85.

Meijboom F.L.B. and Stassen E.N. (2016). The end of animal life: a start of ethical debate. On the role of the human-animal relationship and the plurality of views on the value of animals. In: The end of animal life: a start for ethical debate Ethical and societal considerations on killing animals, Wageningen Academic Publishers, Wageningen, the Netherlands.

Rollin, B.E. (2011). Euthanasia, Moral Stress, and Chronic Illness in Veterinary Medicine, Veterinary Clinics of North America: Small Animal Practice, 41/3, 651-659.

Schukken, Y.H., van Trijp, J.C.M., van Alphen, J.J.M. and Hopster, H. (eds) (2019). The State of the Animal in The Netherlands. Den Haag: Raad voor Dierenaangelegenheden.

Springer, S., F. Jenner, A. Tichy and H. Grimm (2019). Austrian Veterinarians' Attitudes to Euthanasia in Equine Practice. Animals 2019, 9, 44. https://doi.org/10.3390/ani9020044

Yeates, J. (2010). Ethical aspects of euthanasia of owned animals. Practice 2010, 32, 70-73.

Yeates, J.W. Main, D.C.J. (2011), Veterinary opinions on refusing euthanasia: justifications and philosophical frameworks Veterinary Record 168, 263.

# 39. Small animals, big decisions: potential of moral case deliberation for a small animal hospital on the basis of an observational study[1]

*M. Long[1]*, S. Springer[1], I.A. Burgener[2] and H. Grimm[1]*
*[1] Unit of Ethics and Human-Animal-Studies, Messerli Research Institute, University of Veterinary Medicine Vienna, Medical University of Vienna, University of Vienna, Veterinaerplatz 1, 1210 Vienna, Austria; [2]Division of Small Animal Internal Medicine, University of Veterinary Medicine Vienna, Veterinaerplatz 1, 1210 Vienna, Austria; mariessa.long@vetmeduni.ac.at*

## Abstract

Veterinary medical decisions often involve ethically challenging aspects that go beyond purely technical questions. In human medicine, different models and methods of clinical ethics support services (CESS) have been established in order to help healthcare professionals, patients, family members, and other stakeholders to handle these ethical challenges and to improve patient care. So far, only few attempts at implementing CESS in veterinary medicine have been published. The aim of this article is to elaborate on the potential of moral case deliberation (MCD) and CESS in general in a small animal hospital. The first part of this article presents an observational study that was conducted to identify requirements for such a service based on problems veterinarians faced with regards to decision-making processes and cases. The second part of this paper shows how CESS in general can assist in addressing these challenges and why MCD is a particularly promising candidate for a CESS in a small animal hospital.

**Keywords:** clinical ethics support services, clinical ethics consultation, veterinary ethics, clinical ethics

## Introduction: forms of clinical ethics support services

Clinical ethics support services (CESS) aim to support healthcare professionals, patients and family members and other stakeholders with handling ethical challenges in medical decision-making and by that to improve patient care (Fournier, 2016). They have been implemented and developed since the 1970s in human medicine and a variety of models and methods have been established (Fournier, 2016; Molewijk *et al.*, 2016; Rasoal *et al.*, 2017). Veterinary medicine is characterised by ethical challenges of different kinds as well, often related to the tensions arising from the so-called 'veterinary triad' of veterinarian, patient and client (Batchelor and McKeegan, 2012; Kimera and Mlangwa, 2016; Kipperman *et al.*, 2018). Establishing CESS in veterinary medical institutions appears like a logical next step. However, so far only few implementations of CESS in veterinary medicine have been published (Adin *et al.*, 2019; Moses, 2018; Rosoff *et al.*, 2018; Springer *et al.*, 2018).

Following Fournier (2016) approaches to CESS in human medicine can be broadly distinguished into clinical ethics consultation (CEC) and moral case deliberation (MCD). CEC focusses on a justified decision for an ongoing case (Fournier, 2016). MCD aims to train healthcare professionals in moral deliberation by addressing past and ongoing cases and questions that result from clinical practice, in regular meetings with a neutral facilitator who structures the conversations with different methods such as the dilemma method (Stolper *et al.*, 2016) and the Socratic dialogue (Steinkamp and Gordijn, 2003).

[1] This paper presents a part of the as yet unpublished master's thesis 'Exploring the Potential of Clinical Ethics Support Services for a Small Animal Hospital' submitted by M. Long (formerly Stademann) at the University of Veterinary Medicine Vienna which includes a comparison of methods of clinical ethics support services.


EurSafe 2021 – DOI 10.3920/978-90-8686-915-2_39, © Wageningen Academic Publishers 2021

The dilemma method addresses ongoing cases whereas the Socratic dialogue aims to explore concepts and principles more generally (Steinkamp and Gordijn, 2003; Stolper *et al.*, 2016). The facilitator does not provide recommendations or value judgments but guides the participants to develop and reflect upon their own contextual solutions in a dialogical process (see e.g. Porz *et al.*, 2011). Ethical theories, concepts and guidelines are used to inform and question during the reflection process, however, they are not considered as a fixed reference framework that forms the basis for analysing a case (Stolper *et al.*, 2016).

The present paper aims to answer the question of whether CESS in general and more specifically MCD are suitable for addressing (ethical) challenges in the context of a small animal hospital. Firstly, results from an observational study of case discussions are presented highlighting identified problems emerging with regards to cases and decisions. Secondly, it is discussed how CESS in general can assist in addressing these challenges and why MCD is a particularly promising candidate for a CESS in a small animal hospital.

## Observational study in a small animal hospital

To map (ethical) decision-making challenges specific for a small animal hospital setting, a qualitative empirical study was conducted at the Division of Small Animal Internal Medicine at the University of Veterinary Medicine Vienna. The main aim of the study was two answer the following two research questions: (1) Which issues are discussed as challenges in relation to cases or decisions and in relation to the interaction with each other during the case discussions? (2) Which problems become apparent through the observation and reflective analysis of form and content of the case discussions and are not openly discussed by the participants?

### *Method of observation, setting of observation and mode of documentation*

The overt, semi-structured observation by a passive participant observer took place in the natural situation of regular morning and midday meetings on five successive days in December 2019 (Monday to Friday, 10 meetings in total). The morning meetings took place between 8 and 9 am, lasted 25 to 53 min and were held in a conference room. They served the purpose of briefing all attendees on current and newly admitted in-patient cases. The midday meetings took place between 12 pm and 3 pm, lasted 59 min to 2 h 20 min and were held in a recreational and office room. Their purpose were in-depth discussions of in-patient cases. Visible to the participants, the observer silently sat at the side during the meetings and took notes. The observer could see the participants but not all their faces. No video or audio recording was performed. All meetings were conducted in German.

The continuous observation was guided by a set of questions[2] which were answered by the observer after each meeting on the same day, based on memory and the notes taken during the observation. The questions were designed based on literature on possible methods of CESS and with different decision-making procedures in mind. They included the above-mentioned research questions as well as additional questions on the content and decision-making processes (e.g. 'Have all participants shared their opinion when a decision is being made?', 'What kind of reasons are given for opinions?') and on the interaction style and form of the discussion (e.g. 'Who is leading the discussion?', 'How often do participants interrupt each other?'). For the purpose of identifying problems, a 'problem' was understood as the combination of an obstacle and a goal (for a discussion see Grimm, 2010). Problems were identified by the observer in the case discussions either through explicit labelling as a problem by the participants, through the identification of an obstacle towards a goal or the identification of doubts about a goal itself. A second observer joined the first and the fourth morning meeting and subsequently discussed their findings from each session with the primary observer.

[2] The full question catalogue is available on request from the corresponding author.

### *Participants*

Between and during the meetings, participant numbers fluctuated within a bandwidth from 12 to 20 (morning meetings) and 2 to 10 (midday meetings). Midday meetings were attended by veterinarians responsible for in-patient cases at the Division of Small Animal Internal Medicine, including different (post)graduate levels (interns, fellows, residents and always a Senior Clinician, a European and/or American Board Certified Specialist for Small Animal Internal Medicine). In addition to that, students and senior and staff veterinarians of other subspecialties attended morning meetings. Participants were not deliberately recruited or identified by the observer and remained anonymous to the observer apart from their occasional use of first names during meetings. Approval and consent for the observation was attained through the head of the Division of Small Animal Internal Medicine at the University of Veterinary Medicine Vienna.

### *Data analysis*

First, answers to the listed criteria and pre-formulated questions were noted down after meetings and were used as the material for analysis and complemented by re-reading the notes taken during meetings. Second, the problems found were clustered into categories emerging from the data. Third, identified categories were grouped into problem areas. The final problem areas include the findings for morning and midday meetings as well as the findings for both research questions. No quantification of individual identified problems was performed.

### *Results*

Based on the data, nine problem areas were identified and are presented in Table 1.

*Table 1. Problem areas identified during observational study of morning and midday meetings in a small animal hospital with example categories.*

| | Problem area[1] | Example categories |
|---|---|---|
| A | Breed-specific problems | Breed-specific increased risk for cancer |
| B | Lack of medical possibilities | No medical possibility to improve the patient's situation |
| C | Medical uncertainty and equivocality | Missing data and studies on particular questions, unclear medical nomenclature, plurality of veterinarians' opinions |
| D | Necessity to justify clinical decisions professionally | Legal responsibility, tension between participants, criticism of argumentations |
| E | Necessity to appear competent | Pressure to appear competent |
| F | Other medical disciplines, units or veterinarians | Criticism of other veterinarians, missing information from other veterinarians, maintaining functional relationships |
| G | Owner | Relationships and interactions with owners, financial restrictions, owner compliance |
| H | Patient's perspective | Patients' wellbeing affected by being in the hospital and away from familiar environment |
| I | Structural aspects of clinical routine | Time pressure, stress, problems with resources and equipment, interruptions during meetings |

[1] Displayed in alphabetical order.

## Discussion of identified problem areas with regards to CESS and moral case deliberation for a small animal hospital

The following three points resulting from the findings of the observational study will be discussed with regards to what they reveal about the potential of MCD, and CESS more generally, for addressing (ethical) challenges in a small animal hospital: (1) problem areas related to the other stakeholders, namely the owner, other medical disciplines, veterinary units or veterinarians inside and outside the hospital (problem areas F, G); (2). necessity to justify clinical decisions professionally and to appear competent (problem areas D, E); and (3) structural aspects of clinical routine (problem area I). These points were selected for the discussion because they go beyond medical-technical problems in a narrower sense.

### *Problem areas related to the owner, other medical disciplines, units or veterinarians*

Problem areas were found to include other stakeholders, namely the owner, other medical disciplines, veterinary units or veterinarians inside and outside the hospital. Studies have highlighted the impact that owner behaviour and financial limitations can have on veterinarians (see e.g. Kipperman *et al.*, 2018; Kondrup *et al.*, 2016; Moses *et al.*, 2018; Springer *et al.*, 2019). The importance of relationships between referring veterinarians and those in referral centres have also been previously shown (Best *et al.*, 2018). A subset of North American veterinarians who reported disagreements with veterinarians that they shared a case with, have rated these disagreements as more distressing than those with owners (Moses *et al.,* 2018).

CESS have the potential to improve relationships and collaboration between stakeholders. This is especially the case for those methods that attempt to facilitate the exchange and understanding between stakeholders in group meetings, such as MCD. Van der Dam and colleagues (2013) report from nurses participating in MCD that the multidisciplinary group deliberations improved the ability to take other disciplines' perspectives into account.

Other veterinarians' decisions were questioned and criticised during the meetings in the observational study. A possible implication from this is that recommendations (in the form of decisions) from an ethicist are less likely to find acceptance among the veterinarians compared to solutions that were developed by themselves. It is a strong point of MCD that professionals are seen as the experts of their own world who can find ethical solutions through group deliberations (Porz *et al.*, 2011). This is advantageous from a practical perspective as it can increase acceptance of a CESS in veterinary practice, but also from an ethical perspective as there are good reasons to doubt the concept of ethicists as moral experts who are superior at identifying the morally right thing to do (see e.g. Birnbacher, 2012).

### *Necessity to justify clinical decisions professionally and to appear competent*

The two problem areas concerned with the observed need for justifications and competence point toward another possibility for CESS to support veterinarians, namely bringing clarification about which uncertainties and disagreements are of moral and which are of technical nature. An understanding of the nature of the challenge can help to reduce the threshold for admitting and discussing insecurities and disagreements as it becomes clear that disagreements are not necessarily linked to technical competence. Knowing about the moral nature of a challenge does not eliminate the need to justify decisions professionally but it allows the inclusion of ethical arguments into the discussion. This potential was also highlighted by Moses (2018) who reports from her experiences as an ethics consultant for veterinarians in the US that she 'needed to spend as much time explaining what a moral question is and whether the problem at hand is ethical in nature, as on the ethical analysis and recommendations' (Moses, 2018: 68).

### *Structural aspects of clinical routine*

Another finding of the observational study were structural aspects of clinical routine such as time pressure and interruptions during meetings for example by phone calls. Other studies have reported the contribution of a high workload and long working hours to stress for veterinarians in different countries (Bartram *et al.*, 2009; O'Connor, 2019).

Adding a CESS into the routine as an additional obligation might appear counterintuitive. However, a CESS that takes place on a regular basis, can provide valuable space and time to address questions and challenges which are not fully resolved during the usually stressful clinical routine, and prevent escalation. CESS provide the possibility to leave the familiar tracks of discussing diagnostic findings and medical details and instead address underlying ethical questions and disagreements. MCD which is usually implemented with regular meetings, has the additional advantage that it uses a narrative approach starting from the experience of the participants and deliberately including their emotions (Porz *et al.*, 2011; Stolper *et al.*, 2016). This can stimulate reflection through deviating from the usual fact-focussed way of discussing cases.

## Conclusion

Different approaches to CESS in human medicine have been described in the literature with different procedures and focuses. The abundance of ethical challenges in veterinary medicine raises the question of transferring CESS to veterinary medicine.

The observational study in the small animal hospital has shown nine problem areas for veterinarians which are related to other stakeholders but also structural aspects of the clinical routine. The observational study was limited to discussions among veterinarians and did not include the other stakeholders such as veterinary nurses or owners. In addition to that, rarely occurring problems may have been missed. Nevertheless, the findings of the observational study provide reasons for viewing moral case deliberation as a promising method of clinical ethics support services for a small animal hospital, especially with regards to improving mutual and self-understanding, analysing the ethical nature of challenges, the responsibility of participants for developing a solution and providing space and time for reflection. Follow-up empirical studies on the expectations and necessities for a CESS in a small animal hospital could inform a trial of MCD in practice as a promising method.

## References

Adin, C.A., Moga, J.L., Keene, B.W., Fogle, C.A., Hopkinson, H.R., Weyhrauch, C.A., Marks, S.L., Ruderman, R.J. and Rosoff, P.M. (2019). Clinical ethics consultation in a tertiary care veterinary teaching hospital. Journal of the American Veterinary Medical Association 254(1): 52-60.

Bartram, D.J., Yadegarfar, G. and Baldwin, D.S. (2009). Psychosocial working conditions and work-related stressors among UK veterinary surgeons. Occupational Medicine 59(5): 334-341.

Batchelor, C.E.M. and McKeegan, D.E.F. (2012). Survey of the frequency and perceived stressfulness of ethical dilemmas encountered in UK veterinary practice. Veterinary Record 170(1): 19.

Best, C., Coe, J.B., Hewson, J., Meehan, M. and Kelton, D. (2018). Survey of equine referring veterinarians' satisfaction with their most recent equine referral experience. Journal of Veterinary Internal Medicine 32(2): 822-831.

Birnbacher, D. (2012). Can there be such a thing as ethical expertise? Analyse & Kritik 34(2): 237-250.

Fournier, V. (2016). Clinical ethics: Methods. In: H. ten Have (ed.) Encyclopedia of Global Bioethics. Springer International Publishing AG Switzerland, Cham, Switzerland, pp. 553-562.

Grimm, H. (2010). Das moralphilosophische Experiment: John Deweys Methode empirischer Untersuchungen als Modell der problem- und anwendungsorientierten Tierethik. Mohr Siebeck, Tübingen, Germany, pp. 300.

Kimera, S.I. and Mlangwa, J.E.D. (2016). Veterinary ethics. In: ten Have, H. (ed.) Encyclopedia of Global Bioethics. Springer International Publishing AG Switzerland, Cham, Switzerland, pp. 2937-2947.

Kipperman, B., Morris, P. and Rollin, B. (2018). Ethical dilemmas encountered by small animal veterinarians: Characterisation, responses, consequences and beliefs regarding euthanasia. Veterinary Record 182(19): 548.

Kondrup, S.V., Anhøj, K.P., Rødsgaard-Rosenbeck, C., Lund, T.B., Nissen, M.H. and Sandøe, P. (2016). Veterinarian's dilemma: A study of how Danish small animal practitioners handle financially limited clients. Veterinary Record 179(23): 596.

Molewijk, B., Slowther, A. and Aulisio, M. (2016). Clinical ethics: Support. In: ten Have, H. (ed.) Encyclopedia of Global Bioethics. Springer International Publishing AG Switzerland, Cham, Switzerland, pp. 562-570.

Moses, L. (2018). Another experience in resolving veterinary ethical dilemmas: Observations from a veterinarian performing ethics consultation. The American Journal of Bioethics 18(2): 67-69.

Moses, L., Malowney, M.J. and Wesley Boyd, J. (2018). Ethical conflict and moral distress in veterinary practice: A survey of North American veterinarians. Journal of Veterinary Internal Medicine 32(6): 2115-2122.

O'Connor, E. (2019). Sources of work stress in veterinary practice in the UK. Veterinary Record 184(19): 588.

Porz, R., Landeweer, E. and Widdershoven, G. (2011). Theory and practice of clinical ethics support services: Narrative and hermeneutical perspectives. Bioethics 25(7): 354-360.

Rasoal, D., Skovdahl, K., Gifford, M. and Kihlgren, A. (2017). Clinical ethics support for healthcare personnel: An integrative literature review. HEC Forum 29(4): 313-346.

Rosoff, P.M., Moga, J., Keene, B., Adin, C., Fogle, C., Ruderman, R., Hopkinso, H. and Weyhrauch, C. (2018). Resolving ethical dilemmas in a tertiary care veterinary specialty hospital: Adaptation of the human clinical consultation committee model. The American Journal of Bioethics 18(2): 41-53.

Springer, S., Auer, U., Jenner, F. and Grimm, H. (2018). Clinical ethics support services in veterinary practice. Chapter 48. In: Springer, S. and Grimm, H. (eds) Professionals in food chains. Wageningen Academic Publishers, Wageningen, the Netherlands, pp. 308-313. https://doi.org/10.3920/978-90-8686-869-8

Springer, S., Sandøe, P., Bøker Lund, T. and Grimm, H. (2019). 'Patients' interests first, but ...'–Austrian veterinarians' attitudes to moral challenges in modern small animal practice. Animals 9(5): 241.

Steinkamp, N. and Gordijn, B. (2003). Ethical case deliberation on the ward. A comparison of four methods. Medicine, Health Care and Philosophy 6(3): 235-246.

Stolper, M., Molewijk, B. and Widdershoven, G. (2016). Bioethics education in clinical settings: Theory and practice of the dilemma method of moral case deliberation. BMC Medical Ethics 17(1): 45.

Van der Dam, S., Schols, J.M.G.A., Kardol, T.J.M., Molewijk, B.C., Widdershoven, G.A.M. and Abma, T.A. (2013). The discovery of deliberation. From ambiguity to appreciation through the learning process of doing Moral Case Deliberation in Dutch elderly care. Social Science & Medicine 83: 125-132.

# 40. To heal, to advise, to criticise? A German farm veterinarians and their self-perception

*C. Dürnberger*
*Messerli Research Institute, Unit of Ethics and Human-Animal Studies, VetMedUni Vienna, University of Vienna, MedUni Vienna, Veterinärplatz 1, 1210 Vienna, Austria; christian.duernberger@vetmeduni.ac.at*

## Abstract

The job of farm veterinarians is morally challenging. An online survey (n=123) investigated how farm veterinarians in Germany perceive their job. The questionnaire was available online between 1 June and 30 September 2019 and consisted of three parts designed to collect data on the following: (1) demographic data; (2) moral challenges; and (3) future of veterinary medicine in the context of livestock farming. The article at hand focuses on one specific (open-ended) question of this survey, namely how German farm veterinarians describe their job in their own words. This task showed different self-images: the veterinarian as: (1) physician; (2) adviser and attendant for livestock farmers; (3) 'social worker'; (4) advocate for the animals; (5) 'health manager' of a farm; (6) part of livestock farming; and finally (7) entrepreneur. The identified different roles hint at the moral challenges of the job since each role has its own demands and these claims can conflict with each other.

**Keywords:** veterinary ethics, ethics, livestock farming

## Introduction

Livestock farming is currently causing social controversy. Studies show that a very large proportion of the European population not only want high food safety and low food prices (Boogaard *et al.*, 2008), but is also increasingly concerned about the consequences of farm animal husbandry on the climate (crisis) and about animal welfare issues (Special Eurobarometer 473, 2018). The so-called 'animal rights' position even call for the abolition of livestock farming in general, arguing that animals should not be kept and slaughtered for food production (Schmitz, 2016). This field of tension is also shaping the daily job of farm veterinarians. Not only do they have to deal with the changing societal expectations, but also with the demands of animal patients, animal owners and professional obligations (Magalhães-Sant'Ana *et al.*, 2016; Mullan and Fawcett, 2017; Rollin, 2006).

However, there is little literature on how farm veterinarians themselves perceive their profession and its corresponding moral challenges. This research gap was the starting point of an online survey among German farm veterinarians. Key findings of this study have already been presented and discussed in peer-reviewed articles: Two papers focused on the moral challenges (Dürnberger, 2020a,b); one paper described how German farm veterinarians see the future of veterinary training, livestock farming and the profession in general (Dürnberger, 2020c). The article at hand focuses on one specific (open-ended) question of this survey, namely how German farm veterinarians describe their job in their own words: Which roles appear in their descriptions?

## Material and methods

For greater transparency, it is mentioned that the following description of the study's methodology has also already been published in part in the peer-reviewed articles mentioned above. The survey targeted veterinarians who are working (fully or partially) in the livestock sector in Germany. Google was used to


EurSafe 2021 – DOI 10.3920/978-90-8686-915-2_40, © Wageningen Academic Publishers 2021

identify veterinarians or veterinary practices in the livestock sector. This research was considered finished after having found 227 contacts. These contacts were invited by e-mail to take part in the online survey on www.umfrageonline.com. Furthermore, *MSD Tiergesundheit*, the magazine *Der Hoftierarzt* as well as the Federal Association of Practising Veterinarians (*Bundesverband Praktizierender Tierärzte e.V.*) referred to the online survey. Participation in the study was voluntary and anonymous. No incentive was offered. The participants were informed that the results would be published anonymously. The questionnaire was available online between 1 June and 30 September 2019. The survey was conducted in German. When veterinarians are quoted in the text below, these statements were translated from German into English.

The questionnaire of the full survey consisted of three parts designed to collect data on the following: (1) demographic data (8 items: gender, age, working in which federal state, with which animals and since when, self-employed or employed, studied at which university and date of graduation); (2) moral challenges (2 open and 12 closed questions); (3) future of veterinary medicine in the context of livestock farming (6 open and 6 closed questions).

The results presented below focus on one open-ended question and its qualitative evaluation. In the tradition of hermeneutics, such qualitative analyses do not focus on quantitative assertions. Instead they aim at a heuristic exploration of the substantial content of an argument or position. Answers to the open questions in the study were evaluated according to the method of qualitative content analysis, following Mayring (2015) and Kuckartz (2012): (1) an initial round of work on the texts, in which important passages are marked; is followed by (2) the preparation of a first draft setting out a system of main categories. The first test run checks the general adequacy of this system. (3) The material is coded to reflect the categories. (4) All text passages coded with the same category are compiled. (5) Working directly on the raw material, subcategories (if appropriate) are obtained inductively, and further tests are conducted to check and refine the system of main categories and subcategories. Finally, (6) the material as a whole is coded, using the differentiated system. At this point, (7) the discussion of the results can begin. In regard to the open questions, the current survey followed these steps.

The description of the methodology must also provide information about the limitations: Qualitative interviews or focus groups might have been more appropriate formats to explore self-descriptions and moral challenges. The decision to carry out an online survey was taken for pragmatic reasons: Potential participants live all over Germany and have a tight schedule, making it difficult to invite them to semi-structured interviews or to visit them. The present online survey gave them the opportunity to be flexible in terms of time and place to answer the questions.

The study does not provide information on the extent to which the participants represent the total population of German farm veterinarians. Furthermore, a specific selection bias is possible, arguing that veterinarians who are more sensitive about 'ethics' were maybe more willing to complete a questionnaire about the moral challenges of their job. In line with this, it cannot be excluded that the results of this study might represent the attitudes of veterinarians who are more sensitive about 'ethics' and 'moral challenges' than the general veterinary population. Finally, a topic like 'moral challenges' poses the question to what extent the study partially documents socially desirable answers. Although such answers can never be completely ruled out, the study has tried to avoid socially desirable answers by formulating the questions and given statements neutrally and with restraint.

## Results

This paper presents only the answers to a selected question of the study that have not yet been published.

An open-ended question in the study was: 'Describe your job in a few sentences: What do you do in your job?' (n=78). As said in the introduction, this simple sounding question should help to identify different roles and self-images of veterinarians. In the following, the answers of the participants are presented in a structured and summarized way (the following ranking is not weighted quantitatively).

1. Answers with focus on veterinary expertise and veterinary tasks: Participants described their job via outlining their veterinary expertise and certain veterinary tasks (like 'ultrasound', 'pathology' or 'vaccinate') or via the focus of their practice (like 'cattle practice' or 'farm veterinarian with focus on dairy cattle'.)
2. Answers with focus on animal owners: Participants described their job via referring to the farmers they are working for or working with. More precisely, (a) they see themselves as someone who advises farmers. As an example, one participant wrote: 'I advise pig farms on animal welfare, animal health, and management.' A word that was also used in this context, was 'to accompany' (*begleiten*). For example, one participant stated: 'We accompany (family) farms every day in their work with (...) their animals.' The topics here are diverse: The veterinarians advise animal owners not only with regard to animal health and animal welfare, but also, for example, with regard to compliance with the law. Some answers also underlined the necessity of cooperation between veterinarians and farmers. An exemplary response was: 'I examine and treat cattle in close collaboration with the farmer.' (b) Farm veterinarians also described themselves as actors who sometimes have to 'morally support' farmers. As a veterinarian, you have to 'listen to the needs of the farmers'; you have to 'guide and care for people in difficult situations'. (c) As the following point (3) will show, veterinarians also see themselves as critics of farmers, especially when it comes to animal welfare.
3. Answers with focus on animals: Participants described their job via referring to their responsibilities towards the animals. They understand themselves as 'advocate for the animals', that means, for example, that a veterinarian has to 'promote and, in some cases, vigorously implement animal welfare...'. In this context, veterinarians are also something like a 'critical control authority' of the farmers.

   (a) Participating veterinarians described their job with statements like: I am 'treating diseases' of animals. Exemplary answers spoke about 'helping animals', 'I treat sick cows', 'saving lives' or also 'Healing. Trying to achieve maximum cure rates at the lowest possible financial cost.' (b) However, the job is not only about healing, but sometimes also about 'redeeming animals', as could be read in an exemplary answer. The keyword here is 'euthanasia'. (c) Furthermore, participants also mentioned the bigger context of their work, underlining that farm animals are treated for one reason only: So that they can perform and produce. Against this background, an exemplary answer stated: I am 'improving the performance' of animals. (d) Finally, in their answers to this question, participants repeatedly raised the issue of herd health management (in German: *Bestandsbetreuung*) vs curative practice on individual animals. While curative practice focuses on treating sick animals, herd health management means that the veterinarian is mentoring the whole farming business with a view to animal health and other aspects. Exemplary questions could be: What can be improved in feeding or ventilation in the barn? Which considerations are decisive for the farm in terms of genetics? In this context, participants underlined the importance of prevention as essential part of their job. Being a farm veterinarian also means to 'keep animals healthy' and to make sure that they do not fall ill in the first place.
4. Answers with focus on the production of food: Participants described their job referring to the importance of food production. Exemplary answers were: 'I am also significantly involved in ensuring that healthy and high-quality food is produced' or 'We provide food for the world's population.' In this context, also the aim of 'food hygiene' was mentioned.
5. Answers with focus on entrepreneurship: Farm veterinarians described their job via underlining, that they are not only a 'physician/veterinarian', but also an 'entrepreneur'. That means, that they have to pay attention to economic concerns and that they are, in some cases, also 'employers'.

6. Other responses that are more difficult to categorize describe the profession as follows: A farm veterinarian spends a lot of time in the car; the job is physically demanding; as an employee, you have little decision-making power but have to do what the boss says; the job is extremely labour-intensive and the legal requirements are sometimes 'idiotic'.

## Discussion

The whole survey provides an explorative insight into the (moral) challenges and self-understandings of farm veterinarians in Germany. It creates an initial database that allows hypotheses to be framed, points towards further (representative) studies and generates new research questions. Although the paper at hand only presents answers to one selected question of this survey, the results do give some important hints in regard to key self-images of farm veterinarians and corresponding challenges.

The task of describing their job in their own words, showed different self-images of farm veterinarians, more precisely the participants addressed the following roles: The veterinarian as (1) physician, that means as an expert with medical expertise and skills; (2) adviser and attendant for livestock farmers; (3) a kind of 'social worker', that means that a veterinarian is not only confronted with the needs of animals, but sometimes also with the problems and difficulties of the farmers; (4) advocate for the animals and in this regard also a critical control authority of agriculture; (5) a kind of 'health manager' of a farm, focusing not on treating sick (individual) animals but on keeping the whole 'herd' healthy; (6) part of livestock farming, that means, that veterinarians cure and help animals in order to keep them productive for food production; and finally (7) entrepreneur.

The identified different roles make it clear that veterinarians face more than 'just' animal claims (i.e. demands that the animals make on them, such as freedom from suffering) and, in doing so, hint at the moral challenges of the job (cf. Dürnberger, 2020a,b): Farm veterinarians do not understand themselves 'just' as advocates for the animals, but also as adviser and attendant for livestock farmers, entrepreneurs, social workers or part of agriculture. Each role has its own demands and responsibilities – and these claims can conflict with each other. To give an example: As 'advocate for animals', a veterinarian would like to help animals, for example by criticising the owner if necessary; as 'entrepreneur', the veterinarian has to keep in mind that the owner is the customer who pays the bill. Such conflicts between roles and self-understandings could be an ideal starting point for discussions in the context of ethics seminars for veterinarians.

Having said that, the identified roles could help to better understand which topics should play a more important role in veterinary training – at least from the point of view of the participants and at least under one specific condition, see the last paragraph of the article – in order to prepare future farm veterinarians for their job. For example – and as discussed in Dürnberger, 2020c – knowledge about entrepreneurship has been identified in various studies as a crucial competence of veterinarians (Cake *et al.*, 2014). German farm veterinarians join these voices that underline the importance of the corresponding skills. Another educational focus could be a stronger focus on herd health management. The answers indicate that farm veterinarians have less to do with an individual, sick animal than a layperson might think. These results can be interpreted as a typical sign of a structural change of the profession: a shift from curative individual animal care towards herd health management can be observed worldwide in the livestock sector (Gerber *et al.*, 2020). As suggested (Dürnberger, 2020c) this dynamic should be understood as kind of shift – not replacement – of self-understanding, which can be described as follows: a shift from the veterinarian as a 'healer' to a kind of 'health manager'.

However, this analysis with a view to which topics should possibly play a greater role in training must also be critically discussed, because it does not question the current system. Instead of asking which curricula better prepare veterinarians for the current situation in the job, we could and should also ask to what extent the given conditions can be changed so that veterinarians, for example, can actually focus on doing what motivated most of them to take up this profession in the first place: helping animals.

## References

Boogaard, B.K., Oosting, S.J. and Bock, B.B. (2008). Defining sustainability as a socio-cultural concept: Citizen panels visiting dairy farms in the Netherlands. Livestock Science 117: 24-33.

Cake, M.A., Rhind, S.M. and Baillie, S. (2014). The need for business skills in veterinary education. Perception vs evidence. In: Collete, H. (ed.) Veterinary Business and Enterprise. Theoretical foundations and practical cases. Elsevier, pp. 9-20.

Dürnberger, C. (2020a). Am I actually a veterinarian or an economist? Understanding the moral challenges for farm veterinarians in Germany on the basis of a qualitative online survey. Research in Veterinary Science. https://doi.org/10.1016/j.rvsc.2020.09.029

Dürnberger, C. (2020b). I would like to, but I can't. An Online Survey on the Moral Challenges of German Farm Veterinarians. Journal of Agricultural and Environmental Ethics. https://doi.org/10.1007/s10806-020-09833-0

Dürnberger, C. (2020c). The last of us? An online survey among German farm veterinarians about the future of veterinary training, livestock farming and the profession in general. International journal of livestock production. https://doi.org/10.5897/IJLP2020.0697

Gerber, M., Dürr, S. and Bodmer, M. (2020). Survey among livestock veterinarians from the Canton of Fribourg investigating the topics of herd-health management, counselling and the use of antimicrobial drugs. Schweizer Archiv für Tierheilkunde. 162(1):23-36. https://doi.org/10.17236/sat00240

Kuckartz, U. (2012). Qualitative Inhaltsanalyse. Methoden, Praxis, Computerunterstützung. Beltz Juventa, Weinheim and Basel.

Magalhães-Sant'Ana, M., More, S.J., Morton, D.B. and Hanlon, A. (2016). Ethical challenges facing veterinary professionals in Ireland: results from policy Delphi with vignette methodology. Veterinary Record. https://doi.org/10.1136/vr.103786

Mayring, P. (2015). Qualitative Inhaltsanalyse. Grundlagen und Techniken. 12$^{th}$, revised edition. Beltz Verlag, Weinheim.

Mullan, S. and Fawcett, A. (2017). Veterinary ethics: Navigating tough cases. 5M Publishing, Sheffield.

Rollin, B. (2006). An Introduction to Veterinary Medical Ethics: Theory And Cases. Wiley-Blackwell, Oxford.

Schmitz, F. (2016). Animal ethics and human institutions: Integrating animals into political theory. In R Garner, S O'Sullivan (eds), The Political Turn in Animal Ethics (pp. 33-50). Lanham: Rowman & Littlefield.

Special Eurobarometer 473 (2018). Europeans, Agriculture and the CAP. Wave EB88.4. Fieldwork December 2017. Publication February 2018. Available at: http://data.europa.eu/euodp/en/data/dataset/S2161_88_4_473_ENG.

# 41. Vetsplaining – dealing with uncertainty as inescapable part of veterinary practice

*J. Karg* and H. Grimm*
*Unit of Ethics and Human-Animal-Studies, Messerli Research Institute, Vienna, Veterinaerplatz 1, 1210 Vienna, Austria; johanna.karg@vetmeduni.ac.at*

## Abstract

When it comes to decision-making in veterinary practice, veterinarian and animal owner often face each other as competitors: Each of them presumes to be in the position to make the right decision. The veterinarian in the role of the expert could in this case develop a patronising and condescending attitude towards the animal owner, hereafter called 'Vetsplaining'. How can this attitude be legitimised when veterinarian knowledge is based only on probabilities and approximate prognoses? In this paper we argue for narrowing the gap between veterinarian and animal owner by identifying and valuing the uncertainty that decisions in veterinary practice imply. Clinical uncertainty is considered a problem to be either ignored or decimated in the clinical context of veterinary medicine. However recognising and appreciating the unknown could be a useful step in the decision-making process, as we suggest. The veterinarian's admitting that consequences of a decision can never be predicted with certainty and one hundred percent guarantee, stimulates critical and creative thinking. The animal owner is thus actively involved in the process of deliberation, for the benefit of all parties. By respecting the animal owner's autonomy and reaching a decision on a collaborative basis, the responsibility for the animal will be distributed appropriately: The veterinarian shares his/her knowledge about possible negative consequences and the animal owner decides to take the risk or not. In the end, the success of a therapy in veterinary medicine also depends on the cooperation of the animal owner. By actively involving both veterinarian and animal owner in the decision-making process, individual animals and circumstances can be better addressed and each party can fulfil its responsibility towards the animal patient.

**Keywords:** veterinary ethics, clinical decision-making, Vetsplaining, uncertainty

## Introduction

Clinical decision making in veterinary medicine and the resulting interaction with the animal owner are routinely affected by culturally determined codes of 'control' and 'self-aggrandizement' (Clarke and Knights, 2019: 16). These traditional codes are primarily of masculine nature as the veterinary profession in the western world was up to now a predominantly male one. Despite the demographic shift of veterinary medicine towards a female profession, veterinary practice remains permeated by these patterns and structures of masculinity. However we cannot neglect that daily practice of veterinary medicine involves dealing with probabilities and unpredictabilities. In order to still maintain old expectations and role models, veterinarians could fall into the trap of '*Vetsplaining*', that means, similar to the concept of *mansplaining*, ignoring or patronising the other person in a conversation in order to control the situation (Joyce and Huma, 2020). What impact could this behaviour have on each of the three parties involved, animal, veterinarian and animal owner, and also on the final decision? And what other methods of discussion are available to the veterinarian that might lead to more promising results?


EurSafe 2021 – DOI 10.3920/978-90-8686-915-2_41, © Wageningen Academic Publishers 2021

## Animal owner or vet – who is the strongest?

Decision-making is an essential part of veterinary practice and takes place at many different levels (Mossop, 2015). In the classic example of a diseased animal that the veterinarian is confronted with, decision-making starts with considerations about which examinations, apart from the general ones, are sufficient to make a reliable diagnosis. Depending on the veterinary indication a diagnosis implies a specific therapy, measure or advice. When implementing these, the veterinarian's freedom is limited. His or her judgment and the animal owner's ideas about appropriate measures concerning the treatment often diverge widely. The reasons for this are manifold and can rarely be completely grasped right away. Naturally, often both parties would claim to make the best decision: the veterinarian on one side because he or she has the necessary expertise and knowledge and an independent standpoint to decide what is best for the animal, a point on which veterinary actions are mostly oriented (Hammond, 2018). The animal owner on the other side bears the financial burden for the veterinary action. Apart from the animal patient, he or she takes the consequences of this action, also those on the emotional level. Often the animal owner also firmly believes knowing the individual animal better and therefore being in a privileged position to decide on behalf of the animal (Hammond, 2018). Veterinary ethics deals with conflicts like these. How can veterinarians deal with difficult situations and the responsibilities they entail?

Face to face with the animal owner, many veterinarians may now be tempted by 'Vetsplaining', trying to persuade the animal owner. By insisting on being able to make the better decision due to expertise, the veterinarian overrules the animal owner and denies him or her the ability to choose autonomously. Implicitly, 'Vetsplaining' reduces the triad into a dyad that bypasses the owner's role in veterinary decision making. This is unfortunate, since after all, it is the animal owner, who carries the consequences of the veterinary decision, such as costs or burdens linked to the therapy. But isn't it still legitimate to adopt a 'Vetsplaining' attitude when the veterinarian only has the animal's best interests in mind?

The knowledge on which the veterinarian relies during 'Vetsplaining' is one based on empirical findings and statistical probabilities. Whether the chosen therapy leads to the desired success, though, can be predicted only with some probability, but never with complete certainty. Just because one has had good and extensive experience with a veterinary treatment in previous cases does not automatically mean that it is going to happen exactly the same way the next time. Clinical decision-making is always based on probabilities, better known in this context as the probability of prognoses (Ramaekers *et al.*, 2010). Every veterinary action involves risks and an evaluation of these cannot be made with hundred percent guarantee. Even the best veterinarian will not be able to make an empirically secured future prognosis about the treatment of an animal. And whoever bears the responsibility for the decision-making process therefore also bears the risk that unintended consequences will occur with some likelihood. If veterinarians now take on the role of the 'Vetsplainer' and only inform the client what has to be done, they deprive the animal owner of all moral responsibility and take all the responsibility on their shoulders respectively.

'Vetsplaining' seems to be of no advantage to neither one of the two actors: the animal owner seems deprived of his/her say in the game. When he or she makes a decision that has been put into his or her mouth it does not necessarily correspond to his or her judgement. By denying the animal owner a right to a say in this matter, the veterinarian vouches for himself/herself the sole responsibility for the decision about the animal. So, what would be an alternative way for the veterinarian to avoid this opposition with the animal owner and to come to a more collaborative solution?

## Promoting the uncertainty

> Difficulty or obstruction in the way of reaching a belief brings us, however, to a pause. In the suspense of uncertainty, we metaphorically climb a tree; we try to find some standpoint from which we may survey additional facts and, getting a more commanding view of the situation, decide how the facts stand related to one another.
>
> (Dewey, 1986 [1933]: 122)

Uncertainty as a state of doubt and hesitation when confronted with a difficult, unfamiliar situation might stimulate us to reflective thinking, according to the pragmatist John Dewey (Dewey, 1986 [1933]). In this paper we argue that the state of uncertainty concerning the necessarily limited knowledge about possible unintended negative consequences of veterinary intervention that is shared by both veterinarian and animal owner can serve as a unifying and at the same time stimulating measure against 'Vetsplaining'.

Uncertainty is an unavoidable element of veterinary practice, not only because of the mentioned precarious prognosis, but also because of the confrontational veterinary practice itself. Understanding the source of this state offers the opportunity to reduce undesirable effects during the process of decision-making (Hall, 2002). There are various coping strategies for dealing with uncertainty. One possibility might be to ignore or suppress the uncertainty. Within a conversation, people tend to act out a false certainty (Buetow, 2011). In veterinary practice, such behaviour can appear in the form of 'Vetsplaining' in order to maintaining control over the situation and to give oneself and the animal owner a (deceptive) feeling of certainty. Signs of uncertainty are usually perceived as being detrimental to the profession and the professional. Possession of specific knowledge and skills gives professionals such as veterinarians a certain status of trust by society implying the correct application of knowledge and skills (Evetts, 2003). Here, uncertainty is seen as a problem that needs to be countered with plenty of evidence-based medical knowledge. This idea is also widespread in veterinary medicine: a study from Southern England showed that 44% of the participating veterinarians believed that the clients' confidence in them is reduced when veterinarians show verbal or non-verbal expressions of clinical uncertainty (Mellanby *et al.*, 2007).

Another possibility to cope with clinical uncertainty would be to acknowledge it and to get into exchange about it with the animal owner. In the aforementioned British study, the clients of a small animal clinic were also interviewed and could not confirm the veterinarians' impression: 34% of them stated, that signs of uncertainty improved their confidence in the vet and 52% confirmed that it had no effect on it. Only 12% thought it would reduce their confidence (Mellanby *et al.*, 2007). It thus seems as if expressing and addressing uncertainty in veterinary practice does not necessarily imply a loss of the veterinarian's self-confidence nor the clients' trust into the veterinarian's authority. On the contrary, it may be advantageous for the clinician to acknowledge this fact and address the probability of success of a chosen veterinary intervention in conflict with the animal owner. 'Uncertainty is a practical matter' (Dewey, 1984[1929]: 178), says John Dewey. Admitting uncertainty in clinical decision-making to him- or herself may motivate the veterinarian to climb Dewey's metaphorical tree, from which a better view of the situation can be obtained. Maybe from there, the animal owner's concerns can be understood better and the individual animal addressed in a preferential manner. In this way one's preconceived judgement can be challenged. Viewing uncertainty as an inevitable but positive quality of human decision-making and ethical judgment has the potential to foster critical and creative thinking (Buetow, 2011).

For yet another reason, the veterinarian could benefit from openly admitting the uncertainty to the animal owner as well, instead of disguising it through 'Vetsplaining': the animal owner is thus not only confronted with a lack of guarantee for the right decision, he or she will thus also be involved in the decision-making process. As soon as the client perceives that the final decision has not yet been made, he or she is also encouraged to deliberate actively. And this would be consistent, because in the end, it is he or she who is responsible for the decision although information and advice by an expert will help to deal with the consequences. The veterinarian recommends or speaks out clearly in favour of a particular action, but in the end he or she is only an aid to decision-making, not the decision-maker him- or herself. From a systematic point of view, the veterinarian cannot claim full certainty and needs the owner to accept the risk of possible, foreseeable unintended consequences. The veterinarian's responsibility will be fulfilled when his or her professional expertise is communicated to the animal owner in order to set up an action plan. However the veterinarian must be free to investigate and evaluate this request in order to fulfil his or her own moral responsibility towards the animal. Thus, by involving the owner to accept risks and uncertainties, the veterinarian transfers a decisive part of the responsibility to the client and thus ensures an appropriate distribution of burdens. Discussing and determining the present problem on a joint basis promises to give better results and distribution of responsibilities. This way, the veterinarian fulfils the responsibility to inform about foreseeable and unintended consequences. This would involve assessing the probability of these possible negative consequences occurring within his/her means and reporting them to the animal owner. The decision whether or not to assume the risk and thus the responsibility for it and the consequences arising from it, rests, on the other hand, with the animal owner. No one, including the veterinarian, is responsible for unintended consequences that cannot be foreseen despite expert knowledge.

A joint decision even seems to increase the probability of a good prognosis, because when it comes to the therapy of a sick animal, the outcome is also significantly dependent on the goodwill of the animal owner and a veterinary measure can be tailored more individually to both patient and patient owner. Only if the client has the opportunity to participate and thus consider in the discussion which consequences he/she can also bear, it is possible for the animal owner to take responsibility for the decision and everything that goes along with it.

## Conclusion

Veterinary action carries risks. Disguising the uncertainties arising from these risks in the form of 'Vetsplaining' does not seem to be conducive to arrive at a solution with the animal owner and to a commonly shared moral responsibility. Instead, facing clinical uncertainty might stimulate critical thinking and challenge one's own assumptions. Making uncertainties and unpredictabilites of medical action a topic of discussion with the animal owner would have the decisive advantage of a sensible distribution of responsibilities between veterinarian and animal owner. By sharing clinical advice and guidance as well as one's personal assessment and concerns regarding foreseeable risks and consequences, the veterinarian fulfils his/her role as an expert in this field. Based on this communicated knowledge and advice, the responsibility to decide and bear any negative impact, on the other hand, lies also with the animal owner. Only such a decision, which is borne by both parties, can in the end be consistent with their respective responsibilities towards the animal.

## References

Buetow, S. (2011). 'The Virtue of Uncertainty in Health Care.' *Journal of Evaluation in Clinical Practice* 17(5):873-876.

Clarke, C. and Knight, D. (2019). 'Who's a Good Boy Then? Anthropocentric Masculinities in Veterinary Practice.' *Gender, Work and Organization* 26(3):267-287.

Dewey, J. (1984). *The Later Works, 1925-1953 : 4 : 1929 : The Quest for Certainty*. edited by H. F. Simon and J. A. Boydston. Carbondale, Ill: Southern Ill. Univ. Press.

Dewey, J. (1986). *The Later Works, 1925-1953 : 8 : 1933 : Essays and 'How to Think.'* edited by J. A. Boydston. Carbondale, Ill: Southern Ill. Univ. Press.

Evetts, J. (2003). 'The Sociological Analysis of Professionalism: Occupational Change in the Modern World.' *International Sociology* 18(2):395-415.

Hall, KH. (2002). 'Reviewing Intuitive Decision-Making and Uncertainty: The Implications for Medical Education.' *Medical Education* 36(3):216-224.

Hammond, J. (2018). 'Training for Uncertainty in Veterinary Education.' University of Glasgow.

Joyce, J. and Huma, B. (2020). 'Speaking out against Everyday Sexism: Gender and Epistemics in Accusations of 'Mansplaining." *Feminism & Psychology* 0-30.

Mellanby, RJ., Crisp, J., De Palma, G., D. P., Spratt, D. P., Urwin, D., Wright, M. J. H. and Zago, S. (2007). 'Perceptions of Veterinarians and Clients to Expressions of Clinical Uncertainty.' *Journal of Small Animal Practice* 48(1):26-31.

Mossop, L. (2015) 'Editorial: Novice to Expert? Teaching and Development of Moral Reasoning Skills.' Veterinary Record 177(5):122-123.

Ramaekers, S., Kremer, W., Pilot, A., Van Beukelen, P. and Van Keulen, H. (2010). 'Assessment of Competence in Clinical Reasoning and Decision-Making under Uncertainty: The Script Concordance Test Method.' *Assessment and Evaluation in Higher Education* 35(6):661-673.

# 42. Intensive animal agriculture, land-use and biological conservation: converging demands of justice

*A. Wienhues[1]* and S. Hirth[2]*
*[1]University of Zurich, Department of Philosophy & University Research Priority Program (URPP) Global Change and Biodiversity, University of Zurich, UZH Philosophisches Seminar, Zollikerstrasse 117, 8008 Zurich, Switzerland; [2]University of Manchester, Sustainable Consumption Institute, The University of Manchester, Booth St W, Manchester, M15 6PB, United Kingdom; anna.wienhues@uzh.ch*

## Abstract

Intensive animal agriculture's high social-ecological footprint is relevant to debates such as on biological conservation and ecological sustainability. However, it does not feature prominently in the (small but growing) conservation sub-literature on 'just' land-use that considers different aims of biological conservation as well as intra-human justice concerns. For instance, the difference in visions of biodiversity conservation, which are often themselves based on diverging value commitments such as having ecocentric or anthropocentric foundations, can be illustrated by the controversy about the so-called Half Earth proposal. In this context, we would like to stress that these differences should not be overstated. For one, different value commitments can – in some cases – still compromise on similar practical implications such as shrinking intensive animal agriculture. Secondly, whilst it is certainly possible to take either a purely anthropocentric perspective, which only finds instrumental value outside of humanity, or, in contrast, a rather misanthropic perspective which reduces humans to the role of 'ecological sinners', we would like to illustrate that less polarising middle ground positions are also available that acknowledge the importance of intra-human *and* interspecies justice. In other words, taking a planetary justice perspective which is the primary focus of this paper. Here again the reduction of intensive animal agriculture is presented as a possible route of addressing some forms of the respective injustices.

**Keywords:** planetary justice, interspecies justice, de-growth, Half Earth

## Introduction

Whilst (intensive) animal agriculture's unassailably high social-ecological footprint is relevant to the debates on biological conservation, ecological sustainability, food security and biosafety; it does not feature prominently in the (small but growing) conservation sub-literature that considers questions of 'just' land-use in the context of different aims of biological conservation as well as intra-human justice concerns (here we use the term 'animals' as a shorthand for nonhuman animals – with both being preferable to the term 'livestock'). Different agricultural methods, for example dairy and meat production, and their land-use and impact on nature conservation have been analysed in quite some detail within different fields such as land system science (e.g. Foley *et al.*, 2011). However, we have noticed that some more specialised debates on 'just' 'visions of biodiversity conservation' do not stress the importance of decreasing or 'de-growing' animal agriculture enough – at least measured by how we regard its potential relevance for considerations about just conservation. That has the effect that dissimilarities between different approaches are highlighted at the cost of emphasising areas of practical convergence.

For instance, within the category of different visions of biological conservation falls the controversy about the so-called Half Earth proposal (HE thereafter). While the proposal's proponents suggest


EurSafe 2021 – DOI 10.3920/978-90-8686-915-2_42, © Wageningen Academic Publishers 2021

designating half of the planet's surface as conservation areas of different types in order to address current levels of biodiversity loss (e.g. Cafaro *et al.,* 2017; Dinerstein *et al.;* 2017; Wilson, 2016), critics find fault with the lack of attention – or the overly optimistic and reductive character of some proponents' analysis – that concerns about intra-human justice and economic management receive as part of this proposal (e.g. Büscher *et al.,* 2017a,b, Schleicher *et al.,* 2019). Although questions of just land-use are central to and even incorporated in the name of this proposal, animal agriculture's disproportionate land-use does not feature prominently in the debate about it (despite exceptions such as Mehrabi *et al.,* 2018). Therefore, we consider the HE debate a prime example for illustrating how reducing animal agriculture can help to 'partially' alleviate some disagreements between the proponents and critics of this radical proposal (see also Napoletano and Clark, 2020; Wienhues, 2020, who have already offered more nuanced commentaries).

In light of growing pressures to enact more ambitious conservation agendas, inclusive and practicable compromises are sorely needed. Inclusivity, we find, would entail considering different cultural perspectives, ethical stances and values as well as various disciplinary insights. Different visions of biodiversity conservation, such as the different perspectives employed in the HE controversy, build on diverging value commitments, in addition to different ways of conceptually understanding the human-nature relationship. For instance, some proponents build on ecocentric foundations (Kopnina *et al.,* 2018) and others on anthropocentric ones (Büscher and Fletcher, 2019). Yet, we maintain that these differences should not be overstated. For one, different value commitments can – in some cases – still compromise on similar practical implications as we will suggest in the following for the case of de-growing intensive animal agriculture (which does not rely on the 'convergence thesis' (Norton, 1991) being true). Secondly, putting it very simply, between a purely anthropocentric perspective that only finds instrumental value outside of humanity and a rather misanthropic perspective that reduces humans to the role of 'ecological sinners' (which are the extreme ends of a spectrum and are 'rarely' employed argumentatively, and therefore do not represent common disputes about value commitments), a range of less polarising positions are available that acknowledge, for instance, the importance of intra-human 'and' interspecies justice alike.

Having set the scene more widely to illustrate how we situate the relevance of what follows, we cannot discuss all of these different points in the necessary detail. So, after briefly sketching some empirical points on the matter of intensive animal agriculture, we will rather focus in the remainder on one specific 'normative' aspect that speaks in favour of de-growing animal agriculture – the demands of planetary justice. The aim is to provide an initial but incomplete overview of how different justice perspectives (and associated value commitments) can converge on the goal of de-growing animal agriculture. Accordingly, we will also largely set aside any details about the more specialised HE debate. Committed to doing justice to humans and nonhuman beings (Wienhues, 2020), while also acknowledging the capitalist roots of the environmental crisis (Büscher and Fletcher, 2020), we will illustrate how a strong reduction of animal agriculture presents itself as a practical route towards more planetary justice.

## Animal agriculture's biophysicality

By focusing on 'intensive' agricultural practices, we leave open the question of whether other small-scale forms of animal agriculture should remain on some land. To be clear, conversion losses of nutritional energy apply to all modes of producing animal-sourced foods. Our analysis is first and foremost directed at intensive animal husbandry because a variety of otherwise differing positions have the argumentative tools to support a reduction of such 'factory farming' methods. Addressing animal farming's biophysicality is key to the sorely-needed implementation of more ambitious visions of conservation, and to social and economic land-use practices that are just and sustainable.

Biophysicality refers here to the circumstance that as energy conversion from one being to another inevitably entails its dissipative losses to the environment (Kolasi, 2019), plants naturally occupy more biomass on this planet than herbivores, and the latter more than carnivores. Simply put, eating animal-sourced foods is an energetic detour which involves one or more extra levels of dissipative conversion losses as compared to eating plants directly. Against this background, we will take it as given that animal agriculture's disproportionate land and resource use is biophysically undeniable (e.g. Nijdam *et al.*, 2012, Poore and Nemecek, 2018) from which conclusions for biological conservation can be drawn. We are aware that this remains a very simplified premise which omits a range of debates such as on 'land sparing' vs 'land sharing' (e.g. Phalan *et al.*, 2011), 'sustainable intensification' and food security (e.g. Godfray *et al.*, 2010, Searchinger *et al.*, 2018), and the links between dietary and planetary health (e.g. Willett *et al.*, 2019).

## The relationship between planetary justice and animal agriculture

A range of considerations of matters of justice point towards the need to de-grow intensive animal farming. More specifically, this is a concern of 'planetary justice' (Hickey and Robeyns, 2020) which broadly denominates a range of justice considerations in the environmental context, such as intra-human environmental justice between current and towards future human generations as well as interspecies justice. Simplified, the term interspecies justice (also sometimes referred to as ecological justice or animal justice) groups together a range of non-anthropocentric positions that attribute justice claims (of different kinds) to (some or all) nonhuman living beings such as animals (e.g. Baxter, 2005; Cochrane, 2018; Garner, 2013; Wienhues, 2020) and sometimes even to ecological systems (e.g. Schlosberg, 2007). The theoretical investigation into interspecies justice and its relationship with all the relevant considerations of intra-human justice is still ongoing, and a consistent overarching framework of planetary justice is still missing (Dryzek and Pickering, 2019). Yet it constitutes the appropriate normative lens for addressing the different philosophical debates that intersect on the question of intensive animal agriculture and its land-use. Two preliminary observations can already be made in this regard.

Firstly, on the face of it, different normative positions (some of which also invoke justice claims) disagree on many issues, such as whether animal agriculture is intrinsically problematic. For example, accounts of animal rights which seriously question all forms of commercial animal husbandry (e.g. Regan, 2004) or whether only/primarily current forms of intensive farming are morally objectionable. The latter positions often put more emphasis on the lived moral experiences of farmers that keep farmed animals and the cultural and ecological significance of certain ways of farming (e.g. Fairlie, 2010; Smaje, 2020). Important here is that both perspectives have the argumentative tools to critically question intensive animal agricultural practices, despite otherwise diverging commitments (Bossert, 2014).

While there is moral disagreement about the extent to which intensive animal agriculture needs to be reduced, it is also the case that this primarily involves what has been termed 'factory farming' methods involving large-scale 'farms' where everything is geared towards efficiency and increasing outputs by exerting considerable control over animal bodies, which is made possible by large-scale feed crop production and fossil fuels. When developing a theory of justice to 'domesticated' animals in our current 'non-ideal' world we first need to address the most pressing injustices to animals which, in turn, involves the elimination of obvious instances of animal suffering (Garner, 2013). In practice, that implies the rejection of factory farming methods and as a consequence a considerable de-growth of animal agriculture.

Such a claim is, of course, in tension with some perspectives advocating the above-mentioned intensification of food production with ever more efficient feed conversion ratios. However, in the context of claims about de-growing 'animal' agriculture the target of efficiency changes towards efficient agriculture 'as a whole'. Putting energy efficiency in the context of avoiding feed conversion losses in the first place ultimately aims at reducing future scarcity of food (with the aim of food security) and of land (with the aim of habitat conservation, biodiversity protection and a more equitable sharing of limited environmental goods).

Secondly, in contrast to the abstract point about energy efficiency, the land in question is very concrete, referring to actual spaces where people and nonhuman beings lead their lives. What is at stake is usually referred to as cultural landscapes, but we find Val Plumwood's 'collaborate' or 'interactive' landscapes (2006: 125) a more apt description. Her terms acknowledge explicitly the human (and, importantly, not only Western) and nonhuman biotic and abiotic influences that shape the land. So, on the one hand, land matters for human and nonhuman life, while, on the other hand, it remains scarce comparatively to the different needs of humans and nonhuman beings. This, in turn, generates a situation where we need to find a 'just compromise' between different demands of justice, if we take our commitment to intra-human and interspecies justice seriously (Wienhues, 2018, 2020). While the needs of humans and nonhuman beings often conflict, it is also the case that intensive animal agriculture intensifies scarcity of land (relative to all needs) and therefore aggravates conflicts between the satisfaction of different needs and interests. Accordingly, reducing intensive animal agriculture and thereby reducing its land-use could alleviate intra-human and interspecies global injustices in several regards, which can be illustrated briefly by the following two non-exhaustive examples.

Firstly, in the context of interspecies justice, the reduction of intensive animal agriculture can alleviate injustices to 'wild' nonhuman beings by reducing the pressure on their habitats created by conversions into agricultural land. This would be a way of doing less (distributive) injustice to nonhuman beings, if these have entitlements to an adequate amount of habitat, and under the assumption that particularly for animals with large territories agricultural developments in their habitats are putting a strain on their ability to survive (for a general point on animal rights and habitat for wild animals see Donaldson and Kymlicka, 2011). That is particularly relevant against the background that human needs can be satisfied adequately by diets of (at least) low meat and dairy intake (e.g. Willett *et al.,* 2019). Animal agriculture's biophysicality implies a comparatively high land-use which indicates that there is little non-anthropocentrically-based reason to maintain intensive forms of animal agriculture (which make the fast-growing global meat consumption possible; Weis, 2013), particularly in the light of the current high rate of species extinctions of 'wild' nonhuman beings as well as the suffering involved in industrial farming methods.

Secondly, feed crops are grown in order to primarily satisfy a demand for meat and dairy of relatively wealthy people whose consumption not only exceeds sustainable ecological footprints (Weis, 2013), but what could also be considered a just share of the Earth's resources when looking at it from an (anthropocentric) intra-human justice perspective alone. This kind of production and consumption can clearly be problematised if it means that it makes it more difficult for other people to live satisfying lives, for example, when considerable pressure is put on ecosystems and land health in their community. The globalised system of intensive animal agriculture allows for a problematic distancing of its benefiters from some of its environmental consequences. Accordingly, concepts such as 'ecological debt' (e.g. internationally of a state X to a state Y; Goeminne and Paredis, 2010) due to 'ecologically unequal exchange' (Martinez-Alier, 2002) can further illustrate the international dimension of this intra-human justice problem.

## Conclusion

Both of the points on efficiency and land-use leave open how much reduction is possible and desirable (which depends on empirical factors as well as one's moral positioning on animal farming in general). Yet, a defensible planetary justice perspective will require this reduction to be significant. The aims of biodiversity conservation and interspecies justice, on the one hand, and the aims of human food security and intra-human justice on the other, align on the social-ecological problems raised by (intensively kept) farmed animals. While we recognise that other conflicts between these different aims remain, their convergence in the specific context of animal-sourced foods' abundance is important for building political alliances and highlights the central place this question should receive in visions about just conservation. For instance, different positions involved in the HE controversy briefly introduced in the beginning should be able to agree on this goal, despite different argumentative routes for reaching this practical recommendation (whether seeing the route problem in capitalist economic structures and linked intra-human injustices, or highlighting the need of extensive areas for nonhuman habitat protection) and while retaining other substantial empirical and normative disagreement.

## References


Baxter, B. (2005). *A Theory of Ecological Justice*. Taylor and Francis e-library.

Bossert, L. (2014). Tierethik. Die verschiedenen Positionen und ihre Auswirkungen auf die Mensch-nichtmenschliches Tier-Beziehung. *Nachhaltige Lebensstile. Metropolis-Verlag, Marburg*, pp.32-57.

Büscher, B., Fletcher, R. (2019). Towards Convivial Conservation. *Conservation and Society* 17, 283-296.

Büscher, B., Fletcher, R. (2020). *The Conservation Revolution: Radical Ideas for Saving Nature Beyond the Anthropocene*. La Vergne: Verso.

Büscher, B. *et al.* (2017a). 'Half-Earth or Whole Earth? Radical Ideas for Conservation, and Their Implications'. *Oryx*, 51(3): 407-410.

Büscher, B. *et al.* (2017b). Doing Whole Earth justice: a reply to Cafaro *et al. Oryx* 51, 401-401.

Cafaro, P. *et al.* (2017). Letter: If we want a whole Earth, Nature Needs Half: a response to Buescher *et al. Oryx* 1-1.

Cochrane, A. (2018). *Sentientist Politics: A Theory of Global Inter-Species* Justice, Online Edition. ed, Sentientist Politics. Oxford University Press, USA – OSO.

Dinerstein, E. *et al.* (2017). An Ecoregion-Based Approach to Protecting Half the Terrestrial Realm. *BioScience* 67, 534-545.

Donaldson, S., Kymlicka, W. (2011). *Zoopolis: A Political Theory of Animal Rights*. Oxford University Press, Oxford.

Dryzek, J.S., Pickering, J. (2019). *The Politics of the Anthropocene*. Oxford University Press, Oxford.

Fairlie, S. (2010): *Meat: A Benign Extravagance*. Permanent Publications: White River Junction.

Foley, J.A. *et al.* (2011). Solutions for a cultivated planet. *Nature* 478, 337-342.

Garner, R. (2013). *A Theory of Justice for Animals: Animal Rights in a Nonideal World*. Oxford University Press, Oxford.

Godfray, H.C.J. *et al.* (2010). Food Security: The Challenge of Feeding 9 Billion People. *Science* 327(5967): 812-818.

Goeminne, G., Paredis, E. (2010). The concept of ecological debt: some steps towards an enriched sustainability paradigm. *Environment, Development and Sustainability* 12, 691-712.

Hickey, C., Robeyns, I. (2020). Planetary justice: What can we learn from ethics and political philosophy? *Earth System Governance* 100045.

Kolasi, E. (2019). Energy, Economic Growth, and Ecological Crisis. *Monthly Review* 71(2): 29-46.

Kopnina, H., *et al.* (2018). 'The 'future of conservation' debate: Defending ecocentrism and the Nature Needs Half movement.' *Biological Conservation* 217, 140-148.

Martinez-Alier, J. (2002). Th*e Environmentalism of the Poor: A Study of Ecological Conflicts and Valuation*. Edward Elgar Publishing.

Mehrabi, Z. *et al.* (2018). The challenge of feeding the world while conserving half the planet. *Nature Sustainability* 1: 409-412.

Napoletano, B.M., Clark, B. (2020). An Ecological-Marxist Response to the Half-Earth Project. *Conservation and Society* 18, 37.

Nijdam, D., *et al.* (2012) The price of protein: Review of land use and carbon footprints from life cycle assessments of animal food products and their substitutes. *Food Policy* 37(6): 760-770.

Norton, B.G. (1991). *Toward Unity Among Environmentalists*. Oxford University Press, New York.

Phalan, B., *et al.* (2011). Reconciling Food Production and Biodiversity Conservation: Land Sharing and Land Sparing Compared. *Science* 333, 1289-1291.

Plumwood, V. (2006). The Concept of a Cultural Landscape: Nature, Culture and Agency in the Land. *Ethics and the Environment* 11, 115-150.

Poore, J. and Nemecek, T. (2018) Reducing food's environmental impacts through producers and consumers. *Science* 360(6392): 987-992.

Regan, T. (1984). *The Case for Animal Rights*. Routledge, London.

Schleicher, J., *et al.* (2019). Protecting half of the planet could directly affect over one billion people. *Nature Sustainability* 2, 1094-1096.

Schlosberg, D. (2007). *Defining Environmental Justice: Theories, Movements, and Nature*. Oxford University Press, Oxford.

Searchinger, T.R. *et al.* (2018). *Creating a sustainable food future. A menu of solutions to feed nearly 10 billion people by 2050. Synthesis Report*. Washington, DC: World Resources Institute.

Smaje, C. (2020). *A Small Farm Future*. Chelsea Green Publishing: White River Junction and London.

Weis, T. (2013). *The Ecological Hoofprint: The Global Burden of Industrial Livestock*. London and New York: Zed Books.

Wienhues, A. (2018). Situating the Half-Earth proposal in distributive justice: Conditions for just conservation. *Biological Conservation* 228, 44-51.

Wienhues, A. (2020). *Ecological Justice and the Extinction Crisis: Giving Living Beings their Due*. Bristol University Press, Bristol.

Willett, W., *et al.* (2019). Food in the Anthropocene: the EAT–Lancet Commission on healthy diets from sustainable food systems. *The Lancet*.

Wilson, E.O. (2016). *Half-Earth: Our Planet's Fight for Life*. Livelight Publishing Corporation, New York.

# 43. Whose interest matters when regulating animal welfare? A Swedish case study

*F. Lundmark Hedman[1*], C. Berg[1] and M. Stéen[2]*
*[1]Department of Animal Environment and Health, Swedish University of Agricultural Sciences, P.O. Box 234, 53223 Skara, Sweden; [2]The Swedish Centre for Animal Welfare, P.O. Box 7053, 750 07 Uppsala, Sweden; frida.lundmark@slu.se*

## Abstract

Sweden is often mentioned as a leading country when it comes to animal welfare (AW). However, substantial changes have been made in the national AW legislation lately, resulting in criticism. This study analysed how the Swedish Central Competent Authority (CCA) has handled AW research findings, and changes in governmental and stakeholder interests when revising the national legislation between 1988 and 2019. AW legislation amendments for cattle, pigs and horses were scrutinized as well as other governmental documents. The AW regulations have been modified several times since 1988. The main motive presented by the CCA's for an alteration of the legislation was to improve AW. However, the CCA's use of scientific results were surprisingly scarce and inconsistent. The second most common motive was that changes were made in the interest of the owners/industry. The stakeholders' major opportunity to influence new AW regulations turned out to be through an early CCA consultation. After referral management, the CCA did generally not make substantial changes in relation to the original proposal, even if a majority of referral bodies were critical. The industry had the biggest impact, while the opinions from AW, animal rights, and consumer NGOs never per se induced changes. The Government and the Ministry are involved not only in the AW Act and Ordinance but indirectly also at the CCA level. The CCA regulations did mirror the Government's views and wordings, focusing on improving the AW level and making detailed regulations before 2007, and after 2007 recommending that the CCA regulations' AW level shall be maintained but less detailed, more flexible and goal oriented. The governmental directions to the CCA are, and have been, anything but unambiguous. The regulations can hence be perceived as inconsistent. We hence, argue that the politicians need to provide clearer directions and instructions to the CCA.

**Keywords:** amendments, legislation, politics, stakeholders

## Introduction

Sweden is often mentioned as a leading country when it comes to animal welfare (AW). The Central Competent Authority (CCA) in Sweden responsible for issuing the national AW regulations is the Swedish Board of Agriculture (SBA), which is an independent para-governmental entity that can issue binding regulations under the AW Act and Ordinance. This has been the case since 1991 except for the years of 2003-2007, when the Swedish Animal Welfare Agency (SAWA) was responsible. Some of the recent amendments made in the CCA regulations have been criticized, being perceived as relaxing the AW requirements. One criticized amendment involved allowing an earlier weaning age for piglets (i.e. before 28 days of age), a change made to oblige the pig producers.

This study is a part of a larger project, where the development of the Swedish AW legislation from 1988 until 2019 was investigated. The aim of this particular sub-study was to analyse how the CCA handled AW research findings, changes in governmental and stakeholder interests when amending the national regulations.


EurSafe 2021 – DOI 10.3920/978-90-8686-915-2_43, © Wageningen Academic Publishers 2021

## Material and methods

The CCA's AW regulations between 1988 and 2019 on cattle, pigs and horses and their amendments were scrutinized: a total of 28 regulations. The included regulations covered animal management and husbandry systems at farm level, i.e. not transport or slaughter. This resulted in a selection of 77 requirements meeting one or more of the following criteria; the suggested requirement was completely new, it meant major changes for animal husbandry and AW, it generated many comments from the stakeholders, and the requirement was changed considerably between proposal and the final amendment. Each amendment was then studied in relation to how the CCA have motivated the suggested change, what the stakeholders' views were, and how the CCA made alterations depending on the stakeholders' opinions. In addition to this, appropriation directions, assignments and instructions to the CCA were analysed, as relevant governmental investigations during this period, including the governmental bills behind the AW Act of 1988 (SFS, 1988: 534) and of 2019 (SFS, 2018: 1192).

## Results

### *Amendments in animal welfare regulations for cattle, pigs and horses*

The CCA has updated the AW regulations for cattle, pigs and horses several times since 1988; 20 times for cattle, 16 times for pigs and 11 times for horses. Prior to 1989, there were no binding regulations concerning these species, and animal owners had to relate only to the CCA's general advice. Comprehensive AW regulations for horses were not set until 2003. The most comprehensive amendments concerning horses were made in 2007, when the requirements were compiled in a specific regulation for horses, separated from the one for farm animals. In these, SAWA introduced additional requirements on horses' need for social contact, daily exercise in paddocks (or similar area), limitations of the time horses could be tied-up in stalls, and a ban on building new stables with tie-up stalls. In 2018 the SBA revised the horse regulations more structurally than contextually in order to make them more goal oriented, flexible and less detailed. For cattle, amendments have for example been related to details concerning the summer pasture (1995, 1997, 2016), and bans of new building of tie-up stalls (2007). For pigs the most comprehensive amendments were made in 2006, with a focus on the need for straw for all pigs and bedding material for sows before farrowing.

### *Influences from the political sphere*

The CCA in Sweden is a detached expert authority to the Government. Nevertheless, the CCA's activities is governed by the Government through appropriation, directions, instructions, assignments and governmental investigations that sets the framework for the CCA, both contextually and financially. The overall political goal has been, and still is, that the CCA shall ensure good AW. Throughout the years, the CCA has had instructions related to the promotion of an efficient and competitive agricultural industry. However, the focus on improving the competitiveness of the farming industry has increased during the last decade. Before 2007, the CCA was tasked by the government to improve AW, develop stricter regulations and prevent AW problems. Since 2007 the tasks have been to maintain the AW level, making the regulations less detailed, more flexible, goal oriented and cost effective.

In 2015, the Government presented an investigation of how to improve Swedish farmers' competitiveness (SOU, 2015: 15). The conclusion was that the EU level of AW should be the base for the Swedish AW legislation. If any additional national legislation was implemented it should be science-based, and balanced between good AW and the industry's competitiveness with other EU member states. In 2017, the Government presented a national food strategy with goals supporting an AW legislation that supported a competitive, sustainable and increased food production, with requirements that were flexible and goal oriented, without lowering the present AW level (Swedish Government, 2017).

The CCA's regulations mirror the intentions and level of ambition in the AW Act. The AW Acts from both 1988 and 2019 have a preventive purpose, stating that animals shall be protected against unnecessary suffering and shall be able to exhibit natural behaviours. The AW Act of 2019 extends even further with a new aim, ensuring good AW and respect for the individual animal. Hence, the animals shall not only live a life absent of poor welfare, but experience good welfare (Swedish Government, 2018). The bill states that animals are sentient beings and have an intrinsic value, independently of the human purpose of keeping them. However, other interests beside the animals' needs to be taken into account when new regulations are written, and animals are legal objects even if attributed to more than just a material value (Swedish Government, 2018).

### *The CCA's written motivation of amendments*

The most common general motives provided by the CCA at the amendment of a regulation were new knowledge, adaptation to EU legislation or the AW Act, and a need for clearer and more easily interpreted requirements not in disfavour for the farming industry. When SAWA was the responsible CCA, improving AW and paying respect to the animals' natural behaviours, was also mentioned as general motives. Lately, maintaining an acceptable AW level without reducing it has been in focus, i.e. not improving it.

The CCA did most often motivate the amendments with improving animal welfare (71%), or to facilitate for the farmers/horse owners (38%). The CCA did rarely refer to scientific knowledge. For 16% of the suggested requirements the word '*research*' were used, but without any further explanations or references. For another 16%, literature reviews were made; however, these were not mentioned in the written motivation to the amendments. For one requirement only, the motivation to an amendment was substantiated with scientific support.

### *Influences from stakeholders*

The CCA usually consulted stakeholders in advance when planning a regulation amendment or update, giving them an opportunity to present their views. Stakeholders representing the industry, i.e. farmer and horse organisations, were always present at these pre-consultations, and half of the times, they were the only stakeholders present. Other stakeholders occasionally represented were for example the regional competent authorities, researchers, and AW or animal rights NGOs. Normally, the suggestion for a new regulation is sent on a referral round for formal consultation, and for the regulations in this study, 31-79 stakeholders gave their opinions. The consultation respondents were generally positive to 31% of the suggested requirements, and negative or sceptical to 40% of the suggestions, with stakeholders giving opinions on the horse regulations being the most negative. The responses provided varied in quality. In some cases (18%) it was impossible to understand if the stakeholders agreed or disagreed with the suggested change. After the referral round, 32% of the requirements underwent substantial changes compared with the suggested amendment, while 61% were not changed at all. It was usually not clear if the CCA have prioritized some specific stakeholders' opinions before others. However, 14% of the requirements were changed due to opinions from the industry, 5% were changed due to opinions from regional control authorities and 4% were changed due to opinions from researchers. The opinion of AW, animal right or consumer NGOs did never *per se* induce changes.

In our analysis, the industry appeared to be more consistent compared to other stakeholders. For example, the regional CAs often had divergent opinions concerning a proposed amendment. The AW

NGOs neither agreed with each other consistently. Stakeholders were sometimes quite inconsistent in their own answer to the CCA. For example, the horse industry claimed they were positive to flexible and goal oriented regulations, but simultaneously opposed most of them due to difficulties to reach uniform interpretations. Researchers were on one hand critical to keeping horses in tie-up stalls, but on the other hand opposed a ban on this husbandry system.

## Discussion

### *The Government has a large impact*

The regulations' content and structure mirrors the Government's directions. Hence, even if the CCA is a detached authority the governmental politicians play a strong role in the development of the AW regulations at the CCA level. During recent years, the Government has requested more goal-oriented, more flexible and less detailed regulations, with focus on a competitive and increasing domestic food production. Hence, the possibility to wean piglets before 28 days of age was introduced to enable a more efficient production and meet the requests of the industry. Some of the changes in the pasture requirements for dairy cows were also based on requests from industry. The up-to-date horse regulations were made more flexible and goal oriented, on request from the horse-related NGOs, although these organisations then disapproved of most such requirements in the referral round, claiming the increased risk for unequal assessments and interpretations between horse owners and AW inspectors. A concern that is justified by literature (Forsberg, 2011; Johansson, 2006; Viksten *et al.*, 2016).

Today's Governmental directives to the SBA are partly conflicting. The AW regulations should for example be 'flexible but predictable, easy to interpret and enable uniform assessments'. The regulations shall 'prevent poor welfare but focus on the outcomes' (i.e. be goal oriented), and be 'less detailed' but 'detailed'. The regulations shall also mirror the AW Act giving the animals a 'good' AW, which can be interpreted as a higher basic level compared to just avoiding poor welfare, and hence an increased general level of AW compared to previous versions of the AW Act. The regulations shall also be based on scientific knowledge and experience. Simultaneously, other official documents states goals as 'acceptable' and 'maintained' AW level (i.e. not improved AW), and that the AW level must be balanced against the industry competitiveness.

### *Scientific knowledge*

The CCA's reference to scientific knowledge was surprisingly scarce, even when such knowledge was obviously present. Scientific knowledge was also sometimes handled quite inconsequently by the CCA, leading to the impression that the CCA used knowledge in a way that suited them. For example, in the up-dated horse regulation the CCA motivated the elimination of some requirements with the lack of scientific outcomes, even if such outcomes did exist, and simultaneously introduced new requirements that definitely lacked scientific basis.

The AW legislation sets the level of the legally acceptable, i.e. a minimum level. Therefore, when amendments are made, one minimum level replaces another. Not all requirements can be solely based on scientific evidence. For example, science can show that access to pasture for dairy cows reduce the risk of poor welfare (EFSA, 2009), but science cannot show that spending at least six hours per day on pasture is the perfect minimum level, which is required in the Swedish legislation. Furthermore, scientific results are not always unambiguous. A husbandry system may have both advantages and disadvantages (Bernardi *et al.*, 2009), and different professions may interpret a result differently (Bracke *et al.*, 2008). Hence, also other factors than science determine the content and structure of an AW regulation, such as economy, culture, traditions, ethics and the society's expectations (Croney and Millman, 2007;

Lundmark *et al.*, 2014; Sandøe *et al.*, 2004; Yeates *et al.*, 2011), and the Government expect the CCA to make trade-offs between these.

### *Stakeholder impact*

Our results strongly indicate that it is easier for a stakeholder (e.g. the industry) to influence a new regulation in the beginning of the process, than later during the referral round. Among the stakeholders, the industry seemed to have the largest impact on the AW regulations. The industry does represent the persons expected to comply with the new regulations, and hence, it is not surprising that CCA anchors their amendments with them. However, this should not exclude other stakeholders from being invited to an early consultation. The inconsistent stakeholder replies, together with the CCA's (i.e. SBA) statement: 'we only make changes if new knowledge to the issue could be presented', may have contributed to the low level of changes in the regulations after the referral rounds. Even when a majority of the stakeholders were critical (and united), big changes were uncommon, indicating that the CCA had often made the trade-offs before the referral round.

The finding that the horse industry consistently has been critical to any new AW regulations – milder or stricter – is worth dwelling upon, especially since they have always been involved in the early consultations. The SBA stated in the formal consultation in 2018 that 'the industry's views have largely been included'. Nevertheless, this was not the industry's own experience. Is the industry's criticism a result of a perceived rapid development of horse regulations? Or does the criticism mirror the diverse group of people that the horse industry NGOs represents, from hobby pony owners to elite riders within different disciplines.

The stakeholders presented their arguments from their own perspectives, i.e. the regional CA commented on controllability and applicability, the industry commented from the farmers and horse owners' interests, and the AW NGOs commented with the animals in focus. However, there is not necessarily a conflict between the different views and stakeholders. On the contrary, a stakeholder having one perspective in focus does not mean that other perspectives are unimportant *per se*.

## Conclusions

The CCA's handling of scientific results can be improved, and so can the stakeholders' responses to suggested amendments. The CCA generally avoided making substantial changes even when a majority of stakeholders were critical to the adjustments. The industry appears to be the stakeholder with the largest impact, whose views are taken into account during the early pre-consultations. The opinion of AW, animal right or consumer NGOs has never *per se* induced changes during the referral round. The Government does play a huge role in the development of AW regulations, even if the Swedish CCA shall be formally detached from political influence in matters of fact. The CCA's AW work mirrors the Government's views and wordings and throughout the years, the regulations' content and structure have change accordingly. The governmental directions to the CCA are, and have been ambiguous. Hence, it is not surprising if the CCA's handling of AW matters in general, and handling of AW regulations in particular, can be perceived as inconsistent. Therefore, we argue that the politicians need to give the CCA clearer short- and long-term directions.

## References

Bernardi, F., Fregonesi, J., Winckler, C., Veira, D.M. von Keyserlingk, M.A.G. and Weary, D.M. (2009). The stall-design paradox: Neck rails increase lameness but improve udder and stall hygiene. Journal of Dairy Science 92: 3074-3080.

Bracke, M.B.M., Edwards, S.A., Engel, B., Buist, W.G. and Algers, B. (2008). Expert opinion as 'validation' of risk assessment applied to calf welfare. Acta Veterinaria Scandinavica 50.

Croney, C.C. and Millman, S.T. (2007). Board-invited review: The ethical and behavioral bases for farm animal welfare legislation. Journal of Animal Science 85: 556-565.

EFSA (2009). Scientific Opinion on the overall effects of farming systems on dairy cow welfare and disease. The EFSA Journal 1143: 1-38.

Forsberg, E-M. (2011). Inspiring Respect for Animals Through the Law? Current Development in the Norwegian Animal Welfare Legislation. Journal of Agricultural & Environmental Ethics 24: 351-366.

Johansson, V. (2006). Tillsyn och effektivitet [Control and efficiency]. Boréa Bokförlag: Umeå.

Lundmark, F., Berg, C., Schmid, O., Behdadi, D. & Röcklinsberg, H. (2014). Intentions and Values in Animal Welfare Legislation and Standards. Journal of Agricultural & Environmental Ethics, 27, 991-1017.

Swedish Government (2017). Bill 2016/17:104. En livsmedelsstrategi för Sverige – fler jobb och hållbar tillväxt i hela landet [A national food strategy]. Ministry of Enterprise and Innovation. Stockholm. Sweden.

Swedish Government (2018). Bill 2017/18:147. Ny djurskyddslag [A new animal welfare act]. Ministry of Enterprise and Innovation. Stockholm. Sweden.

SandØe, P., Forkman, B. and Christiansen, S.B. (2004). Scientific uncertainty – how should it be handled in relation to scientific advice regarding animal welfare issues? Animal Welfare 13: S121-S126.

SOU 2015:15. Attraktiv, innovativ och hållbar – strategi för en konkurrenskraftig jordbruks- och trädgårdsnäring [Attractive, innovative and sustainable – strategy for a competitive agricultural and horticultural industry]. Swedish Government Official Report. Elanders Sverige AB, Stockholm, 185 pp.

Viksten, S.M., Visser, E.K. and Blokhuis, H.J. (2016). A comparative study of the application of two horse welfare assessment protocols. Acta Agriculturae Scandinavica, Section A – Animal Science 66: 56-65.

Yeates, J.W., Röcklinsberg, H. and Gjerris, M. (2011). Is welfare all that matters? A discussion of what should be included in policy-making regarding animals. Animal Welfare 20: 423-432.

# 44. Citizen views on welfare and rights of farmed animals in food production

*M. Niva*[1*] *and J. Mäkelä*[2]
[1]*Department of Economics and Management, Faculty of Agriculture and Forestry, University of Helsinki, P.O. Box 27, Latokartanonkaari 5, 00014 University of Helsinki, Finland;* [2]*Department of Education, Faculty of Educational Sciences, University of Helsinki, P.O. Box 9, Siltavuorenpenger 5, 00014 University of Helsinki, Finland; mari.niva@helsinki.fi*

## Abstract

Animal production is increasingly contested in food-related discourses and practices due to increasing knowledge about animal welfare, the rise of the animal rights movement, and ecological problems exacerbated by animal production. However, these developments have not resulted in decreases in consumption of animal products even though the number of vegetarians and vegans is gradually rising in many Western countries. In this paper we analyse the meanings, justifications, and ethical perspectives related to animal rights and welfare by omnivores, vegetarians and vegans. The data is based on a survey conducted in co-operation with the Finnish newspaper Helsingin Sanomat, including an open-ended question asking how, in the respondent's opinion, animal welfare and rights should be taken into account in food production. In total 22,803 people completed the survey. The data for this paper comprise 5,933 answers to the question above. The results of qualitative content analysis suggest that animal welfare was regarded as important generally in all groups. However, omnivores' comments relied on a multitude of ethical perspectives and presented a range of views varying from stressing good living conditions and avoidance of (unnecessary) suffering to denial of animal rights. Vegans' comments strongly questioned current animal production. Vegetarians' arguments often used hybrids of utilitarian and animal rights views and stressed welfare and species-specific behaviours, although the value of animal life as such was also recognised. The results evidence contested views on the future of animal production.

**Keywords:** animal rights, animal welfare, qualitative research, omnivores, vegetarians, vegans

## Introduction

In affluent societies animal production is increasingly contested in food-related discourses. This is related to several developments in which, first, the rights and welfare of farmed animals particularly in intensive farming have been questioned. As more is known about animal sentience and mind, it has become increasingly difficult to defend poor living conditions of farmed animals (e.g. Kupsala, 2019). Second, climate change and other ecological problems as well as the health impacts of excessive meat consumption in the global north raise increasing concerns about the ecological, economic and social consequences of animal-based food production.

In affluent countries per capita consumption of meat seems to have reached a saturation level, but with growing populations, global production and consumption of meat, particularly poultry, is increasing (e.g. OECD and FAO, 2020). Meat and other animal products are highly valued, and giving up habitual use of animal-based products is not easy (Niva *et al.*, 2017). In Western countries meat as the centre of the meal is appreciated as tasty and healthy, and is assigned meanings of power and masculinity (e.g. De Bakker and Dagevos, 2012). Earlier quantitative research indicates that people following omnivore, vegetarian or vegan diets apply divergent animal ethics approaches, and that omnivores tend to approach animals from a more mixed ethical perspective compared to vegans and vegetarians (Lund *et al.*, 2016).

In this paper we examine omnivore, vegetarian and vegan perspectives on rights and welfare of farmed animals from a qualitative perspective. We analyse how people following the three diets give meanings and conceptualise rights and welfare issues. In the concluding section, we apply the concepts of justifications and excuses ('accounts', see Scott and Lyman, 1968) to examine what kinds of explanations the respondents used in defending animal production and challenging the paying of attention to animal welfare and rights. In addition, we take a look at the roles that the five ethical perspectives differentiated by Sandøe and Christiansen (2008) – contractarianism (focusing on human self-interest), utilitarianism (taking into account the interests of all affected sentient beings), animal rights view (emphasising respect for life), relational view (stressing human-animal relations), and respect for nature view (protecting species and ecosystems) – play in the accounts presented. We also apply the framework for 'good animal life' by the same authors (Sandøe and Christiansen, 2008) to see how the accounts make use of the ideas of hedonism/pleasure (that good animal life entails as many positive and as few negative experiences for the animal as possible), perfectionism (that the animals are able to realise species-specific potential and to 'live in accordance with their nature', *ibid.*: 36) and preference theory (that the preferences of the animal are satisfied).

## Materials and methods

The data of the study is based on a survey conducted in Finland, carried out in co-operation with the largest Finnish newspaper Helsingin Sanomat, which invited readers to respond to a survey including both structured and open-ended questions related to diets, food consumption and food-related motives and attitudes. The survey was open in March 1-21, 2020, and 22,803 readers completed it. The respondents were not representative of the Finnish population: 72% were women, 67% had at least Bachelor level education and 46% lived in the capital area. One of the open-ended questions asked: 'The use of animals in food production has recently been criticized in public discussion. In your opinion, how should animal well-being and rights be taken into account in food production?' (it was not obligatory to give an answer). It should be noted that the respondents were not provided an explanation about what 'welfare' and 'rights' mean, i.e. the respondents relied on their own understandings of the concepts. This enabled us to examine how the respondents understood the concepts and what kinds of meanings the concepts were assigned.

Respondents' diets were inquired by asking whether they follow a special diet with response options: (1) 'I don't follow at the moment'; (2) 'I try go to this direction'; (3) 'I follow quite closely'; and (4) 'I follow closely'. In this paper, three diet variables are used to categorise the respondents into vegans, vegetarians and omnivores: 'vegan (no animal products at all)', 'vegetarian diet with eggs and/or milk products', and 'diet with no red meat'. Those who selected response option 4 for the vegan diet were defined as vegans, and those who selected option 4 for the vegetarian diet were defined as vegetarians. Finally, those who selected option 1 for vegan and vegetarian diets as well as for a diet without red meat were defined as omnivores. This allowed us to select 'strict' followers of each diet into the analysis. The data used in this paper consist of 5,933 responses to the open question above, of which 3,397 were from omnivores (57%), 1,458 from vegetarians (25%) and 1,078 from vegans (18%).

The length of the responses varied from a few words to a whole paragraph with detailed reflection on animal welfare and rights. A typical length of a response was 1-3 sentences, with vegans (on average 28 words) and vegetarians (27 words) writing somewhat longer comments than omnivores (20 words). We analysed the responses with qualitative content analysis, paying attention to the themes and arguments brought out in each response, and reflecting on them as 'accounts' and as demonstrations of different ethical perspectives and ideas of good animal life. Due to the variety of themes brought out in the responses and lack of space, we describe the most prevalent themes in more detail and give less emphasis to the themes that were mentioned more infrequently. In the following we present the findings of the analysis for each group separately, and finally discuss the differences and similarities of the groups in terms of justifications, ethical perspectives and ideas of good animal life.

## Results

The comments by all three groups included not only direct answers to what was asked ('how should animal welfare and rights be taken into account in food production?') but also comments about the importance of the issue, the acceptability of using and eating animals, the place of animals in society, the subjectivity and rights of animals, and the current state of animal welfare and rights in Finland. There were substantial differences in how these issues were approached in the three groups, and the differences were largely related to what kind of ethical perspective can be seen as the basis of the views.

### *Omnivores: reconciling animal welfare and the eating of animal products*

Among omnivores, a variety of meanings were assigned to animal welfare. 'The importance of welfare issues' was acknowledged in a large majority of the responses. The comments were often declaratory or expressed a deontic modality, stating that animals must/should be treated 'well', 'humanely', 'with respect', or 'properly': 'Animals need to have a good life and they must be treated well to the end' (60-year-old woman).

Many brought out 'elements of what was regarded as good treatment', including, e.g. 'good', 'appropriate', 'natural' or 'sufficient' living conditions, sufficiency of space, freedom of movement, possibility of species-specific behaviour, appropriate or 'natural' food (indeed, instead of 'feed', the respondents most often wrote about 'food' or 'nutrition' for the animals), possibility for grazing, good health, proper medication, short transport distances, painless slaughtering and avoidance of (unnecessary) suffering. Industrial/mass production was problematized, and small-scale, family farming was supported. Organic production was mentioned as a method enabling better welfare; although it was also criticised, often without specifying why. It was noted that better welfare induces costs, and some were willing to pay more if there was a certification or a product label guaranteeing animal welfare (apart from organic production, at the moment such a welfare label scheme does not exist in Finland).

Many of the omnivores' comments on the importance or characteristics of animal welfare were supplemented with a note about the 'benefits accrued for humans' of such measures: the welfare of the animal was believed to lead to: (1) better productivity and success for the farmer ('An animal that is well produces better', wrote a 61-year-old man); or (2) better taste, healthiness and quality of the product ('A well-treated animal tastes better', noted a 31-year-old man). Here, animal welfare was not assigned value in itself but was rather presented as something serving human self-interest and ultimately benefiting the human community, maximising the pleasure and well-being of the eater or the financial gain of the producer. However, it was also noted that only farmers who are well themselves are able to take good care for the animals, and thus the farmers need a decent compensation for their work.

It was often explicitly remarked that farmed animals are 'meant for human consumption'; a state of affairs which the omnivores rarely questioned. Using animals in food production was regarded as acceptable provided that animals are properly taken care of: '[t]his is about farmed animals, and they exist in order to produce food for people. However, their suffering must be minimised' (39-year-old man). A related normalisation of animal production could be seen in presenting meat-eating as 'natural', 'age-old', or part of the 'cycle of nature', and in describing 'humans as "meat-eaters" or "omnivores", I think animal-based food is naturally part of the human diet' (55-year-old man). Often such comments also reflected on animal welfare, but others left the welfare implications of the 'naturalness' of eating animals open.

In omnivores' comments, the focus was indeed on welfare issues, and 'a minority of the responses explicitly mentioned animal rights'. In these comments, animal rights were conceptualised largely in terms of a right to good treatment and good life: 'Animals must have a right to welfare' (32-year-old woman). On the other hand, animal rights were also questioned outright by maintaining that 'animals do not have rights' (several respondents). 'Humanisation' of animals was criticised, and animals were differentiated from humans.

A very prominent thread in the comments included 'constructing an opposition between animal production in Finland and elsewhere'. Finnish animal production was praised as 'good', 'proper', 'ethical' or 'top-quality', whereas the treatment of animals in (many) other countries was regarded as 'poor'. Generally these comments were presented as factual propositions suggesting that the level of animal welfare and the regulatory mechanism governing it are already high enough in Finland and consequently nothing more is needed: 'They [animal welfare and rights] are already well taken into account!' (48-year-old man). Despite its prominence, the view of the properness of Finnish farming practices was not unanimously shared: an opposite perspective suggested that the current 'laws, requirements, control mechanisms and sanctions are not sufficient' to guarantee animal welfare.

Other, minor themes included, e.g. support for reducing animal-based production, descriptions about one's own efforts to eat more ethically (e.g. buying organic, local or domestic food, eating game, or reducing meat eating); comments on the polarisation of the public discourse; and reproaching the critics of animal production, such as vegans, urbanites 'estranged' from food production, and the media, for exaggerating the problems of animal welfare and rights, and for creating unnecessary controversies.

#### *Vegetarians: ambivalence in accepting animal production*

To some extent vegetarians shared many views with omnivores. Animal welfare, its regulation and control were considered to be important and it was stressed that animals must be treated well, 'humanely' or 'with respect', with possibility for species-specific behaviours. Often, however, such comments showed 'a more critical tone compared to omnivores': 'Present day intensive production is completely wrong and extremely cruel. [--] The living conditions of animals should be improved radically' (35-year-old woman). Animal production in general and intensive production in particular were heavily criticised. It was also noted that animal production is principally unethical, but if it cannot be totally given up, animal suffering must be minimised. Organic production and welfare certificates were supported.

Compared to omnivores, vegetarians' comments were more divided in terms of whether the use of animals in food production was regarded in principle acceptable at all. Some, consistently with the reported vegetarian diet, made a 'distinction between eating animals (meat) and eating the foods that animals produce (milk, eggs)', and accepted the latter but not the former. In some comments the perceived moral difference between killing for food and utilising what animals produce was explicitly brought out. Some regarded using animals in meat production as morally wrong, but accepted the eating of game. A few suggested that dairy cows and egg-laying hens should have a right to retire instead of being slaughtered. However, many 'questioned the right of humans to use animals in food production for any purposes, including milk and egg production'. This can be partly explained by the fact a large majority of those who reported to strictly follow a vegetarian diet also reported an interest in vegan diet either by following a vegan diet quite closely or by trying to go to this direction. Whereas among omnivores, it was not unusual to deny that animals have rights, many vegetarians, particularly those with interest in vegan eating, stressed animal rights as a basic principle that must guide food production. Some conceptualised rights as a right to welfare (similarly to omnivores), but more often they were seen as a right not be exploited by humans.

It was noted that animals are sentient beings, and that they should not be treated instrumentally as 'production machines'. It was suggested that 'the target should be gradually ending or at least considerably reducing meat/milk/egg production/eating'. Policies, e.g. stopping or reducing agricultural subsidies or inducing taxes on animal production, were supported to increase the price of meat and to turn meat a luxury food. At the same time, it was noted that farmers need to be supported to find new, plant-based production lines.

#### *Vegans: using animals in food production should be stopped or phased out*

Among the three groups, vegans were, unsurprisingly, most critical towards food production and the eating of animal-based foods. Most vegans started from the premise that since 'the oppression and exploitation of animals by humans cannot be accepted', their use in food production should be stopped. The 'right of humans to raise and kill animals for food was contested', particularly today, when animal-based foods are not needed for survival. It was insisted that 'animals should not be used in food production' and that 'animals should not be eaten'. Cultural preferences, habits and pleasure were not considered to justify the exploitation of animals. From this perspective, 'improving welfare is not enough'.

Others adopted the view that even though the use of animals is wrong, it is not likely that animal production can be totally closed down in the near future (or ever). From this perspective, 'many considered an incremental approach more realistic than a radical turn to plant-based farming and eating'. 'Animals should not be utilized at all in food production, but since there's a long way to go to that, the living conditions of farmed animals should be improved first' (27-year-old woman). Similarly to omnivores and vegetarians, measures for improving animal welfare (e.g. reform of subsidies, taxation, banning cruel practices, informing the public about the conditions at farms, etc.) were suggested, and it was noted that animal production must be reduced. However, the 'ideal' and 'utopia' was a 'vegan society'.

It was argued that animals should be respected as sentient beings, and that they are not supposed to be slaves or means of production but individuals with 'rights equal to humans or pets'. 'Animals are not here for us, but with us' (22-year-old woman). Some, however, considered it as acceptable to keep hens at home yard or a small number of animals in family farms in which good care can be assured. Some accepted hunting and sustainable fishing since they 'do not make animals factories' (27-year-old woman). A minority of the comments focused solely on animal welfare without commenting on the morality of eating animals.

## Discussion and conclusion

The results show that animal welfare gained support in all three groups, while rights were emphasised mainly by vegans and vegetarians. In terms of 'good life' (Sandøe and Christiansen, 2008), all groups supported both the pleasure principle and perfectionism: that through good care, farmed animals' negative experiences must be minimised and positive experiences maximised, and that animals must have a possibility for species-specific behaviours. Here it should be noted that vegans strongly stressed the giving up of animal agriculture and did not always take a stand on animal welfare. Accounts relying directly on preference-theoretical ideas were rarer, although comments stressing that animals should have a 'natural' life and that farming violates that right, as well as those condemning animal farming but accepting the eating of game can be interpreted to support the perfectionist view.

When looking at the justifications and the ethical perspectives applied we can find substantial differences between the groups. Particularly in omnivores' accounts, we can discern 'excuses and justifications', first, for eating meat, and secondly, for the idea that nothing in particular that needs to be done in food production to take animal welfare and rights into account. The main excuses and justifications for meat eating and use of animals in food production included depicting humans as natural meat eaters, presenting meat as a healthy food, denying animal rights, and emphasising the benefits of meat/animal production to Finnish agriculture and farmers. The accounts that challenged the need to pay attention to animal welfare and rights appealed to allegedly well-functioning regulation, responsible practices of Finnish farmers and good state of animal welfare, and similarly to above, denying animal rights. In Scott and Lyman's terms (1968), such accounts demonstrate both denial of victim ('animals have no rights'), denial of injury ('animals are already well treated'), appeal to loyalty ('Finnish farmers act responsibly'), appeal to defeasibility ('regulation works already well'), and appeal to a biological drive ('humans are meat-eaters', 'meat is healthy').

As regards the 'ethical perspectives' (Sandøe and Christiansen, 2008), the differences between the three groups are evident. Omnivores' comments relied on a multitude of ethical perspectives, including a contractual view (focusing on human interests and leaving the interests of other sentient beings aside), a utilitarian perspective (taking animals' interests into account and focusing on reduction of suffering), and a relational view (seeing that the relation between the farmer and the animal is characterised by both good care and eventual killing). Vegetarians often used hybrids of utilitarian and animal rights views, objecting the killing of animals but accepting the production of milk and eggs with very high levels of animal welfare. In our data vegetarian perspectives in many ways aligned with those of vegans, of whom most relied on animal rights perspective but allowed for a gradual process towards a vegan world. There was support in all three groups for a 'reform strategy' (Sandøe and Christiansen, 2008: 73) focusing on improving animal welfare, although in varying degrees; whereas a 'revolutionary change' with the target of giving up animal production, was supported by vegetarians and vegans.

Despite the fact that our data is not generalizable to the Finnish population, the results suggest that there is wide-spread support to paying attention to animal welfare in food production. However, the understandings of what 'welfare' entails, vary. The widespread view among omnivores emphasising that Finnish regulatory and farming practices already secure a high enough level of animal welfare may slow down public demands for improvements. Animal rights are a more contested field compared to animal welfare: from the omnivore perspective, rights are limited to welfare issues, while vegans and to some extent vegetarians conceptualise rights in terms of a right not to be exploited by humans. Such fundamental differences in ideas about the ethics of farming and eating animals are not easily reconciled.

## References

De Bakker, E. and Dagevos, H. (2012). Reducing meat consumption in today's consumer society: questioning the citizen-consumer gap. Journal of Agricultural and Environmental Ethics 25: 877-894.

Kupsala, S. (2019). A sociological study of Finnish attitudes, perceptions and meanings regarding animals in food production. Publications of the University of Eastern Finland. Dissertations in Social Sciences and Business Studies, No 199, 213 p.

Lund, T.B., MacKeegan, D.E.F., Cribbin, C. & Sandøe, P. (2016). Animal ethics profiling of vegetarians, vegans and meat-eaters. Anthrozoös 19: 89-106.

Niva, M., Vainio, A. and Jallinoja, P. (2017). Barriers to increasing plant protein consumption in Western populations. In: Francois, M. (ed.) Vegetarian and plant-based diets in health and disease prevention. Academic Press / Elsevier, London, pp. 157-171.

OECD/FAO (2020). OECD-FAO Agricultural Outlook 2020-2029. FAO and OECD, Rome, 330 p. Available at: http://www.fao.org/documents/card/en/c/ca8861en.

Sandøe, P. & Christiansen, S.B. (2008). Ethics of animal use. Chichester, UK: Blackwell Publishing, 178 pp.

Scott, M.B. and Lyman, S.M. (1968). Accounts. American Sociological Review 33: 46-62.

# 45. What should we breed cattle for – engaging publics in a Democs card game

*D.M. Bruce[1]* and A. Bruce[2]*
*[1]Edinethics Ltd., 11/6 Dundonald Street, Edinburgh EH3 6RZ, United Kingdom; [2]The University of Edinburgh, Old Surgeons' Hall, High School Yards, Edinburgh EH1 1LZ, United Kingdom; info@edinethics.co.uk*

## Abstract

The revolution in genetics has brought the ability to make substantial new developments in breeding livestock. Using advanced genomics many more traits can potentially be taken into account in selective breeding programmes for dairy and beef cattle, including specific disease resistance, methane emissions, feed efficiency, and robustness to environmental stresses. As breeders and farmers can now select for a much wider rage of traits than classical production factors like milk yield or muscle growth, what should the priorities now be? In which directions should we use this new knowledge, in a world where climate change, antimicrobial resistance, biodiversity and even meat consumption itself have become major factors in debate? Such decisions tend to be made with only the major stakeholders round the table. Publics may have different viewpoints and insights about the foods being produced on their behalf, but are seldom involved. As part of the European Commision BovReg project on cattle breeding (www.bovreg.eu), funded within the Horizon 2020 research programme, we have been constructing a Democs card game to explore such questions with small groups of publics in different European countries. The range of information and interrelated ethical and social issues is large and complex, involving genomics, production systems, animal health and welfare, climate change, land use, biodiversity, societal and cultural aspects and world views. In this paper we describe how we negotiated this maze of factors to present questions of breeding priorities and goals, aiming at accurate, ethically balanced and stimulating sets of cards, understandable to lay publics. Covid-19 restrictions have so far prevented playing it normally, face-to-face round a table. The feasibility of an online version is being assessed.

**Keywords:** genomics, priorities, information selection, ethics

## Cattle breeding using advanced genomics

As part of the European Commision BovReg project on the functional genomics of cattle breeding (www.bovreg.eu), funded within the Horizon 2020 research programme, we have created a Democs card game to explain cattle breeding and genomics, and to explore ethical questions about breeding priorities and practices, with small groups of publics in different European countries. In this paper we describe how we constructed the game, aiming to create accurate, ethically balanced and stimulating sets of cards, understandable to lay publics.

The revolution in understanding genetic information across many species has brought also the ability to make substantial new developments in breeding livestock. Since ancient times, farmers have been selectively breeding animals largely based on visible characteristics, like an animal's physical appearance, or on obvious measurable traits such as the number of offspring or milk yield. This selection, combined with natural genetic variation over time, has led to substantially different breeds within the same species, and variations within a given breed. Developments in animal genetics in the 1960's, in the UK and other industrialised nations, led to a gradual displacement of this traditional approach to livestock breeding by selection methods in which scientific measurements and statistical calculation play a major role in influencing the farmer's choice of which males to breed with which females.

In dairy cattle, numerical values are assigned to a wide range of properties of the cow, which include milk yield and composition, fertility, ease of calving, and how placid in temperament the cow is. In the UK in recent years a growing significance has been attached to health and welfare traits such as the incidence of certain well known diseases and conditions such as mastitis and lameness, and also tuberculosis. These numerical data from as many as thirty different factors are fed into very complex statistical models, together with similar data from the siblings and relations of the animals, seeking to identify correlations of the measured factors to the genomes of the animals. This modelling has become steadily more sophisticated as increasing amounts of genomic information about individual animals has been added. In the UK, the combined data about relevant traits and genetics of parents and siblings are rationalised into 'estimated breeding values' and 'predicted transmitting abilities'.

These indices do not necessarily measure the performance of the offspring, so much as the reliability that the traits of a chosen sire will be passed on to the offspring. The aim of these indices is that a farmer can then choose the characteristics which are most suited to his/her situation. For example, if the dairy unit has a high incidence of mastitis, a low score and reduced transmission of this trait might be desirable. If the farmer wishes to expand the herd next year, some milk yield might be sacrificed for higher fertility. Where artificial insemination is standard practice, as in much of the dairy industry, this will involve ordering semen from a catalogue. In the beef sector, many farmers will buy from a regional bull sale. If the index of each bull is available, this can guide the choices made. In more remote areas, selection opportunities are likely to be more limited.

These indices have been developed using complex statistics to identify patterns in the genomes of animals that represent inheritable traits, without knowing how all the individual genes work. The aim of the BovReg project is to analyse the functions of sections of the genome in detail, especially the ways on which genes are regulated, in order to give a better basis for breeding, to widen its scope of traits, and also to include minority and mixed breeds.

The information, values and priorities, and the impetus for this highly technical approach to breeding have come largely from those embedded in the systems of livestock production, and some parts of the food production chain downstream, notably the major supermarket retailers. Breeding companies will develop herds which score well for the traits deemed to be of most importance to their constituency. The role of citizens, outside these systems, is often missing. It was clear from the genetically modified food crisis, when applying genetic technologies to food, that European citizens expect more say in what is being done on their behalf than simply what final purchasing choices they make as consumers.

## Creating a Democs card game on cattle breeding

Public engagement aims at engaging people in reflecting on an issue about which they may not have given much previous thought, but on which they should be given opportunity to do so. Government consultations ask questions aimed at those with specialist knowledge, on relatively short timescales, on which lay people would struggle to frame a response.

The Democs card game is a tool for discussion aimed at informal grassroots groups of 6-8 people. It uses different sets of cards to enable people with no previous knowledge to learn about new technologies and discuss their ethical and social implications. The concept was first developed in 2002 on human embryonic stem cells, and has now been developed on a wide range of issues (Bruce, 2010). Story cards introduce the subject through case studies of imaginary people involved in or impacted by the technology. Information Cards provide necessary factual information as objectively as possible. Issue Cards on the other hand present ethical issues and seek to reflect a fair range of viewpoints, often with some opposed to each other. Players choose which issues and cards to discuss, and frame their own views, drawing upon the suite of cards they selected. There are two outputs. The group is invited to create opinion statements, aiming for consensus as far as possible. Players also then vote as individuals on priorities and questions in cattle breeding, giving reasons in their own words.

The game is intended to be used without experts or facilitation. One person acts as 'dealer', who has read the instructions but knows no more than anyone else about the issues to be discussed. The 'expert' is the cards. This puts a large onus on the creators of the game in the choice of information imparted in the cards and how both factual and ethical issues are expressed. On the more contentious issues, how do the writers of the material handle fairly the opposing views, with disputed assertions even about what is factual, and with profoundly differing value presuppositions? It is also a considerable art to condense the subject of each Information and Issue card into about 35 words, to convey unfamiliar concepts in ways understandable to a lay person. How did we go about selecting what to include in a game, to be accurate, fair, understandable and engaging to help lay people to reflect on the issue?

The first challenge was make a highly sophisticated research field of cattle genomics into something general publics could understand and find interesting. We therefore widened the context to how cattle are bred and the goals to which breeding might best be directed. It was felt this would resonate with citizens, in view of more public interest in where food comes from, evidenced by the number of farming and veterinary programmes on television, the popularity of agricultural shows, and an increased focus on ethical questions about meat eating, linked to diet, health and climate change. This wider scope meant including a complex range of interrelated subjects and issues, involving genomics, production systems, animal health and welfare, but also climate change, land use, biodiversity, societal and cultural aspects, and world views. Our discussions within the work package drew upon a considerable range of ethical and technical expertise on animal biotechnology.

A secondary aim of the game was to bridge a substantial gap in public awareness of how food animals are bred, reared and produced, and the issues this raises. Quite a lot of information cards were necessary to provide basic background information, what animal production systems are used; how animal breeding is currently done and its main goals and target traits; what is genome editing and how does that impact on breeding? We explained that cattle are farmed in many different ways from large dairy units, through medium sized family farms, to small outdoor upland crofts. We explained what is broadly meant by 'intensive' and 'extensive' production. Intensive systems emphasise efficiency, usually in large units where cows are indoors most of the time, and closely monitored for health and performance. Extensive approaches put more stress on giving animals access to the outdoors and as natural a life possible, while using indoor facilities for milking or in bad winter weather. We pointed out that most dairy cattle are Holstein-Friesians, but there are many different beef breeds and some that produce both meat and milk. Minority local cattle breeds struggle to compete with dominant breeds on efficiency, but may have advantages like better tolerance to hot or cold climates, resistance to diseases, or suitability for niche markets like local cheese production.

Having covered these basics, we required a number of cards that between them explained how cattle breeding has developed, and the far reaching changes that the advent of genomics has brought. On the one hand, with improved nutrition and management of livestock and closer veterinary care, milk production per cow has doubled over 40 years. On the other hand there are also significant diseases and welfare problems which can affect livestock, like mastitis, lameness and tuberculosis. These have meant that an earlier focus in selecting for higher production yields has been modulated by including these animal health and welfare factors within the breeding indices. It was important to point this out in the cards, because developments such as these and the restrictions in the use of antibiotics are in farm animals in Europe are not often visible to publics.

Most of this basic information is covered through the 36 Information Cards, discussed in round two of the game. But before this, we have found that the best way to introduce an unfamiliar topic is through people. We created eight Story Cards to be read out in round one, about imaginary people involved in cattle production or affected by it. These are miniature case studies designed to introduce some of the main aspects of cattle production and breeding, and to flag up some of the ethical and social issues it poses. Each story describing an ethical dilemma which the person faces, thus linking also the Issue Cards in round three.

For these we chose:

- a researcher seeking to correlate patterns in cattle genomes with different traits;
- a dairy cattle breeder deciding future priorities amongst an ever widening range of traits;
- an Alpine farmer wanting more consistency in the artisanal cheese his family farm makes;
- a climate scientist advising the Government on how much methane cows emit;
- an environmentalist advocating pasture fed extensive cattle farming even if it costs more;
- a vet and her daily problems of cattle diseases indoors and out, and when to use antibiotics;
- a researcher producing hornless dairy cattle by genome editing, but should we do this;
- a consumer wanting beef with good welfare and low methane but worried about the price.

We started with a much longer list of potential stories, but practicalities limited us to only eight cards. Narrowing the list down was to our final choice reflected a good deal of what we considered were needed both to set the scene and what we considered to be some key ethical issues and viewpoint. The ones we chose were the ones we simply couldn't leave out.

A researcher, breeder and vet were obvious choices, as was an environmentalist wishing to change the prevailing system. We wanted to include a niche market of specialty products from a local breed, to contrast with mainstream intensive dairy. Climate change features in several stories as a major contextual issue, but we felt it necessary to have a story which pointed out the disputed and complex issue of cattle contribution to climate change, to set against the popular notion that giving up meat is sufficient to 'do my bit' for the climate, without needing to use my car any less. The ethical consumer story brings the issues down to earth, faced with how far she is willing to pay for the ethical qualities she would wish.

It was important to include a genome editing story to highlight that, unlike 'conventional' genetic modification, it is already being applied in livestock. The genomic information produced by the BovReg project might also lead to further gene editing applications in future. Several Issue Cards look at genome editing as an illustration of underlying question of how far we should intervene both genetically and functionally in farm animals in the interests of producing food. Should genome editing be used in cattle, and for what purposes, if any?

Ethical and social issues are considered in detail in the 36 Issue Cards. Again what to include was the subject of much discussion in creating the game. Although in the earlier GM debates, selective breeding was promoted as something 'natural' and ethical compared with mixing genes across species, we considered that modern selection can also raise significant issues. To what extent breeding goals should incorporate wider ethical factors than a production orientation has featured from time to time in previous Eursafe meetings and related papers for example, in Olsson *et al.* (2006) and Rocklinsberg *et al.* (2016), but they have not so far been the subject of wide public debate. The game is one step towards such discussion.

We considered how far we should include fundamental ethical question of using animals for food and the current *zeitgeist* to reduce or replace meat in 'Western' diets. We realised that to do justice to these would require a Democs game in itself, whereas our focus in this game was on aims and priorities in cattle breeding. Whether we should use animals for food was beyond our present scope. But the drivers behind such questions indicated some of the ethical issues we should cover in the Issue Cards, namely climate change and methane emissions, animal welfare and modern intensive production, land use (growing crops vs livestock grazing vs 'rewilding'), meat seen as 'unhealthy', anti-microbial use linked to medicine. We also felt it important to raise questions about people's sources of information and their trustworthiness (or otherwise), the role of social media in spreading 'urban myths', who frames the information, and how selective the sources are being for their own ends.

Our central ethical question was what should be the priority traits in breeding, and why, and to what extent these differ from current production priorities and likely future ones? We included the critical view of some that breeding should be 'naturally' done, in the traditional way of characteristics observable without intervention in the animal, not on the basis of technologically measured parameters. We also needed to make clear that breeding is a complex and slow process, in which some traits are easier to improve, some harder, and where changes in one trait might lead to a detriment in another, such as increasing milk yield reducing fertility. We noted the shift in cattle breeding from maximising production traits to a more balanced set of characteristics, and asked what other factors should be included.

We explored to what extent cattle could be bred for environmental benefits. This included breeding for lower methane emissions (if feasible), cattle suitable for pastures requiring less feed, and for using food wastes as part of a 'circular economy' approach to agriculture generally. We also covered possible impacts of climate change on the resilience of breeds, and on feed availability and cost. As part of the context in which breeding is done, we wrote cards exploring the pros and cons of intensive and extensive production. Relevant social factors included the importance of upland cattle in highland and island communities, and of the notion of 'terroir', the place where the meat or dairy produce comes from.

The game is structured by each player in turn sharing cards chosen from their random 'hand' of cards and saying why they chose them. This naturally leads to discussion in the group, in which ideas are supported, altered or challenged. Salient factors are added, from other players' cards, or from players' differing values and opinions. There are two outputs. One is free text statements from the group written on 'Cluster Cards' to record conclusions they have drawn from their discussions, inviting them to cluster the cards relevant to statement they wish to make. The aim as far as possible is to find consensus in these statements, but also to reflect honest differences where it is not. It is up to each group how they do this.

The second output is a voting sheet which first asks about different breeding priorities. 'What traits do you think are the most important for cattle breeders to aim at, given they can't do them all? Here are five improvements in cattle which could be bred. How important (or acceptable) are they? (1) production efficiency; (2) better disease resistance; (3) reduced methane emissions; (4) increasing a cow's productive lifetime; (5) adapted to different types of environment. Space is provided for the player to give their reasons.

Three ethical questions are then asked: (6) Are there moral limits to how far we should adapt cattle by breeding for our own purposes? If so where would you want an ethical line drawn? (7) Should we aim to breed elite highly productive cattle, which depend on stable conditions, or aim for less efficient cattle more robust to varied situations? (8) Is it a good idea to use genomic information in cattle to make desirable changes quickly by genome editing, instead of slowly by cross-breeding?' Finally we ask the ethical consumer's question from Story Card 8. 'How much extra would you be willing to pay for improvements to cattle traits that would make milk or beef more 'ethical', if they also were more expensive? (9) lower methane emissions; (10) healthier meat; (11) disease resistance / less antibiotics; (12) pasture fed cattle?' A range of 0, 2, 10 and 25% is offered against minced beef, steak, milk and a special cheese.

The primary reason for all Democs games is to engage publics with the issues, but in this case we are also collecting data to gauge how publics view priorities in cattle breeding. This raises an important underlying ethical issue. To what extent is the livestock production system open to changes in breeding emphases based on ethical, rather than primarily economic considerations. Typically, cattle breeding values are translated into economic breeding indices, which set a weighting on each of the factors according to their economic return to the farmer. This is one way of taking a range of qualitative value judgements and giving them a numerical score by which they can be weighted against each other. Some strongly ethical factors like animal health have a high economic value as well, because a sick animal is also less productive and a dead one a serious monetary loss. But to incorporate ethical values on an economic basis is both conceptually unsound and likely to miss factors which carry a high ethical value but low economic return. For example, an ethical farmer would not be rewarded economically for having low methane cows, without creating suitable financial incentives.

## Conclusion

The complex process of creating of a Democs card game for public engagement on cattle breeding has been described, explaining the information necessary to engage publics on a relatively unexplored subject, and what were considered the key ethical and social issues on which to seek insights from publics. Covid-19 restrictions have so far prevented playing the game in small groups, but when these are lifted, important feedback is expected from publics on what should be the priorities for breeding, in the light of current wider ethical and environmental values, to set alongside those coming from within the livestock industry.

## Acknowledgement

This project has received funding from the European Union's Horizon 2020 research and innovation programme under grant agreement No 815668.

## References

Bruce, D. (2010) *Playing Democs Game to Explore Synthetic Biology*, Edinethics Ltd [online] Available at: http://www.edinethics.co.uk/synbio/synbio democs report.pdf.

Olsson, A.S., Gamborg, C. and Sandoe, P (2006) Taking ethics into account in farm animal breeding: what can the breeding companies achieve? J. Agric.Envir.Ethics 19:27-46.

Rockinsberg, H., Gamborg, C., Gjerris, M., Rydhmer, L., Tjarnstrom, E. and Wallenbeck, A. (2016). In: Olsson, A.S., Araujo, S.m. and Vieira, M.F. (2016) Food futures: ethics, science and culture, Eursafe 2016, Wageningen Academic Publishers, Wageningen, the Netherlands, pp.61-66.

# 46. Public engagement in biotechnology innovation – the need for research and the role of ethics

*S.G. Carson[1*], B.K. Myskja[1] and A.I. Myhr[2]*
*[1]NTNU – Norwegian University of Science and Technology, 7491 Trondheim, Norway; [2]GenØk-Centre for Biosafety, SIVA Innovasjonssenter, 9294 Tromsø, Norway; siri.granum.carson@ntnu.no*

## Abstract

The promised disruptive potential of genome editing for producing novel agri- and aquaculture products with enhanced qualities has refocused the discussion on non-safety assessments of biotechnological inventions. Since such assessments are based in societal values, the public or their representatives should be involved in handling the normative issues at stake. A starting point of this paper is that such assessments are feasible and justified, given the proper form and level of engagement. Public engagement includes different forms of surveys, focus group interviews, dialogues, and consultations. In this paper we present two recent Norwegian surveys, both with the intention to identify public perspectives on the use of genome editing in animals and plants to be used within agri- and aquaculture. Through an analysis of these two recent surveys, we will show that such approaches tend to fall short when it comes to providing a normative basis for non-safety assessments. For surveys to be appropriate tools for public engagement, it is important that the forming includes a multitude of perspectives on the use of the technology in question. The information material should not be simplified or provide a biased knowledge basis, and the survey should avoid leading questions. By enhancing the research-based framing of public engagement processes, including reflection on and research into the ethical values at stake, these shortcomings could be amended. We argue that in highly contested areas of research such as genome-editing in agri- and aquaculture, public engagement approaches should be based in transparent scientific and normative analyses.

**Keywords:** genome edited organisms, non-safety assessment, social acceptance, ethical reflection, public values

## Introduction: public engagement in technology

To promote public engagement processes in responsible governance of biotechnology is a long-standing tradition. The ideal of involving users in the regulation of technology can be recognized in various theories from stakeholder management (Freeman, 1984), to science communication (Schäfer, 2008), and technology assessment (Guston and Sarewitz, 2001). The reasons for doing so are typically not explicitly stated, but rather implicitly based in ideals of participatory or deliberative democracy. Public engagement is advocated in the context of European science and innovation, for example in a recent report: 'Engaging citizens is a priority across the European Commission and a key component of Horizon Europe' (EC, 2020). The push to include citizens in science builds on the cross-cutting issue of Responsible Research and Innovation (RRI) in Horizon 2020, where RRI is viewed as 'a transparent, interactive process by which societal actors and innovators become mutually responsive to each other' (von Schomberg, 2013).

We may distinguish between three basic objectives for public engagement in biotechnology innovation: 'empowerment, learning and legitimacy' (Hisschemöller and Cuppen, 2015), or alternatively: 'normative, substantive and instrumental' participation (Fiorino, 1990). In other words, participation may serve to:


EurSafe 2021 – DOI 10.3920/978-90-8686-915-2_46, 

- Empower people and put in practice democratic ideals ('normative' – participation as a way to secure that the right decisions are made).
- Gain inputs from citizens' knowledge when facing complex problems ('substantive' –participation as a way to increase our knowledge by including more perspectives).
- Building social legitimacy for the chosen way forward ('instrumental' – participation as a way to secure acceptability for decisions).

These objectives typically co-exist, but in some cases, objectives may contradict, or the emphasize of one over the other may have consequences for the design and application of public engagement methods (Hisschemöller and Cuppen, 2015: 46). While an emphasize on increasing the knowledge basis favours extensive and representative engagement processes, the promotion of social legitimacy may be satisfied with less – and an emphasize on the latter may undermine the objective of building a normative foundation for decision-making in controversial areas (cf. Brunsson, 2003 on 'organized hypocrisy'). The often-implicit coexistence and contradiction between objectives may serve as an explanation for the lack of systematic evaluation of public engagement tools and methods in practice. However, to find the proper level and form of public engagement, it is vital to state the objectives and evaluate the outcomes in relation to the objectives.

Here we precent two public surveys carried out by two different institutions in Norway, both with the objective to identify public perspectives on the use of genome editing in animals and plants to be used within agri- and aquaculture. We focus on shortcomings in these surveys including implications by biased design and analyses and discuss alternatives that provide more balanced and transparent analyses.

## Involving society in biotechnology innovation – the case of genome editing

Genome editing has potential for disruptive changes within agri- and aquaculture, meaning that the implementation of the technology will have significant societal impact. Arguably, this warrants more than mere assessments of risks to environment and human health, by including so-called non-safety concerns in the approval process. These includes questions of societal utility, sustainability, and other value assessments, such as cultural heritage and public morale. Unlike risk assessment, non-safety concerns are usually not quantitative threshold-assessments where a product is either sufficiently safe or not. This has led some to argue that these assessments by necessity are vague, arbitrary, and subjective (Zetterberg and Björnberg, 2017). However, there are ways to overcome these shortcomings through inclusive deliberative processes where all relevant arguments are heard before a decision is made by competent and trusted people (Myskja and Myhr, 2020).

At the outset, it appears as if the best way to ensure that all relevant non-safety issues are covered is by using established risk assessment procedures e.g. expert led and framework based. However, this is not sufficient given that the assessments include issues that involve society at large. The genome-edited product can have implications that will affect the lives in ways that only those potentially affected can foresee. Since the regulatory choices will be value based, the experts have no privileged access to the relevant perspectives and arguments. This calls for public engagement in regulatory discussions as well as in concrete technology development projects, in line with the principles of RRI (Stilgoe *et al.*, 2013). There are several ways to engage the public in the form of communication, consultation or participation (Rowe and Frewer, 2005), including information programs, interviews, focus groups, public meetings, citizens juries and surveys, to mention a few. One standard form of public consultation in regulation of biotechnology has been public surveys.

## Public engagement through surveys

Survey is a method that gives a quantitative input from the public on controversial issues. Unlike qualitative approaches, it is representative and therefore a favoured basis for political decision-making. A well conducted survey gives a reliable indication of the public opinion in the matter at stake. Its main drawback is the lack of depth as it does not provide the argumentative basis for the opinions. It tells what people think is right to do, not 'why' this is the right course of action. This is a crucial weakness, especially in complex issues involving advanced scientific or technological knowledge, as the opinions may be based in misunderstandings, lack of knowledge or direct misinformation. For surveys to be a representation of the considered public view, the background material and questions must include informative presentations of uncontested facts, and must avoid using leading questions, while avoiding confusing and discouraging information overload. In the following we will discuss two recent Norwegian surveys on genome editing and show why they both fail in fulfilling these requirements.

### *The SIFO study*

The first survey is a follow-up study to one conducted in 2017, both by Consumption Research Norway (SIFO), commissioned by the GMO-Network Norway, formerly named the Network for Non-GMO Food and Feed. In the background information for the survey, it says that the questions have been developed in communication with the GMO-Network and their collaborating experts, seeking to ask 'simple, concrete and easily understood', as well as non-leading questions (Bugge, 2020: 3). It starts out with a control of respondents' knowledge followed by series of statements, each introduced by:

> There are many different GMOs. They may have different benefits, disadvantages and possible risks associated with their use. With this in mind, we will present some general statements on the use of genetically modified organisms. Which of the following statements best express your own view of this technology? (Bugge, 2020: 14-15)

The following questions are general in nature, concerning whether GMOs will be necessary to produce enough food, whether it is likely that they will have negative effect on nature and ecosystems, whether it is likely they will constitute a risk to human or animal health and so forth. There is little value in stating that the products will have different benefits and disadvantages, without having this diversity reflected in the questions. Food safety is a general challenge, meaning it is likely that one GMO at some point will constitute a risk to humans or animals. It is not surprising that 49% assents to this statement, even more than was found in the 2017 survey.

There is little reflection on the game-changing potential of genome editing in this survey, although it is mentioned in the explanatory introduction to the survey. The questions overall concern GMOs, which may be understandable given the intention of comparing the results with those in the 2017 survey, in order to find change in opinions. But if we assume that genome editing will change the debate, neglecting how the differences between these and the more traditional GMOs inform the assessments will prevent catching these potential changes. Admittedly, there are two questions concerning the willingness to eat gene-edited pig and salmon, respectively, where the aims of the genetic alterations are included. This is not very helpful, though, as the analysis is restricted to comparing the responses to two specific gene-edited animals, rather than see how the technology and introduced or changed characteristics influence the assessments compared to fast-growing GMO salmon or meat from GMO-fed animals. The salmon is instructive, as 50% would not eat fast-growing GMO salmon, whereas 43% would refuse to eat gene-edited sterile salmon. This is a significant difference, and it would be worth an analysis to find out whether it is the technology or the characteristic that is the main explanation.

### The NBAB survey

The second study is conducted by the Norwegian Biotechnology Advisory Board (NBAB) as part of the research project 'GENEinnovate', funded by the Research Council of Norway. The intention, in addition to making a 'steering tool' for the research project, was to achieve 'a good knowledge base [that] can contribute to an informed public debate and future-oriented policy' (NBAB, 2020: 3). The study combines 'exploratory' focus group interviews, where it is stressed that the 'findings cannot be generalized' (NBAB, 2020: 40), with a survey. The study is designed by members of the NBAB secretariat, but data was collected and compiled by Ipsos, a professional market analysis company. In this context, we will only discuss the presentation of the survey part of the study.

Compared to the SIFO study, this study is very much updated on the reported possibilities by using genome editing. However, merely half of the respondents had heard about this technology before the survey. The questions concerning different technology application are, unlike the SIFO study, not based on existing products, but on hypothetical cases, generally described. People are, for example, asked: 'How positive or negative are you to using gene editing on crop plants ... if the purpose is to: Adapt a crop plant to changing climate, e.g. wheat that better tolerates drought or precipitation?' (NBAB, 2020: 15). This gives a more nuanced material that better reflects the variety of possibilities of the new technology. However, as the applications are more or less hypothetical, the descriptions only deal with their 'intended' effects, which by necessity are positive. The broad public scepticism to agricultural biotechnology, with a few exceptions, does not concern intended effects, but negative side-effects and long-term consequences. This means that the survey questions fail to address the main areas of public concern, be they warranted or not, and skews the response in a technology positive direction.

In the analysis of the relationship between knowledge of the technology and attitudes, the study finds that positive attitude correlates with better knowledge. However, they also find 'the percentage who are 'very negative' about using genome editing for climate adaptation is also highest among those who report having good knowledge about genetic engineering and genetics' (NBAB, 2020: 21). This is explained by showing that in the self-reported good knowledge group, 'those who are very negative about, for example, the use of gene editing for climate adaptation of plants know the least about genetic engineering or have the least trust in science.' (NBAB, 2020: 21). Combining these two respondent groups effectively hides that there is a significant group that lacks trust in science despite possibly having good knowledge of the technology. This again skews the results in a technology positive direction.

The NBAB study includes some leading questions. The respondents are asked to what extent they agree with the statement: 'It can be unethical not to use gene editing in crop plants and livestock if it can contribute towards solving important societal challenges, such as climate adaptation of crops' (NBAB, 2020: 35). 48% agree entirely or somewhat, which is a surprisingly low number, considering the phrasing. It is nearly self-evident that it 'can be' unethical not to use something that may solve important societal challenges. Removing the double negative shows the leading character of the question: 'It can be ethical (or morally right) to use gene editing, etc.'

Unlike the SIFO study, the NBAB study is apparently worked out and analysed by people who have a positive view on genome editing. The conclusion states: 'Overall, our results show that the majority of Norwegian consumers are positive about using gene editing in plants and animals if the purpose can be perceived as beneficial to society and to contribute to sustainable development.' (NBAB, 2020: 40) As we have indicated, this is not supported by a critical reading of the study design and the analysis. It is also claimed in the conclusion that this study, unlike previous studies, shows that the opinions are more nuanced than 'for' or 'against' (NBAB, 2020: 40). However, this is not new as tendency towards nuanced assessments of biotechnology innovations has been repeatedly demonstrated in surveys from the last twenty years (Gaskell *et al.,* 2003, Schaart *et al.,* 2011).

Common for both the SIFO and the NBAB study is the following approach: Ask people about their views and attitudes to (actual or possible) biotechnological innovations, then use their response to evaluate if the technology is in accordance with current public values or not. This approach could provide useful information e.g. regarding the market potential for gene edited products. However, given the complexity of the issues and the limited knowledge of the respondents, it does not provide much in terms of a normative evaluation of the questions at stake. If the objective is to provide a sound normative basis and direction for the future development and regulation of a technology, an alternative approach could be to ask people more open-ended questions about their values in relation to food production, nature, animals, plants, etc.

## The Scylla and Charybdis of public engagement processes

The surveys illustrate the risk of biased design and analyses in 'superficial' public engagement approaches based on quantitative methods. On the other end of the scale, we find extensive participatory processes such as the 'consensus conference', a version of participatory technology assessment involving experts in order to secure that the participants knowledge base is sufficient for well-founded judgement in technically complex questions. The methodology is developed and used by, among others, the Danish Board of Technology (Nielsen *et al.,* 2006) and has attracted considerable attention as an exemplary way to put deliberative ideals of how to govern biotechnology into practice. It does, however, put considerable strain on the participants in terms of investing their time and effort into the process. It is possible to argue such engagement processes in themselves create a bias as they favour the participation of particularly dedicated citizens. For participatory processes to be truly inclusive, they must avoid both the 'Scylla' of too weak and superficial and the 'Charybdis' of too heavy and demanding engagement. In the first case, participants may lack the proper knowledge and be too strongly influenced by prejudice or the way questions are posed. In the last case, one might end up with a kind of 'lay people experts' who no longer represent the public as such.

## Conclusion: the need for research and the role of ethics

To secure sustainable, ethically responsible and socially legitimate use of biotechnology, regulation must be in accordance with public values. The regulation of biotechnological innovation should be knowledge-based while at the same time reflecting our values. In this paper, we have argued that public engagement provides essential input to non-safety assessment of biotechnology innovation, to the extent that the methods for engagement are in line with the stated objectives. The design and analyses of two Norwegian surveys served as examples of how certain forms of engagement may promote bias rather than ethical reflection. To find the right balance between the 'Scylla' of too weak and the 'Charybdis' of too demanding forms of public engagement it is important to reflect on the ideals as well as the objectives for engagement. To achieve this, there is a need for empirical research into the outcome of diverse forms of engagement, as well as normative research into the ethical values and norms at stake in biotechnology innovation and regulation. In the surveys participants were asked to give their opinions on specific issues. Based on their response, deductions were made concerning public attitudes towards genetic engineering. However, if the objective is to promote technology for a good society, public participation should not primarily take the form of polling or evaluation 'after the fact'. Rather, it should take the form of an ongoing dialogue in parallel with (or as part of) the development of new technology. Thus, to the extent that surveys are used as a tool for public engagement, the objective should not be to evaluate the technology by referring to popular opinion. Rather, the objective should be to explore relevant public values as a guide for the regulation of biotechnology. In other words: Rather than attempting to navigate between Scylla and Charybdis, we turn the boat around.

## References

Brunsson, N. (2003). Organized hypocrisy. In: Northern Lights, Czarniawska, B. and Sevón, G.(eds) Northern Lights, Abstrakt Liber Copenhagen Business Press, Oslo, Norway.

Bugge, A.B. (2020). GMO foods or not: Have there been changes in consumers' views on genetically modified foods from 2017 to 2020? SIFO-Report 3-2020. Available at: https://fagarkivet.oslomet.no/handle/20.500.12199/3001.

EC (2020). Citizen science and citizen engagement. Available at: https://op.europa.eu/en/publication-detail/-/publication/c30ddc24-cbc6-11ea-adf7-01aa75ed71a1.

Freeman, R.E. (1984). Strategic management: A stakeholder approach. Pitman, Boston, USA.

Gaskell, G., Allum, N. and Stares, S. (2003). Europeans and Biotechnology. Eurobarometer 58.0. Available at: https://ec.europa.eu/commfrontoffice/publicopinion/archives/ebs/ebs_177_en.pdf.

Guston, D.H. and Sarewitz, D. (2001). Real-time technology assessment. Technology in Society 24: 93-109.

Hisschemöller, M., and Cuppen, E. (2015). Participatory assessment: tools for empowering, learning and legitimating? In: Jordan, A.J. and Turnpenny, J. R. (eds), The tools of policy formulation: Actors, capacities, venues and effects. Edgar Elgar Publishing, Cheltenham, UK. pp. 33-51.

Myskja, B.K. and A.I. Myhr. (2020). Non-safety assessments of genome-edited organisms. Science and Engineering Ethics 26, 2601-2627.

Nielsen, A.P., Hansen, J., Skorupinski, B., Ingensiep, H.W., Baranzke, H., Lassen, J., and Sandoe, P. (2006). Consensus Conference Manual. LEI, onderdeel Wageningen UR, The Netherlands, 40 p.

Norwegian Biotechnology Advisory Board (NBAB) (2020). Norwegian consumers' attitudes towards gene editing in Norwegian agriculture and aquaculture. Available at: https://www.bioteknologiradet.no/filarkiv/2020/04/Report-consumer-attitudes-to-gene-editing-agri-and-aqua-final.pdf.

Rowe, G. and Frewer, L.J. (2005). Typology of Public Engagement Mechanisms. Science, Technology, & Human Values 30: 251-290.

Stilgoe, J., Owen, R., Macnaghten, P. (2013). Developing a framework for responsible innovation. Research Policy 42: 1568-1580.

Von Schomberg, R. (2013). A vision of responsible research and innovation. In: Owen, R., Bessant, J and Heintz, M. (eds) Responsible innovation: Managing the responsible emergence of science and innovation in society. John Wiley & Sons, London, UK. pp. 51-74.

Schäfer, M.S. (2009). From public understanding to public engagement: An empirical assessment of changes in science coverage. Science Communication, 30: 475-505.

Schaart, J.G., Kjellsen, T.D., Mehli, L., Heggem, R., Iversen, T.-H., Schouten, H.J. and Krens, F.A. (2011). Towards the Production of Genetically Modified Strawberries which are Acceptable to Consumers. Genes, Genomes and Genomics 5: 102-107.

Zetterberg, C. and Björnberg, K.E. (2017). Time for a new EU regulatory framework for GM crops? Journal of Agricultural and Environmental Ethics 30: 325-347.

# 47. Openness: terroir and the value of the natural

*L. Ursin*
*Department of Philosophy and Religious Studies, Norwegian University of Science and Technology, 7491 Trondheim, Norway; lars.ursin@ntnu.no*

## Abstract

Why value naturalness? And how? The normativity of naturalness is a tricky topic and concepts like 'nature' and 'natural' are notoriously hard to define. On the one hand, this makes it quite easy to argue that naturalness is devoid of any normative significance, or merely an imprecise term that comprise of several more specific values and norms. On the other hand, the normative and conceptual vagueness of the natural does not seem to cast a damper on the use of naturalness in making normative assessments in daily life and public debate. In this paper, I will discuss the basis of valuing naturalness. My case will be viticulture and winemaking, and the controversial issue of the status of terroir.

**Keywords:** wine, viticulture, naturalness, narrative, authenticity

## Introduction

What is terroir? In the world of wine, terroir is a matter of origin, taste, and authenticity. Terroir is a matter of uniqueness, and to preserve and cultivate this uniqueness into a characteristic and tasteful wine. It's an intriguing idea. But, is it more than an intriguing idea, perfectly adapted to the contemporary ideal of marketing products by appealing to traditions, identity, and authenticity? David Inglis comments:

> A wine deemed to be an authentic 'expression' of *something* – the soil and micro-climate in which it was grown, the environment and culture of the region from which it hails, the genius of the wine-maker who nurtured and brought it into being, the quintessential characteristics of the grape variety it is made from – will likely make much more money than one deemed inauthentic. (Inglis, 2015)

The choice of an authentic and unique wine promises to endow the wine consumer with the same qualities, as well as enabling the consumer to show superior taste, awareness, and appreciation of the local traditions and knowledge that goes into the making of the chosen wine. The authentic wine further allows the consumer to show the ability to taste, decipher and articulate the message of the winemaker, who can cultivate and convey the qualities of the soil to the knowledgeable and receptive consumer. In other words, the concept of terroir is an irresistible selling point.

In such a perspective, the purported authenticity of wines is easy to debunk as a mere myth, conceived by clever winemakers and merchants. This myth is moreover a protectionist market manoeuvre with additional questionable legitimacy if aiming to exclude competitors from making certain kinds of wine, and to promote localism and nationalism among consumers. The notion of 'terroir' is also severely and routinely criticized by scientists as having no chemical or other traceable correlate in wine, for instance by Mark Matthews in his 2016 book with the telling title 'Terroir and Other Myths of Winegrowing' (Matthews, 2016).

Does this kind of critique hit the mark? Is the failure to find consistent chemical data that cause or at least correlate with taste differences between wines a fruitful contribution to the discussion of the significance of soil variations for wine quality? Or, is it a more of a category mistake? Matt Kramer


EurSafe 2021 – DOI 10.3920/978-90-8686-915-2_47, © Wageningen Academic Publishers 2021

regards the myth busting of Matthews as such a mistake. Kramer argues that this kind of critique is an example of 'scientism', by elegantly excluding subjective experiences – like taste:

> Taste some wine. Is a good Chablis really the same as any other Chardonnay grown in a comparably cool climate, never mind whether the soil is chalk or clay or sand? (...) Of course there are reasons: climate, microclimate, elevation, sunlight intensity, wine, rain and yes, soil. Believers in the existence of terroir 'are the first to mention all of these and more. Such differences are collectively called' terroir. (Kramer, 2016)

If taste is excluded from a discussion of terroir, winemakers and wine scientists will talk past each other regarding the ontological and epistemological status of terroir. In line with the perspective of Kramer, Inglis argues for a non-reductionist view of the concept of terroir:

> Terroir is an invention, a label put on to certain things, meaning they are perceived to be authentic. But the things that label refers to – ranging from the slope of a vineyard and the play of sunshine on it, to how grapes grow and when they are picked – are entwined with human semiotics but not completely created by them. (Inglis, 2015)

In other words, there is more to the concept of terroir than social construction. What is, however, the significance of bringing out the qualities of the place, the soil, the vines, the grapes – all these markers of authenticity? If it is more than a myth, more than a gimmick to increase the sales, why all the fuss with bringing out 'natural' properties in the wine?

## Naturalness

The value of the 'natural' is, like the values of the authentic and the terroir, quite controversial. To begin with, to place any special value with something being 'natural' has a hard time to get off the ground, since the concepts of 'nature' and 'naturalness' are very tricky to define. Which phenomena do these concepts include and exclude? In the sceptical view of John Stuart Mill, the concept of nature is either too wide or too narrow to point out a specific set of phenomena in the world: Either nature is taken to comprise of everything that behaves according to natural laws, which means that everything is natural, or nature is taken to comprise of everything that is independent of human intentional interference (Mill, 1969). In the latter sense, nothing on earth in the Anthropocene is natural. Regarding winemaking, a Millan would thus hold that of course any wine is natural, at the same time as no wine is natural. In this way, the notion of 'naturalness' is a non-starter.

To get going, we must somehow decide what does and does not fall under the notion of naturalness. In this line of thought, the notion gets an air of paradox: Naturalness should single out what is given and independent of us, but the notion itself is a cultural construction. As Gregory Kaebnick lets the sceptic sigh: 'Oh, 'nature'. That's a nineteenth-century intervention, you know.' (Kaebnick, 2013:2)

Moreover, any putative 'special value' of naturalness is quite easy to dismiss, as natural events are characterized by moral and aesthetical indifference rather guidance. Humans bring value to the world. Based on our values and needs, there are some phenomena we want to change or remove, like physical obstacles to building projects and germs causing human diseases. Other phenomena we leave to be – and call natural. But, deciding not to act is also an act. It is (to an increasing degree) our decision to leave places and organisms as they are, because this somehow aligns with our values and needs.

In sum, the notion of naturalness seems ill-equipped to have any normative significance. In the case of terroir, this could be brought out in this way: 'You hold that this wine is of superior value, because it expresses the natural properties of the terroir. But, why should I or anyone value anything just because you intentionally refrained from interfering with it?'

Nevertheless, vague and weak as it may be, the normative power of the notion of naturalness has a remarkable popularity and stamina in society. In 2015, the Nuffield Council of Ethics conducted an inquiry into 'what people really mean when they talk about naturalness' (Nuffield, 2015). They found that 'the terms natural, unnatural and nature are often used as placeholders for a range of different values or beliefs that are meaningful and important to people' (Nuffield, 2015). If this is the case, then when we say that something is (un)natural, we simply talk in a too loose way. What we should have said was something more precise.

The Nuffield report lends a hand to all of us here, by dividing the notion of naturalness into five more precise meanings: unnaturalness as disgust and monstrosity, naturalness as the wisdom of nature, naturalness as an empty term, naturalness as natural purpose, and naturalness as the will of God. In their report, the Council recommends that we 'avoid using the terms natural, unnatural and nature without conveying the values and beliefs that underlie them' (Nuffield, 2015).

## Terroir

Now, the proponents of the reality, significance and value of terroir does not as such adhere to the 'natural wine' movement. The proper way of expressing the terroir qualities in wine is not to just stand back and not interfere:

> To let nature express itself does not mean leaving the vats unchecked. Nature cannot express itself on its own; someone has to make it speak, and that requires both great skills and a well-equipped cellar. It is true that practices must be stripped of artifice, but this must be done selectively and skilfully. (Teil, 2012)

Geneviève Teil describes how terroir-conscious winemakers must learn the 'know-how-*not*-to' in order to let the wine express the terroir, but this does not mean that there is not still a lot of 'know-how-to' required:

> The task of 'terroir vintners' is to craft the expression of their terroir. To put it in musical terms, they perform a score composed by the terroir; for terroir to emerge, the performance must allow it to express itself fully rather than conceal it through overpowering practices. (Teil, 2012)

The making of terroir wines thus requires a balancing act of the given and the cultivated. Unlike the protected geographic origin of foodstuffs like marzipan and chocolate, there is on the one hand more to terroir than just craftmanship rooted in local traditions. And, unlike the labelling of foodstuffs like meat and fish by country, there is on the other hand more to terroir than geographic origin. Terroir comprises of both geographic origin and local craftmanship, and moreover poses an essential connection between the two: the aim of the local craftmanship is to bring out flavours stemming from the local soil in the best possible way.

Does it, however, really make sense to say that wine 'express' the terroir? According to Roger Scruton, it does not (Scruton, 2007). Scruton makes a distinction between truly communicative forms of perception, those of sight and hearing, and the non-communicative forms of tasting and hearing. We can represent other things and situations by presenting something visible or audible, like a text, picture, word, or piece of music. By contrast, Scruton holds, we are not able to do so by presenting a smell or a taste. Tastes and smells also lack the required internal order and sequence to make them suitable means of communication.

Is that so? Clearly, it is hard to communicate any complex meaning through a loaf of bread. What about wine? Cain Todd takes issue with Scruton's distinction by describing the highly acquired tastes that the appreciation of fine wine involves. Just like the appreciation of other art forms, like abstract paintings and instrumental music, the connoisseur of fine wine needs to connect with the expressive aspect of wines:

> Tastes can be skilfully combined and structured in wine to constitute, via interaction with us, certain expressive properties. These properties are response-dependent and not fully captured in non-expressive terms. Attributions of expressive properties to wine are thus the result of an imaginative awareness of agency guiding a process resulting in the specific non-expressive features (its tastes, smells, textures) partly constituting the expressive propertie. (Todd, 2012)

Todd's idea is, in other words, that the viticulturalist and winemaker can start out from the specific local conditions where the grapes are grown and harvested, and cultivate the sensory qualities of the resulting wine not only to arrive at pleasant tastes and smells, but in addition to express something by these tastes.

> The expression of terroir, for instance, is the result of the intentional act of the winemaker, and it takes great skill to marry the environment to the particular grape varieties used, to create a unique whole from an array of different parts. (Todd, 2012)

The presence and organization of tastes, smells and textures in the wine is thus expressive, in a similar way as the presence and organization of sounds is expressive in music. But, in contrast to music, as pointed out by Inglis in the quote above, the expression of terroir will inevitably be linked to the properties of a specific biotope. Again, one might say that like a composer should bring out the best of the properties of the instruments, the winemaker aims to bring out the best of the terroir, as crystalized in the grapes.

## Openness

Even if humans have conquered the world and are autonomous beings, there is still a lot of things that we do not control. We are still just a tiny speck in the vast universe. And a tiny virus can create havoc in our societies. Human beings are body and soul, and the major parts of our bodies and minds go on with their daily and vital activities uncontrolled by us: the heart beats, hormones regulate our moods, wounds are healed. Our semi-autonomous bodies are made of the things we eat and drink, and while we can choose what to devour, we have a very vague idea of the complex interaction of the substances we eat and how it is digested and used in our bodies. As human beings, we are one-part nature, and one-part culture.

The human baby is a natural product. It is created through the natural process of conception, gestation, and birth – sometimes with a little help from medicine. The genesis and appearance of a human child stems from the genetics and nurture of its origins and caretakers: its parents. Throughout life, a human being takes care of its body as a natural product, cultivated by its habitat and lifestyle.

Essential to the care of the body is everything that is applied to it or goes into it, like skin care, soap, drugs, foods, and drinks. We have an intimate relation to everything we devour, as we incorporate it. The human body can be embellished and manipulated, but only to a certain extent. With food and drink it is likewise. We appreciate foodstuffs that is natural and has a traceable origin, like ourselves.

Why do we appreciate the natural body and natural foodstuffs? Because it has an openness, something to be discovered, that too much manipulation threatens to hide or ruin. The value of naturalness lies in the limiting of undue demands of taking rational control and the resulting responsibility. (Birnbacher, 2019) In this way, to grow and make foodstuffs share ideals with bringing up a child: to guide and support the realization of the unique potential, and to let it shine. In this way does the narrative, ethos and value of terroir winemaking correspond to the ethos of parenting: to bring out and refine – not erase – the natural.

## References


Birnbacher, D. (2019). Naturalness. Online Encyclopedia Philosophy of Nature.

Clingeleffer, P. (2014). Terroir: The application of an old concept in modern viticulture. Encyclopedia of Agriculture and Food Systems, Academic Press, 277-288.

Inglis, D. (2015). On oenological authenticity: Making wine real and making real wine. M/C Journal, 18(1).

Kaebnick, G.E. (2013). Humans in nature: the world as we find it and the world as we create it. Oxford University Press.

Kramer, M. (2016). It's All Just Myths, You See. Wine Spectator. Available at: https://www.winespectator.com/articles/its-all-just-myths-you-see-53034.

Matthews, M.A. (2016). Terroir and other myths of winegrowing. University of California Press.

Mill, J.S., Robson, J.M., Priestley, F.E.L. & Dryer, D.P. (1969). Essays on Ethics, Religion and Society Essay on Mill's Utilitarianism. Toronto University of Toronto Press London Routledge and Kegan Paul.

Nuffield Council of Bioethics. (2015) (un)natural: Ideas about naturalness in public and political debates about science, technology and medicine. Nuffield Council of Bioethics. London.

Scruton, R. (2007). The philosophy of wine. In: C. Smith Barry (ed.) Questions of Taste: The Philosophy of Wine, Signal Books, Oxford, 1-19.

Teil, G. (2012). No such thing as terroir? Objectivities and the regimes of existence of objects. Science, Technology, & Human Values, 37(5), 478-505.

Todd, C. (2012). Expression and objectivity in the case of wine: Defending the aesthetic terroir of tastes and smells. Rivista di Estetica 51: 95-115.

# 48. Animal business: an exploration of corporate responsibility towards animals

*M.R.E. Janssens*
*Erasmus University Rotterdam, Burg. Oudlaan 50, 3062 PA Rotterdam, the Netherlands; moniquejanssens@ethischbedrijf.nl*

## Abstract

Using insights from business ethics and animal ethics, this paper argues that companies have a strong responsibility towards animals. It argues that animals have a moral status and that moral actors have the moral obligation to take the interests of animals into account. As moral actors, companies should take the interests of animals into account, more specifically their current and future welfare, including continuing a good life. Based on this corporate responsibility, categories of corporate impact on animals in terms of welfare and longevity are offered, with matching normative implications.

**Keywords:** corporate responsibility, animal welfare, moral status, business practice, value chain

## Introduction

Societal and philosophical support for the idea that animals should receive more moral protection is growing. Consumers, investors, and NGOs have become critical of the treatment of animals by corporations and nations (Amos and Sullivan, 2018; Anonymous, 2015). Nevertheless, normative explorations of the responsibility of companies towards animals are scarce (Van Liedekerke and Dubbink, 2008). Therefore, in this paper I will explore the ethical responsibility of companies towards animals.

## The moral status of animals

'To have a moral status is to be morally considerable, or to have moral standing. It is to be an entity towards which moral agents have, or can have, moral obligations', states Warren (1997:3). She argues that sentience of organisms provides a moral status and is therefore a reason not to kill them or cause them pain (Warren, 1997). Rollin (2016) agrees with Warren that sentience is a sufficient condition for moral status. From a Kantian perspective, Korsgaard also agrees to attribute moral status to sentient animals, stating that animals pursue the objects of their inclinations just like humans do, and that both are creatures for whom things can go well or badly (Višak and Garner, 2016). Singer's (1975/2009) utilitarian view on ethical evaluation and decision making is based on a moral status of animals dictated by their sentience. Regan (1983/2004) acknowledges moral status to individuals who have beliefs, desires, preferences, intentionality, memory, and a sense of their future. He defines mammals as subjects-of-a-life and therefore 'moral patients': they have a life that is theirs and can undergo moral acts from moral agents. In other words: there are several ethical approaches that acknowledge a moral status to mammals and in many cases vertebrates, based on their sentience.

## Moral obligations

If sentient animals have a moral status it is an obligation of moral agents to give them moral consideration. To determine how we should act as humans, we need ethical theories. For the purpose of relating to business ethics, specifically those theories acknowledged in business ethics are highlighted briefly: utilitarianism, duty-based ethics and virtue ethics.


Hanna Schübel and Ivo Wallimann-Helmer (eds.) ***Justice and food security in a changing climate***
EurSafe 2021 – DOI 10.3920/978-90-8686-915-2_48, © Wageningen Academic Publishers 2021

Singer's (1975/2009; 1981/2011) utilitarianism argues that the interests of sentient animals as 'their' interests should be taken into account in assessments of right and wrong. Although it is difficult to identify how 'exactly' different interest should be weighed, in many cases, according to Singer, it is not that difficult.

The most well-known duty-based approach to ethics is Kant's deontological theory, which acknowledges no direct human duties regarding animals (Kant, 1785/1902/1996, 1963). This stance had been challenged by several successive duty-based approaches. Regan (1983/2004) offers a duty-based approach to the moral status of animals. He argues that subjects-of-a-life should be treated respectfully and not harmed, which implies direct moral duties of moral agents. Garner (2013) states that at least some animals have an interest in not suffering, which implies that moral agents have duties towards them. Kaldewaij (2013) shows that even starting from the Kantian idea that similarities between rational actors lead to direct duties towards humans, direct duties of humans towards animals are defendable.

A third group of approaches is based on Aristotelian virtue ethics. Nussbaum (2007) argues that, just like humans, animals have a *telos* and should be allowed to exercise certain capabilities so that they can flourish. The core of Rollin's (2006) approach is also the animal's *telos* (Rollin, 2016: 51), from which needs and desires flow. Not satisfying these needs and desires results in poor welfare, which should be avoided.

Apart from these three main approaches in ethics, several authors have taken other routes and reached similar conclusions, e.g. Cochrane (2013), Garner (2013), Donaldson and Kymlicka (2011), and Schmitz (2016): humans should attribute sentient animals a moral status and consider their interests. The broad spectrum of ethical approaches coming to a similar conclusion, offers enough reasons to avoid harming sentient animals.

## The responsibility of companies: animals as stakeholders

Companies have societal responsibilities, as has been argued thoroughly from the point of view of both business ethics (Isaacs, 2011; Kaptein and Wempe, 2002; Kolstad, 2017; Pettit, 2007) and CSR (Van Marrewijk, 2003; Visser, 2014). Within these companies, powerful people can, and should, take individual responsibility and use their power to initiate responsible collective corporate actions. Groups of these powerful individuals (boards, for example) are collective agents. But, Isaacs warns, fuzziness of roles and tasks can pose a challenge to the outcomes. These responsibilities have hardly been applied structurally to the interests of animals (Janssens and Kaptein, 2016; Van Liedekerke and Dubbink, 2008).

Stakeholder theory offers supporting arguments for corporate responsibilities towards animals, because it states that companies have an ethical responsibility to create value for stakeholders (Freeman *et al.*, 2010; Freeman and Velamuri, 2005). A broadly accepted definition of a stakeholder is: any group or individual who can affect or is affected by the achievement of the organization's objectives (Donaldson and Preston, 1995; Freeman, 2010). Animals differ from (other) stakeholders in two possibly relevant ways. The first difference that may be relevant is that they are non-human, but the definition is unclear about whether being human is a necessary feature of stakeholders. The definition does say, however, that it is about groups or individuals who can affect or are affected by the achievement of the organization's objectives, which can be the case for animals. The ethics approaches I discussed above agree (in different words) that sentient animals have interests. Therefore, there is undeniably something at stake for them and they should be included in the stakeholder concept. This stance is supported by Hart and Sharma (2004), who argue in favour of including currently marginalised stakeholders like the poor, the weak, and the non-human, and by Mitchell *et al.* (1997), who classify animals as 'dependent stakeholders'.

This classification of being dependent of others to look after your interests and speak up for you leads us to the second difference: that they cannot speak for themselves (at least not in human language). This can be overcome by letting animal-protection NGOs speak for animals (Carroll, 1991), using scientific knowledge about their preferences (Janssens and Van Wesel, 2018), observing their behaviour (Wemelsfelder *et al.,* 2000) and putting an effort in communicating with them (Meijer, 2017). Webster (2006) describes farm animals as stakeholders of farming, besides farmers and customers.

## The harm of taking life and the replaceability argument

An ongoing discussion in animal ethics is the question of whether taking the life of healthy animals without causing them pain or distress constitutes harm. Regan (1983/2004) is rather clear about this: moral patients have an inherent value, and therefore killing them is a violation of their moral rights. Telos-based approaches are also clear on the issue: Nussbaum (2007) includes continuing life in the list of capabilities we should allow sentient animals; Rollin (2006) takes the position that taking the life of an animal needs thorough justification, which he thinks is lacking in, for example, the case of eating them. At first glance one would think that from a utilitarian perspective, taking life painlessly is not diminishing welfare as long as the animal is not aware of death approaching. No harm has been done before death, and from the moment of death, there is no living animal anymore to have an interest in welfare. This view is opposed by preference-utilitarianism, for which killing is taking away future fulfilment of preferences (Singer, 1993/1999). My conclusion is that killing is a harm if done to a being that at the moment of death has an interest in fulfilling future preferences and enjoying welfare in the future.

The replaceability argument could be brought forward against our case. In a nutshell, it goes like this: Moral actors have a moral duty to maximize overall welfare, but it does not seem to make any difference to which individual this welfare is attached, except that all individuals count equally. This means that taking an animal's life is permissible if the animal is replaced by a newly bred animal with at least the same welfare level (Singer, 1993/1999). Would it then be ethically neutral to kill animals and replace them with equally happy animals?

One of the answers to this question is from Chappell (2013). He argues that an agent with enough life-saving anti-venom for one person would find it horrible to choose which of two poisoned persons to save. Finding this choice horrible is what makes her a moral agent, and it shows that she accords separate value to the interest of each individual, and that those who count are not mere receptacles of welfare. Their interests count equally but as distinctive intrinsic goods. Trade-offs can be made, but a benefit to person A does not absorb a lesser loss by person B. I think Chappell's argument is well grounded, and applies to animals' welfare as well, because animals are individuals too (Braithwaite, 2010; Rollin, 2006). Without analysing the replaceability argument any further, I will assume that killing an animal is an infringement upon the animal's welfare and therefore a harm done to the animal, even if the animal is replaced. If its life is bad, the effort should go into improving the situation.

## Conclusion

My conclusion is that powerful people in companies bear an individual responsibility for the impact of their actions on the interests of animals and that, in addition, the company has an institutional responsibility. This responsibility can be fulfilled by, respectively, actions of individuals within the company and actions of the company as an institution. This leads to the following obligations of companies towards different categories of impacted animals:

- *Animals kept in a negative welfare state, bred and killed for economic reasons* (e.g. industrially kept pigs, broiler chickens, minks, some laboratory animals). Companies who bear direct or chain responsibility for them should consider their living and dying conditions critically, look for ways to solve welfare issues, and consider innovating towards alternate products that make killing animals unnecessary.

- *Animals kept in a positive welfare state, bred and killed for economic reasons* (e.g. free-range pigs and broilers). Companies who bear direct or chain responsibility for them should consider their dying conditions critically, look for ways to solve welfare issues, and consider innovating towards alternate products that make killing animals unnecessary.
- *Animals kept in a negative welfare state, used for economic purposes over a longer period* (e.g. industrial dairy cows and laying hens, circus animals of wild species). Companies who bear direct or chain responsibility for them should consider their living and dying conditions critically and consider innovating towards alternate products that make keeping them unnecessary.
- *Animals kept in a positive welfare state, used for economic purposes over a longer period* (e.g. free-range dairy cows and laying hens, roaming sheep). Companies who bear direct or chain responsibility for them should monitor their welfare to make sure there is no decline, consider their dying conditions critically, and, in case of persistent welfare and killing issues, consider innovating towards alternate products that make keeping them unnecessary.
- *Animals as by-products, killed at an extremely young age for economic reasons* (e.g. male chicks and steers). Companies who bear direct or chain responsibility for them should consider their living and dying conditions critically, take measures to solve welfare issues, and consider innovating towards alternate products or production methods that prevent their coming to life as by-products.
- *Free-living animals, killed or harmed for economic reasons or as a result of economic activity* (e.g. fishes, house mice in buildings, frogs on building sites). Companies who bear direct or chain responsibility for them should consider their impact on these animals critically, and look for measures, alternate processes or products that solve welfare and killing issues.

Critics may argue that the categories are neither complete nor absolute. Intermediate categories can occur (animals taken from the wild to be kept), or animals may change categories (sheep living in a positive state shifting to a negative state when transported for slaughter). In this case, in-between categories or shifts can easily be derived.

## Implications, recommendations and conclusions

I have argued that companies bear responsibilities towards sentient animals in the value chain, which oblige them to consider how to prevent inflicting harm on them, including killing, and to protect their welfare. Nevertheless, in business practice animals are often treated badly by companies (Engster, 2006; Fraser, 2008), or even abused on a large scale (Braithwaite, 2010; Foer, 2009; Leder, 2012; Singer, 1975/2009). A paradigm shift is necessary for businesses to accept that sentient animals have interests that count ethically amongst other interests in and outside the company, and that they should be included in the stakeholder concept. Harming animals is ethically negative and asks for thorough ethical accountability. The impact on animals of companies should be reconsidered and set against the societal necessity or value of the products and services produced. If one includes value for animals into total value, one could say that those who mistreat or kill animals remove value from the value chain rather than add value.

Responsibilities can be limited by vagueness about what can be done, secondly by wrongful social practice combined with widespread ignorance, and thirdly by lack of power (Isaacs, 2011). For most companies, operating in contexts of abundant information and shrinking global distances, the first two limiting factors will play a minor role. The third limitation, lack of power, can at least partly be solved by efforts to cooperate within industries and production chains (Janssens and Van Wesel, 2018).

Aside from death, welfare problems in industrial farming are overwhelming. Measures for enhancing animal welfare will differ between groups of animals. Examples are: cage enrichment, more space, better housing, free-range options, social contacts, safer transport, and humane slaughter methods. Nevertheless, taking animal life will still be removing value. If companies would attribute the interests of animals a relatively heavy weight in their ethical assessments, the far-stretching ideal of a vegan society seems feasible in the future. To reach that point, companies could invest in the development of imitation meat and cultured meat, and promote vegetables, beans, nuts, and plant-based 'dairy'. Economic, political, cultural, and other obstacles should not prevent companies from doing the right thing.

## References


Amos, N. and Sullivan, R. (2018). The Business Benchmark on Farm Animal Welfare; 2017 Report. BBFAW, UK.

Anonymous (2015). Special Eurobarometer 442 Report: Attitudes of Europeans towards Animal Welfare. Available at: http://www.eurogroupforanimals.org/eurobarometer.

Braithwaite, V. (2010). Do fish feel pain? Oxford University Press, UK, 194 pp.

Carroll, A.B. (1979). A Three-Dimensional Conceptual Model of Corporate Social Performance. Academy of Management Review 4: 497-505.

Carroll, A.B. (1991). The pyramid of corporate social responsibility: Toward the moral management of organizational stakeholders. Business Horizons 34(4): 39-48.

Chappell, R.Y. (2013). Value Receptacles. Noûs, 49, 322-332.

Cochrane, A. (2013). From human rights to sentient rights. Critical Review of International Social and Political Philosophy 16: 655-675.

Donaldson, S. and Kymlicka, W. (2011). Zoopolis: A Political Theory of Animal Rights. Oxford University Press, USA, 352 pp.

Donaldson, Th. and Preston, L.E. (1995). The Stakeholder Theory of the Corporation: Concepts, Evidence, and Implications. The Academy of Management Review 20: 65-91.

Engster, D. (2006). Care Ethics and Animal Welfare. Journal of Social Philosophy 37(4): 521-536.

Foer, J.S. (2009). Eating Animals. Little, Brown and Company, USA, 341 pp.

Fraser, D. (2008). Toward a global perspective on farm animal welfare. Applied Animal Behaviour Science 113(4): 330-339.

Freeman, R.E., Harrison, J.S., Wicks, A.C., Bidhan, L.P. and De Colle, S. (2010). Stakeholder Theory: The State of the Art. Cambridge: Cambridge University Press, UK.

Freeman, R.E. and Velamuri, S.R. (2008). A New Approach to CSR: Company Stakeholder Responsibility. Available at: https://ssrn.com/abstract=1186223.

Garner, R. (2013). A theory of Justice for Animals: Animal Rights in a Nonideal World. Oxford University Press, UK, 197 pp.

Hart, S.L. and Sharma, S. (2004). Engaging fringe stakeholders for competitive imagination. Academy of Management Executive 18: 7-18.

Isaacs, T. (2011). Moral responsibility in collective contexts. Oxford University Press, UK, 216 pp.

Kaldewaij, F. (2013). The animal in morality, Justifying duties to animals in Kantian moral philosophy. Utrecht University Press, Netherlands.

Kant, I. (1785/1902/1996). Groundwork for the metaphysics of morals (trans: Gregor, M.). Cambridge University Press, UK.

Kant, I. (1963). Lectures on ethics (trans: Infield, L.). New York: Harper and Row.

Kaptein, M. and Wempe, J. (2002). The balanced company: A theory of corporate integrity. Oxford University Press, UK, 360 pp.

Kolstad, I. (2007). Why Firms Should Not Always Maximize Profits. Journal of Business Ethics 76(2): 137-145.

Leder, D. (2012). Old McDonald's Had a Farm: The Metaphysics of Factory Farming. Journal of Animal Ethics 2(1): 73-86.

Meijer, E. (2017) Political animal voices. University of Amsterdam, Netherlands, 303 pp.

Mitchell, R.K., Agle, B.R. and Wood, D.J. (1997). Toward a Theory of Stakeholder Identification and Salience: Defining the Principle of Who and What Really Counts. Academy of Management Review 22(4): 853-886.

Nussbaum, M. (2007). Frontiers of Justice: Disability, Nationality, Species Membership. Harvard University Press, USA, 487 pp.

Pettit, Ph. (2007). Responsibility Incorporated. Ethics 117(2): 171-201.

Regan, T. (1983/2004). The Case for Animal Rights. University of California Press, USA, 425 pp.

Rollin, B.E. (2006). Animal Rights & Human Morality. Prometheus Books, USA, 400 pp.

Rollin, B.E. (2016). A New Basis for Animal Ethics: Telos and Common Sense. University of Missouri Press, USA, 208 pp.

Schmitz, F. (2016). Animal Ethics and Human Institutions. In: R. Garner and S. O'Sullivan (eds) The Political Turn in Animal Ethics. Rowman & Littlefield, UK/USA, pp. 33-49.

Singer, P. (1975/2009). Animal Liberation. HarperCollins, USA, 311 pp.

Singer, P. (1981/2011). The Expanding Circle: Ethics, Evolution, and Moral Progress. Princeton University Press, USA, 232 pp.

Van Liedekerke, L. and Dubbink, W. (2008). Twenty Years of European Business Ethics: Past Developments and Future Concerns. Journal of Business Ethics 82: 273-280.

Van Marrewijk. M. (2003). Concepts and Definitions of CSR and Corporate Sustainability: Between Agency and Communion. Journals of Business Ethics 44(2): 95-105.

Višak, T. and Garner, R. (2016). Introduction. In: T. Višak, and R. Garner (eds) The ethics of killing animals. Oxford University Press, UK, pp. 1-13.

Visser, W. 2014. CSR 2.0: Transforming Corporate Sustainability and Responsibility. SpringerBriefs in Business, USA, 89 pp.

Warren, M.A. 1997. Moral Status; obligations to persons and other living things. Oxford University Press, UK, 265 pp.

Webster, J. (2006). Ideals and Realities: What Do We Owe to Farm Animals? In: J. Turner and J. D'Silva (eds) Animals, Ethics and Trade: The Challenge of Animal Sentience. Earthscan, UK, pp. 149-158.

Wemelsfelder, F., Hunter, E.A., Mendl, M.T. and Lawrence, A.B. (2000). The spontaneous qualitative assessment of behavioural expressions in pigs: first explorations of a novel methodology for integrative animal welfare measurement. Applied Animal Behaviour Science 67(3): 193-215.

# 49. Visions of gene technology

*S. Kjeldaas* and T. Antonsen*
*GenØk-Centre for Biosafety, SIVA Innovasjonssenter, 9294 Tromsø, Norway; sigfrid.kjeldaas@uit.no*

## Abstract

This article presents visions for the future of gene technology as presented by members of the Norwegian public in response to suggestions by the Norwegian Biotechnology Advisory Board to revise the Norwegian Gene Technology Act. Taking as our starting point the performativity and normativity of techno-scientific visions, we investigate 'what kind of futures' the different actors in the Norwegian debate imagine the new gene technologies to bring forward, particularly in response to challenges to our food system such as sustainability, environmental degradation and climate change. Our findings reveal how different visions emerge from different understandings of environmental or systemic problems of the present. The use of developmental and environmental techno-scientific 'salvation' narratives by proponents of gene technology explains why many agricultural, environmental and scientific organizations refrain from using climate change arguments in their responses, and why they offer less confident visions of the future. We argue that discussions on gene technology's role in our common future cannot proceed until they include the state of the environmental present and its causes.

**Keywords:** GMOs, gene editing, futures, sustainability, biodiversity

## Introduction

There are three main reasons why techno-scientific visions of the future make an interesting study: (1) they may function as rhetorical tools helping actors gather support and resources in ways that alter scientific and political priorities, and may influence democratic processes by defining relevant actors (Davies and Horst, 2016); (2) they reveal the range of imaginaries on the use of technologies: what kinds of problems they will solve and how they will benefit society; (3) they expose the kind of world the different actors imagine the technologies to be part of (Haraway, 1997). If the aim is to bring gene technologies into the common world in ways that ensure that this world is a *good* world for all involved (Latour, 2004), visions of the future provide important windows into the values that shape these imagined worlds.

The Norwegian Gene Technology Act of 1993 (GTA) is unique in an international context in that it applies a precautionary principle to the production and use of genetically modified organisms (GMOs), and demands that sustainability, societal utility and ethics be part of their assessment (§1). This has caused regulation of GMOs in Norway to be restrictive and public debate to include the ethical and cultural aspects of the use of gene technology. In 2017, the Norwegian Biotechnology Advisory Board (NBAB) invited the public to debate their suggestions for revision of the GTA. The revision sought to accommodate the treatment of new genome-edited organisms within the Act and, through this, the future use of new and existing gene technologies. The invitation was presented in the form of a 52-page long hearing report on the state of development of the new gene technologies and on some of the scientific, socio-economic and ethical dilemmas involved in their regulation (NBAB, 2017). 46 Norwegian organizations, research institutions and individuals offered their responses.

We have applied poststructuralist discourse analysis to investigate how the use of concepts and narrative strategies in the hearing responses vary according to the respondents' vocational, political and ideological positions within the field of debate. Our analysis reveals visions of the future to constitute important


EurSafe 2021 – DOI 10.3920/978-90-8686-915-2_49, © Wageningen Academic Publishers 2021

frameworks on which argumentation regarding the use of gene technology – and the legislative revisions needed to promote or discourage this use – depend. What is presented in the following is a summarized account of the content and logic of these visions, and of the conception of the present on which they rely.

In presenting our findings, we would like to specify two things: (1) the hearing invitation asked only for responses to the suggested legislative revisions, to the extent that respondents provided visions of the future, they did so on their own initiative. (2) the future visions presented by the hearing respondents varied less between the different category organizations (umbrella organizations and industrial, agricultural, environmental and/or scientific organizations) than between what we might broadly distinguish as developers/producers of gene technology, and others. Only agricultural organizations, which we identify as potential users of gene technology, deviated from this pattern. In the following, we present the visions of 'producers' and 'non-producers' separately and include the visions of agricultural organizations in the latter category.

## Promising futures and their rhetoric

'Producers' of gene technology and GMOs appear to have clearer, more confident and positive visions for the future than other societal actors. To umbrella organizations of industry and developers of gene technology, the vision of (potentially large-scale) development of gene edited organisms is in itself positive because of the way this is linked to the development of new and expansive global markets in such organisms. Arguing in a variety of ways for how it will contribute to higher and more stable production rates in all parts of the production chain, such actors promote gene technology as the means through which both the sustainability and the overall capacity of food production may be enhanced. In these 'producer' visions, sustainability is associated with production systems and their (biological) carrying capacity. They focus primarily on the economic aspects of sustainability, to the neglect of its social and environmental aspects (Cardonna, 2014).

The future projected by proponents of gene technology operates with modern, abstract conceptions of time and can be characterized as an indistinct future of aspirations. The only time indicators of this future are the full-scale presence of environmental problems like climate change and a fully expanded world population. Against the threat of future actualizations of such looming unknowns, the threat of unknowns associated with the use of gene technologies appear miniscule. Having inscribed the major challenges of our time as problems to be alleviated through the use of gene technology, the only peril to this salvatory vision is the current state of restrictive legislation which may block its manifestation.

## More uncertain futures

Societal actors who do not fit the 'producer' category are more reluctant to offer up visions of the future. To the extent that they do, these visions are both hopeful and pessimistic. Those who offer positive visions present them in the form of politically oriented goals, generally of maintaining the legislative status quo in spite of increasing economic pressures to de-regulate the market in GMOs. The acknowledgement that these *are* political goals, and that they may run counter to economically motivated forms of techno-scientific development, is very likely the reason why these visions are presented with a greater sense of unease.

Among agricultural, environmental and scientific organizations we also find actors who imagine a future in which there is a direct correlation between easily accessible and affordable technology and increasing numbers of GMOs released into the environment. In this more pessimistic vision, a technologically enhanced future is one of enhanced risks (Beck, 1995); one in which there is a need for close and constant surveillance of ecosystems exposed to the direct and indirect effects of agricultural and other production environments using GMOs.

Concern for ecosystems and living beings also inform the understanding of sustainability. One agricultural organization defines sustainability in terms of 'giving plants and animals as good and natural conditions for life as possible,' thereby maintaining their health and vitality. Although there might be economic gains to be had in reducing the use of pesticides and other chemicals, focus is less on the economic aspects of sustainability and more on an ethics of care. Aiming to sustain and develop already existing natural systems of protection internal as well as external to the individual organism, this care for the production organism is associated with an extended sense of care for the ecology of cultured as well as natural ecosystems (Bellacasa, 2017; Preston and Wickson, 2016). This is sustainability of a different and less economically exciting kind than the one promoted by proponents of gene technology; one in which the focus is on the mundane task of maintaining and nurturing already existing ecological relationships (Bellacasa, 2017) and unacknowledged ecosystem services.

## Environmental threat as argumentation for technology

In line with the UN sustainability goals, alleviating the effects of climate change is frequently launched as the ultimate goal of the emerging gene technologies (Det Etiske Råd, 2019; GENEinnovate, 2020). Yet in our data, scientific environments and environmental and agricultural organizations characterized as 'non-producers' appear to be more reluctant than 'producers' of gene technology to apply climate change argumentation in their visions for the future. In order to account for this somewhat surprising finding, we must have a look at the effects of climate change argumentation in the responses of 'producers' of gene technology.

### *The history of civilization*

The department of bioscience at a major Norwegian university exemplifies how narratives of environmental threats can be utilized in support of gene technology. It identifies the 'need for production of safe food' for a growing world population; the necessity of 'fast adaptations' in aqua- and agriculture to counter the effects of climate change; and the 'catastrophic reductions in biodiversity, deforestation and extensive ecosystem changes on a global scale' as challenges that necessitate the use of 'novel technological inventions that to the greatest possible extent make us able to attain sustainable solutions efficiently.'

This and other 'producers' provide a historical narrative which anchors the development of our civilization in agriculture – and turns its history into a history of genetic manipulation of other living beings for our own benefit. The point of this narrative is to communicate the idea that our civilization could not have reached the current state of sophistication (and relative comfort for human beings) without the practice of genetic manipulation – which must be understood as fundamentally part of our cultural history. The evolutionary derailment of other species is, in other words, what has caused the speed and direction of the evolution our own civilization. The environmental challenges facing us at present – particularly those connected with global climate change – constitute a call for even more consciously directed, visionary and inclusive processes of genetic manipulation of living beings if the forward momentum of our civilization is to be maintained.

### *Climate change adaptation*

A Norwegian plant breeder organization articulates how '[g]ene editing will help in our work to develop new varieties adapted to our Nordic climate' as this becomes impacted by climate change. Without access to gene editing and genetically edited material from abroad, it claims, Norwegian agriculture will lose its ability to combine characteristics from good foreign varieties with Norwegian ones and be 'less prepared to develop our own climate and area-specific varieties necessary in the future.'

The argumentation has implications for the way we conceptualize both place-specific adaptation to climate change and naturally occurring genetic material. It indicates that in order to produce place-adapted plants for the future, we have to rely not merely on gene editing of local varieties, but on importing genetically edited varieties from producers abroad. This concept of climate-based place-adaptation relies on breeders' ability to predict future climates and climatic stress factors on our most important commercial crops and to counteract them in well-known and well-controlled production systems. It comes with the presumption that climate change comes in predictable patterns, and devalues considerably the reactionary capacity of 'natural genes' and local gene pools. Valuable genetic material becomes that which has been technologically modified for productive efficiency and climatic resistance by humans – even when this modification has taken place in another country.

### *Biodiversity*

Within this and other narratives of human-directed evolution of other species, more extensive use of gene technology is perceived as contributing to the conservation of biodiversity in two ways. Firstly, by increasing the efficiency of agricultural production systems, the use of gene modified organisms will reduce the need for production acreage and allow not only the preservation but ideally also the opening up of areas in which natural ecosystems may flourish. This vision implies an intensification of production on already existing agricultural land. It is implicitly dependent also on other measures – like political decisions or legislation to regulate and conserve areas of nature untouched by agricultural and other industries. However, as these other preconditions are neither mentioned nor perceptibly a part of the vision itself, the autonomy and importance of science and scientists in leading society's response to the environmental threats of the future is reinforced (c.f. Jasanoff, 2005).

Secondly, gene technology will contribute to biodiversity by 'preventing production species from breeding with natural/wild species, and in this way secure the species diversity of nature' (quote from a university bioscience department). As the design and use of genetic watermarks are vital to this vision, it should be noted that both the hearing invitation report (NBAB, 2017) and other respondents debate the adequacy of this practice, as it entails more extensive modifications than necessary to the genomes of living beings. We find it problematic for its narrow understanding of biodiversity and what is necessary for its protection. Leapfrogging important aspects of 'the variety of life' between naturally occurring individuals and gene pools across populations, communities and ecosystems (Convention on Biological Diversity, 1992), biodiversity becomes in this response a question of non-contamination by organisms produced through industrial farming and gene technology. Gene technology – a natural extension of industrial agriculture – is envisioned as re-establishing the purity of natural ecosystems which agriculture (earlier praised as a natural part of the evolution of our civilization) continuously threatens to contaminate.

### *Separate narratives and narratives of separateness*

In the techno-scientific narrative of civilization, a biodiverse nature is the starting point away from which our technocratic knowledge society and its production organisms have evolved. In future-oriented techno-scientific salvation narrative, it is the endpoint of the evolution of gene technologies, but not in the sense that the gene technologies in themselves contribute to diversification. Rather, biodiversity is understood as that which has remained – and drastically crumbled – beyond the boundaries of civilization and its production systems. By keeping the biodiversity of nature and the production systems of agriculture absolutely separate, gene technology can be launched as the tool through which the new technologies will stop further expansion in land used for food production and hinder contamination by unwanted genetic material modified by humans. By leaving out discussions of what effect large-scale industrial aqua- and agricultural practices have had on naturally occurring ecosystems, plant and animal populations and their vitality, gene technology becomes a guarantor for the protection of untouched tracts of nature and the economic sustainability of efficient production systems becomes the fundament for the kind of ecological sustainability that will lead us through the environmental crisis.

In summary, the environmental argumentation presented by proponents of gene technology involves reductive interpretations of environmental challenges and their causes which diminishes issues of biodiversity to questions of production acreage and non-contamination. It also indirectly reduces the value of natural genetic material as well as of traditional agricultural practices and/or breeding techniques, both of which are understood to be less than optimally productive and too slow in developing the necessary responses to oncoming climate change. Making use of modernist and abstract timescapes, the climate argumentation presented in these hearing responses underscores how gene technology can de-couple targeted organisms both from the constrictions of place and from the timescales of natural organisms and processes. By genetically accelerating 'some' (positive) developmental processes, the argument goes, we will be able to revert other (negative) ones. In the case of climate change the latter, of course, are 'of our own doing'. Rather than recognizing how our civilization's technological-scientific development has led us into the environmental crisis now facing us, proponents of the technology use the time pressures of oncoming climate changes to promote and naturalize more extensive forms and scales of genetic modification of the living beings that constitute our food production systems.

## Environmental threat as argumentation against technology

Contrary to 'producers' of gene technology, agricultural and environmental organizations identify industrial agriculture to be a major cause of problems associated with the current environmental crisis. Because they are part of the same systems of techno-scientific and industrialized production, gene technologies cannot be the solution to these problems. Rather, they pose a threat because of the way they attempt to remedy unethical and unsustainable forms of agriculture and prevent a more profound discussion of alternative solutions to the problems these production systems have caused.

Recognizing the extent to which natural ecosystems are today threatened by the practices of industrial agriculture, respondents in the 'non-producer' category believe these threats will be intensified rather than alleviated if the development of new and easily accessible gene technologies allow the release of novel GMOs into the environment. This discloses their distrust in the idea of genetically induced 'non-contamination' between production sites and surrounding ecosystems. In expressing their concern for the 'balance of nature,' some agricultural, environmental and research organizations reveal their fear that non-local, genetically engineered plants and animals may come to disrupt the intricately woven fabric of relationships that constitute natural ecosystems, many of which are still undetected. These organizations highlight the value, variability and environmental response ability of natural genetic material and naturally occurring ecosystems – and suggest seed banks rather than molecular labs as remedies to agricultural monocultures' problems of genetic homogeneity/lack of genetic variability.

'Non-producer' respondents to the Norwegian hearing in this way harbour an orientation towards relationality and environmental complexity that 'producer' respondents lack. They acknowledge the unintended effects of current production systems as complicating background against which both intended and unintended effects of the new gene technologies must be evaluated. Accordingly, their focus is less on the practical problems to be solved by gene technology in the future than on the environmental problems already facing us. If a future is envisioned, this is one of future generations; an entity partially abstract, partially one into which the concrete and personal ties of the present bind us with a sense of care.

## Conclusion

Our analyses show that different visions of the future are associated with different understandings of fundamental societal or environmental problems. Whether gene technologies act as saviour or villain (the cause of already existing problems) in narratives of the future depends on the different actors' relationship with the present, and how this present is understood to have emerged. We question the practice of projecting visions of a technologically enhanced and sustainable future without first seeking a more nuanced and common understanding of this present. Only on such a basis can gene technologies to be of lasting importance in a new and more sustainable future.

## References

Beck, U. (1995). Ecological Enlightenment. Essays on the Politics of the Risk Society. Trans by Ritter, M.A. Humanities Press, Amherst, N.Y., 159 pp.

Bellacasa, M.P. (2017). Matters of Care: Speculative Ethics in More Than Human Worlds. U of Minnesota P., Minneapolis, M.N. 269 pp.

Cardonna, J.L. (2014). Sustainability: A History. Oxford UP, New York, N.Y., 332.

Convention on Biological Diversity (1992). Available at: https://www.cbd.int/convention/articles/?a=cbd-02.

Davies, S.R and Horst, M. (2016). Science Communication. Culture, Identity and Citizenship. Palgrave, London, UK, 266 pp.

Det Etiske Råd (2019). GMO og etik i en ny tid. Available at: https://www.etiskraad.dk/etiske-temaer/natur-klima-og-foedevarer.

GENEInnovate (2020). Norske forbrukeres holdninger til genredigering i norsk landbruk og akvakultur. Available at: https://www.bioteknologiradet.no/filarkiv/2020/04/Rapport-holdninger-til-genredigering.pdf.

Haraway, D. (1997). Modest_Witness@Second_Millennium.FemaleMan_Meets_OncoMouse.

Routledge, New York, N.Y., 361 pp.

Jasanoff, S. (2005). Designs on Nature: Science and Democracy in Europe and the United States. Princeton U.P., Princeton, N.J., 374 pp.

Latour, B. (2004). Politics of Nature: How to Bring the Sciences into Democracy. Trans by Porter, C. Harvard U.P., Cambridge, M.A., 307 pp.

Ministry of Climate and Environment. Gene Technology Act (English translation). Available at: https://www.regjeringen.no/en/dokumenter/gene-technology-act/id173031/.

Norwegian Biotechnology Advisory Board (2017). Genteknologiloven – invitasjon til offentlig debatt. English translation (2018) and hearing responses available at: https://www.bioteknologiradet.no/2018/12/genteknologiloven/.

Preston, C. and Wickson, F. (2016). Broadening the lens for the governing of emerging technologies: Care ethics and agricultural biotechnology. *Technology in Society* 5: 48-57.

# 50. Hornless cattle – is gene editing the best solution?

*P. Sandøe[1,2*], S. Borchersen[3], W. Dean[1], P. Hyttel[2], L.P. Sørensen[3] and C. Palmer[4]*
*[1]University of Copenhagen, Department of Food and Resource Economics, Rolighedsvej 23, 1958 Frederiksberg C, Denmark; [2]University of Copenhagen, Department of Veterinary and Animal Sciences, Grønnegårdsvej 8, 1870 Frederiksberg C, Denmark; [3]VikingGenetics, Ebeltoftvej 16, Assentoft, 8960 Randers SØ, Denmark; [4]Texas A&M University, Department of Philosophy, College Station, Texas, TX 77843-4237, USA; pes@sund.ku.dk*

## Abstract

Most dairy cows in Europe and the US have no horns. But this is mostly an artefact. Typically, the ability to grow horns is removed by means of so-called disbudding which, even when it is done with the use of local analgesia in combination with painkillers, may have long term negative effects on animal welfare. This suggests the need to seek alternatives. One alternative approach makes use of the genetic disposition not to grow horns (called polledness), which occurs naturally among domestic cattle but is not widespread within the typical dairy breeds. For economic reasons, the achievement of 100% polled dairy cattle through conventional breeding has a very long time-horizon. Gene-editing, most likely by using CRISPR-Cas 9, is an obvious alternative technique that has been shown to work. In this paper we consider whether using gene-editing is preferable to disbudding and conventional breeding from an ethical and a regulatory perspective. We discuss four kinds of ethical concerns: (1) naturalness; (2) respect for animal integrity; (3) animal and human welfare; and (4) human health and safety. Regarding (1) we argue that gene-edited cows are not significantly more unnatural than other modern cattle. Regarding (2) we argue that gene-edited individual cows are not disrespected. Regarding (3) we argue that there may be short-term significant negative effects on animal welfare, but that in the longer term there will be significant positive effects; and regarding (4) we argue that the well-being and safety of those working with the cattle is a strong argument in favour of polled cattle, while potential issues for consumer health are negligible. Our conclusion is that gene-editing dairy cattle for polledness seems to be a potentially acceptable and feasible solution. However, regulatory changes, either in the EU or in the US, will be required before the solution will work in practice. Recent developments in the US indicate that regulatory changes are very likely.

**Keywords:** gene-editing, cattle, ethics, horns, animal welfare

## Introduction

The ancestor of the modern dairy cow, the aurochs, was a horned animal. However, around 80% of dairy cows in Europe and the US do not have horns (Cozzi *et al.*, 2015). This is in most cases an artefact: Typically, the ability to grow horns is removed by means of so-called disbudding, an approach developed in the 1960s and 70s with the introduction of loose-housing systems. The aim is both to prevent cows from hurting each other and to increase the safety of farmers and farm workers. Disbudding is undertaken by means of a hot iron that is placed on the horn buds of calves between 4 and 12 weeks old. Done without analgesia this is painful, but even when local analgesia in combination with painkillers in the form of non-steroidal anti-inflammatory drugs are injected prior to the intervention, as is legally required in some countries, including Denmark, there may still be long term negative effects on animal welfare (Stafford and Mellor, 2011). This suggests that alternative approaches should be considered.

One alternative approach to avoid cows growing horns is genetic. The genetic disposition not to grow horns (called polledness) occurs naturally among domestic cattle. A bull carrying one copy of one of the


Hanna Schübel and Ivo Wallimann-Helmer (eds.) ***Justice and food security in a changing climate***
EurSafe 2021 – DOI 10.3920/978-90-8686-915-2_50, © Wageningen Academic Publishers 2021

three known gene variants for polledness will produce 50% polled offspring, whereas a bull carrying two copies of these gene variants will produce 100% polled offspring. Some beef breeds are already 100% polled, but polledness is less widespread among Holstein Friesians and other typical dairy breeds. One possible way to speed up the process of making dairy cattle polled is by means of gene-editing, a technique that can relatively easily be applied, given the various forms of reproductive biotechnology already used in advanced dairy cattle breeding. In a paper published in 'Nature Biotechnology' (Carlson *et al.*, 2016) this has been shown to work by means of a protein-based gene editing technology called Talens. Today, such a gene-editing approach would most likely be undertaken by means of the CRISPR-Cas 9 technology, either applied directly on *in vitro* fertilized embryos, or on fibroblasts, which are subsequently inserted into enucleated cow eggs by means of nuclear transfer as used for cloning of mammals (Figure 1). Proof of concept of the latter approach was published in 2020 (Schuster *et al.*, 2020).

The question to be discussed in this paper is whether using gene-editing is preferable to disbudding for producing polledness in dairy cattle viewed from an ethical and a regulatory perspective. These perspectives will be addressed in the following sections; and a short synthesis will be made in a concluding section.

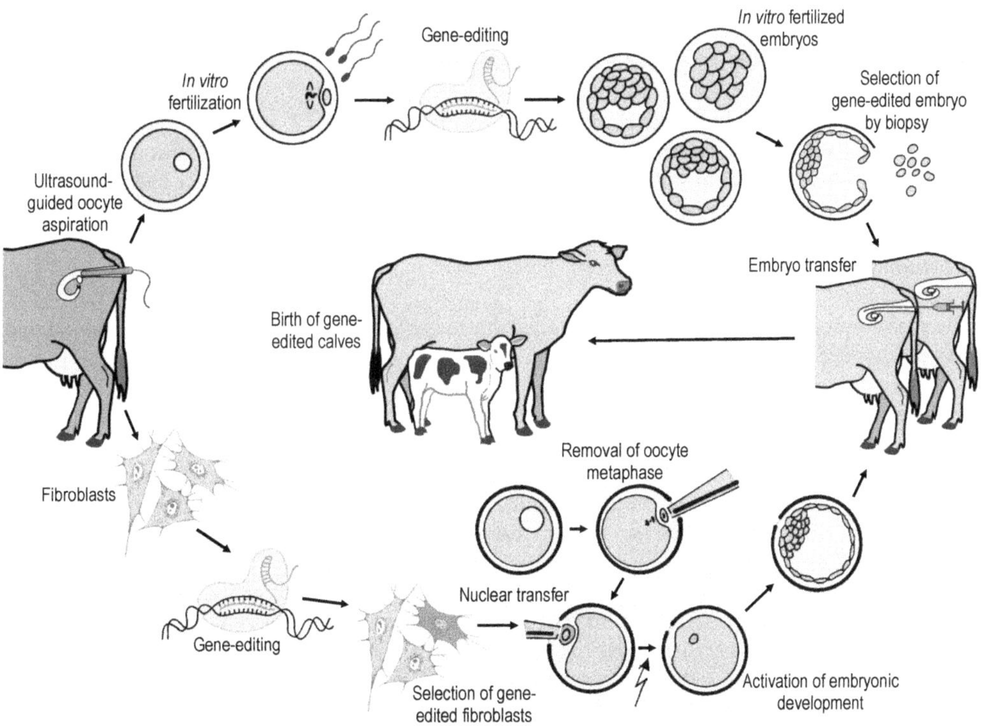


*Figure 1. Overview of the implementation of CRISPR-Cas9 gene editing in cattle. The top route shows gene editing of* in vitro *fertilized embryos with subsequent selection of successfully edited embryos by means of a cell biopsy, followed by transfer of such embryos to recipients. The bottom route shows gene editing of cells, e.g. fibroblasts, selection of successfully edited cells, the use of these cells for production of embryos by somatic cell nuclear transfer, and subsequent transfer of the embryos to recipients.*

Before proceeding, we should point out, as context for this paper, that there are other possibilities here. Cattle could retain their horns. However, in current loose housing systems, having horns may have a negative impact on cattle welfare. In a survey taken by European experts, 57% of the respondents said that the main reason for dehorning is that it reduces the risk of injuries among loose housed cattle (Cozzi *et al.,* 2015). It might be argued that therefore the husbandry systems *per se* need to be changed, rather than using dehorning to allow 'husbandry conditions that are insufficiently adapted to the species-specific needs of cattle' (Knierim *et al.,* 2015). However, realistically, for social and economic reasons, such husbandry changes are extremely unlikely to be implemented any time soon. Polled cattle could, in theory, also be produced by means of traditional selective breeding. However, gene-editing seems vastly superior to traditional breeding in terms of efficiency. The introduction of polledness by means of traditional breeding methods would lead to lost genetic potential or increased inbreeding, because the naturally polled animals to be used will not be elite animals. Therefore, breeding for economically important traits would lose momentum, which in turn would give rise to astronomic costs, measured in terms of lost breeding qualities (Mueller *et al.,* 2019). In practical terms the choice will therefore be between disbudding and polledness achieved by means of gene-editing. We also do not discuss welfare consequences and other ethical issues flowing from the existing use of reproductive technologies in advanced cattle breeding, i.e. artificial insemination, multiple ovulation and embryo transfer, and *in vitro* fertilization. These technologies are already deeply entrenched and widely used, independently of gene-editing for polledness, in the current production of horned cows.

## Ethical perspectives on gene editing for polledness

Here we will discuss four kinds of ethical concerns: (1) naturalness; (2) respect for animal integrity; (3) animal welfare; and (4) human health and safety. We acknowledge that other ethical concerns, for example the role of public consultation in the approval of new technologies (Bruce and Bruce, 2019) should be considered but due to space limitations we do not go into these.

### *1. Naturalness*

Here the key question is: 'Are cows that are polled due to gene editing significantly more unnatural than cows that are dehorned through disbudding?'

One starting point here is that neither kind of cow would have come about 'in nature' understood as 'without human intervention'. Both are domesticated and selectively bred for human purposes; individual cows are brought into existence because humans chose to breed them, in many cases using reproductive technologies. So, the entire context is, in this sense, one of 'unnaturalness'. The question, then, is whether gene editing increases this unnaturalness more than disbudding.

One interpretation of unnaturalness in the context of certain forms of genetic engineering is that outcomes could not happen 'in nature'. So, for instance, if genes found in chicken (but not in cattle) were used to gene-edit cows – say to improve the texture of beef – such genetically modified cows could be argued to be 'unnatural', since these chicken genes would never find their way into cows without human intervention. Such trans-species genetic modifications are referred to as 'transgenic'. However, even if this argument works for chicken-genes, it does not seem to hold true for polledness, which, due to the use of gene variants already found in cattle, is referred to as a non-transgenic modification (Schultz-Bergin, 2018). Hence, the polled gene variants are already in the cow population, and could have become more widespread, for instance by selective breeding rather than gene editing (though practical reasons, as mentioned above, will militate against taking this route). Since polledness is not achieved by transgenesis, and could have occurred without human intervention, it is not more unnatural in this sense than existing selective breeding.

Naturalness might, alternatively, be understood in a slightly different way here – as something like the manifestation of a being's inner nature or *telos*. Growing horns, it might be argued, is natural to dairy cows; preventing their horns from growing may be regarded as 'the loss of the essence of the creature' (Ishii, 2018). One might then think that both disbudding, and gene-editing are equally unnatural. But if what is meant here is what is natural to the particular individual cow, disbudding seems less natural than gene-editing. After all, for a gene-edited cow, what is natural is *not* to grow horns. But for a cow that has not been gene edited, what is natural is to *have* horns, and thus disbudding is interrupting a natural process.

### 2. Integrity

This concept can be understood in rather different ways, but here we take it to refer to something like the wholeness and intactness of the animal (Rutgers and Heeger, 1999). Respecting this integrity, then, requires us not to violate such wholeness and intactness for human instrumental use (Bovenkerk, 2020). On this definition, one might think that both disbudding and gene editing fail to respect animal integrity: in both cases, cows have lost their horns. However, it could also, again, be argued that gene editing does not disrespect particular individuals – with the likely exception of those individuals involved in the development of the technology (though here loss of integrity is not about polledness itself, but about other impacts of experimental procedures). After all, once the gene is established in a population, the individuals that are born are as their genes determine – without horns. They have not lost their horns; these particular individuals were never destined to have any horns. So, their wholeness and intactness has not been violated, and thus their integrity has not been disrespected.

An alternative way of worrying about integrity, though, might be that it's not particular individual cows whose integrity has been disrespected, but rather the whole gene-edited population. Again, this argument is difficult to defend, since the gene variants are naturally occurring. Where populations of cattle are 'naturally' polled, no-one would raise an issue about loss of integrity; and polledness 'could' become more widespread among cattle populations without gene-editing. So, it is hard to see how introducing the gene variants for polledness into a cow population is a problem in this sense, especially as the reason for introducing it is not primarily for human instrumental benefit, but to improve animal welfare.

### 3. Animal welfare

In the short term, there will be negative consequences that go beyond those produced by the existing use of the above-mentioned reproductive technologies in advanced cattle breeding. Due to the potential use of somatic cell nuclear transfer cloning, foster-mothers may go through difficult pregnancies and births; and some calves will be born weak and suffer as a consequence. In the original experiment, creating gene-edited polled offspring using Talens, only two out of 26 embryos transferred ended up as calves alive at day 60 (Carlson *et al.*, 2016); and in a later experiment, based on CRISPR-Cas 9, only two of 12 embryos resulted in pregnancies past the first trimester. One foetus was killed together with the mother for analysis, while the remaining pregnancy was carried to term and delivered one calf, which died shortly after birth (Schuster *et al.*, 2020). Furthermore, the gene editing technique currently poses certain risks in the form of off-target editing effects elsewhere in the genome, some of which can lead to welfare problems.

However, according to published estimates only a small number of gene-edited sires (less than 200) in a breed (e.g. Holstein or Jersey) (Mueller *et al.*, 2019) would be needed in the first decade of a breeding program. After that all breeding bulls would be homozygous polled, and gene editing would no longer be needed to maintain a breed of polled animals. So, in terms of animal welfare there would be a trade-off of a limited number of animals (perhaps a few thousand) suffering to some degree, while a very much larger number of animals (billions, potentially) will avoid painful disbudding procedures. Of course, there are ethical approaches, including many animal rights views, on which sacrificing some sentient animals for welfare gains to many more individuals would be unacceptable (although such views are likely to reject the dairy cattle industry *tout court* on the grounds that animals are instrumentalized and their rights violated). Given the non-ideal assumption stated above, however, the authors of the current paper are sympathetic to a broadly consequentialist argument in favour of gene-editing for polledness as the best alternative in terms of animal welfare.

### 4. Human health and safety

Here there are two main potential concerns. One is the physical health and well-being of farmers and farm workers. The other is consumer health.

In the survey, mentioned earlier, on reasons for dehorning, 24% of the respondents gave lowering the risk of injuries to stockpersons as the primary reason for performing dehorning. A qualitative European study based on group interviews of cattle farmers also found farmer safety to be a key reason for dehorning loose housed cattle (Kling-Eveillard *et al.*, 2015). On the other hand, anecdotal evidence indicates that farmers and farm workers in general do not feel comfortable about disbudding, not only because it is widely perceived, as the mentioned surveys also show, to be painful to the animals, but also because it is seen as a filthy and unpleasant job. Therefore, from a farmers and farm workers' perspective, avoiding the problem altogether through polled animals would clearly be a welcomed solution.

Unintended genetic changes to the gene-edited animals, which are side-effects of the gene-editing, could in theory have negative effects on consumer health. The researchers behind the first two bull-calves, polled by means of gene-editing, in their paper claimed that the calves were 'free of off-target effects' (Carlson *et al.*, 2016). However, years later a stretch of bacterial DNA including a gene conferring antibiotic resistance (raising the possibility of potential impacts on human health) was found in the genome of one of the calves (Regalado, 2019). This was clearly a result of a flaw in the modification process. Moreover, it should be said that there has been no suspicion that this bacterial DNA would have negative effects on human health, and a subsequent very thorough analysis found no other unintended genomic alterations (Young *et al.*, 2020). While off-target effects are possible, and could conceivably affect human health, the risk here is widely thought to be very low.

It is, of course still an open question whether the technology will be accepted by consumers. In light of previous public debates concerning cloning of cows and genetic modification of crops and food animals consumer resistance may be expected. However, there are reasons to think that this will not be a major obstacle in the case of cattle gene-edited for polledness. Firstly, the gene-edited animals themselves most likely will not enter the food chain; they will only serve as breeding animals and, hence, play an intermediate role. After a while there will be no gene-edited breeding animals left. Over time, the issue may therefore become more or less invisible. Secondly, two studies conducted in North America indicate an acceptance of gene-editing of food animals if this serves to improve animal welfare (McConnachie *et al.*, 2019; Ritter *et al.*, 2019).

## Regulatory issues for gene editing for polledness

The only remaining obstacle seems to be regulatory. According to a ruling of the European Court of Justice gene-edited organisms fall under the genetically modified organism (GMO) Directive. Similarly, the US Food and Drug Administration (FDA), which currently is the key regulator, has proposed a mandatory premarket 'new animal drug regulatory evaluation' for all food animals whose genomes have been intentionally altered. Therefore, the introduction of gene-edited cattle into the food chain may require safety testing, which, according to many voices in the industry, will be prohibitively expensive (Eenennaam *et al.*, 2019). Thus, a regulatory change, either in the EU or in the US, will be required before the solution will work in practice.

However, some new US regulations may be forthcoming: According to a notice in the *Federal Register* (Anonymous, 2020) the US Department of Agriculture (USDA) will take on the regulatory approval and food safety processes currently administered by FDA for all agricultural products. USDA currently regulates the use of gene-editing in agricultural plants and considers the introduction of genetic traits into plants as regulatory compliant, when these traits already occur in the species. More specifically, 'USDA is considering an expedited safety review for any genetic modification made that is already known to occur in the gene pool of the species'. It is plausible that the conclusion of this process will lower the bar for regulatory compliance for agricultural animals.

## Conclusion

So, given the world as it is, with an intensive dairy sector that is not going to be discontinued any time soon, dairy cattle gene-edited for polledness seems to be the best solution to achieve hornless cattle. Unlike disbudding and other horn-removing practices, gene-editing for polledness does not give rise to serious long-term welfare or other ethical concerns, and unlike conventional breeding it is economically viable. The only main obstacle, so far, has been regulatory in that gene-edited animals both in the EU and in the US are regulated as GMOs leading to requirements for prohibitively expensive safety testing. However, this may change in light of regulatory changes recently announced in the US. It is unlikely that semen for polled dairy cattle, derived from breeding with gene-edited animals (unlike the gene-edited animals themselves), can and will be prevented from entering the European market.

## References


Anonymous (2020). United States Department of Agriculture. Regulation of the Movement of Animals Modified or Developed by Genetic Engineering. Federal Register 85 (248): 84269-84275.

Bovenkerk, B. (2020). Ethical perspectives on modifying animals: beyond welfare arguments. Animal Frontiers 10: 45-50.

Bruce, A. and Bruce, D. (2019). Genome Editing and Responsible Innovation, Can They Be Reconciled? Journal of Agricultural and Environmental Ethics 32: 769-788.

Carlson, D.F., Lancto, C.A., Zang, B., Kim, E.-S., Walton, M., Oldeschulte, D., Seabury, C., Sonstegard, T.S. and Fahrenkrug, S.C. (2016). Production of hornless dairy cattle from genome-edited cell lines. Nature Biotechnology 34: 479-481.

Cozzi, G., Gottardo, F., Brscic, M., Contiero, B., Irrgang, N., Knierim, U., Pentelescu, O., Windig, J.J., Mirabito, L., Kling Eveillard, F., Dockes, A.C., Veissier, I., Velarde, A., Fuentes, C., Dalmau, A. and Winckler, C. (2015). Dehorning of cattle in the EU Member States: A quantitative survey of the current practices. Livestock Science 179: 4-11.

Eenennaam, A.L.V., Wells, K.D. and Murray, J.D. (2019). Proposed U.S. regulation of gene-edited food animals is not fit for purpose. Npj Science of Food 3: 3.

Ishii, T. (2017). Genome-edited livestock: Ethics and social acceptance. Animal Frontiers 7: 24-32.

Kling-Eveillard, F., Knierim, U., Irrgang, N., Gottardo, F., Ricci, R. and Dockès, A.C. (2015). Attitudes of farmers towards cattle dehorning. Livestock Science 179: 12-21.

Knierim, U., Irrgang, N. and Roth, B.A. (2015). To be or not to be horned – consequences in cattle. Livestock Science 179: 29-37.

McConnachie, E., Hötzel, M.J., Robbins, J.A., Shriver, A., Weary, D.M. and von Keyserlingk, M.A.G. (2019). Public attitudes towards genetically modified polled cattle. PLOS ONE 14: e0216542.

Mueller, M.L., Cole, J.B., Sonstegard, T.S. and Eenennaam, A.L.V. (2019). Comparison of gene editing versus conventional breeding to introgress the POLLED allele into the US dairy cattle population. Journal of Dairy Science 102: 4215-4226.

Regalado, A. (2019). Gene-edited cattle have a major screw up in their DNA. MIT Technology Review. Available at: https://www.technologyreview.com/2019/08/29/65364/recombinetics-gene-edited-hornless-cattle-major-dna-screwup/#Echobox=1567519481.

Ritter, C., Shriver, A., McConnachie, E., Robbins, J., von Keyserlingk, M.A.G. and Weary, D.M. (2019). Public attitudes toward genetic modification in dairy cattle. PLOS ONE 14: e0225372.

Rutgers, L.J.E. and Heeger, F.R. (1999). Inherent worth and respect for animal integrity. In: Dol, M., Fentener van Vlissingen, M., Kasanmoentalib, S., Visser, T. and Zwart, H. (eds) Recognizing the intrinsic value of animals: beyond animal welfare. Van Gorcum, Assen, Netherlands, pp. 41-51.

Schultz-Bergin, M. (2018). Is CRISPR an Ethical Game Changer? Journal of Agricultural and Environmental Ethics 31: 219-238.

Schuster, F., Aldag, P., Frenzel, A., Hadeler, K.-G., Lucas-Hahn, A., Niemann, H. and Petersen, B. (2020). CRISPR/Cas12a mediated knock-in of the Polled Celtic variant to produce a polled genotype in dairy cattle. Scientific Reports 10: 13570.

Stafford, K.J. and Mellor, D.J. (2011). Addressing the pain associated with disbudding and dehorning in cattle. Applied Animal Behaviour Science 135: 226-231.

Young, A.E., Mansour, T.A., McNabb, B.R., Owen, J.R., Trott, J.F., Brown, C.T. and Eenennaam, A.L.V. (2020). Genomic and phenotypic analyses of six offspring of a genome-edited hornless bull. Nature Biotechnology 38: 225-232.

# 51. Genome edited salmon: fish welfare as part of sustainability criteria

*T. Blix* and A.I. Myhr*
*GenØk-Centre for Biosafety, SIVA Innovasjonssenter, 9294 Tromsø, Norway; tba073@uit.no*

## Abstract

Salmon is the most important species in Norwegian aquaculture, and in 2019 1,3 million tonnes of farmed salmon were produced along the Norwegian coast. The rapid growth of the industry comes with several challenges related to environment, economics and animal welfare, all hindering a sustainable production. The newest proposed solution to some of these challenges is to use genome editing. Some examples of research initiatives are sterile salmon, disease resistant salmon, and salmon with increased omega-3 production. In Norway, all organisms generated by gene technology are regulated by the Gene Technology Act of 1993, which, besides safety assessment, demands that ethical justifiability, social utility and contribution to sustainable development to be part of the assessment of GMOs. Operationalisation documents for these non-safety criteria with guidelines have been developed, and furthermore case-specific reports on the contribution to sustainability of different genetically modified plants have been developed. Given the predicted launch of genome edited salmon into international marked, guidelines should be adapted for use on such animals as well. We do also identify that the topic animal welfare is missing in present guidelines and this must be included in the sustainability evaluation. We suggest The Norwegian Animal Welfare Act can be used as a guide for how to evaluate animal welfare, which includes aspects as health, feelings, needs and the intrinsic value of animals.

**Keywords:** animal ethics, salmon, gene technology, GMO regulation, aquaculture

## Introduction

The Norwegian salmon industry has grown rapidly in its short lifespan and has been forespoken additional growth in the coming decades. In 2019 the production reached 1,3 million tonnes (SSB, 2020). At the same time, Norwegian aquaculture is facing numerous challenges like sea lice, infections, escapees, feed availability, and environmental pollution. These issues pose a threat to the development of a sustainable industry. A solution for several of these issues has been proposed – genome editing (GE) of the salmon so that it becomes lice or disease resistant, sterile, or vegetarian. In Norway, all genetically modified organisms (GMOs) are regulated by the Gene Technology Act (GTA) of 1993, which states that a GMO is an organism in which the genetic material has been altered by means of gene or cell technology (Government.no, 2005) Chapter 1, §4). For the time being genome edited organisms in Norway are also regulated as GMOs, but this is under discussion. In the GTA, a requirement is that (Chapter 1, §1):

> [...] the production and use of genetically modified organisms and the production of cloned animals [shall] take place in an ethically justifiable and socially acceptable manner, in accordance with the principle of sustainable development and without adverse effects on health and the environment.

This implies that the assessment must, in addition to risk analysis, include analyses of how the GMO contributes to sustainable development, has societal utility and is ethically justifiable – non-safety criteria. The Norwegian Biotechnology Advisory Board (NBAB) has the responsibility to give advice on


EurSafe 2021 – DOI 10.3920/978-90-8686-915-2_51, 

applications based on these non-safety criteria. The non-safety criteria of the GTA are operationalised in three different guideline documents (NBAB, 2009, 2018, Forsberg *et al.,* 2019). The guidelines operationalizing these criteria have never before been applied to GM animals because no such application has been submitted in Norway.

In this text we present the current guideline document for assessment of contribution to sustainable development (NBAB, 2009). The present guideline has mainly been used on GM plants, adaption needs therefore to be done for animals. There is especially one important topic missing – animal welfare. We elaborate on how this topic can be included in the guidelines, and more specifically for salmon. This includes a discussion on how animal welfare is connected to sustainable development.

## Sustainable development as a non-safety criterion in evaluation of GMOs

In the GTA, sustainable development is mentioned both in the mission paragraph §1 and as an explicit criterion for approval in §10. In 1998, the Norwegian Ministry of Climate and Environment asked for an operationalization guide for the three non-safety criteria. NBAB established a working group and published a guide in 2000 (revised ed. 2009), 'Sustainability, societal utility and ethics in evaluation of genome modified organisms – operationalization of the terms in The Gene Technology Act §§1 and 10' (NBAB, 2009). The document introduces six core ideas of sustainable development, with respective control questions. Core ideas are listed in Table 1. The guideline document has been followed by four reports on how to operationalize the sustainability criterion further for different GM plants (NBAB, 2011, 2014; Catacora-Vargas, 2014; Gillund and Myhr, 2016). Such a framework has not yet been developed for GM animals. This might be because no application for the use of GM animals has yet been received by the Norwegian authorities. However, already in 2013 NBAB made a statement on AquAdvantage, the growth enhanced GM Atlantic salmon which has later been approved for production in the U.S. (FDA, 2019). NBAB emphasized that several issues had not been evaluated by the FDA for the GM salmon, e.g. ethics and societal impact and effects outside the U.S./Canada. They also pointed out that FDA did not consider a) malformations as a welfare issue, and b) issues related to only selecting for growth in livestock (NBAB, 2013).

## The guideline document and operationalization for plants

The core ideas and control questions in the original guideline document (NBAB, 2009) are, together with topics from the reports on GM plants, also relevant for an assessment of genome edited salmon. Table A presents the original guidelines (NBAB, 2009) together with the topics used in the four frameworks for evaluation of the contribution to sustainable development of GM plants, combined. The latter have been slightly adopted to GE salmon. This table thus illustrates how a framework for GE salmon could be designed. Although Table A cover a range of topics relevant for different GM products, we find it necessary not only to adapt the questions to living GE animals, but also to specify the core ideas and evaluation questions, and draw on other relevant strategies, approaches and regulation as for example the UN sustainable development goals (SDGs) (UN, 2015), The Norwegian Animal Welfare Act (AWA), and the World Animal Health Organization (OIE) – The Aquatic Code. Further elaboration is also needed for topics like traceability through the food web and in the environment, impact on wild salmon as well as biodiversity in general, ownership and liability, and coastal as well as wild salmon-based community development. Within the scope of this article and context, we will elaborate on the most prominent core idea 'missing' in the guidelines and in the former reports, namely animal welfare.

*Table 1. Core ideas (NBAB, 2009) and relevant topics (Catacora-Vargas, 2014; Gillund and Myhr, 2016; NBAB, 2011, 2014) for assessment of sustainable development by salmon.*

| Original guideline document | Operationalization of guidelines: report on plants adapted to salmon | |
|---|---|---|
| Core ideas | Pillars | Topics |
| Global effects | Ecology and environment | The genetically modified organism |
| Ecological limits | | Interaction between the GM and the environment |
| Basic human needs | | Gene flow to wild relatives |
| Distribution between generations | | Preservation of biological diversity in ocean and rivers |
| Distribution between rich and poor | | Resistance in salmon to diseases and parasites |
| | | Comparison with control salmon (farmed) |
| (For all core ideas: do these effects differ between production and use?) | | Safety of human health and the environment over time |
| | Economy and society | The right to sufficient, safe and healthy food |
| | | Animal welfare[1] |
| | | Living conditions and profitability for fish farmers and coastal communities in short and long terms |
| | | Biodiversity and genetic resources for food and aquaculture |
| | | Independent risk assessment |
| | | Freedom to choose a different aquaculture system in the future |

[1] Animal welfare is a topic in NBAB (2013) and Catagoras-Vargas (2014). However, this was regarding the use of the evaluated plant for animal feed, not for the GM animal itself.

## The novel core idea: animal welfare

### *How can animal welfare be included in the framework?*

Welfare is a range of states from good to poor, and this state represent how the animal is '[...] faring' (Broom, 2010). An animal's welfare is an evaluation of its needs based on the biological state and the feelings of the animal. Animal welfare can also be seen as a multidimensional concept, considering how it includes both physical, measurable status and mental, less available status. This needs to be taken into consideration when designing the evaluation framework for the welfare of GE salmon. Which aspects that must be considered in the assessment of the welfare of such an animal depends on the characteristics of the changes introduced in the modified animal, and the genotypic and phenotypic effects of this change. There are, however, different criteria for fish welfare in guiding documents, illustrated by comparing the Norwegian Animal Welfare Act to the OIE Aquatic Code.

OIE work under the phrase 'protecting animals, preserving our future' (OIE, n.d.) and aquatic and terrestrial animals are designated separate rights and thus status. The Terrestrial Code describe standards for welfare of terrestrial animals including the five freedoms stating what harms and negative impact animals should be protected from. Here welfare and health are separated, indicating an awareness of mental condition. The Aquatic Code, on the other hand focuses on disease prevention (OIE, n.d., OIE, 2019a,b). This does not necessarily diminish the importance of good welfare, but it may exclude the consideration that fish have feelings and needs. It also excludes allowing fish the five freedoms from hunger, malnutrition and thirst; heat, stress and physical discomfort; pain, injury and disease; and freedom to express normal patterns of behaviour.

The AWA states that 'all' animals should be treated well and be protected against dangers, stress, pain and suffering; that they shall not be restricted in the ability to function or in their quality of life, and that their natural behaviour shall be allowed (Government.no, 2009). This implies that if the new genotype created by GE in a salmon leads to a phenotype generating pain, stress or suffering, this is not acceptable. The genotype shall neither inflict on the needs of the fish, lead to an increased feeling of fear or leaving the animal more exposed to disease. The salmon should be allowed and able to fare well and engage in natural behaviour. Further, for animals used in research, the concept of the 3 R´s is guiding the treatment of these animals. 'Refine, reduce and replace' can guide healthy activities in animal protein production as well. This concept will be important in an evaluation of welfare of GE salmon because the research animals needed for developing the commercial fish have to be included in the evaluation. This is also acknowledged in the GTA as it covers both the production, and use/release of GMOs.

### What is the relation between animal welfare and sustainable development?

The welfare of a GM salmon cannot be poorer than the un-modified relative (Broom, 2010), and welfare must therefore be thoroughly evaluated for the GE salmon. Poor animal welfare will lead to poor production efficiency – which is not sustainable (Tarazona *et al.,* 2020), neither to the environment nor to the economy. It can also be argued that poor animal welfare is ethically unacceptable. Broom (2010) emphasises that a sustainable system must be morally acceptable and considers animal welfare as '[...] an aspect of care, sustainability and food quality required by the public' (Broom, 2010).

In addition, it can be argued that good welfare is important for biodiversity, human health and energy consumption (Keeling *et al.,* 2019). Recent work by Keeling *et al.* (2019) studied how animal welfare is connected to the 17 SDGs, which are based on sustainable development as '[...] development that meets the needs of the present without compromising the ability of future generations to meet their own needs' (Brundtland *et al.,* 1987). They revealed that animal welfare is less important to sustainable development than it is to animal welfare. Animal welfare is not mentioned in the SDGs, and Keeling *et al.* (2019) emphasises the need for more research on this connection. Torpman and Röcklinsberg (2020) have discussed how and to what extent the SDGs should also concern animals. They concluded that some of the SDGs actually do concern animals directly (sentience-centered), while others don't (human-centered). Looking at the sentience-centered SDGs we want to especially emphasise goals 6, 13, 14 and 15 as these have also been designated as contributing to biospheric sustainability in a model presented by Stockholm Resilience Center (2016).

### Intrinsic value – a part of Norwegian assessment of animal welfare

The AWA has a statement which is equally applicable for both terrestrial and aquatic animals. It states that all animals have intrinsic value independent of their value to humans (Chapter 1 §3, Government.no, 2009). The statement refers to how animals should be treated, not necessarily what welfare is. This must therefore also be included in the assessment of genome edited salmon, to ensure that the evaluation is following all aspects of the AWA. The AWA does not only align with Broom (2001, 2010) in the need for protecting animals and making sure they can fare well, but it also gives them moral status. Compared to other regulations and standards, e.g. the Aquatic Code of the OIE, this recognition for fish is rare. As described in earlier work by Trøite and Myskja (2018), intrinsic value in the AWA has a symbolic role. Using moral concepts as symbols only, with no actual effects on regulation, has two consequences. Either the concept is diminished in meaning, or it is promoted as a direction and can aid '[...] long-term political change, despite [its] mere symbolic function' (Trøite and Myskja, 2018). When implementing animal welfare in the evaluation of contribution to sustainability of GE animals, the concept of intrinsic value cannot be overlooked. We believe that there must be a clear decision on whether intrinsic value should have a guiding role promoting for how we want to understand the moral status of animals, or whether the moral concept should not be included as it only has symbolic value, and hence no direct effect on regulation. This decision should follow the definitions of sustainability given by Brundtland *et al.* (1987) and Broom (2001). Performing genome editing which harms the animal's integrity by diminishing its moral status, will according to the definition of Broom (2001) not be sustainable exactly because it is not morally acceptable. An example of this is presented by Forsberg (2008), where she uses as a case a GM animal modified for purposes of entertainment for humans to state that 'Even if it would have no negative welfare consequences for either the modified animal or the animals that were used to develop this variant, it would violate the integrity of the animal'. Based on this it can be argued that in the assessment of contribution to sustainable development one should evaluate whether the genetic change(s) diminishes the moral status of the animal by harming its integrity.

## Conclusions

The requirement of sustainable development of the aquaculture sector may open up for the introduction and use of GE fish. This may address some of the sector's present challenges, but also introduces new questions with regard to sustainability. Guidelines developed by the NBAB functions well as a starting point for the assessment of sustainability of GE salmon but needs further elaboration so that the topics can be adapted to salmon and be used for a case-by-case evaluation. We have suggested one novel core idea – animal welfare. In addition to evaluation of health aspects – physical and mental welfare, the evaluation of a GE fish must therefore include issues related to whether or not the genetic changes affect the intrinsic value of the animal. Further work must approach what such changes are, this includes elaboration of what changes that are morally acceptable and should include engagement with stakeholders.

## References

Broom, D.M. (2001). Does present legislation help animal welfare? In Sustainable Animal Production, Landbauforschung Völkenrode, 222.

Broom, D. M. (2010). Animal Welfare: An Aspect of Care, Sustainability, and Food Quality Required oy The Public. Journal Of Veterinary Medical Education, 37 pp.

Brundtland, G. H., Khalid, M., Agnelli, S., Al-Athel, S. A., Chidzero, B. T. G., Fadika, L. M., Hauff, V., Lang, I., Shijun, M., De Botero, M. M., Singh, N., Nogueira-Neto, P., Okita, S., Ramphal, S. S., Ruckelshaus, W. D., Sahnoun, M., Salim, E., Shaib, B., Sokolov, V., Stanovnik, J., Strong, M. and Macneill, J. (1987). Our Common Future. Oxford University Press.

Catacora-Vargas, G. (2014). Sustainability assessment of genetically modified herbicide tolerant crops. the case of Intacta(Tm) Roundup Ready(Tm)2 Pro soybean farming in Brazil in light of The Norwegian Gene Technology Act. Genøk - Center For Biosafety. Available from https://genok.no/arkiv/4214/.

FDA (U.S. Food and Drug Administration) (2019). Questions and Answers on FDA's Deactivation of an Import Alert for Genetically Engineered Salmon. Available at: https://www.fda.gov/animal-veterinary/animals-intentional-genomic-alterations/questions-and-answers-fdas-deactivation-import-alert-genetically-engineered-salmon.

Forsberg, E.-M. (2008). Principles of animal ethics in Scandinavian animal welfare legislation and governance. Work Research Institute, Oslo. Available at https://fagarkivet.oslomet.no/handle/20.500.12199/6361.

Forsberg, E.-M., Hofmann, B., Kaiser, M., Myskja, B., Strand, R., and Ursin, L. (2019). Vurderingskriteriet etikk - Veileder for operasjonalisering av vurderingskriteriet etikk i genteknologiloven (in Norwegian).

Gillund, F. and Myhr, A. I. (2016). Important considerations for sustainability, social utility and ethical assessment of late blight resistant GM Potato - Biosafety Report 2016/01. Genøk. Available at: https://genok.com/arkiv/6143/.

Government.no (2009). Animal Welfare Act. Available at: https://www.regjeringen.no/en/dokumenter/animal-welfare-act/id571188/.

Government.no. (2005). Gene Technology Act. Available at: https://www.regjeringen.no/en/dokumenter/gene-technology-act/id173031/.

Keeling, L., Tunón, H., Olmos Antillón, G., Berg, C., Jones, M. Stuardo, L., Swanson, J., Wallenbeck, A., Winckler, C. and Blokhuis, H. (2019). Animal Welfare And The United Nations Sustainable Development Goals. Frontiers In Veterinary Science 6(336).

NBAB (Norwegian Biotechnology Advisory Board) (2009). Sustainability benefit to the community and ethics in the assessment of genetically modified organisms: implementation of the concepts set out in sections 1 and 10 of the Norwegian Gene Technology Act. 2nd Revised Edition 2009. Available at: https://www.bioteknologiradet.no/filarkiv/2010/07/2009_11_18_diskusjonsnotat_baerekraft_engelsk.pdf.

NBAB (2011). Insektresistente genmodifiserte planter og bærekraft (in Norwegian). Available at: http://www.bioteknologiradet.no/filarkiv/2011/06/rapport_baerekraft_110627_web.pdf.

NBAB (2013). Opinion on the assessment of Aquadvantage Salmon. Available at: http://www.bioteknologiradet.no/filarkiv/2013/06/Opinion_AquAdvantage_NBAB.pdf.

NBAB (2014). Herbicide-Resistant Genetically Modified Plants and Sustainability. Available at: http://www.bioteknologiradet.no/filarkiv/2014/09/Herbicide-resistant_genetically_modified_plants_and_sustainability_NBAB.pdf.

NBAB (2018). Samfunnsnytte og genmodifiserte organismar (in Norwegian). Available at: https://www.miljodirektoratet.no/globalassets/publikasjoner/m977/m977.pdf.

OIE (World Organization for Animal Health) (2019a). Aquatic code. Available at: https://www.oie.int/en/standard-setting/aquatic-code/access-online/?htmfile=chapitre_welfare_introduction.htm.

OIE (2019b). Terrestrial code. Available at: https://www.oie.int/en/standard-setting/terrestrial-code/access-online/?htmfile=chapitre_aw_introduction.htm.

OIE (nd). What is animal welfare. Available at: https://www.oie.int/en/animal-welfare/animal-welfare-at-a-glance/.

SSB (Statistics Norway) (2020). Aquaculture (Discontinued). Available at: https://www.ssb.no/en/jord-skog-jakt-og-fiskeri/statistikker/fiskeoppdrett/aar.

Stockholm Resilience Center (2016) How food connects all the SDGs, EAT Forum. Available at: https://www.stockholmresilience.org/research/research-news/2016-06-14-how-food-connects-all-the-sdgs.html.

Tarazona, A. M., Ceballos, M. C. and Broom, D. M. (2020). Human Relationships With Domestic And Other Animals: One Health, One Welfare, One Biology. Animals 10(1): 43.

Torpman, O., and Röcklinsberg, H. (2021). Reinterpreting the SDGs: Taking Animals into Direct Consideration. Sustainability, 13(2).

Trøite, M. F. and Myskja, B. (2018). Legal protection of animal intrinsic value - mere words? In: Springer, S. and Grimm, H. (eds.) Professionals in food chains. Wageningen Academic Publishers, pp. 369-374.

UN (United Nations) (2015). Transforming Our World: The 2030 Agenda for Sustainable Development. UN General Assembly. Available at: https://sdgs.un.org/2030agenda.

# 52. An agent-centred approach to innovation for 21$^{st}$ century challenges of agriculture

*Z.H. Robaey*[1] *and P. Sandin*[2]
[1]*Wageningen University and Research, Gebouw 201, 6706 KN Wageningen, the Netherlands;* [2]*Swedish University of Agricultural Sciences, P.O. Box 7070, Uppsala, Sweden; zoe.robaey@wur.nl*

## Abstract

Innovation is necessary to deal with challenges that climate change brings for agriculture, such as droughts, floods, pests and pathogens that enter new climatic regions, and challenges relating to the labour force. There is a dominant narrative that science and technology are the locus of innovation, and that the solutions developed can change systems. Indeed, history shows how the Green Revolution started a massive change in practices worldwide and gave science and technology the main role. Innovation, however, also happens outside universities or industry research. Particularly in agriculture, it happens in the field through farmers' experimentation (Hansson, 2019). This opens up the idea of who is an innovator, and it is crucial to tackling climate challenges to agriculture. In this paper, we investigate the role of curiosity as a virtue of innovators. Instead of looking at systems of innovation, or at innovations themselves, we add to the growing scholarship on virtues of innovators. Among those virtues, curiosity is prominent. We argue that different contexts require different forms of the innovator virtues, and we conceptually investigate how such virtues ought to be cultivated in order to contribute to solving 21$^{st}$ century challenges for agriculture.

**Keywords:** innovation, virtues, curiosity

## Introduction

Innovation is necessary to deal with challenges that climate change brings for agriculture, such as droughts, floods, pests and pathogens that enter new climatic regions, and challenges relating to the labour force. There is a dominant narrative that science and technology are the locus of innovation, and that the solutions developed can change systems. Indeed, history shows how the Green Revolution started a massive change in practices worldwide and gave science and technology the main role. What is the role of science and innovation? There is a recognized division of labour in that that science is knowledge driven, whereas innovation is need driven, and where innovation makes use of science in an applied manner, thereby relating to engineering disciplines. In other words, science pursues knowledge, and its application to useful ends is innovation. Helga Nowotny (2008), for instance, speaks of science having an insatiable curiosity, and Godin (2008) notes that as Romans saw curiosity as a vice, it then became a virtue, or a desirable cognitive ability in modern science. This modern view on science and innovation has been predominant and has dictated many of the policies supporting the development of innovations. For instance, the 'innovation paradigm' has arguably affected the role genetically engineered crops came to play in the global food system (Schurman and Munro, 2010). Another example is the enthusiasm for use of CRISPR technologies for gene editing plants for non-browning traits, or to facilitate mechanical harvesting (Clapp, 2020).

Moving forward with tackling agricultural challenges under intensified and unpredictable climatic conditions, the roles of science and innovation need to be further conceptualized for policies to open up the realm of possibilities and support relevant interventions. In this paper, we problematize the idea of curiosity as something pertaining to scientific practice only. We discuss recent literature conceptualizing


EurSafe 2021 – DOI 10.3920/978-90-8686-915-2_52, © Wageningen Academic Publishers 2021

curiosity, and sketch implications of opening up the discussion on curiosity to foster empirical research on inclusive innovation for the challenges of 21$^{st}$ century in agriculture.

## Curiosity in scientific and innovation practices

What kind of role curiosity has been attributed to in the practices of science and innovation?

First, curiosity in scientific practice is thought to work in two ways: curiosity for the pursuit of knowledge, but also curiosity for achieving good. Considering how public funds are spent on research and innovation, there are increasing demands to clearly connect fundamental research agendas to societally relevant issues. This is an instance of an essentially Baconian view of the role, value, and *raison d'être* of science: Science generates knowledge, knowledge is power, and this power is to be put to good use (on some dominant perception of 'good'). One out of many examples of this is the presentation of the Dutch National Research agenda that developed as a result of public consultation in main routes of investigation that are directly relevant to societal challenges, envisaging to provide a 'contribution to the global knowledge society of tomorrow, in which 'new knowledge flows freely from researcher to user' and where new questions from practice and society quickly and automatically find their way into new research' (NWA, 2021, emphasis added). Another sign of this is the emerging field of Responsible Research and Innovation (RRI), which suggests that research and innovation are not inherently useful to society. Instead, to have societal relevance, RRI activities will help align research and innovation with society's needs, interests, and values. The field of RRI engages equally with innovations and fundamental research (Stilgoe *et al.,* 2013). For instance, recent research on the Synthetic Cell which is still at an early stage, engages in RRI processes call for more participation (Habets *et al.,* 2020).

Second, curiosity can be found in other practices than scientific research. It is difficult to conflate all innovation practices in one, but we would like to illustrate the role of curiosity in the practice of design. Steen (2013) underlines specific qualities of what he calls 'cooperative curiosity', as being 'being open and receptive towards other people and their experiences, and towards one's own experiences and learning' in the context of the practice of participatory design practices. Similar analytical and empirical work in other innovation practices deserve further attention. At least experience suggests that that curiosity has an important role to play in science and innovation within academia, industry, farming, and citizen projects.

Together, this underlines that curiosity in scientific and innovations practices 'should have certain normative qualities'. We find that these are often implied in various practices and require further discussion, which we elaborate on in the next section.

## Curiosity as a virtue

In the previous section, we underlined how the role of science and innovation is increasingly to serve the needs and interests of the public, and that traditional institutions like universities, research centres, and businesses are not the only loci of science and technology. If we start looking broadly for where innovation happens, we also need to have a way of discussing the normative qualities of curiosity for innovators in order to support their efforts. In doing so, we shift focus from institutions to individual innovators. This is not to deny the importance of institutions – to the contrary, developing a better understanding of individuals can in turn inform demands on institutions.

In the philosophical literature, the discussion on curiosity distinguishes between curiosity as a 'state' and a 'trait', respectively. Focussing on curiosity as a virtue means inquiring into the latter, since merely being in a particular state does not necessarily mean that one possesses a stable disposition to be curious.

(We take it that possession of such a stable disposition is a necessary but not sufficient condition for having a certain virtue.) It is also clear from this discussion that curiosity not always must be regarded as a virtue. It can also be a vice (Manson, 2012), for instance if the object of curiosity is inappropriate, or if the exercise of curiosity is incompatible with exercise of other virtues.

Recently, Lewis Ross has argued that curiosity as a virtue should be understood in the following way, 'The virtue of curiosity is the disposition to experience appropriately discerning, exacting and timely curiosity, motivated by a non-instrumental appreciation of epistemic goods.' (Ross, 2020: 116) This means that an agent possesses virtuous curiosity if:

- the extent that their curiosity is directed at the right object(s);
- arises in appropriate situations; and, finally
- is not too easily satisfied.

So far we agree with Ross. However, it is hard to see that the condition of non-instrumentality is necessary. Consider a farmer experimenting with plant breeding, curious to see whether certain seeds will germinate and develop into plants with desirable characteristics (Hansson, 2019). She is curious – but instrumentally motivated by a desire to improve the crops needed for the continued existence and well-being of her community. Arguably, this is a display of the virtue of curiosity. It is essential, however, that the farmer is exercising the curiosity within a virtuous practice, such as a sustainable farming system. If she were part of a non-virtuous practice (let us assume a non-sustainable, environmentally destructive and socially disruptive agricultural system), the story is very different. On this account, curiosity becomes a 'role-relative' virtue.

This discussion is largely inspired by Hansson's account of farmer's experimentation (2019) and, as mentioned in the previous section, the authors find that curiosity is a recurring trait that scientists and innovators in different contexts mention. While this investigation is conceptual, the authors invite furthering conceptual and empirical research on this virtue of innovators. By analysing the virtues of agents, we can open up the realm of who is an innovator by considering qualities, instead of loci of practice. As such, citizens, farmers, and other agents with a given role, can be considered innovators within their practice.

## Conclusion: inclusive innovation

Opening up the idea of who is an innovator may be crucial to tackling climate challenges to agriculture and avoiding technological bandaids. We suggest (after Ross, 2020) that curiosity as a virtue in innovation is:

- A stable trait of character, in the form of a disposition to be curious about the right object(s), in appropriate situations, and that is not too easily satisfied.

But we also note that:

- The rightness of the objects and the appropriateness of the situation is dependent on curiosity being exercised within a *virtuous practice*, for instance, an innovation system within sustainable agriculture

This understanding of curiosity is compatible with an agent's acting at least partly for instrumental reasons, given that those reasons are located within the virtuous practice. It is also consistent with an Aristotelian interpretation of the virtue as a 'Golden' Mean between two extremes of deficit and excess. A final requirement for curiosity to be a virtue is that its exercise is compatible with the exercise of other virtues, whether those are specific to the virtuous practice or not. If curiosity were to hamper one's exercise of, say, justice, we can question whether it is a virtue (in that situation for that specific agent).

This conceptual investigation aims to pave the way for further empirical research and informing policies supporting agricultural innovations in several different contexts. By recognizing various forms of practicing curiosity, we suggest that policies dealing with agricultural innovations should support various forms of formulating solutions and thereby supporting inclusive innovations. This means that we will have come full circle: back from individuals to virtuous institutions. Properties of virtuous practices for agricultural innovations, and the institutions dedicated to upholding them, is a field of inquiry warranting further studies (conceptual and empirical) in order to deal with challenges of our current predicament.

## References

Clapp, J., Ruder, S.-L., 2020. Precision Technologies for Agriculture: Digital Farming, Gene-Edited Crops, and the Politics of Sustainability. Global Environmental Politics 20, 49-69,

Godin, Benoit. 2015. *Innovation Contested: The Idea of Innovation Over the Centuries*. Routlegde. https://www.routledge.com/Innovation-Contested-The-Idea-of-Innovation-Over-the-Centuries/Godin/p/book/9780415727204.

Habets, Michelle G.J.L., Hub A.E. Zwart, and Rinie van Est. 2020. 'Why the Synthetic Cell Needs Democratic Governance'. *Trends in Biotechnology* 0 (0). https://doi.org/10.1016/j.tibtech.2020.11.006

Hansson, S.O. Farmers' experiments and scientific methodology. Euro Jnl Phil Sci 9, 32 (2019). https://doi.org/10.1007/s13194-019-0255-7

Manson, N.C. 2012. Epistemic restraint and the vice of curiosity. Philosophy 87, No. 340, 239-259

Nowotny, Helga. 2008. *Insatiable Curiosity | The MIT Press*. MIT Press. https://mitpress.mit.edu/books/insatiable-curiosity.

NWA. 2021. https://www.nwo.nl/en/researchprogrammes/dutch-research-agenda.

Ross, L. 2020. The virtue of curiosity., Episteme, 17, 1 (2020) 105-120.

Schurman, R & Munro, W.A. 2010. Fighting for the future of food: activists versus agribusiness in the struggle over biotechnology. Minneapolis, MN: University of Minnesota Press.

Steen, Marc. 2013. 'Virtues in Participatory Design: Cooperation, Curiosity, Creativity, Empowerment and Reflexivity'. *Science and Engineering Ethics* 19 (3): 945-62. https://doi.org/10.1007/s11948-012-9380-9

Stilgoe, Jack, Richard Owen, and Phil Macnaghten. 2013. 'Developing a Framework for Responsible Innovation'. *Research Policy* 42 (9): 1568-1580. https://doi.org/10.1016/j.respol.2013.05.008

# 53. Vegan food system and biodiversity: an ethical analysis

*M. Oksanen*[1*] *and T. Kortetmäki*[2]
[1]*University of Eastern Finland, Department of Social Sciences, P.O. Box 1627, 70211 Kuopio, Finland;*
[2]*University of Jyväskylä, Department of Social Sciences and Philosophy, P.O. Box 35, 40014 Jyväskylä, Finland; moksanen@uef.fi*

## Abstract

The purpose of the paper is to look at the linkages between diets, biodiversity, and food system from the vegan angle (without however committing to veganism). Food production and consumption are important drivers of biodiversity loss both directly using components of biodiversity as a harvested resource and indirectly through the land-use impacts (and pollution and climate change to a slightly lesser extent). For a vegan, the destruction of non-sentient biodiversity can be wrong through the loss of well-being it causes to sentient animals and humans. Since the primary reason for agriculture related clear-cutting is the need for increased feed production and thus to support the animal agriculture, there are strong and obvious indications that a vegan diet is more biodiversity-friendly than average omnivorous diets. The paper consists of the following sections. First, we identify four assumptions that instruct the analysis. Second, we describe how the biodiversity concern for animals has historically been resource-motivated. Third, we compare the idea that meat-eating generates an enormous number of sentient animals (the so-called logic of the larder) with the idea that the conservation of biodiversity rests on resourcist ideas (what we call the logic of the ostrich). These logics are discussed with relation to how they link biodiversity conservation to diets and food ethics, a linkage that is also shortly addressed from vegan and vegetarian angles. Fourth, after characterising the previous approaches as anecdotal, we take a broader, a food-system perspective and consider how the systemic approach differs in linking biodiversity, diets, and food ethics. Our analysis is empirically informed philosophy.

**Keywords:** food, meat-eating, species loss, sentientism, resourcism

## Introduction

Food production and consumption are important drivers of biodiversity loss both directly using components of biodiversity as a harvested resource and indirectly through the land-use impacts (and pollution and climate change to a slightly lesser extent). An example of the former is overfishing and industrial-like extraction of marine resources and the extensive use of game meat (Machovina *et al.*, 2015); and of the latter is the clear-cutting rainforests for livestock pastures and for soya and corn fields for feed (Williams *et al.*, 2020) and the related climatic impacts. Animal-based protein production is estimated to provide 18% of our calories and 37% of proteins yet it accounts for most food system originated harm to ecological systems and biodiversity, using ~83% of the world's farmland, mainly due to feed production (Poore and Nemecek, 2018). Thus, if there is less demand for meat, especially for beef and pork, then there is less demand for feed and less pressure on biodiverse land and habitat loss and fragmentation (Machovina *et al.*, 2015). Or, as George Monbiot (2016) put it, 'Rainforests, savannahs, wetlands, magnificent wildlife can live alongside us, but not alongside our current [omnivorist] diet.' Though the above-mentioned articles can suggest that diets rich in meat should be replaced with plant-based diets, or even vegetarianism or veganism (e.g. Food for Thought campaign, 2021), often the stances are articulated more moderately and suggested consuming meat from more biodiversity-friendly species (meat with lesser land-use impacts).


EurSafe 2021 – DOI 10.3920/978-90-8686-915-2_53, © Wageningen Academic Publishers 2021

For a vegan, the destruction of non-sentient biodiversity can be wrong through the loss of well-being it causes to sentient animals and humans. Since the primary reason for agriculture related clear-cutting is the need for increased feed production and thus to support the animal agriculture, there are strong and obvious indications that a vegan diet is more biodiversity-friendly than average omnivorous diets.

This idea of environmental, including biodiversity, friendliness of vegetarianism has been promoted in environmental ethics for a quite some time. A case in point is Paul Taylor (1986) who famously argued for vegetarianism although defending biocentrism. His theory attributes an equal moral status to all living organisms, humans, animals, and plants, and thus it does not approach vegetarianism solely based on concerns for sentient animals or human health or food justice. Because humans have to eat to survive, Taylor (1986: 293) asks, 'why should eating plants be ethically more desirable than eating animals?' He first recognises the badness of inflicting suffering on sentient animals by raising and killing them for food. The second point, however, is the main point resting on wilderness ideology: so the proportion of wild (non-cultivated) nature to cultivated nature should be larger than it is currently (notably, the situation has worsened significantly in 35 years that have passed since Taylor's argument). The transition to vegetarianism should be an objective because it drastically reduces the amount of cultivated land needed for human food production (Taylor, 1986: 296).

However, in light of the emergence of biodiversity-focussed conservation in the 1980s and culminating in the Rio Convention on Biodiversity in 1992, the appeal of veganism for biodiversity reasons remains unclear. Firstly, the human-dependent biodiversity (domestic animal breeds, crop varieties, etc.) was counted in the concept of biodiversity and the main concern was their sustenance as components of biodiversity, neither animal welfare nor their rights. Thus, domestic varieties of cattle and sheep, bred and adapted to local climes and food systems, became positive contributors to biodiversity (in their own right, but not necessarily regarding their impacts on the overall biodiversity) and there was a concern for their survival in the face of more efficient new breeds. Second, discussion on human diets and biodiversity of that time was, however, somewhat anecdotal (from the perspective of food systems) and focused on examples of ecological consequences of, for example, sustainable harvesting of particular animal species and culling of invasive animals that could threaten native populations. We will contrast this 'anecdotal view' with a more systemic view on diets and biodiversity to illustrate the difference between these viewpoints; the contrast also has implications for dietary ethics in relation to environmental problems including biodiversity loss.

This paper consists of the following sections. First, we identify four assumptions that instruct the analysis. Second, we describe how the biodiversity concern for animals has historically been resource-motivated. Third, we compare the idea that meat-eating generates an enormous number of sentient animals (the so-called logic of the larder) with the idea that the conservation of biodiversity rests on resourcist ideas (what we call the logic of the ostrich). These logics are discussed with relation to how they link biodiversity conservation to diets and food ethics, a linkage that is also shortly addressed from vegetarian and vegan angles. Fourth, after characterising the previous approaches as anecdotal, we take a broader, a food-system perspective and consider how the systemic approach differs in linking biodiversity, diets, and food ethics. Our analysis is empirically informed philosophy.

## Four assumptions

To begin with, let us spell out four assumptions on human needs, veganism, biodiversity and its current state and the impact of diets and the food system. First, humans need food. Food scarcity and the difficulties in feeding sufficiently everyone is a major global issue with biodiversity implications. Second, there is diversity in living and non-living nature and this diversity is of 'some' (moral) 'importance' even to an extreme individualist for whom only individuals are of intrinsic value. Biodiversity loss may not be a major problem for an individualist for whom the treatment of individuals determines the moral value of action, at least insofar as the biodiversity loss does not imply greater harm to individuals (vis-à-vis the situation of no biodiversity loss) now or in the future. Admittedly, there may not be representatives of this position although certain ideas, both sentientist and biocentric 'all animals/organisms are equal' positions, move towards it. In their view, biodiversity may be recognised and valued but only for instrumental and derivative reasons: biodiversity is important because it enables individual living and flourishing and is therefore worth protecting. This is relevant to food ethics because veganism, in its non-prudential and non-anthropocentric forms, is typically based on the principles of anti-speciesism and moral individualism that set the contours of morally appropriate treatment of any sentient organism (Lamey, 2019). The defenders of these principles have also been critical of biodiversity policies that violate those principles and involve killing or harming animal individuals to protect endangered species (e.g. Rawles, 2004).

The third assumption is, following Sarkar (2017), that a defensible concept of biodiversity 'must play some role in conservation biology, perhaps most saliently in systematic conservation planning [--].' Given that this conventional approach to biodiversity has focussed on non-individual entities, chiefly on species because they are measurable, moral individualism and related vegan diet can be overlooked. In addition, the variation within species and between ecosystems has been targeted. In the conventional approach to biodiversity, the role of individual organisms tends to vary from harmful (specimens of invasive alien species) to minuscule (members of the species of least concern on the IUCN priority listing) and crucial (representatives of endangered species and unique places). Thus, in the conventional approach to biodiversity, the key criterion for the right treatment is species membership. Again, there is a possibility of conflict between sentientists and the proponents of biodiversity conservation.

The fourth assumption is rather an empirical fact based on observations and assumptions in environmental sciences: biodiversity is significantly diminishing due to human action. An important driver for this loss is the current (global) food system, that is, what, where and how food is produced, processed, sold, and consumed.

Both certain biodiversity conservation practices (e.g. the culling of invasive alien animals) and the justification of biodiversity conservation (the focus on populations and lineages rather than, and sometimes at the cost of, individuals) have been problematic to a principled vegan whose focus is on individual sentient beings and their wellbeing. Veganism exists in many forms that have different implications to biodiversity consideration. Nevertheless, the loss of biodiversity is a matter of concern to a vegan both because sentient animals are included in biodiversity and may suffer from the impacts of biodiversity loss, and for various other reasons.

## Animal biodiversity as a resource

The Rio Convention on Biological Diversity (1992) promotes the idea of combining use and conservation to provide both a funding and a motive for conservation. Thus, the loss of biodiversity can be best prevented by means of constructive use in which human interests and the preservation of biodiversity are integrated. The rudiments of this stance are summarized by Freese (1997: 2): 'The logic behind this proposition is that revenues generated by the commercial, consumptive use of wild species will provide economic incentives for sound management of the harvested population(s).' There is more than that in this suggestion: actually, we come to the paradoxical situation that it is the human use that can contribute to the maintenance of the viability of populations for resources purposes (within the suitable institutional settings). The idea has quickly gained popular support as a reasonable compromise between environmental preservation and the meeting of human needs. Such a rationale seems to underlie the Rio Convention on Biological Diversity (CBD) to some extent. In § 6, the objective is to 'integrate, as far as possible and as appropriate, the conservation and sustainable use of biological diversity into relevant sectoral or cross-sectoral plans, programmes and policies' and § 10 is about 'sustainable use of components of biological diversity'.

Perhaps the most significant person to fuel the resource approach has been E. O. Wilson, who famously states that 'Biodiversity is our most valuable but least appreciated resource' (Wilson, 1994: 269). Wilson speaks about 'a revolution in conservation thinking', about 'New Environmentalism' that has shifted the focus to 'the practical value of wild species' that he assumes could 'give the invisible hand of free-market economics a green thumb' (*ibid.*: 270-77). His proposal on the utilization of Amazon river turtles as food is indicative of this tactics:

> The seven know species [of the genus Podocnemis] are highly regarded as a protein source by local people. The meat is of excellent quality and the base of a pleasing native cuisine. As the river banks have been more thickly settled, the turtles have been overhunted and several species are now endangered. But they are also easy to cultivate. Each female lays a clutch of up to 150 eggs, and the young grow rapidly. One species, the giant *Podocnemis expansa*, reaches a length of nearly a meter and a weight of 50 kilograms. (*ibid.*: 282)

Moreover, the production of protein by farming these turtles may make the conversion of rainforests into pasturelands economically inefficient, as turtles can be farmed on natural floodplains. As Wilson says, the turtle may produce meat 'more than 400 times the yield of cattle raised in nearby pasture cut from surrounding forests.' There are more similar tendencies, in which the domestication (or semi-domestication) of the species has aided its survival and recovery; one may consider the ostrich in Africa, the buffalo and the alligator in America, or some species of crocodiles, some of them being farmed in other continents. From a vegan perspective, the model Wilson offers, or more generally the resourcist view on biodiversity, seems objectionable for two main reasons. First, the attitude to animals is instrumentalist; animals may serve some non-economic purposes (spiritual or cultural), they are to satisfy basic dietary needs. Second, their treatment is tied to their status as designated through the biodiversity conservation policies. Moreover, it shares, as will be explored next, some features with the logic of the larder argument.

## The logic of the larder and the logic of the ostrich

What Lamey (2019) considers 'a long-standing view' holds that it is in the interest of an animal to exist for human purposes, for instance as a food source, because without a human purpose the animal would not exist. Thus, if we were following strictly a vegan diet, chicken, cow, sheep, and other domesticated animals would not exist, and the non-existence is worse than their existence regardless of the conditions of their existence. A century ago, the British animal rights writer Henry S. Salt named this reasoning as the 'logic of the larder'. Somewhat similar reasoning appears in the 'use it or lose it' debate accounted in the previous section: without use, people lack motives for maintaining certain biological lineages and/or protecting them against other aims in trade-off situations. Let us call the latter argument as the 'logic of the ostrich' to commemorate the once almost extinct species of the bird that is no longer threatened and whose range has expanded to new continents, mostly farmed to provide eggs and meat for humans. There is even a long-standing feral population in Australia.

The logic of the larder addresses individuals and whether it is a morally good thing to bring into existence creatures whose lives are not going to be decent. Accordingly, there is a discussion on the ethics of reproduction in which some are supporting the imperative of avoiding bad outcomes (have a life full of suffering) and others considering that bad life is better than no life at all (see Lamey, 2019). In the logic of the larder, the species membership tends to play no role, it is about sentient individuals, about their existence and quality of life.

In the logic of the ostrich, the focus is on species and/or on within species variation such as wild populations and farmed breeds and landraces. The conservation of farmed breeds (landraces) essentially follows the logic of the larder, as without sufficient incentives (such as food or workforce) to bring animals into existence and to keep them existing the variety is at risk of disappearance. There has in recent decades been a general trend that the less efficient (productive) varieties become replaced with more efficient or otherwise suitable varieties. The disappearance of traditional varieties has been quick, and therefore seen as a matter of concern; concerns about the impacts of climate change on the predictability of future farming conditions have also increased the interest in reviving the more robust varieties back into use. For example, as a part of biodiversity policies, there has been an interest to conserve the three breeds of Finn cattle, adapted to cold and harsh northern conditions. Food ethics that embrace a vegan diet and view any use that can require the killing of an animal as wrong implies risking these breeds: without sustainable use of animals, there are either no incentives to conserve these animals or the incentives are inadequate (i.e. can be overridden by other motives).

The scope of concern can be rather narrow and focus on a few domesticate species and subspecies, but as the example of the ostrich and the Amazon river turtle indicate, the range of animals is not limited to traditional domesticates. Within the conventional approach to biodiversity, the expansion of the scope of edible species of animals and plants can be greeted as welcome (as far as it does not endanger diversity in wildlife) to conserve biodiversity.

The logic of the ostrich can be recognized also at the levels of specific communities, where there can be strong cultural and economic interests in conservation. Namely, biodiversity conservation has invested a lot of effort in the integration of the cultural, economic, and ecological dimensions of sustainability. Especially, the indigenous perspectives have been important, in particular because many of the biodiversity hotspots are there where indigenous peoples live and act as the guardians of biodiversity. The vegan bid, as far as we follow the vegan principles literally, is to give up many of the ancient, culturally-specific practices that involve using animals for food. A case in point is traditional whaling and eating of whale meat (on discussion, see Posey, 1999).

## Vegan food system

While vegans clearly object the logic of the larder, how do they response to the logic of the ostrich where the eating of animal-based foods is justified by the argument that it promotes the protection of the species (whose members are eaten)? Here's one possibility: The logic of the ostrich is anecdotal and addresses isolated cases without providing us with a fuller picture of the biodiversity footprint of the whole food system. Because the currently dominant patterns and modes of producing animal-based food, especially livestock, and overharvesting of aquatic animals, are major drivers of biodiversity loss, biodiversity action goes hand in hand with the need to significantly reduce meat and animal-based food consumption in average diets and therefore in food systems.

As we began with referring to studies on the ecological impacts of the current food system activities, a vegan food system would likely result in more benign impacts on biodiversity. A vegan food system has at least the following implications for biodiversity from the empirical viewpoint: (1) it does not involve raising animals for food (or feed); (2) it does not involve harvesting or hunting animals for food; (3) because of no need for feed production, its land use impacts are a fraction compared with those of the present food systems and it would probably require no land clearance (but instead allow ecological restoration); (4) it would have smaller impact on climate change; (5) it would not give economic/diet-based incentives for preserving a particular species.

The logic of the ostrich emphasises the possible cost of the vegan food system for biodiversity: it risks of losing domesticated varieties and some singular species. As it can be seen, the cases brought up by the logic of the ostrich are anecdotal by addressing the fifth point above, without paying attention to the four other, more wide-ranging points. For example, the logic of the ostrich neglects the question of the 'impacts of producing and eating' the specimens of the given species: if ostrich farming (or whatever the species meriting protection by this strategy) has large-scale ecological impacts and becomes a widely practised mode of production, it may be the case that the preservation of the ostrich comes at the cost of losing several other species.

The above system-level implications mean that a fully vegan food system would not in itself be ideal (optimal) from the biodiversity point of view because of the fifth implication. There might be cases where local or seasonal omnivorism supports biodiversity. On the other hand, biodiversity considerations rarely ask what would be 'ideal' from the biodiversity point of view but what would be 'sufficient' to stop biodiversity loss. Thus, a key question is how comprehensive a move to a vegan food system should be or how vegan would an environmentally ideal food system be? These imply questions for the individual level dietary choices: should the ideal diet concern every normal person, everywhere and all the time, or should it be sensitive to exceptions that allow using animals for food? These questions are at the heart of the food and resource sovereignty locally and nationally.

## References


Food for Thought Campaign (2021). Why vegan. Available at: https://www.foodforthoughtcampaign.org/why-vegan.

Freese, C.H. (1997). The 'Use It or Lose It' Debate. Issues of a Conservation Paradox. In: Harvesting Wild Species. Freese. C.H. (ed.): The Johns Hopkins University Press, Baltimore, pp. 1-48.

Lamey, A. (2019). Duty and the Beast. Cambridge University Press, Cambridge, UK, 262 pp.

Machovina, B., Feeley, K.J. and Ripple, W.J. (2015). Biodiversity conservation: The key is reducing meat consumption. Science of the Total Environment 536: 419-431.

Monbiot, G. (2016). 'I've converted to veganism to reduce my impact on the living world.' Available at: https://www.theguardian.com/commentisfree/2016/aug/09/vegan-corrupt-food-system-meat-dairy.

Poore, J. and Nemecek, T. (2018). Reducing food's environmental impacts through producers and consumers. Science June 1, 2018: 987-992.

Posey, D.A. (ed.) (1999). Cultural and Spiritual Values of Biodiversity. UNEP, Nairobi; IT publication, London, UK, 731 pp.

Rawles, K. (2004). Biological Diversity and Conservation Policy. In: Oksanen, M. and Pietarinen, J. (eds) Philosophy and Biodiversity. Cambridge University Press, Cambridge, UK, pp. 199-216.

Sarkar, S. (2017). Approaches to Biodiversity. In: Garson, J., Plutynski, A. and Sarkar, S. (eds) The Routledge Handbook of Philosophy of Biodiversity. Routledge, London, UK, pp. 43-55.

Taylor, P.W. (1986). Respect for Nature. Princeton University Press, Princeton, USA, 329 pp.

Williams, D.R., Clark, M., *et al.* (2020). Proactive conservation to prevent habitat losses to agricultural expansion. Nature Sustainability. https://doi.org/10.1038/s41893-020-00656-5

Wilson, E.O. (1994). The Diversity of Life. Penguin, Harmondsworth, UK, 406 pp.

# Section 5.
# Methodology and further challenges to environmental ethics

# 54. Food security and symbolic legislation in Switzerland: a false sense of security?

C.E. Blattner[1]* and O. Ammann[2]
[1]University of Bern, Faculty of Law, Schanzeneckstrasse 1, 3012 Bern, Switzerland; [2]University of Zurich, Faculty of Law, Rämistrasse 74/64, 8001 Zurich, Switzerland; charlotte.blattner@oefre.unibe.ch

## Abstract

In 2017, 78.7% of the Swiss people voted to enshrine the concept of 'food security' in the Federal Constitution. Originally prompted by agricultural interest groups for reasons of protectionism and then revamped by the Swiss Parliament, the new article 104a includes a wide array of demands for food policy, including protection of agricultural land, local production, conservation of natural resources and their effective use, responsiveness to market demand, and trade relations contributing to sustainable development. As a one-of-its-kind constitutional norm on food security, the now four-year-old article still raises questions about its precise scope and normative content. In particular, it is often said that the norm is largely symbolic. We shed light on these developments by examining the emergence of the norm and embedding it in the broader international discourse on food policy. We show that Switzerland's understanding of food security is greatly flawed, as it seeks to secure its own access to goods without regard to their environmental footprint and effect on human rights abroad. In line with Switzerland's climate and human rights commitments, we propose new ways to interpret food security, including moving away from market needs, distinguishing foodtypes based on whether they thwart food security in the long term, paying attention to food security elsewhere, and giving effect to considerations of distributive justice.

**Keywords:** constitutional law, direct democracy, agriculture, Swiss Constitution

## Introduction

Switzerland is in the midst of an intense debate about food and agriculture. Between 2018 and 2020, no less than three popular initiatives related to food and agricultural policy were voted upon.[1] In 2017, with a record approval of 78.7%, Swiss voters agreed to include a new provision on food security in the Constitution. While this new article 104a was drafted by the Swiss Parliament, the proposal to adopt a constitutional norm on food security came from the Swiss Farmers' Union (SFA). This proposal was largely driven by protectionist motives, despite the fact that high-quality food in Switzerland is highly accessible and household spending on food is, in comparative perspective, very low. Unlike billions of other people, the Swiss have, throughout the last century, not experienced food insecurity at any point in time (the last famine in this country took place in 1918 during the onset of the Spanish flu). How come, then, that close to 80% of voters recently chose to adopt a new constitutional provision on food security?

## Article 104a of the Swiss Constitution

On 24 September 2017, an overwhelming majority of Swiss voters (78.7%) and all Swiss cantons accepted the federal decree on food security, thereby agreeing to enshrine the concept of food security in article 104a of the Swiss Constitution. This provision reads as follows:

[1] I.e. the 'Fair Food Initiative', the 'Food Sovereignty Initiative', and the 'Horned Cow Initiative'.


EurSafe 2021 – DOI 10.3920/978-90-8686-915-2_54, © Wageningen Academic Publishers 2021

> In order to guarantee the supply of food to the population, the Confederation shall create the conditions required for: (a) safeguarding the basis for agricultural production, and agricultural land in particular; (b) food production that is adapted to local conditions and which uses natural resources efficiently; (c) an agriculture and food sector that responds to market requirements; (d) cross-border trade relations that contribute to the sustainable development of the agriculture and food sector; (e) using food in a way that conserves natural resources.

The constitutional norm is the result of a longer back-and-forth between the SFA's initial proposal (which was later withdrawn), the Federal Council's direct counter-proposal in favour of strengthening domestic production (which was eventually abandoned), and the direct counter-proposal by the Swiss Parliament. The latter provision, which is the one that was finally put to vote, emphasizes the need to protect agricultural land, while taking the entire food production chain into account and being more market-oriented. Political parties and interest groups almost unanimously recommended accepting article 104a, but for different reasons: conservative farmers hoped for protectionism, while mixed progressive and liberal farmers hoped for fair trade and sustainable production. Due to its vague wording and the lack of knowledge about its precise scope and normative demands (especially vis-à-vis the legislature), article 104a was supported by a majority along this spectrum, as it 'can be invoked by almost any interest group to support its policy preferences' (Hettich, 2019: N 8; see also Schöchli, 2017).

Article 104a introduces several new terms into the Constitution, i.e. 'food security', 'food sector', 'trade relations', and the requirements that food production/use be 'adapted to local conditions', 'uses natural resources efficiently', and 'conserves natural resources' (Biaggini, 2017: 896, N. 2). This suggests that the provision has autonomous normative content. Yet, commentators described the vote as a 'ballot about nothing' and as having symbolic value only (Schöchli, 2017). Thus far, this assessment is correct to the extent that article 104a has not led to new laws, statutory amendments, new or amended subsidies, or the like. Yet, judgment about the norm being symbolic is premature: in the parliamentary debates, it was stated that the new article 104a 'complements' existing constitutional provisions, namely articles 104, 75, and 102 (Council of States, 2016: 900, 901). Moreover, several MPs across the political spectrum submitted procedural requests to clarify the normative implications of article 104a (e.g. Baumann, 2020; Bourgeois, 2020; Grin, 2017; Schneider Schüttel, 2018; Rytz, 2018).

In Parliament, article 104a was described as 'complementary' to article 104 (National Council, 2017: N 212). While some argue that article 104a is a 'duplicate' of article 104 as both deal with food supply, sustainability, market orientation, and the protection of agricultural land (Renz, 2017), this view ignores the distinctive terminology used by article 104a. Contrary to article 104, it is not limited to domestic production and applies to the entire food chain; further, its notion of market orientation refers not only to agriculture but to the entire food industry.

Article 75(1) provides that the Confederation has the power to 'lay down principles on spatial planning' in order to guarantee 'the appropriate and economic use of the land and its properly ordered settlement'. The Federal Council has noted that this provision implicitly includes the protection of agricultural land (2015, 10), which article 104a highlights explicitly. This focus on the protection of agricultural land should, however, not be conflated with spatial planning, which article 104a does not touch on (National Council, 2017: N 212, 214, 216 f).

Article 102 provides for the supply of essential goods and services 'in the event of the threat of politico-military strife or war, or of severe shortages that the economy cannot by itself counteract' (article 102(1)). Unlike article 104a, it is tailored to respond to acute crises rather than dealing with recurring, longer-term issues (Council of States, 2016: 900, 907).

Thus, article 104a does not merely replicate existing constitutional norms. The Federal Council and Parliament repeated many times that it creates 'an overall concept for food security' (e.g. CS-CEAT, 2016: 10) – albeit a concept that remains highly indeterminate at this point in time.

## The problem with article 104a of the Swiss Constitution

As of 2018, the most recent estimate, Switzerland's degree of self-sufficiency was at 58%, a percentage which has been steady since several decades (FAOG, 2020: 114). Due to limited availability of land and a comparatively high number of inhabitants in relation to its size, Switzerland is uniquely dependent on imports. With feed imports of over 1 million tons a year, and substantial parts of its land used for animal production, there is no way Switzerland could become any more food secure. Splitting up self-sufficiency by plant and animal products brings this dilemma to light: per 2018 (and similarly for the previous years), self-sufficiency for animal products was at 100%, while for plant products, it amounted to 40% (FAOG, 2020: 115).

The specific steps envisaged by the Swiss agricultural policy after 2022 to meet its rising demand for food is to limit the supply of raw protein to animals, to reduce ammonia, to let animals be outside more, and to ensure 'a longer use' of cows used for milk production (Federal Council, 2020: 11). The credo is 'let's do more with less,' showing that food security is primarily understood as producer security (Häberli, 2015: 172). Switzerland thereby follows a long tradition in food policy discourse according to which the central problem of food (in)security is (shortage of) food supply. However, today, assumptions that scarcity and population growth cause hunger, and that food production will eradicate hunger, are slowly withering. For, despite rapid advances in agriculture with high-yielding hybrids, chemical fertilizers, powerful pesticides, and extensive irrigation happening half a century ago, for a significant number of people, producing and accessing enough food remains life's central challenge (Burke and Lobell, 2010: 13; Holt-Giménez, 2019: 1).

Departing from the earlier 'productionist policy paradigm' that shaped food policy, the most authoritative and still widely used definition of food security was enshrined in the Declaration on World Food Security in 1996, defining it as 'a situation that exists when all people at all times have physical, social, and economic access to sufficient, safe, and nutritious food that meets their dietary needs and food preferences for an active and healthy life' (World Food Summit, 1996: para. 1). Food security thus is about 'having, on an individual level, the food one needs and wants' (Burke and Lobell, 2010: 14), while taking into account broader policies in agro-biodiversity, natural resource decline and depletion, as well as sociocultural complexities and emerging food needs. For people to be food secure, the following four pillars must be fulfilled: (1) 'food availability' of sufficient quantities of food of appropriate quality, supplied through domestic production or imports; (2) 'food access' by individuals to adequate resources (entitlements) for acquiring appropriate foods for a nutritious diet; (3) 'utilization of food' through adequate diet, clean water, sanitation, and health care to reach a state of nutritional well-being where all physiological needs are met; (4) 'stability', meaning that a population, household or individual must have access to adequate food at all times and should not risk losing access to food as a consequence of sudden shocks (e.g. an economic or climatic crisis) or cyclical events (e.g. seasonal food insecurity) (World Food Summit, 1996).

Despite decade-long international efforts to achieve 'zero hunger' (SDG No. 2), the world is anywhere but on track to reach this goal. Updated methods taking into account household data show that the 'decades-long decline in hunger in the world (...) had unfortunately ended' (FAO *et al.,* 2020: xviii). As of 2020, nearly 690 million people are hungry (i.e. suffering from undernourishment), which is 8.9% of the world population (id. at 3, 4). This number was up by 10 million people in the last year, and by nearly 60 million in the past five years (id.). By 2030, the number of hungry people is expected to exceed 840 million (id. 8). Climate change significantly exacerbates existing pressures on the global food system due to rising temperatures, increased temperature variability, greater frequency of draughts, heavy precipitations and other extreme weather events, rising sea levels, and the salinization of arable land and freshwater – leading to crop failures, yield losses, new pathogens and diseases, animal health issues, compromised forest management, and losses in fisheries, especially in the Global South (FAO *et al.*, 2020: 5; IPCC, 2019: 10; IPCC, 2018: 9; Tirado *et al.*, 2009: 129, 133). According to the latest IPCC Synthesis Report, 'All aspects of food security are potentially affected by climate change, including food production, access, use and price stability (high confidence)' (IPCC, 2014: 69; also FAO, 2016).

Switzerland is significantly adding to the problem of food insecurity, securing its population's unhindered excessive use of resources while exacerbating food security elsewhere. The demand of consumers for grain and soy-fed animal proteins (meat, eggs, and milk) is, in qualitative and quantitative perspective, one of the biggest threats for food security and efficiency, putting pressure on global food systems, the environment, and public health (Blattner and Ammann, 2019). Food produced in the 'Global South' is preferentially given to export countries in the North (mostly as feed for animal production), resulting in a lack of food for local communities. Further, subsistence farmers are pushed to live and work on barren land in mountains terrains for large agricultural producers to let their livestock graze on the flat plains (Gaidetzka, 1997: 13). By excessively supporting their own producers, well-off Northern countries significantly increase global food prizes, which reduces access to food worldwide. Overall, increasing production – despite the fact that there is in fact enough food for everybody – neither increases food security in Switzerland, nor ends hunger elsewhere. Instead, it will push our planet beyond its ecological limits and destroy the lives and livelihoods of billions of people (Holt-Giménez, 2019: 8).

With its inequitable and extractivist food system, Switzerland threatens to violate other people's right to food, which is protected by the Universal Declaration of Human Rights (art. 25(1)), and guaranteed by various international human rights treaties, most comprehensively by the International Covenant on Economic, Social, and Cultural Rights (ICESCR). In art. 11 ICESCR, parties 'recognize the right of everyone to ... adequate food' (para. 1) and 'improve methods of production ... of food,' *inter alia* 'by developing or reforming agrarian systems in such a way as to achieve the most efficient development and utilization of natural resources' (para. 2). States have a duty of progressive realization with respect to this right (art. 2(1)), and they cannot discriminate against specific groups of individuals in this context (art. 2(2)). To be clear, ensuring the right to food and freedom from hunger principally rests with national governments, and considering food security elsewhere is often seen as voluntary. However, under international law, States' duties with respect to the right to food are owed not only to people within their territories, but also outside. The minimum content of extraterritorial obligations is the duty to respect, meaning that no State is allowed to violate the right to adequate food of a person living in another country. Thus, States must refrain from impairing food security in other countries (Häberli, 2015: 149; Paasch, 2010: 21-22).

## New ways to interpret 'food security' in Switzerland

One cannot look at a country's level of food security without considering food security globally, for all States either have to increase their own production or import goods (and Switzerland has, as mentioned, limited choice in this regard). At the occasion of the World Food Summit in 1996, the parties – including Switzerland – committed to implement the annexed World Food Summit Plan of Action. In this Plan, they underlined, among others, the importance of conservation and sustainable use of natural resources (such as land, water, and forests) to achieve this goal (comment one to objective 1.2, para. 15). For this purpose, taking into account 'the present and future needs of the people as well as the natural resources potential and limitations' (comment three, para. 27) is key. Eighteen years later, at the occasion of the Second International Conference, Switzerland, among other States, adopted the Rome Declaration on Nutrition, recognizing 'the need to address the impacts of climate change and other environmental factors on food security and nutrition, in particular on the quantity, quality and diversity of food produced, taking appropriate action to tackle negative effects' (FAO and WHO, 2014: para. 8). As the host state of the UN, the WHO, and other international organizations concerned about food security, Switzerland should lead by example, pay attention to food security elsewhere, and give effect to considerations of distributive justice.

In the Paris Agreement (2015), parties recognized the 'fundamental priority of safeguarding food security and ending hunger, and the particular vulnerabilities of food production systems to the adverse impacts of climate change' (preamble, recital 9). The fact that inequitable and extractivist agriculture drives global food insecurity and climate change is accepted by major global players in food discourse. As the UN Environmental Programme stated in 2010: 'Impacts from agriculture are expected to increase substantially due to population growth increasing consumption of animal products'. It is recognized that '[u]nlike fossil fuels, it is difficult to look for alternatives: people have to eat' (82). However, agriculture also has unique potential to contribute to stabilizing the world's climate and securing food security worldwide (IPCC, 2014: 100; FAO, 2016: 5). As part of its mitigation duties under the Paris Agreement (2015, article 4), Switzerland must work towards these goals, through changes in consumption patterns and dietary changes (IPCC, 2014: 100) by 'reducing demand for emission- and resource-intensive food products' (FAO, 2016: 10), which includes – most prominently – animal-sourced foods (FAO, 2016: xv; IPCC, 2019: 24; UNEP, 2010: 82). Consumption patterns and dietary habits must move toward more balanced and healthier diets featuring 'plant-based foods, such as those based on coarse grains, legumes, fruits and vegetables, nuts and seeds' (IPCC, 2019: 24; see also FAO, 2020: xix). The German and Austrian food authorities are already appealing to the public to change their consumption habits away from animal products (German Council for Sustainable Development, 2018; Kumpitsch, 2014: 8), and the EU is funding alternatives to meat as part of its sustainability strategy of the Green Deal, providing 10 billion Euro to promote research on how to source alternative proteins and to increase their availability.

## Conclusion

The new article 104a of the Swiss Constitution on food security emerged from protectionist ambitions of the agricultural lobby, which the Swiss Parliament toned down by drafting an alternative provision. Due to its vagueness, the provision received broad political support, but is largely viewed as 'symbolic' and toothless normatively speaking. Yet as we have argued, article 104a introduces new concepts into the Swiss Constitution and has a distinctive normative content. What is deeply problematic is that article 104a provides, both literally and figuratively, a false sense of security to Swiss citizens. It cannot increase food security domestically due to Switzerland's dependency on feed imports. Worse even, it misunderstands food security for protectionism and unbridled consumerism, thereby threatening food security elsewhere.

While Switzerland remain trapped in the old producitivist approach to food security, global food policy has taken huge leaps in terms of how food security is understood: from supply to needs, from quantity to quality, from calorie intake to nutrient sufficiency, from production to distribution. Further, there is increasing acknowledgement of the need to address the effects of food security policies on people abroad, e.g. by recognizing the vicious circle of 'bad foodtypes' that contribute to climate change and that exacerbate existing stressors to secure food (Häberli, 2015: 175). As such, overconsumption and overproduction are emerging as the central challenge of our times (Holt-Giménez, 2019: 5), considering that hunger does not just happen but is 'manmade' (Gaidetzka, 1997: 13; Reimann, 2000: 90). Instead of relying on an anachronistic and inequitable concept of food security, Switzerland should reinterpret its new constitutional norm to abide by its climate commitments and obligations under international human rights law.

## References

Baumann, K. (2020). Interpellation no. 20.4176: Absatzförderung für Schweizer Fleisch im Widerspruch zu den Verfassungszielen zu der Ernährungssicherheit?.

Blattner, C. and Ammann, O. (2019). Agricultural exceptionalism and industrial animal food production: Exploring the human rights nexus. Journal of Food law & Policy 15(2): 92-151.

Bourgeois, J. (2020). Interpellation no. 20.3521: Im Ausland verursachte Umweltbelastung.

Burke, B. and Lobell, D. (2010). Climate effects on food security: An overview. In: Lobell, D. and Burke, M. (eds) Climate Change and Food Security: Adapting Agriculture to a Warmer World. Springer, Dodrecht *et al.*, pp. 13-58.

Council of States (2016). Winter Session, Second Meeting, November 26. 15.050, AB 2016 S 900 ff.

Council of States, Committee for Economic Affairs and Taxation (2016). 15.050 Für Ernährungssicherheit. Volksinitiative, Gegenentwurf und Fristverlängerung. Bericht [*cited* CS-CEAT].

FAO (2016). The state of food and agriculture: Climate change, agriculture, and food security. FAO, Rome.

FAO (2020). The state of food and agriculture: Overcoming water challenges in agriculture. FAO, Rome.

FAO and WHO (2014). Second International Conference on Nutrition: Rome Declaration on Nutrition. Rome, 19-21 November 2014.

FAO, IFAD, UNICEF, WFP and WHO (2020). The state of food security and nutrition in the world 2020. FAO, Rome.

Federal Council (2015). Erläuternder Bericht zum Gegenentwurf des Bundesrates zur Eidgenössischen Volksinitiative 'Für Ernährungssicherheit' vom 14. Januar 2015.

Federal Council (2020). Agrarpolitik 22+: Botschaft des Bundesrates zur Weiterentwicklung der Agrarpolitik nach 2022 (AP22+).

Federal Office for Agriculture FOAG (2020). Agricultural report 2020.

Gaidetzka, P. (1997). Ernährung – Ein Recht für alle? In: Chandy K.T. *et al.* (eds) Ernährung – Ein Recht für alle. Horlemann, Aachen, pp. 7-16.

German Council for Sustainable Development (2018). Der Nachhaltige Warenkorb_Ihr Einstieg.

Grin, J.-P. (2017). Interpellation no. 17.4173: Strategie des Bundesrates bezüglich der Agrarpolitik 2018-2021 und darüber hinaus.

Häberli, C. (2015). Ernährungssicherheit in der Schweiz. In: Norer, R. (ed.) Landwirtschaft und Verfassungsrecht – Initiativen, Zielbestimmungen, rechtlicher Gehalt. Dike, Zürich/St. Gallen, Schriften zum Recht des ländlichen Raums/Collection de Droit Rural 9: 145-175.

Hettich, P. (2019). Art. 1 LwG. In: Norer, R. (ed.), Stämpflis Handkommentar: Landwirtschaftsgesetz. Bern, Stämpfli, pp. 44-56.

Holt-Giménez, E. (2019). Can we feed the world without destroying it? Polity, Cambridge UK.

IPCC (2014). Climate change 2014 synthesis report. IPCC, Geneva.

IPCC (2018). Global warming of 1.5°C. Summary for policymakers. IPCC, Geneva.

IPCC (2019). Climate change and land. Summary for policymakers. IPCC, Geneva.

Kumpitsch, O. (2014). Klimagerechte Ernährung: Ein Beitrag zur Reduktion von Treibhausgasemissionen. GRIN, Norderstedt, Germany.

National Council (2017). Spring Session, Seventh Meeting, March 7. 15.050, AB 2017 N 212 ff.

Paasch, A. (2010). Kampf gegen den Hunger – Paradigmen im Widerstreit. In: Gabbert, K. *et al.* (eds) Jahrbuch Lateinamerika: Analysen und Berichte 33, Über Lebensmittel. Westfälisches Dampfboot, Münster, pp. 16-35.

Reimann, C. (2000). Ernährungssicherheit im Völkerrecht. Boorberg, Stuttgart.

Renz, F. (2017). Bauer pfeift, Politiker tanzt, Tages-Anzeiger, August 17.

Rytz, R. (2018). Motion no. 18.4394: Freihandelsabkommen zwischen EFTA und Mercosur. Verbindliches Nachhaltigkeitskapitel.

Schneider Schüttel, U. (2018). Interpellation no. 18.4243: Umsetzung der standortangepassten Produktion gemäss Artikel 104a der Bundesverfassung.

Schöchli, H. (2017). Ernährungssicherheit: Der Urnengang über das Nichts. NZZ, June 29.

Tirado, C., Cohen, M.J., Aberman, N.-L., Thompson, B. (2009). The impact of climate change on nutrition. In: Clapp, J. and Cohen, M.J. (eds) The Global Food Crisis: Governance Challenges and Opportunities. Wilfrid Laurier UP, Waterloo, Canada, pp. 129-44.

UNEP (2010). Assessing the environmental impacts of consumption and production: Priority products and materials. UNEP, Nairobi.

World Food Summit (1996). Rome Declaration on World Food Security. Rome, Italy.

# 55. Rights for rivers

*M. Hansche*[1] *and S. Meisch*[1,2*]
[1]*International Centre for Ethics in the Sciences and Humanities (IZEW), University of Tübingen, Wilhelmstr. 19, 72074 Tübingen, Germany;* [2]*Centre for the Study of the Sciences and the Humanities (SVT), University of Bergen, Parkveien 9, 5020 Bergen, Norway; simon.meisch@uni-tuebingen.de*

## Abstract

For some years now, a growing environmental movement has been demanding that rivers be given rights or the status of legal persons. Proponents argue that this approach would better protect river ecosystems. In addition, it resonates with calls to give indigenous peoples more say in the management of their livelihoods. Various countries (e.g. New Zealand or Colombia) have already created corresponding legal frameworks. This paper will first deal with inherent tensions between ethical justifications of rights for rivers (ecocentrism vs corrective justice). Second, it will look at the case of the Whanganui River (NZ) which has the most elaborated legal framework of a river right. Concludingly, it argues that the case for rights of rivers raise challenging questions for various domains of applied ethics (such as agriculture, food or conservation ethics).

**Keywords:** environmental ethics, indigenous rights, ecocentrism, the Whanganui River

## Introduction

A growing global movement demands rights for non-human nature in general and rivers in particular (Benöhr and Lynch, 2018; Chapron *et al.*, 2019; Strang, 2020a,b; Wilson and Lee, 2019). Currently, some countries assign rivers the status of legal person (cf. Chapron *et al.*, 2019). The most prominent case is the Whanganui River in New Zealand. In Colombia, the Supreme Court ruled that the Rio Atrato has legal rights. In Ecuador and Bolivia, nature and thus rivers have the status of legal persons, while in Bangladesh all rivers enjoy such a status. In India, the attempt to grant rights to rivers was stopped by the Supreme Court.

Calls to grant rivers a legal status usually refer on two different arguments. First, 'ecocentric perspectives' argue for a greater protection of rivers from harmful human interventions. Second, by drawing on concepts of 'corrective justice' (on concepts of justice, cf. Kalfagianni and Meisch, 2020), other perspectives call for the participation of indigenous peoples in issues concerning them and their livelihoods. In the past, they and their concerns have generally been marginalised in many settler societies (O'Donnell and Macpherson, 2018; Strang, 2020b). Both arguments are often mentioned together. However, there are inherent contradictions between both lines of argumentation, which, in addition, become manifest in practical contexts of nature use and conservation (e.g. Bremer and Funtowicz, 2015; Yates *et al.*, 2017).

Despite these contradictions, advocates of rights for rivers often seem evasive or ambivalent when it comes to ethical justifications, as can be seen quite well in Chapron *et al.* (2019: 1392): 'Whether nature has moral rights is likely to remain debated, but nature clearly can have legal rights [...] Although moral considerations often influence the development of legal rights (and vice versa), legal rights need not have a moral basis. The law can give rights to all kinds of entities if it finds reason to do so.'

In this paper, we believe it is necessary to address these normative challenges. First, we address inherent tension between normative justifications of rights for rivers. Subsequently, we discuss a prominent case, the Whanganui River (New Zealand), and speculate how the legal personhood of a river might affect


EurSafe 2021 – DOI 10.3920/978-90-8686-915-2_55, © Wageningen Academic Publishers 2021

agriculture and food production. This case, we argue concludingly, raises challenging questions for further research in applied ethics.

## The dilemma of ecocentricm

Advocates of rights for rivers refer to various ecocentric positions (Chapron *et al.*, 2019; Strang, 2020a,b; Wilson and Lee, 2019), drawing on descriptions grounded in the critical social science analysis, according to which Western societies 'entrenched, over centuries, a dualistic worldview in which (feminised) Nature is 'other' to (masculinised) human Culture' (Strang, 2020b: 115). By contrast, they call for attributing 'Nature' an 'intrinsic value' (Chapron *et al.*, 2019: 1392). In this way, river rights advocates embrace ecocentric arguments (on ecocentrism, cf. Dierks, 2016; Eser and Potthast, 1999). This environmental ethical approach rejects anthropocentric approaches, which it accuses of an instrumental stance to nature. In contrast, ecocentrists advocate a nature-centred ethics and argue for the moral primacy of holistic entities – such as species or ecosystems – over individual beings. Thus, ecocentrism even transcends other non-anthropocentric approaches such as sentientism and biocentrism that seek to give direct moral consideration to individual entities (such as animals or plants). Philosophers associated with this position are Aldo Leopold or J. Baird Callicott.

Ecocentrism has also been subject to criticism. Two aspects are directly relevant to the debate of rights for rivers. The first concerns the epistemological problem of determining the moral subject. The second relates to the place of humans in ecocentric approaches. From both, tensions can arise, in theory and practice, between ecocentric rationales and those seeking to give indigenous communities greater say over their own affairs.

First, as ecosystems are mental constructs, questions about systemic boundaries emerge. Ecosystems are neither stable, nor is it an undisputable given where a system ends and another begins. This creates the challenge of determining what a system (e.g. a river) is that should merit intrinsic value and, thus, (legal) rights. River rights advocates seem more concerned with what rights rivers should have than with how exactly to delineate river systems. The challenge becomes apparent by considering the following description: 'Rivers change progressively over their course from headwaters to mouth, from steep streams that are narrow and turbulent to wider, deeper, often meandering channels.' (Middleton, 2012). Rivers are dynamic systems. So, the question arises as to what belongs to a river: its catchment area? Its banks? Tributaries? Floodplains? The animals and plants living in it? Certain amounts of rainfall? Feeding glaciers? What happens when a river changes its course? These questions are only few examples – and depending on the respective answers, it soon becomes clear how rights collisions might arise. So, clarifying what is meant by river is not trivial, neither in scientific nor in ethical terms. After all, the scientific criteria of what constitutes the river that should receive rights are based on value judgments. Such judgments need to be made transparent and ethically justified.

Second, ecocentrism does not ascribe any special or even higher value to humans within entities such as ecosystems. Various ecocentric positions even have a reputation of being misanthropic (cf. Dierks, 2016), that is, of seeing humans as a disruptive element, to be removed if necessary. This perspective has, in itself, been criticised as standing, almost inevitably, in contrast to a human rights regime. In addition, some ecocentric approaches advocate positions often called 'ecototalitarianism' or 'ecofascism'. Obviously, the use of such labels is highly questionable. What becomes evident, though, is that holistic positions assigning a higher value to a system than to individuals must provoke considerable ethical inquiries – and rightly so.

In the case of rights for rivers ecocentric positions risk running into a dilemma. In fact, a major driver for introducing river rights is to give indigenous peoples a greater say in the management of their livelihoods.

Paradoxically, an ecocentric interpretation of river rights could make this more difficult if 'Nature' were to be protected from 'Humans' (O'Donnell and Macpherson, 2018).

Challenges also arise in practice, as environmental and conservation policies are often driven by ecocentric intuitions (Bremer and Funtowicz, 2015; Yates *et al.*, 2017). From a philosophical perspective, one can of course ask to what extent these practical contexts are really based on ethical ecocentrism. However, they do share the conviction that ecosystems need to be protected from humans. In addition, there is the scientific pedigree of ecocentrism and the conviction that ecologists, conservationists, and other scientific disciplines know best how to deal with and protect ecosystems such as river or water systems. Thus, even when indigenous (as well as other local) communities are involved in resource management, the difficulty arises that their knowledge and values are often discounted as unscientific and therefore as less relevant (Bremer and Funtowicz, 2015; Yates *et al.*, 2017).

## The case of *Te Awa Tupua (Whanganui River) in* New Zealand

Above, we have drawn attention to the danger of inherent contradictions in the justification of rights for rivers. Below, we look at the *Te Awa Tupua (Whanganui River)* on the Northern Island of New Zealand which is the most prominent and elaborated case of a legal personhood of a river and also the one attracting the most academic interest so far. Hence, the example can indicate in what ways the above-mentioned contradictions actually arise in practice.

### *Legal and political provisions*

In March 2017, Te Awa Tupua was granted the status of a legal person in the 'Te Awa Tupua (Whanganui River Claims Settlement) Act, 2017'. This provides the river with some legal rights, including a right to 'enter [...] contracts, own [..] property, [or] sue and be sued in court' (O'Donnell *et al.*, 2020: 409; Talbot-Jones and Bennett, 2019: 3). In other words, such a step gives the river the legal tools to protect itself, and 'expands legal systems to include consideration of the needs and rights of nature' (O'Donnell, 2018). Conceptually, this was regarded as a step to overcoming anthropocentric understandings of law, to recognising nature as more than just a factor for benefits to humans and to accounting legally for damage done to the environment (O'Donnell and Talbot-Jones, 2018). Yet, a closer look reveals that such ecocentric arguments were less important for initiating the Te Awa Tupua Act. Instead, the ethical arguments justifying the status of Te Awa Tupua as a legal person are rather based on socio-ethical considerations. The Act especially refers to the indigenous law and worldviews of the local Maori, the Whanganui *iwi*. In their understanding, the river and the *iwi* are the same, reflecting the physical and metaphysical importance of the river to the *iwi* – hence, contradicting Western legal understanding of nature as a dichotomous counterpart to the human (O'Donnell, 2018). Consequently, recognising Te Awa Tupua as a legal person, primarily reflects the increasing acceptance of indigenous worldviews within the legal system rather than ecological or conservationist ambitions. This becomes evident as the Act contains an extensive part of apology and acknowledgements towards the Maori *iwi* (Te Awa Tupua Act, 2017), classifying the Act as a case of corrective 'and' distributive justice (Kalfagianni and Meisch, 2020). Other international cases of rights for rivers support this observation by showing the strong role of indigenous thought in the drafting and implementation of such acts (O'Donnell *et al.*, 2020).

The legal person of Te Awa Tupua is defined by its non-physical and cultural components as well as its physical components as 'the space of land that the waters of the Whanganui river cover at its fullest flow without overtopping its banks and [...] the subsoil, the plants attached to the bed, the space occupied by the water, and the airspace above the water' (Te Awa Tupua Act, 2017: Part 1, Section 7). So, the Act is basically about a property right, i.e. the river owns itself. However, there are conceptual path dependencies inbuilt in this construction. First, there is a long tradition in the Anglo-Saxon legal systems of dealing with property issues. It is now applied to a river – thus, creating interesting regulations. For instance, the river owns the airspace above its surface. One could speculate over potential hydrological or spiritual reasons for that; however, it is a feature of Anglo-Saxon property rights regimes that a property holder also owns the air column above its property. Second, the law refers to former Crown property and not to the entire river. So, strictly speaking, the river does not belong to itself outright.

The Act created the Te Pou Tupua as guardian 'required to act and speak to the benefit of the river's health and well-being' (O'Donnell and Talbot-Jones, 2018). This legal body consists of one representative of the *iwi* and one of the Crown (Te Awa Tupua Act, 2017: Part 2, Section 3). Consequently, Te Pou Tupua plays a central role as it legally represents the river, allowing to practically appeal to the rights given to the river. Additionally, the Act creates different collaboration groups and committees for the development of the river management. This includes Te Kopuka, a forum of various stakeholders, including representatives of the *iwi*, governments, business, and others. They discuss and develop a concept for Te Awa Tupua regarding all its hydrological and hydro-social meanings (Kauffman and Martin, 2018: 7), recognising the importance of environmental protection but also other aspects of public interest like the usage for hydroelectricity (O'Donnell and Macpherson, 2018: 37; O'Donnell and Talbot-Jones, 2018; Talbot-Jones and Bennett, 2019: 3). Next to Maori traditions, this forum is meant to bring together different interest groups to ensure environmental protection, while still recognising and mediating between the different interests of fields such as economy, health, and ecology (Kang, 2019: 643-644).

### *Relevance for food production*

How then does recognising a river as a legal person affect sectors such as agriculture and food production that are highly dependent on water? As developments in New Zealand are quite recent, we cannot tell much for certain. From the legal provisions, some educated guesses are possible. The legal status of Te Awa Tupua hardly touches agriculture and food production. For instance, the use of water (O'Donnell and Macpherson, 2018: 40) as well as aquatic life (Talbot-Jones and Bennett, 2019: 3) are little affected by the Act, making fishery generally possible. While direct restrictions for food production are unlikely, they might occur indirectly when a certain consumption by the food sector affects other, protected elements of the river or Te Tou Tupua (O'Donnell and Macpherson, 2018: 37). Despite potential limitations, it appears more likely that holistic river management concepts and cooperation between different stakeholders might incentivise, or even demand more sustainable approaches to food production.

## Conclusion

In our paper, we looked at the phenomenon of rights for rivers. Such rights are demanded by a global movement hoping to stop the destruction of river systems and to give indigenous peoples a greater say in determining their own livelihoods. We first showed that epistemological and ethical challenges arise. They result from tensions between justifications referring to ecocentrism and corrective justice. In the worst case, they would stand at odds to each other. Second, we looked at the most prominent case of a right for rivers. It turned out that the creation a legal personhood for the Whanganui River was not motivated by environmental, but social ethics. Since the *iwi* see themselves as part of the river, the right to the river could not rest with them but had to rest with the river. Respecting *iwi* cosmology, the 'Te Awa Tupua (Whanganui River Claims Settlement) Act' settled the epistemological issue of what the river is. It also created institutions of joint management. Meanwhile, it also turned out that, and how, the Act merges indigenous worldviews with Anglo-Saxon legal traditions (especially concerning property rights) (Salmond, 2014). For the future, it remains to be seen how ecocentric and indigenous framing interact with each other and how they are dealt with in specific situations.

If the current trend gathers momentum, the discussion on rights for rivers might become an object to study normative contradictions and tensions between different public policies. Cases in other countries reveal the complex interplays between right for rivers (or nature) and environmental protection legislations. Often environmental protection could well be possible even without river rights if existing conservation laws would be respected. For instance, the first successful case of a lawsuit concerning rights for nature, the case of Vilcabamba river in Ecuador, was only necessary because local governments had previously ignored existing environmental law (Corte Provincial de Justicia de Loja, 2011). In addition, rights for rivers 'may well create more problems than it solves' (O'Donnell, 2018) by shifting the focus from a social responsibility to protect the environment to a self-responsibility of nature (Eckstein *et al.*, 2019: 811) or by creating the image of environment and humans competing over water (O'Donnell, 2018), thus 'undermin[ing] the cultural narratives that support environmental protection' (O'Donnell and Macpherson, 2018: 41). Hence, without considering appropriate cultural and social preconditions, rights for rivers could pose a threat to previous achievements in environmental protection. Since agriculture and food production are major water consumers, one of the key issues will be how much water they use and in what ways.

For normative ethics, exciting challenges arise, as different cosmologies and ontologies about what water or rivers are (Yates *et al.*, 2017) need to be brought together in a way that enables joint action. According to O'Neill (2009: 230), deploying normative principles 'is a matter of constructing, supporting and working within structures and policies, institutions and cultures that buttress the feasibility of the joint enactment of the principles to which agents are committed'. In the case of different and potentially incommensurable ontologies, as in the case of rights for rivers, this becomes an exciting challenge for applied ethics of different domains (such as environment, food, agriculture, water, energy). This concerns ethical reasonings as well as the role of applied ethics within practical contexts (Meisch, 2018, 2019).

## Acknowledgement

We would like to thank Leonie Bossert, Lieske Voget-Kleschin and two anonymous reviewers for their comments on earlier versions of this paper. Simon Meisch acknowledges funding for his CANALS project from the European Union's Horizon 2020 framework programme under the Marie Skłodowska-Curie grant agreement No 895008.

## References

Benöhr, J. and Lynch, P.J. (2018). Should Rivers Have Rights? A Growing Movement Says It's About Time. Yale Environment 360 (August 14, 2018).

Bremer, S. and Funtowicz, S. (2015). Negotiating a place for sustainability science: Narratives from the Waikaraka Estuary in New Zealand. Environmental Science and Policy 53(Part A): 47-59.

Chapron, G., Epstein, Y. and López-Bao, J.V. (2019). A rights revolution for nature. Science 363(6434): 1392-1393.

Corte Provincial de Justicia de Loja (2011). Juicio No. 11121-2011-0010, 30 de Marzo del 2011.

Dierks, J. (2016). Ökozentrik. In: Ott, K., Dierks, J. and Voget-Kleschin, L. (eds) Umweltethik. Stuttgart: Metzler, pp. 169-177.

Eckstein, G., D'Andrea, A., Marshall, V., O'Donnell, E., Talbot-Jones, J., Curran, D. and O'Bryan, K. (2019). Conferring legal personality on the world's rivers: A brief intellectual assessment. Water International 44(6-7): 804-829.

Eser, U. and Potthast, T. (1999). Naturschutzethik: Eine Einführung für die Praxis. Nomos, Baden-Baden.

Kalfagianni, A. and Meisch, S. (2020). Epistemological and ethical understandings of access and allocation in Earth System Governance. International Environmental Agreements 20: 203-221.

Kang, K. (2019). On the problem of the justification of river rights. Water International 44(6-7): 667-683.

Kauffman, C.M. and Martin, P.L. (2018). When Rivers Have Rights: Case Comparison of New Zealand, Colombia, and India. International Studies Association Annual Conference, San Francisco, April 4, 2018.

Meisch, S. (2018). Water ethics – lessons from post-normal science. In: Springer, S. and Grimm, H. (eds) Professionals in food chains. Wageningen Academic Publishers, Wageningen, pp. 459-464.

Meisch, S. (2019). *I Want to Tell You a Story*. How Narrative Water Ethics Contributes to Re-theorizing Water Politics. Water 11(4): 631. https://doi.org/10.3390/w11040631

Middleton, N. (2012). Rivers: A Very Short Introduction. Oxford University Press, Oxford.

O'Donnell, E., Poelina, A., Pelizzon, A., and Clark, C. (2020). Stop Burying the Lede: The Essential Role of Indigenous Law(s) in Creating Rights of Nature. Transnational Environmental Law 9(3): 403-427.

O'Donnell, E. (2018). Legal rights for rivers: More power, less protection? Available at: https://globalwaterforum.org/2018/04/23/legal-rights-for-rivers-more-power-less-protection/.

O'Donnell, E. and Macpherson, E. (2018). Voice, power and legitimacy: the role of the legal person in river management in New Zealand, Chile and Australia. Australasian Journal of Water Ressources 23(1): 35-44.

O'Donnell, E. and Talbot-Jones, J. (2018). Creating legal rights for rivers: lessons from Australia, New Zealand, and India. Ecology and Society 23(1): 7.

O'Neill, O. (2009). Applied Ethics: Naturalism, Normativity and Public Policy. Journal of Applied Philosophy 26: 219-230.

Salmond, A. (2014). Tears of Rangi. Water, power, and people in New Zealand. Journal of Ethnographic Theory 4(3): 285-309.

Strang, V. (2020a). Re-imagining the River: New Environmental Ethics in Human Engagements with Water. One Earth 2(3): 204-206.

Strang, V. (2020b). The Rights of the River: Water, Culture and Ecological Justice. In: Kopnina, H. and Washington, H. (eds). Conservation. Springer, Cham, pp. 105-119.

Talbot-Jones, J. and Bennett, J. (2019). Toward a property rights theory of legal rights for rivers. Ecological Economics 164: 106352.

Te Awa Tupua (Whanganui River Claims Settlement) Act 2017, New Zealand.

Wilson, G. and Lee, D.M. (2019). Rights of rivers enter the mainstream. The Ecological Citizen 2: 183-187

Yates, J., Harris, L. and Wilson, N. (2017). Multiple ontologies of water: Politics, conflict and implications for governance. Environment and Planning D: Society and Space 35(5): 797-815.

# 56. More than life-sustaining resources – on the integrity argument for natural resources

*S.I. Espinosa Flor*
*University of Vienna, Universitätsstraße 7, 1010 Vienna, Austria; sarah.espinosa@univie.ac.at*

## Abstract

The things that matter most to us are usually those that have some deep meaning and special connection to us. Either because we find them beautiful, because they bring back good memories, or simply because they are things, whose existence we cherish (even if cannot fully explain why). There are also those things that are needed to live and therefore, we value more than any other. Water, food and air are good examples of these things. They are natural resources intrinsically necessary to sustain life and to which all humans want to have access. However, this may not be the only reason why these resources are valuable or worthy of respect. In this paper, I shall defend the view that certain natural resources can be considered valuable not only because they are life-sustaining resources or because they bring direct and countable benefits to people. The 'integrity' of these resources, from which many goods and services are derived, can be considered valuable regardless of how necessary it is to sustain human life. Based on this idea, it is expected to question those forms of access and use of natural resources that are beneficial to humans, but that cause serious damage to the environment or that fail to respect the life of other non-human living beings.

**Keywords:** scarcity, resource exploitation, sustainability, non-instrumental values, environmental ethics

## Water, soil, air...

Natural resources encompass many things, but their analysis becomes especially important when it comes to those resources that represent a basic necessity for humans, for example, clean water, unpolluted air and adequate food. In this sense, the drive to access certain ecosystems is based on non-trivial claims that people have on these ecosystems, as they provide them with basic goods and services to live. In other words, there seems to be good reasons to exploit especially those resources, whose use and/or extraction directly and undeniably benefits human beings. Thus, the use and/or extraction of said natural resources only becomes problematic when their sustainability is threatened or they become ultimately exhausted. However and to the extent that the use and/or extraction of natural resources satisfies non-contingent human needs, the exploitation of those resources is justified; opening up the possibility of depleting them if such exploitation is not carried out in a sustainable manner (whatever this means).

This line of reasoning is compatible with basic resource rights theories (Shue, 2020) and natural resource economics (Perman *et al.*, 2003) and is therefore especially important in emphasising the fair share each person should have to vital resources and for that, the need to think about and design models of 'efficient and optimal exploitation' of natural resources. However, this argument ignores other perspectives and considerations for valuing natural resources not only in terms of human profit (Kallhoff, 2017) and is silent on the normative dimensions of the concept of 'exploitation' of non-human parties, including natural resources. For this reason, this paper attempts to analyse one possibility that may be compatible both with the sustainable use of life-sustaining resources and other considerations in valuing those resources for reasons that go beyond human interests. In this way it is also expected to be critical of the concept of exploitation of natural resources as an unproblematic type of human-nature relationship. This possibility is based on the 'integrity' of ecosystems and their mutual dependence on humans (Poteete *et*


EurSafe 2021 – DOI 10.3920/978-90-8686-915-2_56, © Wageningen Academic Publishers 2021

*al.*, 2010). I shall begin by outlining those resources needed to sustain life and then move on to develop what I call the integrity argument for natural resources.

Given the amount of goods and services that ecosystems such as forests, fisheries, water systems and even air provide to humans, it does not seem that not accessing or using them is a viable option. It is however true that there are various ways to access and make use of these resources and that not all goods and services provided by ecosystems meet what could be arguably called vital human needs. The strategies that limit or regulate these actions are those that ultimately prevent, delay or lead to the depletion of a natural resource and their analysis encompasses more than can be covered in this paper. Nevertheless, it can said that most life-sustaining resources, such as water and soil, are renewable resources, i.e. they are resources that comprise biotic, plant and animal populations and therefore have the capacity to grow in size over time through biological reproduction (Perman, 2003: 11). They renew and sustain themselves as long as their biological reproduction processes are not affected by some external forces such as human over-exploitation. As a result, renewable resources are defined by being exhaustible if they are harvested for too long at a rate that exceeds their regeneration capabilities (*ibid.*); making them scarce and potentially unable to satisfy human and non-human needs.

Resource scarcity is therefore an issue closely related to sustainability; both to the capacity of a resource to sustain itself over time, and to the capacity of today's societies to sustain the basic conditions of human life and well-being. Consequently, the fact that the ability of many renewable resources to sustain, grow and meet long-term human needs is currently under threat, puts particular pressure on the international political agenda and classic economic models to focus on different paths to alleviate scarcity and promote sustainability (*ibid.*: 14). However, it can be specially difficult to think about and design ways to achieve this purpose, since many of these resources, such as water and air, have the characteristic of common goods and therefore, may be particularly subject to depletion if their use or access is not regulated on a way that reflect collective interests (Ostrom, 1990). As famously illustrated by Garrett Hardin, depletion scenarios tend to occur when there are no control and cooperation mechanisms to regulate common goods free access (Hardin, 1968). Free and unrestricted access jeopardises the sustainability of these resources and, with it, their ability to meet future needs. In the face of this, depletion scenarios can only be avoided if individuals and communities choose to subscribe to norms that define best practices for addressing access to and use of a common good (Kallhoff, 2017: 419).

## Non-instrumental values?

Throughout history, countless and diverse academic efforts have been articulated to address and reflect on the value of non-human nature. Particularly in recent decades, many environmental philosophers and ethicists have tried to reflect on different approaches to understanding why non-human nature, and all that encompasses it, should be considered valuable beyond human benefit (Brennan and Lo, 2020). It is especially in the face of the effects of human overpopulation, environmental pollution, overexploitation of resources, extinction of species and increased emissions of greenhouse gases into the atmosphere that the debate on the non-instrumental value of nature takes on special importance. One of its purposes being to draw attention to less anthropocentric perspectives on nature, seeking to challenge, among other things, the assumption that natural resources are exclusively means to certain human ends, thus having only instrumental value; an assumption at the heart of many economic calculations applied since the 18th century and of classical and neoclassical economics traditions (Smith, 1766; Marx, 1867). Admittedly, it may be difficult to think of the non-instrumental value of resources such as water, soil and forests, independent of their attribution by humans.

Even if we argue that a resource is valued for aesthetic reasons such as admiration versus the sublime (Kallhoff, 2017: 418f.), this valuation will depend on what kind of experience a person has when faced with said resources. In this sense, the value of this resource will depend on what an experience brings about to a person, even if this value cannot be objectively calculated. For example, I may not value a forest because of the wood I can get from it to build a nice bedside table, but because I find it amazingly beautiful and therefore worth preserving. This is a trivial example, but it should help illustrate how difficult it can be to think about the value of certain natural resources and ecosystems independently of human acknowledgement. Nature is valuable for people, whether their lives depend on it or not. As a result, it is easier to see why and to what extent some renewable resources may be more valuable to people than others. The difficult part comes when one tries to argue that while these resources are indeed very valuable in providing the means for the satisfaction of human needs and desires, this is not the only reason they may be valuable.

If it is plausible to argue in favour of something like a non-instrumental value of natural resources, – not necessarily in terms of intrinsic value in the parallel sense of humans' intrinsic value – then it may be easier to recognise why certain practices justified on the basis of benefit and human interest may be somewhat problematic. This becomes clear when the exploitation of a natural resource is taken to limits that threaten the existence of that resource and with it the possibility of continuing to use it. More specifically, exploitation is considered problematic when profit and human interest are compromised due to scarcity or depletion of a resource, but not before that. Nevertheless, many environmental relevant issues like those concerning the resilience of natural resources subject to human use and extraction, are not the result of overexploitation or even climate change, but of mutually incompatible or directly dangerous and polluting practices (*ibid.*: 421). In other words, many environmental relevant issues are harmful not only to humans, but also to ecosystems and living beings whose existence and growth is affected by human action.

In this regard, I do not aim at reconstructing the arguments and main debates on something like natural resources' intrinsic value, but rather to emphasise that some renewable resources can be more than a source of livelihood and that this is because they can be valuable for reasons that exceed the direct and countable benefits they bring to human beings. This means coming to reason about experiences with nature (*ibid.*: 418f.) and about properties of natural resources that are worth of acknowledgment and respect because of what they are and not only because of how necessary they are to people. An important clarification is that, unlike people, whose most uncontroversial understanding of intrinsic value is based on the premise that each person is valuable in itself, i.e. independent of external ascription, the non-instrumental value of natural resources must be recognized by humans in order to avoid harmful actions towards them. Especially those actions that may not necessarily be harmful to humans, or at least not directly, but primarily affect non-human parties and entities.

As noted above, these actions can occur when the use and/or extraction of a natural resource is sufficiently unsustainable to be challenged on its own (when over-exploitation is not necessary to cause serious harm to non-human nature). It seems therefore important to delve into those natural resource exploitation actions that do not include environmentally relevant issues in their cost-benefit analysis and that, by not assigning a non-instrumental value to the environmental capital from which many goods and services derive, ignore many of the damages they cause. A step in this direction would be to question not only the conceptual bases underpinning these actions, but also to become aware of the mutual dependence between natural resources and human beings (Poteete *et al.*, 2010). In addition, to take seriously into account how fragile, unique and complex certain ecosystems are. Thus, to promote those practices and institutions that do not lead to the lack of provision of basic services and goods for human beings, but also to the irreversible extinction of species, to environmental pollution and to the increase of greenhouse gas emissions into the atmosphere.

## The integrity argument

One way to make sense of the non-instrumental value of natural resources is to appeal to some essential characteristics of natural resources that are worthy of respect or simply to the mere existence of these resources. Natural resources can be worthy of respect because they are valuable in their own right and also because of the relationship they have with humans and other living beings. An attitude of respect can arise when natural resources are not seen simply as means to achieve specific ends such as subsistence, but when their existence triggers feelings such as satisfaction, admiration and even fear (Kallhoff, 2017: 418). But it does not take existential experiences for people to recognise the need to respect natural resources beyond the benefits they bring to humans. The prospect of actions that are not harmful to humans (they might even be beneficial to them), but may be harmful to other living beings, may be reason enough to find these actions problematic.

For example, there are non-human entities that qualify as life-sustaining resources but whose existence does not (at least directly) benefit human beings, such as the hot springs of Yellowstone or the Galapagos Islands. In the case of Yellowstone, these are springs that are close to boiling point and extremely acidic, yet are home to microbes (the bacterium Thermus aquaticus) that produce certain pigments that give the springs bright and unusual colours. It is thus clear that the springs have an instrumental value for microbes, but not for humans, as they can even be deadly for them. Therefore, the disappearance of these springs not only represents no harm to humans, but may be the unintended consequence of an action beneficial to them (let us imagine, for example, that these springs turn out to be a great source of energy and that it is possible to exploit them). However, actions that can lead to the disappearance of the springs could rationally be considered problematic. Let me elaborate.

Witnessing the colours of the springs and having experienced how beautiful and also threatening they can be, can generate in oneself a feeling of awe and admiration that translates into an attitude of deep respect towards them. Even if you never get to experience what it is like to be in Yellowstone and see the colourful spectacle that these microbes produce, you may still experience an inexplicable feeling of sadness at the prospect of the springs ceasing to exist because of human action. A feeling that would generate an attitude of rejection towards this action (even if you do nothing about it). This may be even clearer in the case of the Galapagos Islands. These islands are basically lava formations that emerged on top of still active volcanoes, being in itself an inhospitable place to live. However, they have the right conditions to support the life of countless unique species in the world, i.e. they are vital resources for many non-human parties and entities, while they are not directly so for humans. Even so, their disappearance would be regrettable even for those who do not benefit at all from their existence; many would reject actions that would endanger the islands, even if their lives were not affected in the least by their disappearance.

Therefore, when it comes to life-sustaining resources, I will join those who argue that the integrity of an ecosystem should be understood as a fundamental value that deserves moral respect (Costanza *et al.*, 1998: Kallhoff, 2017). First, natural resource integrity is defined as having an empirical side in terms of essential ecofunctions or habitat, biological properties and ecosystem processes (Costanza *et al.*, 1998). Second, and more importantly for humans, it is from these biological properties or ecosystem processes that goods (such as food) and services (such as waste assimilation), – directly and indirectly beneficial to humans – are derived. Natural resources that span entire ecosystems have salient properties or processes necessary to sustain human and non-human life, but their existence and long-term sustainability will depend largely on how people use them. Consequently, I shall argue that these empirical aspects, that is, some properties or processes from which ecosystem goods and services are derived, as well as their interdependence with humans and other living beings, constitute the integrity of a resource. If it is plausible to accept this idea, then we could say, in general terms, that an (exploitative) action that harms or does not respond adequately to the integrity of a natural resource, can and should be considered problematic.

Basically, the use and/or extraction of some resource in order to profit from it, seems to imply that such actions do not cause any serious damage to some essential characteristic of the object being exploited or that such actions do not fail to respect its mere existence. One could say that by exploiting a forest to benefit for example from the wood of its trees, an essential part of that forest is automatically considered and treated as a product that can be marketed, bought and sold. This resource is being considered and treated, therefore, as a commodity that serves to satisfy human needs and desires and nothing else. This view is particularly relevant with regard to human-nature interactions, since the exploitation of natural resources for profit implies that such actions do not fail to respect any non-instrumental value of those resources. However, not because exploitative actions are carried out without disrespecting any non-instrumental value, but because these resources do not seem to have any. If there is no non-instrumental value that can be disrespected, then there can be no wrongful action towards the integrity of natural resources. If some non-instrumental value of natural resources is recognised, it is possible to think about how and to what extent to access and making use of them, without risking their ability to grow and sustain themselves over time.

## Conclusion

There are several approaches and attempts to show why ecosystems and non-human nature in general is and should be valued regardless of how beneficial and useful it is to humans. This paper focuses on how necessary certain natural resources are to sustain human life, but tries to deepen some arguments to consider and recognise also the non-instrumental value of these resources. In particular, it aims to show that the integrity of natural resources, understood as the empirical aspects from which goods and services flow, and whose existence is affected by human actions, could also be valuable in itself and for the sake of other living beings that depend on it. If this proves to be plausible, then it may be easier to challenge exploitative actions that are harmful not only to humans but also to countless non-human entities and ecosystems across the globe.

## References

Brennan, A. and Lo, Y.-S. (2020). Environmental Ethics. In Zalta, E.N. (ed.) *The Stanford Encyclopedia of Philosophy*. Metaphysics Research Lab, Stanford University, winter 2020 ed.

Cafaro, P. and Sandler, R.D. (2010). *Virtue ethics and the environment*. Springer.

Costanza, R., d'Arge, R., De Groot, R., Farber, S., Grasso, M., Hannon, B., Limburg, K., Naeem, S., O'Neill, R.V., Paruelo, J., *et al.* (1998). The value of the world's ecosystem services and natural capital. *Ecological economics*, 25(1):3-15.

Engel, J.R., Westra, L., and Bosselmann, K. (2009). *Democracy, Ecological Integrity and International Law*. Cambridge Scholars Publishing.

Garrett, H. (1968). The tragedy of the commons. *Science*, *162*(3859), 1243-1248.

Kallhoff, A. (2017). Water ethics: Toward ecological cooperation. *The Oxford Handbook of Environmental Ethics*, pages 416-426.

Leopold, A. (1949). A sand county almanac and sketches here and there. Oxford university press. *New York, NY.*

Marx, K. (1867). Capital, vol. 1. Penguin, London, 1992.

Ostrom, E. (1990). *Governing the commons: The evolution of institutions for collective action*. Cambridge university press.

Perman, R., Ma, Y., McGilvray, J. & Common, M. (2003). *Natural resource and environmental economics*. Pearson Education.

Poteete, A.R., Janssen, M.A. and Ostrom, E. (2010). *Working together: collective action, the commons, and multiple methods in practice*. Princeton University Press.

Sandler, R.L.(2009). *Character and environment: A virtue-oriented approach to environmental ethics.* Columbia University Press.
Shue, H. (2020). *Basic rights: Subsistence, affluence, and US foreign policy*. Princeton University Press.
Smith, A. (1776). *An inquiry into the nature and causes of the wealth of nations: Book II, Chapter III.* London: printed for W. Strahan; and T. Cadell.

# 57. The ethical sustainability matrix: a practical tool for assessment of GMOs including genome-edited organisms

*T. Dassler** and A.I. Myhr*
*GenØk-Centre for Biosafety, SIVA innovasjons-senter, 9294 Tromsø, Norway; tim.dassler@uit.no*

## Abstract

An ethical sustainability matrix can serve as a useful tool for finding and assessing relevant aspects of genetically modified organism (GMOs) regarding their sustainability and ethical justifiability. Here we present a restructured version of the ethical matrix originally proposed by Mepham and discuss its usefulness in technology assessment of GM and genome-edited (GE) organisms, as well as in GMO regulations that include non-safety factors. We reorganised the original ethical matrix by taking in ethical insights from virtue ethics and ethics of care in order to reduce the risk of bias. Sustainability aspects inspired by public values such as the UN sustainable development goals were also included. The proposed ethical sustainability matrix also takes into account the different types of (non)knowledge, important ethical traditions and a broad set of values, stakeholders and public opinions and emotions. Using an actual case on the GE sterile salmon, we show how the ethical sustainability matrix can be used to identify challenges and assess contribution to sustainable development, as well as its ethical justifiability and social acceptability. Thus, making it a useful assessment aid for authorities, producers, regulators and the public alike.

**Keywords:** genome editing, trust, non-safety, care ethics, social acceptability

## Introduction: challenges with the original ethical matrix

The original ethical matrix (EM) was developed by Mepham (1996) to help decision makers to consider ethical issues raised by biotechnologies especially in the agri-food sector. The strengths of using an ethical matrix as a decision support framework is that it exposes case-relevant and ethically relevant aspects, includes multiple perspectives, ethical arguments and important values at stake (Kaiser *et al.*, 2007).

In applying the EM to the assessment of safety and non-safety criteria and the ethical justifiability and social acceptability of genetically modified (GM) including genome-edited (GE) organisms we encountered four major challenges for the original ethical matrix:
1. How to include relevant and representative values from various moral sources and perspectives of common morality in addition to consequentialism and deontology?
2. How to include a comprehensive understanding of sustainability encompassing both environmental protection as well as social and economic development of present and future generations?
3. How to include public opinions and how people feel about gene technology?
4. How to deal with the complexity arising from 1., 2. and 3, as well as the various dimensions and stakeholders the matrix needs to address.

These challenges were identified through an analysis of European GMO legislation, regulation and policy objectives. We investigated various sources of public morals and societal norms such as European and national law, including reports from government agencies, advisory boards and public surveys. Our study was especially influenced by the German Law of Nature Protection, the German Gene Technology Law, as well as the Norwegian Gene Technology Act (GTA) which in addition to not posing risk to


EurSafe 2021 – DOI 10.3920/978-90-8686-915-2_57, 

health and environment requires GMOs to be ethically justifiable for them to be approved. The GTA also requires a positive assessment of sustainability and societal utility.

### *Ethical principlism and its limitations*

Mepham's original EM draws on *prima facie* ethical principles that are derived from Beauchamp's and Childress' (1994) biomedical ethics. *Prima facie* principles are rules of action that are valid at 'first sight' and create a duty to act. They facilitate ethical decision-making by allowing weaker principles to be overruled by stronger ones in particular cases or contexts. Although this kind of ethical principlism has pragmatic advantages due to its being able to circumvent complex and highly theoretical moral discussions, one of its challenges is that it may lead to a biased outcome, for example, due to its limited framing and understanding of ethical decision making, as well as unfair weighing of various stakeholders' interests.

The *prima facie* principles of Mepham's EM are primarily derived from deontological and consequentialist theory (Mepham, 2010). Mepham draws on these principles following Beauchamp and Childress (1994) who argue that they are grounded in a common morality. This approach effectively leaves out two hugely influential philosophical traditions. Virtue ethics and ethics of care. The history of the ethics of care exemplifies that relying on ethical principlism that is grounded in purely consequentialist and deontological principles is at risk of neglecting incompatible and conflicting moral views and principles, as well as underestimating the significance of context and relations to solving ethical problems (Dassler, 2016). Furthermore, the *prima facie* principles tend to be blind for 'different' ways of solving ethical issues (Gilligan, 1982). Care ethics and virtue ethics add important moral insights and a broader perspective on moral problems and stakeholder views to draw on and should thus be included in the matrix to reduce the risk of bias. The inclusion of a broad and representative set of thick and thin values and principles important to public morals into the matrix is also important for creating trust and legitimacy regarding the assessment of a specific technology such as GMOs.

Therefore, we suggest that there is a need for a more inclusive account of common morality, which is not restricted to *prima facie* principles, but builds on ethical values that serve 'as resource of well-established moral insights and experiences, which have proved generally valid but neither sufficient nor infallible' (Rauprich, 2008, p. 43). A more inclusive account of common morality would also leave room to include different ethical perspectives and traditions such as care ethics and virtue ethics.

### *Sustainability is about more than fairness and biota*

Caring for a sustainable future means that we may need to balance ethical principles such as autonomy and wellbeing of existing generations against the autonomy and wellbeing of future generations and the survival of planetary ecosystems. Mepham's EM adapted to the evaluation of GMOs places sustainability as an issue of biota and fairness, that is, it is concerned with the sustainability of biotic populations (Mepham, 2000). Even though this addresses the importance of a healthy environment to achieve sustainable development, it does not capture the two other significant dimensions; social and economic sustainability and their relation to environmental protection. Thus, the EM becomes of limited use in cases when authorities need to justify that the protection of the environment for future generations may be more important than increasing welfare and economic activity today. Since the concept of sustainability encompasses all the relevant environmental, social and economic dimensions across space and time, it is about more than fairness with regard to biota. It is about how the environment is interlinked with social and economic development and how to achieve such development without exceeding planetary boundaries. It is as much about wellbeing, welfare, choice, or fair trade as it is about conservation, biodiversity, and the intrinsic value of nature.

Through the global community's commitment to the United Nations Sustainable Development Goals (SDGs) the idea of sustainability has become an important public value and part of a global common morality. Consequently, sustainability has started to function as a guiding principle for political, institutional, business and even private conduct and should thus be added as an additional principle to the EM. This presupposes the above mentioned more inclusive account of common morality that allows for the inclusion of sustainability as a public value or guiding principle. If we agree that sustainability should be included into the matrix, where should we place it? It is neither a *prima facie* principle, nor an interest group.

### *Adding more complexity: how to deal with emotions and opinions?*

The third challenge which we want our matrix to be able to deal with for it to be a useful assessment tool is inclusion of how different interest groups such as producers, authorities, regulators, and especially the public think, feel and react to GMOs. Mepham's EM emphasizes the role of ethics as the ground for political motivation. This is legitimate when understanding ethics as dealing with the question 'What should I do?'. But there are certainly other factors than purely ethical ones motivating our actions and decisions. In a non-ideal world this is true both for private as well as for political decisions. Political decisions are influenced and motivated by all kinds of empirical and non-empirical knowledge, interests held by various stakeholders, as well as the public's feelings and opinions about political topics. The case of GE and GMO approval is no exception. Their assessment is neither decided purely on ethical nor on empirical grounds. Rather, attitudes towards GMOs, especially GMOs developed to be used in food production, involve strong feelings of safety and health as well as identity and culture (Carson, 2018; Myskja and Myhr, 2020). Public values, social and cultural aspects, empirical knowledge, as well as public opinions, feelings and emotions may all influence decisions regarding (non)approval and thus need to be included into the matrix. This further adds to the matrix's complexity and we need to find a way to deal with this complexity for the matrix to be a useful decision aid and not turn into a complex and impractical monster.

## How to address these challenges – introducing the ethical sustainability matrix

Above, we identified four major challenges for the original ethical matrix. To manage these challenges, we analysed the blueprint of an ethical matrix for the assessment of GE and GM organisms developed for the Norwegian Environment Agency (Forsberg *et al.*, 2019) that builds upon the original EM by Mepham. This matrix is very comprehensive and includes both sustainability as well as a broader selection of public values and societal norms. But it is also a highly complex matrix consisting of four stakeholder dimensions where both sustainability and future generations, country of production, the national level, as well as animal well-being and the environment are included alongside twelve ethical principles and values: care, trust, beneficence, no-harm, well-being, fairness, autonomy, stewardship, solidarity, naturalness, respect for dignity and integrity, as well as the precautionary principle. The matrix also includes suggestions for more than 100 guiding questions and is thus considered too complex by the Norwegian Biotechnology Advisory Board (2020). We used this matrix as a starting point for developing an ethical sustainability matrix (ESM), but in order to ensure applicability to the user context which is often characterized by lack of resources and limited expertise with ethical assessment, we had to reduce the matrix's complexity without the matrix becoming banal or losing sight of what is at stake.

Based on the work done by the Norwegian Environmental Agency (2020) on the assessment of ethical justifiability of GMOs and an analysis of European GMO legislation, regulation and policy objectives as well as German and Norwegian sources of public morals and societal norms, we propose to reorganize this complex matrix by moving the stakeholder and value dimensions inside the matrix. We also recommend reorganizing the question of environmental protection and economic as well as societal development into a sustainability dimension. Furthermore, we suggest that assessment of GMOs' social acceptability and ethical justifiability ought to draw on various ethical perspectives and address not only the ethical principles and values that guide the assessment, but also what types of knowledge we have about the organism and its impacts as well as how various stakeholders on different levels feel about this. To reduce complexity, we propose to organize these questions into a justifiability or trust dimension. We here lean on Carson (2018) who emphasizes the importance of trust in addressing both knowledge, emotions, and ethical values. In drawing on Luhmann's (2000) discussion of trust, Carson proposes that the complexity of decision-making situations can be reduced by using trust to manage lack of knowledge as uncertainty as well as communication between different systems (both internally and externally). This understanding of trust goes back to Aristotle's views on what kind of dimensions of a decision or an argument that must be addressed for it to be trustworthy, justifiable and legitimate. Following Carson, we suggest trust as a means for managing the complexity of the assessment regarding ethical justifiability and social acceptability of a GMO application. We thus end up with a matrix consisting of three dimensions; justifiability (trust), sustainability, stakeholder, as presented in Table 1.

Three ESM dimensions: sustainability (social, economic and environmental dimensions); stakeholder (interest groups differentiated by global, national, local/regional); justifiability (trust as strategy to manage complexity).

The ESM will provide us with a reduced number of $3^3 = 27$ guiding questions that are meant to identify challenges to ethical justifiability and social acceptability of GMOs. Let us consider some aspects of these questions in relation to a specific case: the GE sterile salmon.

### *Knowledge – Environment – Global*

What do we know about the GE sterile salmon's environmental impacts on the global level? The GE sterile salmon may have globally (and locally) beneficial impacts on the environment due to potential lower $CO_2$ emissions as well as higher protein efficiency compared to traditional meat production. This may make it seem more sustainable and thus more attractive for the aquaculture industry and consumers and consequently increase the volume of production. Increased production may, however, scale up already identified risks and create additional negative environmental local impacts.

### *Knowledge – Environment – Local*

We can change the question to consider environmental, social and economic impacts on the global, national and local level. This may show that there might be different advantages and drawbacks depending on the level we are addressing. The objective of GE sterile salmon is to avoid spread of genes to wild populations, which would be a positive environmental effect on the local and regional level. But, if solving the problem of spread to wild populations would increase local communities' production in sea-cages, there could also be a negative effect due to increased impact on benthic zone. Especially when considering the combined effect of all local production sites in a region.

*Table 1. The ethical sustainability matrix (ESM) for evaluation of ethical justifiability and social acceptability of a GMO application (adapted from Carson, 2018).*

| | | | |
|---|---|---|---|
| Justifiability | (non)knowledge | values | emotions |
| Sustainability | environment | social | economic |
| Stakeholder | global | national | local/regional |

### Emotions – Social – Local

Large scale salmon farming has the potential to change traditions and the coastal culture from people being fisherman to becoming fish farmers. In Norway, this has already happened to a large degree. Suddenly the livelihood of fishing communities is dependent on employment by the aquaculture sector. How would the use of GE Salmon affect this development, for example, by speeding up the present trend where large multinational corporations (MNCs) are taking over the sector. Would GE salmon, which is a patentable technology, create even stronger dependencies on MNCs? Such dependence and impacts on cultural and social identity may invoke strong feelings and emotions and thus not be considered a desirable outcome by local communities and their political representatives.

### Values – Economic – National

If GE sterile salmon farming is seen as a sustainable source of creating economic development and providing jobs by the national authorities thus becoming a priority policy objective. Can consumers trust the producers and the authorities to balance their interests of economic and social development with protection of the environment and public health? Will this in turn impact consumers' interest in buying the salmon, as survey findings show that there is public scepticism?

### Values – Social – Local/National

What values and social norms on the local and national level are important/in conflict with (non) approval of the GE application? This addresses the question whether (non)approval of a sterile salmon will contribute toward development that is desirable for society, whether it conforms to social norms and fulfils certain ethical minimum requirements and thus is socially acceptable. Is it morally desirable to further instrumentalize salmon as a mere means for human utility and consumption, for example, as something that can be changed by genome editing to solely serve our needs?

## Advantages of the ESM

As seen in the case above, one of the biggest advantages of the ESM is that it enables its user to formulate questions that capture complex ethical issues, their relation to various stakeholders and how these are connected to the three sustainability dimensions. It also illustrates that different types of knowledge, ethical values and emotions are non-subjective, but interconnected in a complex web of relations and thus interdependent.

The ESM is not meant to replace health and environmental assessment but includes it as an important part of a comprehensive evaluation of social acceptability and ethical justifiability. Accordingly, it enables its user to consider various types of knowledge, both safety and non-safety criteria, as well as ethical values and the feelings and emotions of those who are affected by (non-)approval of a GE or other GMOs. The ESM does this by a process of question formulation that will identify relevant stakeholders' interests, reasons for concern, lack of knowledge, lack of transparency, or conflicts with ethical and societal norms. Furthermore, the ESM is a framework that includes public morals in addition to all three sustainability dimensions; environmental concerns and socio-economic aspects. In cases where GMO approval in European Union or European Economic Area member states conflicts with national policy, societal norms, and the protection of the environment these aspects can provide legal reasons for opt-outs. The ESM also gives us a common framework and language to communicate assessments and decisions regarding (non-)approval of GE and other GMOs and can thus be used for considering wider non statutory aspects in assessment and management of ethical justifiability and social acceptability with regard to both safety and non-safety criteria by regulatory authorities, producers and the public. Finally, the ethical dimension of the ESM is not biased toward principlism or limited to certain ethical modes of decision making but draws on the richness of a variety of ethical traditions in addition to being open for learning from experiences with GMOs.

## Concluding remarks

If an ethical matrix is to give us adequate guidance regarding political decisions it must address many levels of policy objectives, values, stakeholders, and types of knowledge, thus increasing complexity of the assessment. Here we have suggested trust as key to reduce complexity. This has the advantage that the ESM is kept open and adaptable to a wide variety of cases, scenarios and contexts. Finally, it is important to emphasize that the ESM hinges on the idea that technological innovation does not happen in a vacuum and that it should make a positive contribution to society, the economy and the planet. GE and other GMOs are thus subject to a social contract and must be measured against their social acceptability and ethical justifiability. In order to see whether the ESM is useful as a decision support framework we need producers, assessors and political decision makers to apply, test and critically discuss the ESM.

## References

Beauchamp, T.L., and Childress, J.F. (1994). Principles of biomedical ethics. Fourth edition. Oxford University Press, Oxford, UK, 546 pp.

Carson, S.G. (2018). The corporate legitimacy matrix – a framework to analyze complex business-society relations. Philosophy of Management 18: 169-187.

Dassler, T. (2016). From care ethics to care politics. Toward a political notion of care. UiT – Norges arktiske universitet, Tromsø, Norway, 109 pp.

Forsberg, E.-M., Hofmann, B., Kaiser, M., Myskja, B.K., Strand, R. & Ursin, L. (2019). Vurderingskriteriet etikk: Veileder for operasjonalisering av vurderingskriteriet etikk i genteknologiloven. Cristin-resultat-ID: 1807181

Kaiser, M., Millar, K., Thorstensen, E. & Tomkins, S. (2007). Developing the ethical matrix as a decision support framework: GM fish as a case study. Journal of Agricultural and Environmental Ethics 20: 65-80.

Luhmann, N. (2000). Vertrauen: Ein Mechanismus der Reduktion sozialer Komplexität. (Vol. 2185). Lucius & Lucius, Stuttgart, Germany, 140 pp.

Mepham, B. (1996). Ethical analysis of food biotechnologies: An evaluative framework. In: Mepham, B. (ed.). Food Ethics, Routledge, London, UK, pp. 101-119.

Mepham, B. (2000). A framework for the ethical analysis of novel foods: The ethical matrix. Journal of Agricultural and Environmental Ethics 12: 165-176.

Mepham B. (2010) The ethical matrix as a tool in policy interventions: The obesity crisis. In: Gottwald, F.T., Ingensiep, H., Meinhardt M. (eds) Food Ethics. Springer, New York, USA, pp. 17-30.

Mepham, B. (2013). Ethical principles and the ethical matrix. In: Clark, J.P., Ritson, C. (eds) Practical Ethics for Food Professionals. Wiley-Blackwell, Oxford, UK, pp. 39-56.

Myskja, B.K. and Myhr, AI. (2020). Non-safety assessments of genome-edited organisms: Should they be included in regulation? Science and Engineering Ethics 26: 2601-2627.

Norwegian Biotechnology Advisory Board. (2020). Uttalelse om rapporten 'Veileder for operasjonalisering av vurderingskriteriet etikk i genteknologiloven'. Available at: https://www.bioteknologiradet.no/filarkiv/2010/07/2020-05-14-veileder-for-etikk-kriteriet-i-genteknologiloven.pdf.

Norwegian Environment Agency. (2020). Rammeverk for Bioteknologirådets vurderinger av etisk forsvarlighet etter genteknologiloven. Trondheim, Norway.

Rauprich, O. (2008). Common morality: Comment on Beauchamp and Childress. Theoretical Medicine and Bioethics 29: 43-71.

# 58. 'What is wrong with the EAT Lancet report?'

*M. Kaiser*
*Centre for the Study of the Sciences and the Humanities, University of Bergen, Parkveien 9, 5020 Bergen, Norway; matthias.kaiser@uib.no*

## Abstract

The much celebrated EAT Lancet report can be seen as an example of a global empirical model, with a top-down governance implication. This paper argues that as a consequence of the chosen framework and approach, the EAT Lancet report displays crucial shortcomings as brokerage between science and policy which limit its practical applicability: first, it is not populated by real people and their lifeworld; second, the report fails on scientific rigor in crucial points and puts forward politically alarmist claims; third, the report ignores crucial uncertainties in indices and quantified targets; fourthly, the report proposes a top-down solution, even when all developmental studies argue for the crucial importance of bottom-up governance; and lastly, the report pretends to provide useful strategies, when in reality most of them result in simple truisms with little empirical impact or strategic guidance. The paper will detail these four criticisms.

**Keywords:** EAT-Lancet, food system, uncertainty, brokerage, governance, transdisciplinarity

## Introduction

In principle, there are two ways to understand the world around us: the search for conceptual foundations on the one hand, and the search for global empirical models (or theories) on the other. The first approach is traditionally mainly pursued by philosophy, while the second is pursued in the empirical sciences. Arguably, both have their advantages and their shortcomings. The first approach typically will bring us closer to conceptual clarity, and understanding of logical and pragmatic interdependencies. When we talk about food and food-systems, such an approach would, for instance, try to elucidate the conceptual differences between 'food' and 'nutrition'; similarly the conceptual differences between 'wellbeing' and 'health', or their negative counterparts like 'distress' and 'disease'. Such conceptual clarification can elucidate the cornerstones and basic categories which help us define a food-system. The second approach seeks to chart causal interdependencies, drivers and inhibitors in the dynamic representation of our food system. In order to do this reliably, empirical input is needed. We 'validate' our models and system representations by seeking a good match with reliable empirical data.

However, the knowledge that we produce within the walls of academia has practical implications. Knowledge gives us a handle on how we can change our practices, how we can produce new instruments to reach certain goals and objectives, how we can solve practical challenges and manage complex problems. This aspect of knowledge production relates to policy, it produces science-for-policy. While theorists like Thomas Kuhn (1962) contributed to our understanding of academic knowledge with 'purely' epistemic interests, science-for-policy extends this with a higher complexity, as outlined by Funtowicz and Ravetz (1992). We move from 'normal science' to 'post-normal science'.

As many scholars have realized, our food-systems need a radical overhaul, a new paradigm. Numerous reports and global assessments concluded that the global food system *per se* is not sustainable and in need of radical change (FAO, 2018; Costello *et al.,* 2019; SAPEA, 2020; Odegard, 2014). A recent SAPEA report (2020) stated boldly: '"business as usual" is no longer a viable option and radical change is required' (*ibid.*: 21). Any policy maker is therefore well justified in putting forward the question: What


EurSafe 2021 – DOI 10.3920/978-90-8686-915-2_58, © Wageningen Academic Publishers 2021

can we do about this problem of our current food system, how can we improve human and planetary health in regard to food?

It is no trivial insight to realize that such a demand for policy advice makes us step out of the purely academic exchanges, contestations, and model constructions. The scientist needs to help policy by inserting the best knowledge available that may help policy makers out of their conundrum. And here the scientist has to face two new problems. The problem of good governance, and the problem of ethical and useful policy advice.

The principles of good governance have been widely embraced: legitimacy, accountability, participation, fairness, equity, transparency, and value pluralism (Kaiser *et al.*, 2021; Van Doeveren, 2011). The mention of participation is commonly taken to imply a principally bottom-up approach, basically also because top-down approaches have been shown to lack robustness and compliance from the populace. Food systems are anchored differently and have multiple characteristics if we move across scale and scope of the system.

Yet, even when we are aware of the challenges that good governance provides us with, we may not yet have the right form of presentation of our knowledge to the decision makers. This is admittedly a pragmatic consideration, but nonetheless highly significant in terms of the quality of the outcome. And here is the rub: the best academic knowledge, even when it passes all expert peer review, might lead to bad policies. This is not because the knowledge in itself can be 'false' after all, but it is because the knowledge comes packaged in a way which signals false reliability when in reality it is contextualized in a way which misses the essentials of the reality we want to target. In other words, there is an interface between academia and policy which is crucial for the policy quality of the knowledge, i.e. being fit-for-purpose. Being unaware of the specific demands of this interface can be seen as a failure of our ethics of science. Even 'good science' can become 'irresponsible science' in this setting.

We are talking about a boundary function between academia and policy which can be described as 'brokerage'. Roger Pielke (2007) has differentiated between 'the honest broker' and 'issue advocate'. Issue advocates are typically guided by selective value orientations and are focused on certain aspects of the knowledge which at times may lead to unbalanced representations of the *status questionis* in the sciences. Scientific advocacy can have a positive function in society when e.g. certain issues are not sufficiently in the spotlight of policies.

The honest broker has a different boundary function. Pielke (2007) describes the honest broker as a person or group of persons who put personal biases and values aside in order to assist policymakers in making choices between options, generally by providing clarity on the evidence. The role of honest broker leaves policymaking to the policymaker, and political decisions to the politician, minimizing any impact of the broker's own biases. Generally, this means providing options rather than a single path ahead. The honest broker needs to minimize own bias, have an overview of the science, present information in decision-relevant format, and admit all presuppositions and assumptions of the option space. Communicating caveats, uncertainties, (un-)reliability of data, and implicit values is an inherent part of the appropriate communication package (Gluckman *et al.*, 2021).

The temptation for scientists can be to act as issue advocates under the guise of being an honest broker. This is because the appearance of an honest broker opens the ear of the decision maker to a larger extent than the issue advocate who basically competes with a great number of other lobbyists.

## The EAT Lancet report

The Eat Lancet report (Willett *et al.*, 2019a,b) purports to describe a 'safe operating space' for the global food systems of the world under the conditions of the SDGs and the Paris Agreement. The declared aim of the report is to do for food what the Paris Agreement (and the IPCC) has done for the climate, namely to spell out the targets to be reached and to provide the scientific basis for political action.

I want to state clearly at the outset that I do not have any difficulties with accepting the premises of this report, namely that a radical change of our food systems is necessary for the sake of human and planetary health. This includes the general observation that, among others, reductions in meat consumption and production seems necessary to reach several of the SDGs. I not only believe that this insight is well-anchored in scientific insights, I also believe that our most widely shared ethical principles would endorse these general goals.

But the novelty of the Eat Lancet report lies not in the global goals it describes, but rather in the precise targets and the prescribed strategies it recommends to reach the overall global goals. This has to be embedded in its appeal to the political and strategic context. As consequence of the general scientific report, it includes briefing papers for cities, farmers, health-care professionals, food service professionals, policy makers, and for everyone. This underlines the ambition of the authors to make their analysis practical for wide and global implementation at various scales. I conclude from this that the report aspires to the role of being a broker. Given the distinctions made by Pielke, this raises the question whether it should be categorized among the honest brokers or rather among the issue advocates. Being categorized as the one or the other is not without consequences, since it brings in standards of presentations which in turn reflect on the ethical responsibility of the report.

My guess is that the authors would self-categorize the report as functioning as the honest broker. The repeated references to scientific consensus, to scientific models and data, to objective and nearly certain findings make it natural to examine the report as an attempt to take on the role of honest broker as the default. Several statements indicate this much: 'This Commission is an independent scientific body using the latest available science to make a global assessment of the food system and set global scientific targets for healthy diets and sustainable food production. These targets form the first attempt to provide scientific guidance for a transformation towards healthy diets from sustainable food systems' (*ibid.*: 452). I will therefore proceed to critically discuss the report under this category.

### *The report is not populated by real people*

Food is not simply a naturally occurring system in the world; it is centred around people and their activities. With this in mind, we realize that people figure in the food systems described in the EAT Lancet report only in two indirect ways, namely as producers (farmers and fishers) and as consumers in regard human health. The report is explicit about only looking at two selected endpoints, the 'production' and the 'consumption' of food. In this manner it indirectly reflects the traditional depiction of the food value chain, briefly the linear chain from production via processing via distribution and retailers to consumption. Yet, recent reports have noted the inherent insufficiency in this picture to chart the underlying complexities. The food system interacts with a variety of other systems in various ways, thus it needs to be embedded within various other systems as e.g. environment, politics, health, society, and economy (cf. SAPEA (2020).

The EAT Lancet report is, however, taking the centrality of human agency in a rather cavalier manner: 'Transformation can only be achieved with all actors in all parts of the food system working collectively towards this transformation. Furthermore, we acknowledge that food systems also affect society, culture, economy, and animal welfare. However, given the breadth and depth of the topics discussed, many important issues could not be discussed. These and other issues should be considered to achieve healthy diets from sustainable food systems' (*ibid.*: 452). The important point to remark here is that the transformation aimed at in the report can indeed only be achieved with human agency. I would add here the following statement: 'All serious and encompassing socio-cultural transition processes start in the minds of people!' If one takes this seriously, then three things follow. First, one needs to explicitly address the values and cultural identities of people. Second, given the complexity of sustainable development and its interconnections to social realities, multiple societal pathways to multiple endpoints may exist. Third, some of the most crucial nodes in the food system are power differentials between social actors. Therefore, the report turns the crucial relationships on its head. It acknowledges the impact of the food systems on cultures and other social realities, rather than describing the food system as the outcome of these cultural and socio-economic systems. This leads to basic blind spots in the description of the food systems. What are the main drivers in the food system? I would surmise that sources of significant economic and political power and the interests associated with them act as main drivers for the food system we have. Yet, we do not find these drivers addressed in the report. But it is here where the current debates are most intense. Do we need to move from large scale global industrial production systems to small scale and regional productions? Can we maintain monocultures with industrial herbicides and pesticides, or do we need to move to some variant of organic production?

We are already aware of the inequities following the global trade of food. After all, many people take up the topic of food sovereignty precisely because we have become aware of the power differentials in regard our food systems. This latter point has direct implications for governance. If all change needs to start in the minds of people and their interactions, then we need to opt for bottom-up strategies as the main road to robust transformations. Surely, participatory and empowering processes as bottom-up governance cannot mean simply implementing goals and targets which are set top-down, even including the top in this case coming from science.

In sum, I want to maintain that the food system needs to be human-centred. The fact that in the report it is not, reflects assumedly the bias in the selection of what the authors regard as relevant expertise. A vast amount of social science apparently does not qualify (cf. Béné *et al.*, 2020).

### *Questionable scientific rigor?*

The report sets forward claims which have the ring of being alarmist, for instance the claim that one could save 11 million lives annually if one follows the recommendations of the report. This is obviously not based on an observed (historical) fact, but rather on predictions from highly abstract modelling, in this case a combination of three major studies.

Again, I do believe it is indeed scientifically shown that our nutritional profiles relate to our morbidity and mortality. Furthermore, it is assumedly true that many people eat diets which are not beneficial for their health and wellbeing. For this, I believe, one may have relatively large consensus among the scientific experts on nutrition and health. Therefore, the intention of the report is good. The question is, however, how to arrive at these targets which provide the safe operating space the report aims at. The immediate next question could be why do we not have these targets already, given the immense amount of research on nutrition and health. Would not every country have an interest in promoting such targets for its own population when it is obvious that safer nutrition could not only reduce mortality among its populace, but also reduce health care costs? The country that I live in, Norway, has for a long time been struggling with the question whether the consumption of aquaculture is associated with health risks and / or health benefits. As a matter of fact, it is a constant issue of science-based debates between Norwegian experts and experts from other countries (Russia, EFSA, etc.).

There are reasons for scepticism. Firstly, the EAT Lancet report operates with averages and estimated midpoints for the reference diet. The average is based on an intake of 2,500 kcal/day which corresponds to the energy needs of a 70 kg man aged 30 years and a 60 kg woman aged 30 years with moderate to high physical activity. The scientific report is assumedly well aware about the constraining implicit assumptions as reference is made to age and ethnic / cultural differences. The question is still who answers to the average? Will it not be misleading if applied across the board? Here the report uses the other strategy of working with midpoints in their reference diet. Given that many recommendations seem to vary considerably (e.g. the poultry recommendation is in a range 0-58 g/day, and a midpoint is chosen of 29 g/day), the problem of who this recommendation applies to is amplified. Furthermore, the report is aware that most of the studies are using data from the USA and Europe, highlighting in particular studies of the Mediterranean diet, one may suspect a systematic bias in the data material informing the target values (cf. also Willett *et al.*, 2019c; Zagmutt *et al.*, 2019).

Finally, there is the question of the reliability of the meta-studies they cite. John Ioannidis concluded already in 2013: 'Almost every single nutrient imaginable has peer reviewed publications associating it with almost any outcome' (*ibid.*: 1). The Boston Cook book study: Schoenfeld and Ioannidis (2013) found that a random selection of 50 ingredients of the Boston cook book produced simultaneously positive and negative associations to cancer risks for each and every of the ingredients. These studies (Brown *et al.*, 2014; Ioannidis, 2013; Schoenfeld and Ioannidis, 2013) raise the question if the target values of the Eat lancet report can claim to be based on wide scientific consensus. Perhaps they are appropriate for an issue advocate, but they seem problematic for an honest broker.

### *Ignoring uncertainties and values?*

The EAT lancet report claims to be reflective of inherent uncertainties in the data material. The conclusion they draw from this is that they try to avoid extreme values and rather hit some assumed midpoint of moderate size that would still satisfy the objectives of health and sustainability. They also claim to be reflective of a broad range of alternative scenarios and sensitivity analysis.

I would grant that a global study like this has to make compromises and that applying sensitivity analysis is in general a good approach. However, the question which bugs me is whether the accumulated uncertainties in a study of this scope and scale in the end not undermine its claim of practical guidance. This is the well-known difficulty of all meta-studies, and of data collections in different studies and different countries with different sample populations. In order to arrive at quantified comparisons, one has to build uncertainties upon uncertainties. Any global index operates with subjectively chosen categories, which again have a set of chosen indicators, which often consist of another set of chosen sub-indicators, which assume comparable data across all samples (Kaiser *et al.* in press).

The complexity which follows the construction of such global targets necessitates value-based decision making on methods and simplifying strategies to counter the inherent uncertainties. For decision-makers in policy, the knowledge of and critical insight into these uncertainties and values is typically as important as the positively asserted results. A policy maker is accountable for the decision in more ways than a scientist is. In sum, I would tend to argue that the Eat Lancet report lacks the transparency necessary for the intended audience to act responsibly on the recommendations.

### *Are the recommended strategies useful, well-founded and original?*

The Eat Lancet report has a number of policy recommendations, even some specified for different user-groups. There are five general strategies they recommend: (1) 'seek international and national commitment to shift towards healthy diets!'; (2) 'reorient agricultural priorities from producing large quantities of food to producing healthy food!'; (3) 'sustainably intensify food production, generating high-quality output'; (4) 'strong and coordinated governance of land and oceans'; (5) 'at least halve food loss and waste, in line with global SDGs!'. Reading those, one may well ask: And what's new? Indeed, these recommendations, read in isolation, sound like some generally recognized truisms.

Thus, more detail comes under them in explanatory texts. And here my impression is that the basic message is similar to: 'it all depends'. For instance, in relation to animal production they not only mention animal welfare and antimicrobial resistance, but they also continue: 'Discussion about what large quantities of food and healthy food means in different contexts is emerging. We support the continuation of this holistic assessment of context-specific trade-offs or win-wins that could arise from animal production' (Willett *et al.*, 2019a: 481). Isn't this precisely what is bothering many of us, that these discussions are highly contentious, and seldom adjusted to local contexts, and by far not involving all relevant actors, and leaving out civil society? I would also add that these assessments are carried out in disregard to the value-landscapes of the people in many regions.

International trade is dealt with in a special text box. However, here one also merely finds a list of some pros and some cons. 'Evidence indicates that policies that liberalize trade has mixed outcomes' (*ibid.*: 483). What should those countries do that experience an increase in their food insecurity through a global system of free trade? A honest broker might not come up with strong recommendations, but an honest broker should present the spectrum of concrete policy options with their pros and cons. In terms of strategic advice it is simply not sufficient to outline a dilemma.

The recommendations for specific user-groups contain some helpful guidance, e.g. the advice to consumers ('everyone') to plan the week ahead. Useful, I think, but hardly new. And then repeating what others already have concluded: many in wealthier countries should eat less red meat and dairy.

In sum, I see a discrepancy between the boldness of sharply quantified targets on the one hand, and the commonplace recommendations on the other. Because of this gap it is also difficult to unite around common international strategies.

## Conclusion

I tried to argue that in spite of my basic agreement with the general outlook on the problems of our food systems of the EAT lancet report, I do not believe they provide the right medicine to heal this system. They do what scientists typically can do, namely provide numbers and models, and they do this because this is what they can. But their mistake might be that: (1) they argue with confidence that they represent a purely 'objective' account of the problem and its solution, while subjective judgements permeate the whole report; and that (2) they believe that their global analysis based on quantified targets is a good starting point for policy design. One might ask if not some simple heuristics, adjusted to local and regional conditions, and a more transdisciplinary effort from bottom-up might have better chances of transforming our food systems (Kaiser *et al.*, 2021). In any case, appropriating the flag of honest broker under these conditions might stretch our ethics of science.

## References

Béné, C., Fanzo, J., Haddad, L., Hawkes, C., Caron, P., Vermeulen, S., ... & Oosterveer, P. (2020). Five priorities to operationalize the EAT–Lancet Commission report. Nature Food, 1(8), 457-459.

Brown, A.W., Ioannidis, J.P., Cope, M.B., Bier, D.M. & Allison, D.B. (2014). Unscientific beliefs about scientific topics in nutrition. Advances in Nutrition, 5(5), 563-565.

Costello, C., L. Cao, S. Gelcich *et al.* (2019). The Future of Food from the Sea. Washington, DC:World Resources Institute. Available online at www.oceanpanel.org/future-food-sea (accessed 20.09.2020).

FAO, F. (2018). The future of food and agriculture – alternative pathways to 2050.

Funtowicz, S. & Ravetz, J. 1993. 'Science for the Post-Normal Age.' Futures 26, no. 6: 568-82.

Gluckman, P., Bardsley, A. & Kaiser, M.. (2021). 'Brokerage at the science-policy interface: from conceptual framework to practical guidance'. Humanities and Social Sciences Communications 8:84. https://doi.org/10.1057/s41599-021-00756-3

Ioannidis, J. (2013). Implausible results in human nutrition research. BMJ 347: f6698. https://doi.org/10.1136/bmj.f6698

Kaiser, M., Goldson, S., Buklijas, T., Gluckman, P., Allen, K., Bardsley, A., Lam, M.E. (2021). 'Towards Post-Pandemic Sustainable and Ethical Food Systems'. Food Ethics, 6:4. https://doi.org/10.1007/s41055-020-00084-3

Kaiser, M., Chen, A.T.Y. & Gluckman, P. (in press). 'Should policy makers trust composite indices? A commentary on the pitfalls of inappropriate indices for policy formation'. Health Research Policy and Systems.

Kuhn, T.S. (1962 [2012]). The structure of scientific revolutions. University of Chicago Press.

Odegard, I.Y.R. & Van der Voet, E. (2014). The future of food – scenarios and the effect on natural resource use in agriculture in 2050. Ecological Economics, 97, 51-59.

Parsons, K., Hawkes, C. & Wells, R. (2019). Understanding the food system: Why it matters for food policy. Available at: https://www.city.ac.uk/__data/assets/pdf_file/0008/471599/7643_Brief-2_What-is-the-food-system-A-food-policy-perspective_WEB_SP.pdf.

Pielke, R. (2007). The Honest Broker. Cambridge University Press.

SAPEA Evidence Review Report no. 7 (2020). A Sustainable Food System for the European Union. SAPEA Berlin. https://doi.org/10.26356/sustainablefood

Schoenfeld, J.D. & Ioannidis, J.P. (2013). Is everything we eat associated with cancer? A systematic cookbook review. The American journal of clinical nutrition, 97(1), 127-134.

Van Doeveren, V. (2011) Rethinking Good Governance, Public Integrity, 13:4, 301-318.

Willett, W., Rockström, J., Loken, B., Springmann, M., Lang, T., Vermeulen, S., ... & Murray, C.J. (2019a). Food in the Anthropocene: the EAT–Lancet Commission on healthy diets from sustainable food systems. *The Lancet*, *393*(10170), 447-492.

Willett, W., Rockström, J., Loken, B., Springmann, M., Lang, T., Vermeulen, S., ... & Murray, C.J. (2019b). EAT-Lancet Commission Report Summary. Available at: https://eatforum.org/eat-lancet-commission/eat-lancet-commission-summary-report/.

Willett, W., Rockström, J. & Loken, B. (2019c). The EAT–Lancet Commission: a flawed approach? – Authors' reply. The Lancet, 394(10204), 1141-1142.

Zagmutt, F.J., Pouzou, J.G. & Costard, S. (2019). The EAT–Lancet Commission: a flawed approach?. The Lancet, 394(10204), 1140-1141.

# 59. Empirical ethics is not a magic bullet for applied ethicists[1]

*R.B. Mikkelsen[1*], P. Sandøe[1,2] and M. Gjerris[2]*
*[1]University of Copenhagen, Department of Food and Resource Economics, Rolighedsvej 23, 1958 Frederiksberg C, Denmark; [2]University of Copenhagen, Department of Veterinary and Animal Sciences, Grønnegårdsvej 8, 1870 Frederiksberg C, Denmark; rbm@ifro.ku.dk*

## Abstract

Ethicists are often included as part of interdisciplinary research projects when the research is deemed ethically controversial, e.g. in research aimed at developing biotechnologies to be applied in food production. The stated or implied expectations are often that ethicists can deliver a stamp of moral approval or advice on how to avoid moral disapproval. We argue that these expectations ask too much of applied ethics. A normative judgment requires a combination of ethical analysis and a normative perspective, but it is not possible for the ethicist to decide whether one normative starting point is objectively better than another. In light of this, some ethicists have turned to empirical ethics to find a normativity that is independent of the ethicist's particular biases. In this paper we assess the dialogical empirical ethics methodology developed by Widdershoven *et al.* as an example of an approach that claims to develop an emerging normativity through interaction with the stakeholders in the process. However, we find that this approach inevitably introduces ethicist-dependent normativity that cannot be relied upon as the sort of objective answer that is sought by those posing the question. The focus on generating normativity undermines the methodology when it is not successful, and in fact does not represent the most important contribution of such methods when they are successful. When methods are directed away from their attempt to 'generate' normativity, and instead onto their ability to 'translate' normativity into a given context, they represent a valuable tool in providing the type of guidance that is increasingly asked of ethicists. Understanding and pursuing this alternative aim does not diminish the importance of including ethicists in controversial research projects, but it changes their role: Instead of providing normative judgment, they should be used to make the values that are relevant to the research project transparent, determine how those values and commitments could be expressed in the given context, and facilitate dialogues between stakeholders.

**Keywords:** ethics consulting, ethics methodology, bioethics, normativity

## Introduction

Ethicists are often included as part of interdisciplinary research projects when the research is deemed ethically controversial, e.g. in research aimed at developing biotechnologies to be applied in food production (Jao *et al.*, 2020; Mertz *et al.*, 2020). The stated or implied expectations are often that ethicists can deliver a stamp of moral approval or advice on how to avoid moral disapproval. In other cases, they are asked to provide guidance on specific actions that are to be carried out in practice, either in the local context, or in broader policy perspectives. In the following, we will refer to the act of providing such moral evaluation or action guiding recommendations as providing an 'ethics solution' to an issue.

We argue that this expectation asks too much of applied ethics. The field of applied ethics is diverse both in opinions and commitments, and when single ethicists or groups of ethicists are included in projects, they cannot be expected to be able to provide ethical guidance that is comprehensive enough to include

[1] This paper is in part a result of earlier work done within the Horizon 2020 project 'MycoSynVac' (www.MycoSynVac.eu), and we would like to thank our collaborators on this project.


EurSafe 2021 – DOI 10.3920/978-90-8686-915-2_59, © Wageningen Academic Publishers 2021

all these diverse views, without the conclusion becoming so broad that it becomes irrelevant in terms of guiding specific actions or the conclusion being dependent on some normativity that cannot with any kind of honesty be said to be 'objective'.

Of course, there may be cases where different, potentially conflicting, ethical perspectives combined with a factual analysis give overlapping advice. In these cases, an ethicist might be able to deliver a broadly socially acceptable answer, but not even 'a majority vote' cast by ethical theories necessarily imbues objectivity on an opinion. However, often there is no consensus between the major normative perspectives. Another exception to our claim consists of cases where ethicists provide ethical consults in hospitals, in various ethical committees, etc., where there are often existing guidelines on the subject that allow ethicists to provide more direct recommendations. So this paper will concern itself with cases where ethicists are asked to provide an 'ethical solution' within areas where there are no pre-existing guidelines and standards of practice that can be readily applied.

A normative judgment of the type requested in an 'ethics solution' requires two separate elements: a combination of a) a concrete ethical analysis of the issue at hand and b) an application of a normativity to the given issue. In current methodologies, empirical ethics attempts to contribute to both of these elements, which we find overstretches the justified reach of the methodologies proposed in ways that can undermine their findings. Our central concern is that it is not possible for any given ethicist to objectively determine whether one normativity is better than another. Some of the current methodologies within empirical ethics attempt to overcome this issue by employing justifications for their normativity that is separate from the ethicist on the surface, but we find that they are in fact introducing the same 'ethicist-bias' issue in more subtle ways (Carter, 2020; Jao *et al.*, 2020; Mertz *et al.*, 2020; Wöhlke and Schicktanz, 2019). Whether the normativity comes from the ethicist or some sort of process involving the general public or a specific set of stakeholders it is still normativity and thus not an objective account of 'what is right'.

Consequently, current empirical ethics methodologies fail to provide convincing answers to one of the elements needed to deliver the ethics solution asked of the ethicist (they fail to deliver a normativity that can be applied to the issue at hand), and thus are not satisfactory methodologies for providing answers to the type of issue that they purport to address. Interestingly, we find that such methods run into issues exactly because they are attempting to deliver on the normative side of the question (so to speak), and that the attempt to establish this normativity is also a flawed division of labour. Thus, even if ethicists claim to use new empirical tools to imbue some independently justified value base into their ethical evaluation, they are in fact still supplying a subjective normativity through their methodology. In democratic societies it falls within the domain of democratic institutions (or funding bodies) to set the normative agenda in research – and take responsibility for it as an opinion that however well- or ill-informed can be discussed as any other normative prescription.

## An example

The authors of this paper have been a part of the H2020 project 'MycoSynVac' which serves as a useful example of such a case. In this project, we were tasked with providing ethical evaluation and assessment (an ethics solution) to a research project intended to develop a synthetic vaccine for use in food production animals in order to combat diseases that currently are difficult or impossible to treat. Here, some of the decisions within the normative element of our work were to a large extent decided by the context of the task, including sociological studies of stakeholder views that preceded the normative element of our work. The project had been funded by the EU, and by researching the sentiments and opinions of both laypersons, vaccine experts, synthetic biology experts and stakeholders, a quite clear normative background was set out. This included the starting point that such a vaccine is *prima facie* morally acceptable but is less desirable than a 'natural' vaccine, and that the potential for increases in animal welfare is essential, along with a focus on ensuring a fair distribution of the goods derived from the development of the vaccine.

However, translating these broader normative 'standards' into specific recommendation and actions remained difficult, representing the real challenge in providing the ethics solution requested of us, and it is an area where better methodological tools would be useful. Thus, the essential challenge of this type of work is not 'generating' normativity, but rather 'translating' normativity into useful prescriptions in practice. So, where some current empirical ethics methodologies focus on generating normativity (and often stumble in our view), the real usefulness of such methodologies, (in the context of providing the ethics solution needed in this and many other cases), lies in their ability to translate a rather nebulous set of normative values into practical and actionable guidance and assessment.

## An empirical ethics that generates normativity

While more than one empirical ethics methodology potentially suffers from the issue described in this paper, we will discuss one of the prominent empirical ethics methodologies in more detail in order to support our claim. This is the dialogical empirical ethics methodology described in (Widdershoven *et al.*, 2009).

Widdershoven *et al.* described a 'dialogical' empirical ethics methodology in their 2009 paper. The approach supposedly generated normative conclusions through dialogue with relevant stakeholders, more specifically through focus group interviews that produce a fusion of horizons. Essentially, the ethicist works together with participants to generate the normative conclusions of a study. As an example, Landeweer *et al.* (2011) used the methodology to justify their claim that there had been an improvement in the moral views of participants in their study on coercion in psychiatry. In both Widdershoven *et al.* (2009) and Landeweer *et al.* (2011), the authors argue that the fostering of moral deliberation between study participants can lead to deeper moral reflections about difficult ethical cases that participants encounter in practice, and that this type of dialogue can lead to changes in their moral stance. This new considered stance that is shared among the stakeholders is then claimed to be an ethically superior stance.

However, claiming that there is a change in stakeholders' moral stances and that the new stance is more considered is only a very preliminary step in claiming any normative justification for this new stance beyond that which the participants had before engaging in the process. If the process is left at this stage, the authors themselves acknowledge that they are only justified to claim that: 'Intuitively we consider the new perspective developed in the case story as qualitatively better than the views the nurses shared before the focus group started' (Landeweer *et al.*, 2011). However, this does not carry much moral weight. Intuitions should be used with caution when legitimizing moral stances. In order to be able to credibly argue that this new stance that has emerged from the process is a more justified normative conclusion, there must be some feature about the claim which this justification can be attributed to beside it being 'more considered'.

## Ethicist bias and claims of 'moral improvement'

The authors recognize this limitation and proceed to argue that moral conclusions cannot only be changed, but that they 'can be improved' (Landeweer *et al.*, 2011). But 'improve' is of course a key word in this context: What does it mean to improve an ethical decision? It would seem that the through their answer to this question the authors claim to have a methodology which can generate normativity that is ideally free of ethicist-bias.

One interpretation of the claims in the paper is that the nurses in the study initially were making sub-optimal decisions and that the study shows that by engaging in the process described in the paper, they are able to improve these decisions from a normative viewpoint. On this interpretation, the paper has a rather limited conclusion, but crucially, this conclusion would also rely on some ethical standard which is not established within the paper. Essentially, if the claim is interpreted narrowly, the authors are claiming that they have not established a normativity but rather have developed a method that allows participants in their study to align more naturally with a pre-determined normativity by following the prescribed method. This would clearly fall victim to the exact issue that is at hand in this paper: the methodology relies on an externally supplied normative standard in order to evaluate moral improvement in the practical context.

Thus, the claim that the method can generate normativity that is in some sense 'free' of the ethicist is not achieved, and the method fails to generate both elements mentioned in the introduction that are needed to satisfactorily provide an ethical solution to the case (i.e. an ethical analysis of the context of the case as well as a normativity that can be applied to the case in question). In other words, the paper turns out to be an empirical paper 'about' ethics, rather than an empirical ethics paper. Furthermore, as the authors do not appeal to any established normativity in the paper, it is a presentation of a method that claims to improve decision-making, without transparent standards for how 'improvement' is measured. More could be said on such an interpretation of the paper, but it is likely that a stronger interpretation is a fairer representation of the authors intent, and as such, we will now move onto assessing the paper in this light.

We interpret the paper as having the more ambitious goal of seeking to present and employ a methodology that can in itself reliably generate improvements upon current ethical decisions without including an external normative perspective. In other words, the claim must be that there is some feature of the methodology that provides justification for the claim that a normative conclusion P that has been reached by engaging in the process described in the paper is reliably more normatively justified than a conclusion Q that has been reached prior to or without engaging in this process. All this allegedly independent of the normativity of the ethicist herself. Support for this interpretation can be found within the paper itself as the authors state that they want to 'determine whether the changes in moral views of the nurses were actually an example of moral development, improving the moral quality of the decisions concerning seclusion at the wards?' (Landeweer *et al.*, 2011). When providing a positive answer to this question, they appeal to the dialogical method established in the earlier paper (Widdershoven *et al.*, 2009).

However, the question remains, how moral justification can be generated by the methodology itself. In the two papers discussed here, the argument is claimed to be based in part on the hermeneutic philosophy of Hans-Georg Gadamer (Widdershoven *et al.*, 2009: 238). The argument essentially is that dialogue about morally complex issues can lead to development of new perspectives, as well as a shared and developed understanding of the world and questions at hand. Through the process, a fusion of horizons happens, and the participants are left with a 'better, fuller understanding of moral situations' (Widdershoven *et al.*, 2009). Consequently, participants are able to see new solutions to difficult moral problems and are better able to morally evaluate solutions in general, and this ideally leads to agreement on more normatively justified decisions.

This approach has some similarities to a theory of justification that is native to the ethics field, namely the reflectivc equilibrium made famous by John Rawls in his 'A Theory of Justice' (1999) (first published in 1971), and later expanded upon by Norman Daniels (1979). However, the wide reflective equilibrium concerns itself with finding justification (traditionally in some form of coherentism) viewed from the perspective of the individual engaging in the exercise, whereas the methodology developed in

(Widdershoven *et al.*, 2009) appears to rely on an interactive process between individuals that allow insight into some normative features that are external to the individual.

Thus the approach can purportedly reliably improve upon the normative judgments of the participants in a way so that not only coherence is achieved within the perspective of an individual but so that intersubjective development is achieved. So, in essence, by joining the methodological approach with a hermeneutic meta-ethical commitment, the authors are able to claim moral authority for the normative conclusions reached in this type of study. If you accept the premises, the authors can legitimately claim that conclusions regarding difficult moral cases can be improved by engaging in this process, and that the justification for this claim is a function of the process itself.

But whether to accept the premises is exactly the issue at hand: On the surface, the appeal of this methodology is that it seems to be able to provide both elements of the answer often sought from ethicists: (1) it provides concrete analysis of the issue at hand through deliberation with relevant stakeholders and ethical expertise; and (2) it provides an ethicist-neutral normativity which can provide guidance on the specific issue. However, it should be clear that the normativity that is at the foundation of the conclusion is in no way neutral to the biases introduced by the ethicist. Rather, the conclusion is held up by a meta-ethical commitment to an absolutist reading of the hermeneutic philosophy of Gadamer, which cannot be said to be broadly shared within the ethics community, nor if it were, would represent any objective normativity.

## A way forward for ethics solutions

In the above, we have presented two interpretations of what a prominent empirical ethics methodology can provide when searching for normatively justified solutions to a specific case. If the hope is that such methodologies can provide an improvement to how ethics solutions are carried out by ethicists when called upon by decisionmakers to provide guidance or input on policy or practice with regard to increased normative justifiability through ethicist-neutrality, we find that the methods fail. On the narrow interpretation, the method requires an externally provided normative standard in order to be able to get the claims of 'moral improvement' off the ground. On the ambitious interpretation, the method requires a commitment to dialogical methods that can deliver universally valid results in order to show 'moral improvement'. In both cases, the method fails to deliver a perspective that is sufficiently ethicist-independent, insofar as it implies a fairly specific set of values or 'justification by process' normativity that cannot be objectively justified as being preferable to other normativities.

So, should empirical ethics methodologies like the one described here be dismissed as effective tools to provide ethics solutions of the type more and more frequently requested by decision-makers? In our view, this is not the case. Rather, we believe that such methods have become vulnerable to this form of criticism exactly because they attempt to deliver more than is possible. If their ambition is scaled back such that they do not claim to be able to generate objective normativity (especially independently of the ethicist), they can serve as one element of a successful strategy to provide ethical solutions going forward. By adapting the methodology to a more narrow interpretation, what one is left with is a useful tool for assessing the context within which one is attempting to generate and express ethical advice that is useful in practice. In this case, no normativity is provided by the method itself, but neither is such a normativity necessary. In most cases, some normative standards are set by the context within which the ethicists are asked to provide their analysis, like in the case of MycoSynVac. Often, funders, politicians, stakeholders, or general academic ethics consensus has provided a broad normative standard that will be necessary to rely on in the practical context. In such cases, methods like the one described here are useful tools to translate such a normativity to a set of practical recommendations that can be implemented. In this way, empirical ethics methodologies can provide significant value by way of their ability to bridge the gap between broader normative standards, and the type of specificity required from ethical solutions. Especially if the ethicists remember to emphasize the limits of their guidance with regard to normativity.

## Conclusion

It is becoming more common for ethicists to be asked to provide ethical solutions by funders and stakeholders that provide an objective normative evaluation of e.g. a technology. However, providing such a solution is a challenge for ethicists because ethics by its very nature cannot provide objective answers to normative questions. Many empirical ethics methods purport to be able to solve this problem by providing two elements: An ethicist-independent normativity as well as a way to translate this normativity into practice. In (Widdershoven *et al.*, 2009) the authors provide a somewhat integrated strategy for achieving both of these aims. However, we find that the method fails to deliver on the first element: It cannot deliver an ethicist independent normativity. But, in our view, this is not necessary for such methods to be useful. In practice, broad normative standards are most often set by the context within which the ethicist is working. We argue that by removing focus from the ability to deliver objective normativity, and instead focusing on translating a transparent normativity into useful ethical advice, not only are many of the methodological weaknesses avoided, but the method is also better positioned to provide an important tool for ethicists going forward.

## References

Carter, S.M. (2020). Methodological challenges in deliberative empirical ethics. *Journal of Medical Ethics*, *46*(6): 382-383. https://doi.org/10.1136/medethics-2020-106185

Daniels, N. (1979). Wide Reflective Equilibrium and Theory Acceptance in Ethics. *The* Journal of Philosophy 76(5): 256-282. https://doi.org/10.2307/2025881

Davies, R., Ives, J. and Dunn, M. (2015). A systematic review of empirical bioethics methodologies. BMC Medical Ethics 16(1). https://doi.org/10.1186/s12910-015-0010-3

Jao, I., Marsh, V., Chi, P.C., Kapulu, M., Hamaluba, M., Molyneux, S., ... Kamuya, D. (2020). Deliberately infecting healthy volunteers with malaria parasites: Perceptions and experiences of participants and other stakeholders in a Kenyan-based malaria infection study. *Bioethics*, *34*(8): 819-832. https://doi.org/10.1111/bioe.12781

Landeweer, E.G., Abma, T.A., and Widdershoven, G.A. (2011). Moral margins concerning the use of coercion in psychiatry. Nursing Ethics 18(3): 304-316. https://doi.org/10.1177/0969733011400301

Mertz, M., Nobile, H. & Kahrass, H. (2020). Systematic reviews of empirical literature on bioethical topics: Results from a meta-review. *Nursing Ethics*, *27*(4): 960-978. https://doi.org/10.1177/0969733020907935

Nichols, P. (2012). Wide reflective equilibrium as a method of justification in bioethics. Theoretical Medicine and Bioethics 33(5): 325-341. https://doi.org/10.1007/s11017-012-9223-3

Rawls, J. (1999). A Theory of Justice. Revised Edition. Harvard University Press, Cambridge, Mass., USA.

Widdershoven, G., Abma, T. & Molewijk, B. (2009). Empirical Ethics as Dialogical Practice. Bioethics 23(4): 236-248. https://doi.org/10.1111/j.1467-8519.2009.01712.x

Wöhlke, S. & Schicktanz, S. (2019). Special Issue: Why Ethically Reflect on Empirical Studies in Empirical Ethics? Case Studies and Commentaries. *Journal of Empirical Research on Human Research Ethics*, *14*(5): 424-427. https://doi.org/10.1177/1556264619862395

# 60. Coating moral misconduct – the motivational function of shame in ethical debates

*K. Dieck* and H. Grimm*
*Unit of Ethics and Human-Animal-Studies, Messerli Research Institute, University of Veterinary Medicine, Vienna, Medical University of Vienna, University of Vienna, 1210 Vienna, Austria; katharina.dieck@vetmeduni.ac.at*

## Abstract

The fur debate of the late 1980s and 1990s in North America has been remarkable in many ways. It was of such magnitude that it almost brought a whole flourishing industry sector to a standstill within just two decades. Still today, the arguments of the fur opponents are relevant insofar that fur breeding bans are already introduced or at least discussed in many countries. Compared to other forms of animal husbandry, the debate over fur thus receives impressive attention from the media and political side. One element that particularly characterized the fur debate is fur-shaming: the practice of inducing shame in fur wearing people through public, offensive confrontations. The aim of this paper is to examine fur-shaming in its relation to the fur fashion, which changed drastically, when fur became more and more controversial. Fur started to be processed in such a way that it was made unrecognizable as such to the environment. It disappeared inside the coat as lining or was shorn and dyed in such a way that it could no longer be distinguished from fake fur. The public confrontations were changed by the shift in fur fashion, however the moral conflict was only hidden in the lining. Through the non-discursive measures of the fur-fashion-industry in this debate, new questions can be raised regarding the motivational function of shame: How can shame be understood to motivate a certain behaviour without having adapted corresponding moral beliefs? By incorporating recent findings of moral intuitionist theories and social functionalist accounts, the overall aim of this paper is to shed light on this question. Thus, this paper offers a perspective on the fur debate as a moral discourse, which can be relevant for a deeper understanding of the role moral emotions play in ethical debates.

**Keywords:** fur, shame, moral discourse, functionalism

## Introduction: the fur industry in the public eye

Unlike many other economic sectors based on animal products, the fur industry has faced an incisive moral debate in western countries. The fur debate has spread through the whole western world and has been remarkable in many ways. It diminished international fur trade significantly and shaped the public perception of fur until today. As a representative study in Austria showed, 68% of the population were in favour of a legal fur-ban and 78% regarded is as right to publicly denounce fashion stores that sell fur (Balluch, 2013). A fashion element had undergone a discursive transformation – what had been a symbol of luxury once, has emerged as a morally dubious, shameful commitment.

How decisive the public perception of fur can be in a political dimension has recently been revealed by the corona crisis: When a mutated form of coronavirus had been found on mink farms in November 2020, Denmark decided to cull all its mink for a mutation might have risked the efficiency of a future vaccine. Up to 17 million animals were slaughtered, regardless whether they were infected or healthy, thereby devastating EU's then largest fur producer (BBC, 2020). Since the fur industry is considered a rather unpopular economic sector not only in Denmark, voices have been raised that the quick decision to cull Denmark's minks was at least partly motivated by collective aversion to its fur industry (SZ, 2020).


EurSafe 2021 – DOI 10.3920/978-90-8686-915-2_60, © Wageningen Academic Publishers 2021

Denmark thus adds its name to the list of European countries that have already introduced a legal ban on fur farming (Balluch, 2015: 281).

This recent debate, however, confirms the trend that discussions about fur receive impressive attention from the media and political side compared to other forms of animal husbandry and the number of animals kept there. Fur production seems to meet a special interest in public discourse. But any collective aversion and moral interest in fur has been established in a discourse culture about fur, whose roots go back to the 1920s, when debates about exotic furs emerged. It reached the mainstream in the 1980s through massive campaigning of the anti-fur movement, which consisted of animal-rights' activists as well as animal welfare activists. They proclaimed several reasons for resisting fur consumption, animal welfare issues due to housing conditions as the most prominent one. Regardless whether fur was farmed or trapped, activists maintained that 'all methods of obtaining fur [were] cruel' (Farley and Farley, 2014: 152), thereby inferring fur-wearers to be cruel. This moral flaw in character was then rendered as something shameful, something to be publicly exposed.

This paper will provide a closer look at the fur-debate in North America, as it is the most documented. To date, there have been no cross-national studies concerning debates about fur. North America's debate is therefore assumed to be paradigmatic, there are reciprocal references with other countries in the sources as well. Consequently, this paper also hints at aspects of the European fur debate.

It sketches how these changes in the public perception of fur came about, which arguments pervaded the debate and which measures were adopted by the anti-fur movement. The focus, however, will not be on any ethical dimensions or argumentations in the fur-debate, but on aspects targeting a moral psychological dimension of actively induced shame in the debate: The strategy of fur-shaming is proposed to have triggered a design revolution within the fur industry, thereby displaying moral mechanisms made visible via possibilities the matter of fashion provides. This paper thus reconstructs the fur debate in terms of shame as a moral emotion that fulfils a function in shaping moral discourse. Relevant findings of this interpretation are then integrated within a moral intuitionist and functionalist framework as presented in the moral psychological research by Haidt (e.g. Haidt, 2001, 2007; Haidt and Kesebir, 2010; Keltner and Haidt, 2016), to shed further light on the question: How does an active inducing of moral emotions impact moral discourse?

## The fur debate in North America

Before the arising of the anti-fur movement, North America's fur industry could look back on a long-standing legacy of success. In the mid-1980s, the fur industry had 15 years of steadily increasing sales behind it, and with labour markets opening up to women and cheaper production methods, the rise of the fur was further promoted. Once an unattainable symbol for wealth, fur could then be produced in a way which made it affordable for America's middle-class working women (Hochswender, 1989; Morris, 1986). The future looked promising for the fur industry. After the peak of the fur debate, however, the US fur market had lost over a third of its sales in the beginning of 1990 (Burns, 1991). In some European countries, this economic sector has even been brought to a complete standstill by then. How could these massive effects on a flourishing economic sector have been brought about?

The fur opposition tries to reinterpret positive characteristics associated with fur in various print- and media-campaigns, in which cruel aspects of the production process and the suffering of animals are highlighted. Especially PETA activists were at the centre of the debate in the media for their attention-grabbing performances, such as burning or burying fur in public places (*ibid.*: 258). At the same time, activists tried to create new associations of fur with cruelty through print campaigns. By disseminating images depicting living conditions and killing methods of fur animals, the living animal behind the fur was brought to attention, thus negating the usual separation of animal and product (Olson and Goodnight, 1994: 260). Their strategy was to stigmatize fur, according to a representative of 'Trans Species Unlimited': 'We decided to make fur seem vulgar, a symbol of someone who is tasteless and uneducated, so that people would feel they were being admired if they didn't wear fur' (Kasindorf, cited in Olson and Goodnight, 1994: 262). The public display of fur, like smoking, should only be tolerated, but no longer appreciated – an act that leaves one slightly ashamed and is therefore not worth the effort.

### *Fur-shaming*

One method that particularly characterized the fur debate then is fur-shaming: the practice of inducing shame in fur wearing people through public, offensive confrontations. The aim of the act of fur-shaming was to communicate that wearing fur is a moral transgression, that someone should be ashamed to commit. In order to promote the transformation of fur as a status symbol into something vulgar and cruel in the public perception, the fur opposition worked with the concept of shame on several levels.

In 1988, for example, the Humane Society published a large-scale print advertising campaign called 'The Shame of Fur Campaign' (Olson and Goodnight, 1994: 263). It appealed directly to the sense of shame of potential fur buyers, for example in a pamphlet that advertised: 'It's a shame to wear fur! People should be ashamed to wear fur. When you choose to wear fur, animals suffer and die needlessly' (*ibid.*). These and similar publicity offensives set the stage for the prominent role of induced shame in the debate.

However, direct encounters between individuals or groups of activists with fur wearers pervaded the discourse as well and had an especially prominent role at public events such as fashion weeks. Activists positioned themselves in public places to confront individual fur wearers about their fur consumption. PETA in particular often drew attention to itself and its cause through such actions, such as in 1989 when a group of activists gathered in front of fur shops in Manhattan and shouted 'Shame! Shame! Look at that Fur!' (Olson and Goodnight, 1994: 263). It was also common to throw vegan pies, red paint, dead raccoons or various other objects at fur wearers in public places. If a fur wearer was actually hit, her outward appearance was usually ruined and the moral stain clearly marked.

The culture of fur-shaming that developed during the fur debate of the 1980s and 1990s has at least been successful insofar as fur wearers were usually aware of the controversial nature of the product. In 1989, for example, 'Time' featured a woman who, according to her own account, had been delighted to buy a silver fox coat in 1984, but who emphasized the insecurities that the arguments of the fur opposition had triggered in her: 'I've been struggling with the dilemma of buying fur. [...] I like the look, but I feel really guilty' (Reed, 1989).

### *Reaction of fur industry*

The protests against fur remained unacknowledged by the fur industry for a long time (Olson and Goodnight, 1994: 258). When the sales figures fell so drastically that the activists' claims could no longer be ignored, the fur industry invested in various print campaigns, that communicated humane treatment of the fur animals during the production process and linked fur to sustainability (*ibid.*: 265).

However, one remarkable strategy of the industry consisted in reinventing the design principles of fur. As the New York Times stated 1992: 'Buffeted by a sluggish economy, animal-rights protesters with their spray cans and threats, not to mention a warm winter in many parts of the country, furriers have decided to fight back. Their weapon: fashion' (Morris, 1992a).

Since the early 1990s, fur has increasingly been processed in such a way that it was only visible to the wearer. As a lining in winter coats, fur could be shielded from unwanted glances. Likewise, many fur accessories such as collars and trimmings were no longer sewn on tightly but were so functionally processed that the wearer could easily remove and reattach them on her own. On the catwalks of the autumn-winter collections, reversible garments were often seen as well: 'There are also a number of reversible coats that keep their fur side under wraps when the weather or the neighbourhood is bad' (Morris, 1992a). These reversible coats had fur on one side and usually high-quality wool fabric on the other allowing the wearer to decide each day whether she wanted to show or hide the fur.

At the same time, faux and real fur developed to a point where they were indistinguishable from each other: In the early 1990s, the so-called 'fun furs' that had enjoyed great popularity in the 1970s returned (Farley and Farley, 2014: 150). They can be characterized as "Frankly fake' furs ... not only imitate the animal kingdom, but poke fun at it. They have been dyed in fanciful colours and made with exaggerated markings' (Anonymous, 1950). Due to the now disreputable status of real fur, the need for faux fur seemed to have grown. Even though fur imitations have been around for much longer than the debate (cf. Farley and Farley, 2014), it was now the real fur that was designed to imitate fake fur. No fur product whatsoever should remind of its animal origin.

In 1994, there was a multitude of faux fur coats on the catwalks: 'No need to bother about euphemisms – those shaggy fur coats are frankly fake. [...] Designers are making no attempt to refer to them as real' (Morris, 1994). Its bright colours, short hair and fuzzy look that characterized faux-fur's optic, visually mocked real fur. So, no one would mistake it for the real thing, rather seeing one's dismissal of fur practices displayed in a parodic fashion. Even though real fur obviously continued to have qualitative advantages for many people, the garish fake fur had one advantage over it: It provided the wearer with the feeling of security of not being confronted by activists on the street and not having anything morally disreputable associated with oneself.

The fur industry took advantage of the growing need for fake fur and the return of fun furs. Thus, real fur was increasingly processed in such a way that it could no longer be distinguished from the flashy fake furs. The fur was shorn, trimmed and dyed in 'non-realistic colours like pink, it belied its sheepskin origin' (*ibid.*). Consequently, the animal origin was no longer discernible, undermining the connection between animal and product for which the fur opposition created awareness.

The changes in fur fashion during the fur debate meant that by hiding and camouflaging real fur, the line between fake and real fur became so blurred that it was no longer possible to distinguish between the two in public spaces. In this way, fur advocates managed to keep the risk of potential confrontations by activists low by removing the potential target from sight. However, the moral conflict was only hidden in the lining.

## Shame and functionalism

For this paper, the fur debate is interpreted as a moral discourse, in which a former symbol of wealth and luxury has been reconceptualized in terms of its production methods and therefore transferred into a morally dismissible area. Public fur-shaming is therein proposed as a strategy, that was vital in bringing about the shift in public conception of fur. In this form, shame had a certain function in moral discourse. Active shaming triggered the design transformation in fur fashion. By hiding the real fur, the new designs concealed potential individual moral misconducts.

### *Functionalism: Jonathan Haidt*

Looking at the relation between the development of fur and shame in moral discourse, a moral-psychological account may help to clarify the functional relation of shame and morality. In defining morality, Haidt focuses primarily on the social, interpersonal mechanisms that shape the way we live together. He working with a social-functional framework, which explains the phenomenon of morality through its function and not on the basis of content. Accordingly, Haidt defines morality as follows:

> Moral systems are interlocking sets of values, virtues, norms, practices, identities, institutions, technologies, and evolved psychological mechanisms that work together to suppress or regulate selfishness and make cooperative social life possible.
> (Haidt and Kesebir, 2010: 800)

This definition thus allows for moral pluralism, the idea that there are different, mutually incompatible but morally justifiable ways of organising a society (*ibid.*).

In this functionalist account, moral emotions serve certain functions and constitute our moral intuitions (Haidt, 2003; Keltner and Haidt, 2016). They help us navigate the social world with its various moral norms and guide our actions. Regarding Haidt, it is scarcely conscious considerations about what is right, but moral emotions that move us to moral actions. He ascribes a special value to moral emotions when growing into a social environment: Moral emotions condition us and it is only through them that we learn norms and develop an understanding of morality (2001: 824). Therefore, moral emotions such as shame are a powerful social mechanism to align oneself regarding moral contents. If my action does not conform to a shared moral norm in my community, other persons will probably address this transgression. Consequently, I am most likely to react with a corresponding moral emotion, usually shame or guilt. The moral ideas that are brought to me through my social environment are thus internalised and form a strong motivation to adapt my behaviour. This mechanism also functions in an anticipated form: The mere idea of a certain action that is considered immoral in my community can fill me with shame and thus provide motivation not to commit any transgressions.

### *Shame in the fur debate*

To examine fashion as an object of moral transgression in a debate is insightful, for it uses visible material to create three-dimensional objects. It constructs an inside and an outside, thereby displaying or hiding relevant aspects. The moral transgression can be camouflaged and made indiscernible as to provide a simple solution to moral tensions.

The developments in fur fashion reveal something about the workings of shame as a crucial moral mechanism. The product that was the target of the anti-fur-movement remained the same throughout the debate as did the demand for it. The change of behaviour is thus not likely to be caused by a moral point taken: Fur wearers did not want to relinquish fur altogether but its public display, thereby avoiding being subject to shaming practices.

Thus, there was no genuine alignment of norms as proposed by the functionalist account, for people kept wearing fur. They did not themselves regard fur as something truly dismissible or shameful, or they would not have worn it. Still, the fur wearers' behaviour is altered without internalizing the new norm, since solely the target of moral criticism is made invisible. This suggests that shaming practices gave sufficient reason to change one's course of action.

By appealing to the concept of shame in public space, one directs public attention to something morally relevant, which may serve as a strong individual motivation. Fur-shaming constituted not only an uncomfortable situation for a shamed individual, but through the public gaze and the ongoing debate was increasingly interpreted as something of moral significance. So, even though there was no alignment of norms, active shaming served to alter behaviour in such a way that the moral conflict was eliminated, thereby enhancing the possibility of cooperative social life.

## Conclusion

Looking at the fur debate, the role of shame can be seen as supporting the functionalist notion, that emotions are not only in one's head, but a matter of relation and interaction with other persons (Keltner and Haidt, 2007). Still, shame as discussed here suggests an even wider interpersonal scope of moral emotions. It works to alter behaviour to reduce moral conflict even when shame is only induced from the outside but not actually felt for the reason that one is urged to adopt. This brings another important characteristic of shame to light, that adds to a moral functionalist account: Shame can not only be brought about resulting from oneself transgressing one's own norms. It can be brought about by other persons actively alluring to shame in characterizing one's behaviour. This constitutes another dimension of shame, i.e. a specific form of unwanted public attention. Shame can thus be also seen as located in various interpersonal dimensions and as a tool for navigating pluralist societies by reducing social conflict without the alignment of moral norms.

For future research within functionalist approaches in moral psychology, it might be illuminating to examine whether the interpersonal and inherently active scope of shame in the fur debate is a specific characteristic of moral emotions. Incorporating these findings further into a functionalist account, might shed light on the social functions of other moral emotions within moral discourse. For practical ethics, it may be insightful to analyse moral discourses in terms of moral emotions, considering their pervading impact on decision-making.

## References

Balluch, M. (2013). IFES – Studie zum Tierschutz. Wien, Institut für empirische Sozialfor-schung GmbH.

Balluch, M. (2015). Pelz. In Ferrari, A., Petrus, K. (eds) Lexikon der Mensch-Tier-Beziehung. Bielefeld, Transcript

BBC (2020). Coronavirus: Denmark shaken by cull of millions of mink. Available at: https://www.bbc.com/news/world-europe-54890229

Burns, J.F. (1991). Fur industry shrinking with no end in sight. New York Times Special February 26, 1991

Farley, G. & Farley, J. (2014). Treatment of animals In Farley, J. & Hill, C. (eds) Sustainable Fashion: Past, Present and Future. London, Bloomsbury.

Haidt, J. (2001). The Emotional Dog and its Rational Tail: A Social Intuist Approach to Moral Judgement. Psychological Review 108(4): 814-834.

Haidt, J. (2003). The Moral Emotions. In Davidson, R.J., Scherer, K.R., Goldsmith, H.H. (eds) Handbook of affective sciences. Oxford, Oxford University Press, pp. 852-870.

Haidt, J. (2007). The New Synthesis in Moral Psychology. Science 316: 998-1002.

Haidt, J. & Kesebir, S. (2010). Morality. In Fiske, S., Gilbert, D., Lindzey, G. (eds) Handbook of Social Psychology, 5th ed., Hobeken, NJ, Wiley, pp. 797-832.

Hochswender, W. (1989). As the Image of Fur Suffers, So Does Profit. New York Times March 14, 1989.

Keltner, D. & Haidt, J. (2016). Social Functions of Emotions at Four Levels of Analysis. Cognition & Emotion 13: 505-521.

Morris, B. (1986). The furrier's secret: A 15-year boom. New York Times June 12, 1986.

Morris, B. (1992). Furriers' weapon: sprightly design. New York Times May 12, 1992.

Morris, B. (1994). Confounding fur and faux. New York Times April 26, 1994.

Anonymous (1950). Fake Fur Becomes a Novelty Fabric. New York Times September 13, 1950, p. 42.
Olson, K.M. und Goodnight, T. (1994). Entanglements of consumption, cruelty, privacy, and fashion: the social controversy over fur. The Quarterly Journal of Speech 80: 249-276.
Reed, J.D. (1989). The furor over wearing furs. Time December 18, 1989.
Süddeutsche Zeitung (2020). Der Fall der Felle. Available at: https://projekte.sueddeutsche.de/artikel/politik/nerze-in-daenemark-der-fall-der-felle-e909969/?reduced=true.

# 61. Normativity in applied ethics teaching: not to have, nice to have, or need to have?

*C. Gamborg* and M. Gjerris*
*Department of Food and Resource Economics, University of Copenhagen, Rolighedsvej 23, 1958 Frederiksberg C, Denmark; chg@ifro.ku.dk*

## Abstract

Arguing for mitigation/adaptation to counter risks to food security, agricultural production and environmental protection caused by climate change is, although based on (sound) science, at the core a value-based argument. Therefore, at many universities, applied science students now have mandatory and/or elective BSc and MSc courses which put their discipline into a wider social context, including ethical analysis and reflection. Such courses thus deal with normative questions. However, it is necessary to ponder what to do with the normativity in applied ethics teaching, including how to deal with the values of the teacher whilst ensuring the best learning outcome. We analyse the role of normativity in ethics teaching drawing upon, respectively, the stance on conflict of interests found in codes on responsible conduct in research (RCR) and an adapted version of Pielke's science-policy interface framework to bring out different types of ethics teachers. We find that not disclosing one's normative position as teacher within applied ethics can be seen as lack of transparency, and may result in being, or being perceived of being, an act of (hidden) persuasion. We argue that teacher's normative positions should be disclosed, to some degree, to enact, what we call responsible conduct in teaching (RCT). We conclude that there are a number of challenges associated with the normativity of teachers in relation to RCT and best learning outcome and suggest that having a teaching team that represent different normative positions may be of benefit to students and deflect criticisms of bias.

**Keywords:** responsible conduct of teaching, science-policy interface, responsible conduct of research, best learning outcome

## Introduction

Strategies to counter the effects of climate change on food security, agricultural production and environmental protection include mitigation (minimizing possible impacts) and adaptation (reducing the negative effects), and seek out opportunities. Coming professionals and actors in the food chains are often taught about these strategies and their implementation as part of their university curriculum. In such contexts, science and technology are often promoted as important tools to help society in a situation where climate change, loss of biodiversity, widespread hunger and pollution threaten the lives of billions of people now and in the future.

Arguing for the mitigation of climate change or adaption to changes to safeguard food security and ensure environmental protection evidently builds on insights from e.g. biology and food science, but is at the core a value-based argument. In this context, almost any choice of action has effects that raises ethical questions related to e.g. social equity or environmental justice. It also relates to how to prioritize efforts in a situation where 'facts [are] uncertain, values in dispute, stakes high and decisions urgent' (Funtowicz and Ravetz, 1993: 744).

Moreover, future professionals studying and working with climate, food, nature and environment will often be faced with disagreement and dilemmas instead of clear-cut decisions between e.g. protection


EurSafe 2021 – DOI 10.3920/978-90-8686-915-2_61, © Wageningen Academic Publishers 2021

of the natural environment and the realization of human interests (or the interests of some at least); or whether societies should aim for environmental or ecological justice, taking into account also non-human species? In response to these challenges, these coming professionals and actors within the food sector (from farming to governance to research) are often introduced to courses putting their knowledge into a wider societal context, including reflection on values and how these values should be handled (ethical theories) in their curricula. For example, being educated to work with food production technologies, an ethical question is whether it is morally permissible to eat food based on animals, and if so, under what conditions and why; or if one specialises in food science, a question could be whether it is morally acceptable to use palm oil in food products – and when/when not?

Inclusion of applied ethical issues in applied science curricula have come in waves. The first struck the world of medicine in the 1970s, the second biotechnology in the 1990s and since the 2000s food production and climate and biodiversity management. Rationales probably range from a genuine recognition of ethics as part of finding socially robust solutions, over to a kind of greenwashing of certain professions, production paradigms and/or technologies. Whatever the rationale, applied ethics acts as the bridge between practical problems and conceptual critical analysis, analysing the presuppositions of different kinds of arguments being used in academic and societal discourses on e.g. climate, society, nature and the environment.

The authors of this paper teach ethics at a (mostly) natural science faculty at BSc, MSc and PhD level to students of whom many after graduation will contribute with scientific and technological knowledge to address the abovementioned societal challenges. Teaching ethics in such a setting obviously raises many questions (Gjerris, 2006; Stumann and Gamborg, 2013). Here, we will focus on two intertwined questions: What should the aim of such teaching be, and what to do with the normativity of the teacher in relation to the taught subjects and cases used? In short: how to deal with the values of the teacher to ensure the best learning outcome?

The issue of how to deal with the normativity of researchers is integral to being a researcher when reporting on possible conflicts of interests when communicating research in both the academic and the public context and when engaging in advisory work in both private and public policy processes. However, it is noteworthy that normativity in teaching has not been the subject to the same scrutiny as the other tasks – even though teaching is a huge part of the professional life of many university researchers. We will analyse the role of normativity in ethics teaching in the context described above, drawing upon, respectively, the stance on conflict of interests found in codes on Responsible Conduct in Research (RCR) and a science-policy framework developed by Roger Pielke (2007). At the same time, we develop a sketch of what the goals of said teaching should be. We ask: should the normativity of the researcher as teacher in ethics teaching be seen as something – as far as possible – to shun, as a regretful necessity or a positive contribution to reach the best learning outcome?

## Normativity

Let us begin by acknowledging that although environmental decision-making, *sensu lato*, can be based on science, it cannot be left to science alone:

> If you leave decisions about nature and environment to 'the experts' within science and technology, it does not mean that these decisions will be objective and neutral. It only means that the values and ethical assumptions that will shape the decisions are the ones held by the experts. (Des Jardin, 2013: 7)

The same can be said in relation to teaching applied ethics to applied science students at university level (by teaching, we also mean supervision on projects, theses, etc.), as values influence pretty much all aspects of a course. Here, we will flag three aspects: (1) 'course design and planning', that is decisions on how to prioritize the course: selection of theories, concepts, texts, and cases; and how the role of ethics is presented overall (meta-ethical questions) like: is ethics a 'hammer' or a 'torch' (see e.g. Gjerris, 2009). (2) 'course execution' e.g. choice of lecture format, guest lecturers; whether the emphasis is on discussions, exercises, or project/group work, etc. (3) 'course examination', e.g. formulation of exam questions, choice of exam format; and assessment of student performance. Thus, we can link normativity to course organization, didactics and curriculum. The first place to find inspiration on how to deal with this normativity will be in the concept of Responsible Conduct of Research.

## Normativity and responsible conduct of research (RCR) in relation to teaching

Since the 1970s, there has been a growing interest in responsible conduct of research (here seen as the same as research integrity) both within the research community and society as such caused by, among other things, a number of high-profiled scandals (Jensen *et al.,* 2020). In a Danish context, this has resulted in 'The Danish Code of Conduct for Research Integrity' where three principles that should govern all aspects of research are stated: Honesty, transparency and accountability. This mirrors to a large degree other international endeavours such as e.g. The European Code of Conduct for Research Integrity (ALLEA, 2017) and The Singapore Statement on Research Integrity (WCRI, 2010). By following these principles when planning, performing, and communicating research the aim is to ensure the trustworthiness, credibility and reliability of research, researchers, and research institutions. In turn, this should result in better research as well as continued societal support for research (Ministry of Higher Education and Science, 2014).

One area where the importance of these principles is very visible when discussing the normativity of researchers is 'conflicts of interests'. Most will readily agree that economic interests can influence the way research is planned, performed and/or communicated. However, it seems just as obvious that values can play a role e.g. if an ethicist provides an analysis of the ethical issues related to stem cell research and at the same time holds values that are against the destruction of early foetuses. This is not the place to take the discussion of which conflicts of interests should be reported (Jensen *et al.*, 2020). Yet, it shows that if the overall aim is to follow acknowledged principles of RCR then one needs to consider how to be honest, transparent and accountable in one's communication of research, both in an academic and a public setting. This also goes for the communication of science that takes part when researchers are involved as advisors in policy processes.

An ideal of university teaching is that it should be research based. It seems to us that reflection on responsible conduct is just as relevant when teaching, not least when seen as a form of research communication. Thus, the RCR principles of honesty, transparency and accountability can serve to maintain trust in the knowledge that the teacher present to the students, and trust in the teacher and/or the university as an institution. Another justification is that it reflects seeing students as citizens, respecting their integrity and the democratic processes as such, as well as treating students as ends in themselves. This opens up for a discussion of what role the teacher should be and how the inherent normativity should be dealt with.

## Normativity and teacher roles

We will describe four archetypes of 'the ethics teacher' based on Pielke's (2007) framework for science-policy interface that deals with the issue of normativity. According to Pielke's model, there are four possible main roles of the researcher in the interface with policy and politics (here only briefly described). (1) 'Pure scientist', assuming a distanced role, being passive or pursuing no targeted policy communication, not helping applying results to specific societal problems and indifferent about the use of her research in policy making. (2) 'Science arbiter', eliciting no specific recommendations to policy makers, consider herself a resource, giving answers when (factual) questions are asked by policy makers, but is not active in framing the issues. (3) 'Issue advocate', providing specific recommendations, and, importantly, making the case for one alternative over another based on own values. Finally, (4) 'Honest broker', serving balanced information about the range of options facing the policy maker, and clarifying choices to facilitate the process of decision-making. In real life, in different situations or across time, the same person might adopt one or more roles. Pielke identifies issues of normativity in all four roles. However, the fourth type, the honest broker, which Pielke endorses, seems to be the one closest to the RCR principles described before and the best way to ensure a legitimate democratic process.

We can translate the four roles into the domain of teaching applied ethics to applied science students: (1) 'The neutral presenter', not offering any normative ethical position, not targeting communication to the applied science audience, and indifferent about the usefulness for the students' future context. Focus will be on presenting ethical theories or history of philosophy. (2) 'The ethics arbiter', providing no specific normative position as the basis for teaching, and expressing no opinion on own ethics preferences. Teaching here becomes kind of a Q&A session with the teacher not providing any framework to help the students create a comprehensive understanding of the field. (3) 'The issue advocate', having an articulated position that is promoted, and giving specific recommendation on the choice of ethical positions to students. Goal of teaching is to instil specific normativity in students, e.g. in the shape of own values or a professional or institutionally instilled code of conduct. (4) 'The honest broker', presenting a broad portfolio of positions and values which are perceived to be relevant for the students and their future work, giving balanced information about the positions' strength, weaknesses and applicability seeking to help the students shape their own informed opinions, as well as challenging positions to make students reflect on their own view.

We can – admittedly somewhat speculative – try to link these types with how normativity is seen in teaching. Here, most likely the neutral presenter would see normativity in teaching as something to stay completely clear off, insisting on that he or she and the applied science students remain objective. The ethics arbiter would presumably see normativity as a regret, perhaps realizing the inevitability of normativity but only let it surface, if directly approached. The issue advocate would likely view normativity as unproblematic as the goal is to make the students adopt specific values, be it those of the teacher or more institutionally based values. The honest broker would, having the end goal of making students reflect, not be interested in forcing his or her own normativity on the students. Nonetheless, she will have to figure out how to deal with it as it will inevitably be part of the teaching process from course organization, to didactical choices, choice of curriculum and so on, as we argued earlier in this paper.

Adopting these roles or types of the ethics teacher – regardless of how realistic they are seen to be – present some challenges in relation to normativity, the RCR principles and as seen from a didactic perspective. The 'neutral presenter' runs the risk failing to engage the students as there is no attempt to show the relevancy of ethics to their context. More importantly in this context, there is a risk of lack of transparency, e.g. if failing to account for the selection of texts or explain that the history of philosophy looks different depending on where it is told from. At a more profound level, this type begs the question of whether it is at all possible to be neutral or objective, and by trying to come across as such, it may be interpreted as a lack of honesty. The 'ethics arbiter' (perhaps unintentionally) risks limiting the menu of ethics options offered and considered. Moreover, students may be never challenged on their value assumptions and the teacher's normativity will partly shape her answers, whereby certain positions will carry an undue weight. Hereby, she risks slipping into the role of the issue advocate, and this may be seen as a lack of honesty and transparency. The 'issue advocate's' own normative views may fill everything, as the goal is to teach students 'the right view'. Normativity might also come through in a more subtle way, e.g. by (mis)representing her favoured ethical theories, through text or case choices and unfairly representing alternative options. It may also be difficult for students to judge when the teacher is actively trying to persuade students to a certain view or merely provoke as a vehicle for learning. Thus, there are issues of missing transparency and credibility when/if it not made clear where the line is between ethics teaching and personal opinions, or, even worse, if it goes to serve her special interests. Finally, although we find that the honest broker is the right role model for an applied ethics teacher, challenges with normativity are still abundant. Ensuring balance and avoiding any form of bias is extremely demanding – and may be seen as an unattainable ideal rather than reflecting actual practice. As such, she may (unwillingly) slip into to the role of issue advocate. This, obviously, will a problem in relation to accountability and transparency. Moreover, avoiding presenting own views to avoid charges of this – to not seem unduly provocative, or to appear as balanced and credible as possible – may in the end not foster enough critical reflection among students, leading to students failing to both catch interest and develop their own views.

Inevitably, normativity is present in all of the four idealized roles. This should not come as a surprise as all communication is, to some extent, coloured by values. However, by carefully formulating and presenting what the strived for learning outcomes are while maintaining the ideals of transparency, honesty and accountability as the basis for Responsible Conduct of Teaching this challenge can, at least to some extent, be addressed. In the final section, we will come with suggestions to learning outcomes.

## What to aim for – best learning outcome

Based on the above analysis and based on our experience as applied ethics teachers for almost two decades as well as leaning on an ideal of teaching of developing and supporting independent, informed, empowered citizens and professionals, we suggest the following learning outcomes when teaching applied ethics to applied science students.

### *Knowledge*

- understand the relationship between ethics and science;
- understand different applied ethical theories and concepts;
- understand the complexity and value-ladenness of concepts such a, sustainability, integrity and nature.

### *Skills*

- analyse discourses and policies relating to one's profession/academic discipline for implicit ethical aspects;
- analyse societal ethical dilemmas related to one's profession/academic discipline and suggest ways to include them meaningfully them in decision-making processes;
- use knowledge of ethics in relation to other curricular activities, including MSc thesis.

### *Competencies*

- develop and maintain an informed ethical perspective regarding professional/academic discipline issues, and practice;
- critically reflect on different views/discourses on the use of the profession/academic discipline;
- structure and participate constructively in discussions between different value perspectives related to one's professional issues/academic discipline.

Clearly, these learning outcomes are broadly formulated, but it should be evident that the overarching goal is to challenge students, provide them with opportunities for critical reflection, and supply them with knowledge and 'tools' to engage with value-based issues. Finally, we find it is vital that they realize how normativity is present in their own professions/academic disciplines.

In our view, achieving these goals requires a plethora of didactical and value-based choices. Two important aspects, as we see it, which are part of responsible conduct in teaching in applied ethics at university level, are: (1) 'Normativity disclosure'. The teacher's own normative position should be disclosed through active use as an inspiring provocation/devil's advocate approach to ensure student engagement and maintain honesty and transparency. Moreover, although aiming at a balance in course organization, didactics and curriculum, dialogue with students on these issues should always be encouraged to safeguard transparency and accountability. (2) 'Heterogonous teaching teams'. Teaching in teams where different normative positions are present can be of benefit to students, both in terms of achieving the best learning outcome and to be able to highlight unwarranted normative biases in all aspects of teaching. Through this we believe that applied science students will be better equipped to identify, understand and engage with questions about ethics and values in relation to their professional agendas, e.g. in relation to identifying well-founded, transparent and socially robust strategies with regard to climate change – not least in relation to food security, agricultural production and environmental protection.

## References

ALLEA (2017): *The European Code of Conduct for Research Integrity. Revised Edition.* Berlin: ALLEA.

Des Jardins J.R. (2013): *Environmental Ethics: An Introduction to Environmental Philosophy, Fifth Edition.* Belmont, Calif: Wadsworth.

Funtowicz S.O. & Ravetz J. (1993): Science for the Post-normal Age. *Futures* 25(7): 739-755.

Gjerris, M. (2006): Breakfast at Milliways – The Didactic Challenges of Bioethics. *Dansk Universitetspædagogisk Tidsskrift* 1 (2): 45-52.

Gjerris M. (2009): This is not a Hammer: on Ethics and Technology, in Bedau MA and Parke EC (eds): *The Ethics of Protocells: Moral and Social Implications of Creating Life in the Laboratory.* Cambridge: MIT Press 291-305.

Jensen K.K., Andersen M.M., Whiteley L. & Sandøe P. (eds) (2020): *RCR – A Danish Textbook for Courses in Responsible Conduct of Research (4th ed.).* Frederiksberg: Department of Food and Resource Economics, University of Copenhagen.

Ministry of Higher Education and Science (2014): *The Danish Code of Conduct for Research Integrity.* Copenhagen: Ministry of Higher Education and Science.

Pielke R.A. Jr (2007): *The Honest Broker: Making Sense of Science in Policy and Politics.* Cambridge: CUP.

Stumann C.B. & Gamborg C. (2013): Reconsidering Social Science Theories in Natural Resource Management Continuing Professional Education. *Environmental Education Research* 20(4): 496-525.

WCRI (2010): *Singapore Statement on Research Integrity.* World Conference on Research Integrity.

# Section 6.
# Covid-19: new directions for ethics and food security?

# 62. Chat with your vet! Covid 19 and veterinary tele-medicine

*K.L. Weich*
*Messerli Research Institute, Unit of Ethics and Human-animal-studies, University of Vienna, Medical University of Vienna, University of Veterinary Medicine, Vienna, Austria; kerstin.weich@vetmeduni.ac.at*

## Abstract

Tele-Medicine had been marginalized in European veterinary medicine. While fostered in human medicine as part of a future-oriented digitalization, in veterinary medicine legal and ethical constraints impeded its implementation. In May 2020, the head of the Austrian veterinary chamber explains this situation by arguing that the direct physical examination cannot be substituted in veterinary medicine, as animal patients do not express their complaints by themselves. This places the direct reference to a fundamental difference between humans and animals, also referred to as 'anthropological difference' at the centre of the debate: The speechlessness of animals is used as an explanation for the rejection of tele-medicine in veterinary medicine. The momentum brought into the veterinary profession's policy-making by the Covid-19 crisis challenges this argumentation: guidelines on a possible integration of telemedical procedures into veterinary care have been published, thus initiating the professional acknowledgement and regulation of veterinary tele-medicine. As an effect, the many facets of the already established veterinary tele-medicine come to the foreground – and with them some serious failings regarding their technical-digital, legal and economic conditions. While supporting the call for a due investment in these regards, this paper argues, that the relevance of the 'anthropological difference' in this development should not be overlooked. In a concluding step, the implications of this argumentative framework become explored on a conceptual landscape. With Derrida, the debate of tele-medicine in veterinary settings can clearly be depicted as a case in which the question of the animal arises. Why should the insistence on animals as being speechless, not able to speak for themselves, impede veterinary tele-care? And how does the implementation of veterinary tele-medicine transform the idea of a 'speechless' animal patient? The speechlessness of 'the animal' is translated into the major concern for the potential loss of the 'physical examination'. This paper argues, that this an invitation to engage in more detail with the specific conditions of veterinary diagnosis and the notion of the (human and non-human) patient.

**Keywords:** digitalisation, anthropological difference, diagnosis, animal patient

## Covid-19 meets veterinary tele-health

E-health and tele-medicine have been fostered in European human medicine during the last two decades as part of a future-oriented digitalization (European Commission, 2004: WHO, 2010 EU). Especially in regard of the so-called 'age time-bomb', which stands for the increasement of elderly people in industrialized nations, e-health is being promoted as a fundamental means for providing solutions (Botrugno, 2019). In contrast, legal and ethical constraints were impeding its implementation in veterinary medicine (Held, 2020b; Stefan-Gromen, 2020; Magalhães-Sant'Ana *et al.*, 2020). Accordingly, in many European countries, tele-health and tele-medicine fall outside of what is considered acceptable veterinary practice by the professional regulator. However, the Covid-19 crisis has effected a change: guidelines on a possible integration of telemedical procedures into veterinary care have been published in several European countries with reference to their transitional nature and the requirement of an ethical evaluation (Reimers, 2020; Magalhães-Sant'Ana *et al.*, 2020). The most recent FVE position paper regarding veterinary tele-medicine caused a lot of discussions between the different members. The finally published position of the FVE is similar to the AVMA, recommending the members to allow the use of tele-medicine and set standards for its regulated use in the context of a veterinarian-


EurSafe 2021 – DOI 10.3920/978-90-8686-915-2_62, © Wageningen Academic Publishers 2021

client-patient-relationship (VCPR) (FVE, 2020; AVMA, 2020). The VCPR has proven to be a crucial criterion for defining veterinary tele-medicine in contrast to the broader terms of e-health or tele-health. In other words: the VCPR allows for distinguishing professional veterinary service from the broader (tele-)exchange on animal health, like a (private or commercial) website providing information e.g. on health-aspects of animal diets. Simultaneously, the VCPR itself is expected to change significantly in the course of digitalization. This renders the theme of VCPR as a pre-requisite for veterinary diagnosis especially relevant for the consideration of the ongoing transformation (BVA News Council, 17.12.2020; Limb, 2020).

Although tele-medicine has been excluded from professional regulation, digitalization – as the developments in the information and communication technologies (ICTs)– has not stopped outside the doors of the veterinary profession (Mars and Auer, 2006; Papageorgous *et al.,* 2001), although remaining underrepresented in veterinary education (Watson *et al.*, 2019). Benefits and threats have been considered (Chitty, 2019; Richards, 2019), and both more sceptical (Kogan *et al.,* 2007) as well as more optimistic opinions have been published (Hess, 2017). Specific tele-medical practices, already consolidated in different veterinary specializations, have been subjected to assessment (Bishop *et al.,* 2018; Roca and McCarthy, 2019). Remote veterinary services are well distributed among the so-called 'four pillars of telemedicine' – namely remote consulting, remote diagnosis, remote prescribing and third-party generated medical data (FVE, 2020).

Covid-19 crisis thus did not create a situation calling for a decision for or against tele-medicine. Instead, the many facets of the already established veterinary tele-medicine come to the foreground – and with them some serious failings. The profession's reluctance towards the acknowledgement of tele-medicine now seems to take its toll. Turned into a more positive perspective, one could also say, that the momentum brought into the veterinary profession's policy-making by the Covid-19 crisis will finally result in obvious and overdue next steps. Veterinary tele-medicine is lagging behind human tele-medicine, especially concerning the economic, technical-digital and legal conditions.

## Veterinary tele-medicine lag

The reimbursement of remote veterinary services also pushes for reconsideration. Empirical research has shown that a de-valuation of veterinary services may come along with its digitalization (Richards, 2019). This concern has to be taking seriously, as a well-known side-effect of digitalization. In a broader perspective, the push of Covid-19 to the recognition of veterinary tele-medicine asks for a reconsideration of the relevance of communication, information exchange and shared decision-making in veterinary services (Küper, 2021). Currently, e.g. in the German fee schedule for veterinarians, the single rate for a 15-minute consultation (both present and remote) is 7,04 €. The complexity of commodifying (remote) veterinary advice lies outside of the scope of this paper. But this information points out that the economic regulation is a key element for maintaining professional standards within a market situation, in which more and more (extra-professional) players successfully profit from a digitalized animal health market (Reimers, 2020).

Concerning the lag in technical-digital conditions in veterinary tele-health, the example of already established digital forms of vet-to-vet-counselling on clinical cases helps to identify next steps. In a recent Delphi-study on veterinary tele-medicine in Portugal, there is a report on one nation-wide veterinarian Facebook-group, rated as useful and supportive. Also, in the German-speaking part of Europe, many Facebook-groups for vet-to-vet-counselling do exist, which suggests a similar development to the Portuguese case. The Policy Delphi Study conducted in Portugal gives evidence, that veterinarians already profit from the remote exchange on diagnosis, indications and outcomes related to respective ongoing VCPRs. Simultaneously, the digital mediation of veterinary professional care raises concerns related to values like privacy, data-protection and the maintenance of professional standards (Magalhães-Sant'Ana *et al.*, 2020). At this point, the discussion of human tele-medicine provides an important lesson: Tele-medicine is to be considered as a transformation in working realities which will only succeed when work and money are invested into its development (Botrugno, 2019). Applied to the former example of vet-to-vet counselling using Facebook, the serious transformation towards tele-medicine would call for investment in the development of independent digital platforms and tools that foster the benefits of virtual counselling and exchange while fully securing professional standards, data privacy and complying to societal demands on veterinary professionalism.

Covid-19 has shown that despite the restrictions for veterinary tele-health, some forms of veterinary tele-health have been already established. During the pandemic crisis, the profession did profit from these veterinary experiences and skills, effecting a change in the profession's policy regarding tele-medicine. Next steps are directed to the making up of lost ground as well as to the exploitation of tele-medicine's potential for progress and improvement. Not only empirical research on already established forms of veterinary tele-medicine, but also the advance in human tele-medicine will contribute to fact-based and future-oriented policy-making. For the remaining part of this short paper, I will argue, that the due investment in constructive implementation of ICT should also include an investment in conceptual research.

## Veterinary tele-health and the question of the animal

Since the ongoing developments demand fact-based policies, the focus easily shifts away from the causes for the current gap in tele-medicine to the search to remedy its effects. Thus, the fundamental theoretical question of why tele-medicine has been fostered in human medicine while it has been restricted and excluded from medicine directed to animals seems to lose relevance vis-a-vis the practical task of identifying next steps for its implementation. This observation can be connected to an insight brought forward by the French philosopher Jacques Derrida (2002): he claims that one important precept of our general thinking is a fundamental difference between humans and animals, which is so profound that it becomes virtually invisible to us. In effect, all our reasoning and acting about animals starts from this distinction but rarely manages to analyse its character and justification. This foundational but unreflected distinction, also referred to as 'anthropological difference', unsurprisingly informs the ongoing debate of tele-medicine. Consequently, its articulation and analysis will help to better understand this debate as well as veterinary medicine's reluctance towards tele-medicine. In May 2020, the head of the Austrian veterinary chamber explained the unequal situation in veterinary and human tele-medicine by referring to the human-animal distinction as the allegedly obvious difference between both medical realms. More specifically, he argued, that the direct physical examination could not be substituted in veterinary medicine, since animal patients were not capable of expressing their complaints by themselves (Stefan-Gromen, 2020). A Portuguese veterinary officer, who affirms this argument, is cited in the already mentioned survey on tele-medicine: '[Clinical examination of the animal is essential in 99% of cases.] Unlike humans, the animal is unable to describe its symptoms and animal owners are unable to interpret them.' (Magalhães-Sant'Ana *et al.*, 2020: 3-4).

As an effect, tele-medicine becomes one of the many cases in which the question of the animal arises. It is the task of medical philosophy to prevent its fast disappearance by scrutinizing the specific intertwinement of the medical and the animal in the debate of tele-medicine and unfolding its impact. In a concluding step, I contour possible directions in this research in order to stimulate this endeavour.

## Exploring conceptual presumptions and implications

Before Covid-19, veterinary medicine mostly discarded tele-medicine. But the worldwide pandemic and ongoing political measures to counter it (i.e.: social distancing, quarantine, lockdown) forced it to re-consider the actual and potential use of tele-medicine. Thus, Covid-19 has brought unrest into a medical landscape so far being characterized by a strong 'yes' to human tele-medicine and a decisive 'no' to veterinary tele-medicine.

A simple 'no' as an answer, because of dealing with animals instead of humans, is no longer helpful. At this point the notion of the animal itself is turned from an answer to a question. Why should the insistence on animals as being speechless, not able to speak for themselves, impede veterinary tele-care? Wasn't 'the animal' already 'speechless' before the digital revolution took place? Wasn't the animal patient always reliant on vets and companions, articulating and interpreting (and technically mediating) its condition? The differing opinions on veterinary tele-medicine are rooted in the concern, that significant features of veterinary care become endangered. This concern is focused on the potential loss of the 'physical examination' of the animal. Veterinary tele-medicine is fine, as long as this direct contact between vet and animal patient has taken place, but it's dangerous whenever this element in veterinary care – and more precisely in veterinary diagnosis – is omitted. This is not an argument against this concern, but an invitation to engage in more detail with the specific conditions of veterinary diagnosis. What exactly is endangered here in comparison to human medicine? Which specific information-flow or knowledge-gaining has been developed in veterinary diagnosis, as medicine dealing with non-human patients? The pursuit of these and related questions promises to challenge the human-animal-border by making visible, that and how this border is constantly being breached and transformed in the medical realm.

In a broader perspective, the picture of an already advanced human tele-medicine in contrast to an underdeveloped veterinary medicine becomes connected to the question of the animal and how it is perceived and treated through the medical system. As depicted above, this picture supports a call to veterinary profession for catching up with the digital development. But how far and on which levels can this call be generalized? On an empirical and practical level, the idea of decreasing differences by pursuing a similar development as human medicine seems well justified and a useful orientation. Of course, one cannot simply take the developments in human tele-medicine and apply them as a general template to veterinary tele-medicine. But the careful investigations of similarities and differences between human and animal medicine when embracing the digital transformation seems to take place in a general frame which suggests a need for veterinary medicine to 'humanize'. We should be careful for the impacts of this configuration on a more conceptual level. For this purpose, it could be asked, whether the finding that the need for veterinary tele-medicine has pushed forward the idea of 'humanizing' the profession in this regard could be turned around. Is it possible to understand the debate on veterinary tele-medicine not as a cause, but as an effect of this idea? Representatives of the veterinary profession's regulatory authorities have defended the 'delay' in tele-medicine, arguing that it endangered the care for animal patients (e.g. for Austria: Stefan-Gromen, 2020): Because veterinarians care for their animal patients, tele-medicine could not be an option. Notably, this concern as well as its (empirical) invalidation is brought forward not only under the pressure of Covid-19, but also mostly in the field of companion animal medicine. Again, in how far does the idea of veterinary progress by 'humanizing' structure the debate? Human tele-medicine is fostered, its benefits and harms are carefully researched and debated. Now, we do face a similar challenge in veterinary tele-health and we feel that we stand at the beginning of a complex process, although there is no need to invent the wheel again. Caring for the conceptual presumptions and implications of this transformation, the question arises, how the notions of the patient and the animal become connected and transformed in this process. Which kinds of animal patients do raise concern and discussion? Is there a narrowing-down to the field of companion animal medicine taking place? Is this because of tele-medicine being an issue for human patients? Is tele-medicine accordingly in veterinary medicine only an issue in the companion animal medicine, being the 'most humanized' sector? The progress and the big developments in implementing ICTs in the veterinary management of livestock did not raise a similar debate (Neethirajan, 2017; Vial *et al.*, 2018). On which assumptions is this justified? Or is an exclusion taking place here, which is based on the unquestioned reproduction of a general anthropological difference, which is translated to a difference between 'real, human-like' animal patients (e.g. companion animals) and other medicalized animals (e.g. livestock or lab animals)? And if so, how was this exclusion actually put into practice? The exploration of the conditions of excluding and including certain animal patients to the debate on tele-medicine forms a significant agenda for the upcoming veterinary humanities as well as for research in medical philosophy applying a more-than-human perspective.

## References

AVMA. *Model Veterinary Practice Act.* (2020). Available online at: https://www.avma.org/resources-tools/avma-policies/model-veterinary-practice-act.

Bishop G.T., Evans B.A., Kyle K.L., *et al.* (2018). Owner satisfaction with use of videoconferencing for recheck examinations following routine surgical sterilization in dogs. J Am Vet Med Assoc 253:1151-1157.

Botrugno, C. (2019). Towards an ethics for telehealth. Nurse Ethics 26, 357-367. https://doi.org/10.1177/0969733017705004

BVA News Council (2020). Under Care and Remote Provision of Vet Services. *Veterinary Record* 187.12 (2020): 506.

Chitty, J. (2019). Telemedicine – the opportunity in the threat. Veterinary Record 184, 322-322. https://doi.org/10.1136/vr.l1072

Derrida, J. (2002). Das Tier, das ich also bin. Wien: Passagen.

FVE position and recommendations on the use of telemedicine, 2020. Available online at: https://fve.org/cms/wp-content/uploads/049-Telemedicine_position-paper_adopted-Nov-GA-3.pdf.

Hess L. (2017). Telemedicine: The Future of Veterinary Practice. J Avian Med Surg. 2017 Jun;31(2):165-171.

Kogan L., Ruch-Gallie R., Schoenfeld-Tacher R. (2020). Veterinarians' use and perceptions of information and communication technologies. Med Res Arch. 2016;4(2):1-29.

Limb, M. (2020). Emerging ideas on changes to 'under our care. VetRec, 30 May/6 June: 550-551

Küper A, Merle R. (2021). Partners in sickness and in health? Relationship-centered veterinary care and self-educated pet owners in Germany: A Structural Equation Model Front. Vet. Sci. | https://doi.org/10.3389/fvets.2020.605631

Magalhães-Sant'Ana, M., Peleteiro, M.C., Stilwell, G. (2020). Opinions of Portuguese Veterinarians on Telemedicine – A Policy Delphi Study. Front. Vet. Sci. 7, 549. https://doi.org/10.3389/fvets.2020.00549

Neethirajan, S. (2017). Recent advances in wearable sensors for animal health management. Sensing and Bio-Sensing Research 12, 15-29. https://doi.org/10.1016/j.sbsr.2016.11.004

Papageorges M., *et al.* (2001). How to make telemedicine work. Clin Tech Small Anim Pract. May;16(2):111-114.

Richards, I. (2019). Telemedicine – the threat in the opportunity. Veterinary Record 185, 25-25. https://doi.org/10.1136/vr.l4535

Reimers, J. (2020). Corona ebnet Weg für Tele-Medizin. VetImpulse 29/9: 1-3.

Roca, R.Y., McCarthy, R.J. (2019). Impact of Telemedicine on the Traditional Veterinarian-Client-Patient Relationship. Topics in Companion Animal Medicine 37, 100359. http://dx.doi.org/10.1016/j.tcam.2019.100359

Stefan-Gromen S. (2020). Tierärztliche Online-Sprechstunden im Fokus der Kritik. Available online at: https://www.ots.at/presseaussendung/OTS_20200521_OTS0049/tieraerztliche-online-sprechstunden-im-fokus-der-kritik.

Vial, F., Székács, A., Quealy, S. (2018). Editorial: Digital Transformation of Animal Health Data: Proceedings of the AHEAD 2017 Workshop. Front. Vet. Sci. 5, 111. https://doi.org/10.3389/fvets.2018.00111

Watson, K., Wells, J., Sharma, M., Robertson, S., Dascanio, J., Johnson, J.W., Davis, R.E., Nahar, V.K. (2019). A survey of knowledge and use of telehealth among veterinarians. BMC Vet Res 15, 474. https://doi.org/10.1186/s12917-019-2219-8

World Health Organization (2010). Telemedicine. Opportunities and developments in member states. Report on the second global survey on e-Health, Geneva, 2010. Available at: http://www.who.int/goe/publications/goe_telemedicine_2010.pdf.

# 63. Epidemics and food security: the duties of local and international communities

*A.K. Martin*
*Philosophisches Seminar, University of Basel, Steinengraben 5, 4051 Basel, Switzerland; angela.martin@unibas.ch*

## Abstract

Over 60% of all epidemics have a zoonotic origin, that is, they result from the transmission of infectious diseases from animals to humans. The spill-over of diseases often happens because humans exploit and use animals. In this article, I outline the four most common interfaces that favour the emergence and spread of zoonotic infectious diseases: wildlife hunting, small-scale farming, industrialised farming practices and live animal markets. I analyse which practices serve human food security – and thus have a non-trivial purpose – and which predominantly have an economic purpose or serve as a symbol of wealth. I conclude that many practices that increase the risk for zoonotic infectious disease outbreaks actually do not contribute to the survival and food security of humans. I make two arguments in turn. First, I argue that in cases where the consumption and use of animal products does not contribute to the food security of a population, then this population has a duty to abstain from them, since they impose a grave and avoidable risk to themselves as well as to innocent third-parties. However, some communities must sometimes rely on practices that increase the risk of emergence and transmission of zoonotic infectious diseases, because they have no healthy alternatives. That is, the food security of the local population depends on the consumption and use of risky animal products. The second argument I advance is that, in such scenarios, the international community has a duty to provide the communities concerned with alternative food options, as well as economic and educational opportunities and technologies, in order to reduce the spill-over risk of infectious diseases. Given that abstention from such practices contributes significantly to the public good and benefits the international community, the latter has a corresponding duty to provide local communities which abandon such practices with alternatives.

**Keywords:** zoonotic infectious diseases, risk ethics, political duties, ethics of meat consumption

## Introduction

Epidemics and pandemics can cause huge economic losses and great suffering – not only in the regions where they originate, but also in distant countries. This has been shown most prominently by the ongoing Covid-19 pandemic, which has caused economic losses of a still-unknown extent to many governments, businesses and individuals. It has also led to the death of over 1.5 million people, as well as causing long-lasting health issues in millions of people. Furthermore, it has caused intangible suffering all over the world in the form of lost family members and friends. Finally, the mental health of many individuals has substantially deteriorated (Cullen *et al.*, 2020; Kwong *et al.*, 2020), due to local lockdowns and reduced social contact, among other factors.

In what follows, I evaluate the key drivers of zoonotic disease transmission from an ethical perspective, taking into account the human interests at stake. Broadly speaking, there are four common modes of transmission of zoonotic disease: (1) the hunting and consumption of wild animals (often referred to as 'bush meat'); (2) backyard farming, that is, small family farms; (3) the intensive farming of livestock or wild animals; and (4) live-animal markets (sometimes referred to as 'wet markets') (Espinosa *et al.*, 2020; Horby *et al.*, 2014; Magouras *et al.*, 2020).


EurSafe 2021 – DOI 10.3920/978-90-8686-915-2_63, 

I will begin by showing that practices that commonly increase the transmission risk of zoonotic infectious diseases are often not crucial for the survival and food security of the population concerned. Then, basing myself on an argument made by Jones (2020), I will defend the following claim: moral agents have a duty to avoid imposing unnecessary and serious risks on non-consenting third-parties. If there is no nutritional or economic necessity for a local population X to rely on products that substantially increase the risk of zoonotic infectious disease outbreaks, then such practices are avoidable and unnecessary. This provides population X with a strong reason to avoid such practices – for their own sake, but also for the sake of all those who could be detrimentally affected by a potential epidemic.

However, some uses of animals that substantially increase the risk of epidemics may be necessary for the economic survival, food security and health of a local population. In such cases, according to the second thesis I defend in this article, the international community has a duty to provide this population with alternatives – e.g. economic resources, food, or educational training – allowing them to lead a healthy life and earn a livelihood in an alternative, low-risk way. Finally, I will defend this claim against the objection that this could create harmful dependencies.

## The drivers of zoonotic infectious disease transmission

It is estimated that over 60% of all infectious diseases in humans and 75% of all newly emerging infectious diseases have a zoonotic origin, that is, their root cause is a transmission of disease from nonhuman animals (hereafter: animals) to humans (Jones *et al.*, 2013; Vorou *et al.*, 2007). The reservoirs of these diseases mostly lie in wild animals, but domesticated animals frequently serve as a bridge for disease transmission to humans. As the United Nations Environment Programme (2016) notes:

> Never before have so many animals been kept by so many people – and never before have so many opportunities existed for pathogens to pass from wild and domestic animals through the biophysical environment to affect people causing zoonotic diseases or zoonoses.

Roughly speaking, we can distinguish four main interfaces that drive the emergence and spread of zoonotic disease (Cutler *et al.*, 2010; Espinosa *et al.*, 2020; Magouras *et al.*, 2020):

1. Wildlife hunting: In some South Asian and African countries, local populations hunt, transport, butcher and consume wild animals on a large scale, which creates a substantial risk for the cross-species transmission of infectious diseases (Magouras *et al.*, 2020; Wolfe *et al.*, 2005). It is estimated that over 5 million tons of wild mammal meat feed millions of people in Neotropical and Afrotropical forests every year (Fa *et al.*, 2002). In addition to being a source of meat, bushmeat hunting generates income (Kümpel *et al.*, 2010; Wright and Priston, 2010), animal trophies (Lindsey *et al.*, 2007) and animal-based medicines (Alves and Alves, 2011).
2. Backyard farming: Small-scale backyard or family farming is a key pillar of food security in countries that lack easy access to plant-based protein (Espinosa *et al.*, 2020). Due to the often close proximity between farmed animals and wild animals, backyard farming carries a substantial risk for epidemic outbreaks (Henning *et al.*, 2011; Wang *et al.*, 2013).
3. Intensive farming: Over the last seventy years, small-scale farming has given way to massive facilities which can contain thousands of animals (both wild and domesticated) in crowded spaces (Graham *et al.*, 2008). The expansion and intensification of human agricultural practices (amongst other things due to increasing demand for meat) promotes encroachment into the habitats of wild animals, which in turn leads to ecosystem changes, bringing humans, wild animals and livestock into closer proximity. Dense living conditions and lack of genetic diversity, in turn, facilitate the rapid dissemination of diseases amongst farmed animals (Jones *et al.*, 2013; Horby *et al.*, 2014; Espinosa *et al.*, 2020).

4. Live animal markets: Some markets, particularly in certain Asian countries, sell live animals, both domesticated and wild (Magouras *et al.,* 2020). These markets often operate under poor hygienic conditions, which increases cross-contamination risks (Lo *et al.,* 2019; Sekoai *et al.,* 2020). In many cases, the consumption of wild animals sold on such markets does not seem to reflect nutritional or economic necessity, but rather amounts to 'a fashionable lifestyle and symbol of elite status' (Li Zhang and Yin, 2014), which is correlated to higher income and higher educational attainment.

There are other factors that increase the risk of the emergence and spread of zoonotic infectious diseases, such as living with domesticated animals. However, those listed here are among the most prominent ones. These uses of animals could be considered as morally problematic with regard to the welfare of the animals concerned: for example, dense living conditions on farms, without any opportunity to engage in species-typical behaviour, could be seen as detrimental to animal welfare. However, I will leave such considerations aside in this article. In what follows, my arguments are entirely based on the risks posed to human health by such practices.

## The duty to avoid the non-necessary use of animals

The four practices outlined above carry a substantial risk for the emergence and spread of zoonotic infectious diseases and may result in epidemics. Their spread may, in turn, affect not only the local population, but also individuals in other parts of the world (as it is the case of the current Covid-19 pandemic). That is, these activities may cause substantial harm to the health, well-being and economic welfare of millions of people.

The fact of exposing others to unwanted and unsolicited risks that are detrimental to their health, well-being and economic welfare is morally relevant. Jones (2020) illustrates this point with an analogy: drink driving is morally wrong for three reasons. First, it poses an 'exceptionally high risk' of harm to others. Second, this risk is 'unnecessary', as there are alternatives to drink driving (driving sober, letting someone else drive or taking public transit). Third, the risk it carries is 'grave': drink driving can result in severe injuries and death, both to the driver and to others.

According to Jones, the same principle applies to the consumption of meat: in many cases, the consumption of meat imposes an exceptionally high, grave and unnecessary risk on others (as there are, in many cases, alternatives to meat). He gives three justifications for this claim: first, we have a duty to prevent individual harm to others if we can – we should not impose grave risks on others without their consent; second, we have a duty to avoid being complicit in collective harm – that is, we should refrain from consuming meat products that come from an interface with an increased risk for zoonotic disease emergence and spread; and third, there are considerations of fairness: not consuming meat deriving from risky sources for infectious diseases represent a huge contribution to the public good, insofar as the emergence and spread of such diseases is prevented.

Not all uses of animals that are correlated with an increased risk of EID transmission enhance the food security of the local population. As shown above, wildlife hunting, for example, is often an income-generating activity, and thus only indirectly linked to food security (Kümpel *et al.,* 2010; Wright and Priston, 2010). The consumption of wild animals from so-called wet markets is frequently a symbol of wealth, and unrelated to nutritional or economic necessity (Li Zhang and Yin, 2014). Additionally, many people currently consume more meat than they should from a health perspective, especially in industrialised countries (Battaglia Richi *et al.,* 2015; González *et al.,* 2020; Walker *et al.,* 2005). If we can refrain from activities and behaviours that impose risks on others, because there are lower-risk or no-risk alternatives, then we have a duty to do so.

Admittedly, some of the practices outlined above have a long tradition and significant cultural value. However, this value is not absolute, especially when it puts the health and welfare of others at risk and alternatives exist. That is, there is a duty to abstain from such activities when they do not contribute to food security and thus to the direct survival of the local population. The reason for this is that the maintenance of cultural traditions is less important from an ethical perspective than reducing potentially fatal risks for others.

## The duties of the international community in no-alternative scenarios

There may be populations that rely on meat products to meet their nutritional needs. Imagine a scenario in which there is no alternative for a population or community X to products that increase the risk of infectious zoonotic disease outbreaks. What should be done in such cases?

The international community 'benefits' if community X reduces its consumption of products that increase the risk of emergence and spread of infectious diseases. Abstaining from such products thus contributes substantially to the public good, insofar as it reduces risk for everyone – not only for the local population, but also the international community. However, it would be disproportionate to require from community X that it sacrifice its personal well-being and health 'just' in order to reduce risk for the community and for others: that is, the welfare of such populations is more important than diminishing a 'mere' risk. However, given that they have no alternative food options, and given that, all things considered, it is preferable for all those who are potentially affected that EID events not take place, the international community (such as the governments of wealthier countries) should step in: they should provide community X with the means for risk-reduction which would otherwise not be feasible. In practice, this means that the international community should provide population X with no-risk or lower risk, affordable and healthy food alternatives (such as plant-based proteins or staple foods), but also with training opportunities and economic resources that would allow population X to turn to food and income options that do not present a risk for both themselves and others. For example, non-meat proteins could be promoted and support for more sustainable agricultural practices could be offered (Wright and Priston, 2010).

One might object here that this would lead to dependency on the part of local communities and states, which is morally problematic. If a population becomes dependent on the international community for basic goods, such as food, then the population loses its independence and becomes more vulnerable to exploitation, because the relationship is asymmetric.

Two points are worth noting here. First, as outlined above, I do not necessarily advocate a transfer of food alternatives alone. Rather, local populations can also benefit from educational training and offers of financial and other resources allowing them to turn to practices that do not carry a high risk for the emergence and spread of zoonotic infectious diseases. This would reduce their dependency on the international community for their food security. Second, the relationship between community X and the international community need not be asymmetric. Those providing a population X with alternatives to risky meat products 'benefit' from the sacrifices of the local population who use other food sources – and so does the local population, which faces a lower risk for disease outbreaks. That is, their relationship is not a one-way street, since both sides benefit from each other.

## Conclusion

In this article, I made two arguments. First, I argued in favour of a duty to avoid products that carry a high risk for zoonotic infectious disease outbreaks, whenever the consumption of such products is unnecessary. Second, I argued that, in cases where the avoidance of such products is not possible (for example, due to a lack of alternatives), there is a duty on the part of the international community to provide local populations with alternatives – with respect not only to food, but also to economic opportunities and training. By adopting such an approach, the risk of zoonotic infectious disease outbreaks may be lowered substantially – which is beneficial not only to the community in which the outbreak was likely to happen, but also to people living far away who might have been negatively affected by a potential epidemic.

## References


Alves, R.R.N. and Alves, H.N. (2011). The faunal drugstore: animal-based remedies used in traditional medicines in Latin America. Journal of Ethnobiology and Ethnomedicine 7(9).

Battaglia Richi, E., Baumer, B., Conrad, B., Darioli, R., Schmid, A. and Keller, U. (2015). Health risks associated with meat consumption: A Review of Epidemiological Studies. 85(1-2): 70-78.

Cullen, W., Gulati, G. and Kelly, B.D. (2020). Mental health in the COVID-19 pandemic. QJM: Monthly Journal of the Association of Physicians 113(5): 311-312.

Cutler, S.J., Fooks, A.R. and van der Poel, W.H.M. (2010). Public health threat of new, reemerging, and neglected zoonoses in the industrialized world. Emerging infectious diseases 16(1): 1-7.

Espinosa, R., Tago, D. and Treich, N. (2020). Infectious Diseases and Meat Production. Environmental & Resource Economics 76: 1019-1044.

González, N., Marquès, M., Nadal, M. and Domingo, J.L. (2020). Meat consumption: Which are the current global risks? A review of recent (2010-2020) evidences. Food Research International 137: 109341.

Graham, J.P., Leibler, J.H., Price, L.B., Otte, J.M., Pfeiffer, D.U., Tiensin, T. and Silbergeld, E.K. (2008). The animal-human interface and infectious disease in industrial food animal production: rethinking biosecurity and biocontainment. Public Health Reports (Washington, D.C.) 123(3): 282-299.

Henning, J., Henning, K.A., Morton, J.M., Long, N.T., Ha, N.T., Le Vu, T., Vu, P.P., Hoa, D.M. and Meers, J. (2011). Highly pathogenic avian influenza (H5N1) in ducks and in-contact chickens in backyard and smallholder commercial duck farms in Viet Nam. Preventive Veterinary Medicine 101(3): 229-240.

Horby, P.W., Hoa, N.T., Pfeiffer, D.U. and Wertheim, H.F.L. (2014). Drivers of Emerging Zoonotic Infectious Diseases. In A. Yamada, L.H. Kahn, B. Kaplan, T.P. Monath, J. Woodall and L. Conti (eds) Confronting Emerging Zoonoses, Tokyo: Springer Japan, 13-26.

Jones, B. (2020). Eating meat and not vaccinating: In defense of the analogy. Bioethics (online first).

Jones, B.A., Grace, D., Kock, R., Alonso, S., Rushton, J., Said, M.Y., McKeever, D., Mutua, F., Young, J., McDermott, J. and Pfeiffer, D.U. (2013). Zoonosis emergence linked to agricultural intensification and environmental change. Proceedings of the National Academy of Sciences of the United States of America 110(21): 8399-8404.

Kümpel, N.F., Milner-Gulland, E.J., Cowlishaw, G. and Rowcliffe, J.M. (2010). Incentives for hunting: The role of bushmeat in the household economy in rural Equatorial Guinea. Human Ecology 38(2): 251-264.

Kwong, A.S.F., Pearson, R.M., Adams, M.J., Northstone, K., Tilling, K., Smith, D., Fawns-Ritchie, C., Bould, H., Warne, N., Zammit, S., Gunnell, D., Moran, P., Micali, N., Reichenberg, A., Hickman, M., Rai, D., Haworth, S., Campbell, A., Altschul, D., Flaig, R., McIntosh, A.M., Lawlor, D.A., Porteous, D. and Timpson, N.J. (2020). Mental health before and during COVID-19 in two longitudinal UK population cohorts. The British Journal of Psychiatry: The Journal of Mental Science. (Online first)

Li Zhang and Yin, F. (2014). Wildlife consumption and conservation awareness in China: a long way to go. Biodiversity and Conservation 23(9): 2371-2381.

Lindsey, P.A., Roulet, P.A. and Romañach, S.S. (2007). Economic and conservation significance of the trophy hunting industry in sub-Saharan Africa. Biological Conservation 134(4): 455-469.

Lo, M.Y., Ngan, W.Y., Tsun, S.M., Hsing, H.-L., Lau, K.T., Hung, H.P., Chan, S.L., Lai, Y.Y., Yao, Y., Pu, Y. and Habimana, O. (2019). A Field Study Into Hong Kong's Wet Markets: Raised Questions Into the Hygienic Maintenance of Meat Contact Surfaces and the Dissemination of Microorganisms Associated With Nosocomial Infections. Frontiers in Microbiology 10: 2618.

Magouras, I., Brookes, V.J., Jori, F., Martin, A., Pfeiffer, D.U. and Dürr, S. (2020). Emerging zoonotic diseases: Should we rethink the animal-human interface? Frontiers in Veterinary Science 7: 582743.

Sekoai, P.T., Feng, S., Zhou, W., Ngan, W.Y., Pu, Y., Yao, Y., Pan, J. and Habimana, O. (2020). Insights into the microbiological safety of wooden cutting boards used for meat processing in Hong Kong's wet markets: A focus on food-contact surfaces, cross-contamination and the efficacy of traditional hygiene practices. Microorganisms 8(4): 579.

United Nations Environment Programme (2016). UNEP frontiers 2016 report: Emerging issues of environment concern. Nairobi, Kenya: UNEP.

Vorou, R.M., Papavassiliou, V.G. and Tsiodras, S. (2007). Emerging zoonoses and vector-borne infections affecting humans in Europe. Epidemiology and Infection 135(8):1231-47.

Walker, P., Rhubart-Berg, P., McKenzie, S., Kelling, K. and Lawrence, R.S. (2005). Public health implications of meat production and consumption. Public Health Nutrition 8(4): 348-356.

Wang, Y., Jiang, Z., Jin, Z., Tan, H. and Xu, B. (2013). Risk factors for infectious diseases in backyard poultry farms in the Poyang Lake area, China. PLOS ONE 8(6): e67366.

Wolfe, N., Daszak, P., Kilpatrick, A.M. and Burke, D. (2005). Bushmeat hunting, deforestation, and prediction of zoonotic disease. Emerging Infectious Disease Journal 11(12): 1822-1827.

Wright, J.H. and Priston, N.E. (2010). Hunting and trapping in Lebialem Division, Cameroon: bushmeat harvesting practices and human reliance. Endangered Species Research 11: 1-12.

# 64. The ethics of intensive agriculture: lessons from the Covid-19 pandemic

*O. Torpman* and H. Röcklinsberg*
*Department of Animal Environment and Health, Swedish University of Agricultural Sciences, P.O. Box 7068, 750 07 Uppsala, Sweden; olle.torpman@slu.se*

## Abstract

Modern intensive agriculture, including both forestry and intensive animal farming, imposes an increased risk of cross-species diseases, so-called zoonoses. When animals are kept in small spaces, and when deforestation pushes wild animals closer to humans, there is a non-negligible risk that zoonotic diseases are developed and spread to humans. While much has been said about the ethical aspects of these types of intensive agriculture in general, less has been said about the ethical aspects of its risks related to zoonoses in particular. The aim of this paper is to highlight some ethical challenges for intensive agriculture that come with such risks, with a special interest in animal farming. In doing so, we analyse the implications from some major moral perspectives. Our hypothesis is that the risks imposed by intensive agriculture, with respect to diseases that affect humans, imply that adjustments are needed. In effect, we argue that this requires substantial animal welfare improvements in intensive animal farming. We moreover argue that this conclusion can be reached without taking animals themselves into direct consideration.

**Keywords:** animal ethics, intensive animal farming, SARS-CoV-2, zoonoses

## Introduction

Today's Western diet relies on intensive agriculture, which is in turn is shown to give rise to biodiversity loss (Bajzelj *et al.*, 2014; Röös *et al.*, 2016), have a negative impact on the climate (IPCC, 2014; Rockström *et al.*, 2009; Steinfeld *et al.*, 2006), and, known since many years, to have a negative impact on animal welfare (Brambell, 1965; Harrison, 1964). It is also established that if intensive agriculture, especially animal based food production, relies on high levels of antibiotic use, which contributes to increased bacterial resistance threatening modern health care (Bengtsson and Greko, 2014). Numerous ethical studies have shown how these aspects of intensive animal industry affect its justifiability, and that current methods of intensive agriculture therefore need to be improved in several respects.

It has recently been observed that intensive agriculture increases the risk of diseases, which affect both humans and animals (Dhingra *et al.*, 2018; Wallace, 2016). For instance, modern intensive agriculture tends to drive deforestation, which in turn pushes wild animals closer to humans, implying an increased risk of spreading zoonotic diseases. The same risks have been observed in intensive animal farming, where animals are kept close together in small spaces. Some examples of diseases that have been developed and spread thus are African Swine Fever, porcine reproductive and respiratory syndrome (PRRS), bovine spongiform encephalopathy (BSE) (related to Creutzfeldt-Jacob Disease in humans), N5H1 (Asian Avian Influenza) and, most recently, Covid-19 (SARS-CoV 2). However, the ethical implications of the spread of diseases for intensive agriculture need to be discussed, and the aim of this paper is to uncover some ethical implications of intensive agriculture in relation to transmissions of diseases.

Our focus is on effects of prevalence of *zoonoses*, which are typically defined as infectious diseases that have jumped from non-human animals to humans. For the sake of argument, we will in this paper assume an anthropocentric perspective, which is in line with the outset of the United Nations' Sustainable


EurSafe 2021 – DOI 10.3920/978-90-8686-915-2_64, © Wageningen Academic Publishers 2021

Development Goals (for criticism, see Torpman and Röcklinsberg, 2021; Visseren-Hamakers, 2020). We shall thus argue that the current pandemic gives us reason to substantially rethink the way current agriculture exploits the animals and the environment, even if we think only about humans and thus disregard any direct concern for the animals or the environment. We will, however, say something about what a non-anthropocentric perspective would imply regarding the same issue.

## The ethics of intensive agriculture

One of the first wake up calls for reflection on the industrialization of farming practices was Ruth Harrison's description of 'factory farming' in her book Animal Machines (Harrison, 1964). Her book led to a committee working on the issue of animal welfare, and publication of the so called Brambell report and the formulation of the 'five freedoms' (Brambell, 1965; Farm Animal Welfare council, 1993) also visible in today's legislation in the EU (Council Directive 98/58/EC; European Convention for the Protection of Animals kept for Farming Purposes).

From an ethical point of view, intensive agriculture has been criticized by both animal and environmental ethicists. Among the first were the philosophers Tom Regan and Peter Singer who edited a book together in 1976 (Regan and Singer, 1976). Although holding different moral theories (animal rights and utilitarianism, respectively), they agree that sentience is the most convincing criterion for moral standing, and hence argue that farm animals should be given direct ethical consideration. Criticism of intensive farming is thus not a new phenomenon, but sometimes considered as 'too radical'. However, given the current global pandemic with the spread of SARS-CoV-2 from animals to humans (OIE, 2021; Oude *et al.,* 2021), it is hard to claim that such reflections are radical. Rather, it is highly relevant not only to scrutinize the empirical evidence for the causes and effects related to intensive agriculture and dissemination routes, but also to investigate the ethical implications of the spreading of these diseases. This is not least due to its risks of negative effects on humans.

It has been argued that the current global Covid-19 pandemic was foreseeable, that its occurrence is related to the shift from small-farming to an industrialized format in China, and that the specialized breeding with small genetic variations between individuals create a perfect context for a virus to spread (Dhingra *et al.,* 2018; Wallace, 2016). Even so, it is impossible to know of all effects in advance. This normally leaves the ethical elaboration with a focus on what is already known – such as issues related to facts concerning animal welfare, human health, food safety, and farmers' profitability. These factors are known to be negatively affected by standard farming practices, which hence make these vulnerable to new challenges.

## Effects of intensive agriculture on prevalence on zoonoses

In previous research, much focus has been put on the effects of the use of antibiotics in animal farming, and how this has negative effects on human well-being in the long run. Most noticeably, the heavy use of antibiotics in animal farming leads to antibiotic resistance that affects human health care negatively (Bengtsson and Greko, 2014; Goodard *et al.,* 2017). Even if previous diseases (such as SARS, MERS, the Mad Cow Disease, the Avian Flu/N5H1, PRRS, etc.) have been studied to a greater extent than more recent diseases, the ethical implications of these diseases regarding farming practices has generally not been in focus. Presumably, this is because these diseases have threatened human lives, which has taken focus from their overall ethical implications – including long term social, environmental, economic, and animal welfare effects (Laxminarayan *et al.,* 2013).

A literature search on Web of Science (January 2021) combining farming, husbandry, industrial animal products with names of the relevant diseases (avian flu (or N5H1), African Swine Fever (ASF), BSE and Creutzfeldt-Jacob Disease (one by one)) led to only a few articles. Regarding risk for BSE in cattle (and Creutzfeldt-Jacob Disease in humans) a couple of older studies of databases for production levels and farm sizes in France show a correlation between intensive farming system and increased risk of BSE (Bonnardière *et al.*, 2006, Ducrot *et al.,* 2006). There was an enhanced BSE risk for the high producing herds as compared to the least producing herds but no link between herd size and BSE, apart from higher risk of at least one affected cow related to higher herd sizes (Ducrot, 2006). It is hence possible to see a connection between BSE and production level, which in turn can be described as one essential factor in intensive farming. As to the question of how consumers or citizens perceive the relation of BSE, healthy food and farming practices, Sibley concluded (2001) that consumers and government demand safe food through animal health and welfare, which increase veterinary involvement on livestock farms. Interestingly, none of these studies seem to have been cited, which might mirror a limited interest about the link between farming systems (including feed regimes) and outbreaks of diseases. A search for the other diseases mentioned above ended in no, very few or irrelevant results, which might be due to the fact that the question of farming system and health (both animal and human) is an emotional and politically sensitive issue.

Taking the abovementioned factors together (the SARS-CoV2 challenge to global health, its origin in intensive forestry, the potential connection between disease prevalence and industrial farming methods (dairy/BSE), the susceptibility of industrial farms for virus and the risk of intensifying up of the virus's virulence) inevitably calls for ethical reflection. In what follows, we will make a brief ethical assessment of the risks of diseases due to intensive agriculture. In doing so, we shall look at this particular issue from three major ethical perspectives: consequentialism, deontology, and virtue ethics.

## An ethical assessment

### *Consequentialism, intensive animal farming, and the risks of zoonoses*

From a consequentialist perspective, intensive animal farming can be justified only if its overall consequences are better than its alternatives. It is typically very hard to determine whether a choice of action is justified on consequentialist grounds, since it is typically very hard to determine its consequences and their net value. However, even prior to the Covid-19 pandemic, the justifiability of intensive agriculture was questioned – on the grounds that it involves animal suffering and makes a substantial contribution to climate change. This critique is most often directed against intensive animal farming, such as, e.g. the meat, dairy, and fur industries. Now, in the midst of the pandemic (February 2021), more than 2 million humans have died, and numerous citizens are living in lockdowns and are suffering from restrictions of democratic rights. On the face of it, it is quite clear that the negative aspects of intensive agriculture are thus more serious than previously thought, and that it would constitute an additional reason for consequentialist to condemn such intensive farming methods.

Of course, if it could be ascertained that intensive agriculture would produce more utility than its alternatives, then it could be justified on consequentialist grounds. But given its negative impacts on both animal welfare and climate change, such a justification seems farfetched: At the very least, it seems that some adjustments regarding current agricultural practices would be even better. We do not yet know whether the long-term effects of the pandemic will lead to a future that is overall better than it would have been had the pandemic never occurred. In that case, one could perhaps argue, from a consequentialist perspective, that intensive agriculture was after all justified, for the reason that it taught us something new. Still, this does not provide a justification for continuing intensive animal farming, since the lesson has now been learned.

### *Deontology, intensive animal farming, and the risks of zoonoses*

In contrast to consequentialism, deontological theories do not justify an act in relation its consequences, but to the intention' behind the act, and certain duties, such as to not lie, not kill, etc. At bottom, these duties requires that we respect individual people's lives and their autonomy. In line with the restriction made above, we shall focus on anthropocentric versions of these views. A deontological approach would typically approve of animal farming industries, unless restrictions would lead to improved respect of the lives and autonomy of the people involved in those industries. The current pandemic seems to be a clear example of a case where the intensive agriculture imposes risks to people's lives and autonomy and hence a rough assessment suggests that restrictions to intensive agriculture can be recommended – or even required – on the basis of deontological morality.

Some would perhaps argue that intentions and foreseeability play crucial roles in deontological ethics, and that the pandemic was neither an intended nor a foreseeable outcome of intensive agriculture. For this reason, one might argue, intensive farming has not involved any moral wrongdoing. But even though it is quite clear that no one has intended such an outcome, it could be questioned whether the risks of those outcomes were not foreseeable. However, since we have come to learn about these related risks, deontological ethics would not allow this type of intensive agriculture to continue unchanged.

### *Virtue ethics, intensive agriculture, and the risks of zoonoses*

From a virtue ethics perspective, what matters is not to follow a certain principle (duties or maximizing good consequences), but rather the virtuous motives behind the choice of action. Acting out of virtuous motives is often taken as a prerequisite for a flourishing human life and reach wisdom, which is in itself taken to be the final goal of a moral life. As this implies, an action can be justified on virtue ethical grounds only if its performance can be motivated out of virtues – such as temperance, courage, benevolence, justice, etc. This also means that an act cannot be justified if it is based on non-virtuous (i.e. vicious) character traits. As before, we shall for the sake of argument stick to an anthropocentric perspective, and so we shall focus on virtuous attitudes towards human beings.

Intensive agriculture cannot 'in principle' be ruled out from an anthropocentric virtue ethical perspective, since there is nothing vicious about producing food for a hungry human population. (This is true also for consequentialism and deontology, since neither of them implies that there is something wrong with intensive agriculture *per se*.) To the contrary, this might seem quite benevolent and just. However, now that we know about the effects of 'intensive' agriculture in particular, the verdict from a virtue ethics perspective is changed for such types of agriculture. The reason is that intensive animal farming methods, and exploitation of natural habitats of wild animals, can no longer continue without being based on motives such as greed and egoism rather than benevolence and justice. Moreover, intensive animal farming, is problematic since it relies on character traits in humans that are vicious – which will in turn risk hindering the flourishing of these humans, with negative effects on those affected by them. Therefore, the current pandemic has turned current intensive animal farming, and its risks of zoonoses, into something that must be condemned even from an anthropocentric virtue ethics.

Although this reasoning is very brief, all three theories propose a redirection to a more sustainable agriculture for the future. Interestingly, one can see an overlapping consensus with respect to intensive animal farming in the particular case of its risks of zoonoses.

## What about the animals and the environment?

In much national and EU legislation, as well as in ethical literature in the field, sentient animals are regarded as morally considerable, and hence ethically relevant for their own sake. We sympathize with this 'sentientist' view and will briefly comment possible implications from a non-anthropocentric perspective.

On a sentientist consequentialist framework, also harms and suffering caused on animals due to intensive agriculture would be considered, including consequences of political measures to tackle the corona virus. For ex., millions of minks and other animals have been euthanized in order to mitigate the risks of mutations and further spreading to humans. A sentientist deontological framework, on the other hand, would probably come close to the approach developed by Donaldson and Kymlicka (2013) who advocate animals' positive rights and suggest we ascribe them citizen rights, much the same as children or elderly who have nationality, and popular sovereignty, but only indirectly democratic agency. This would be taken care of by a system of representatives. Applied to the issue of industrial animal farming, the outcome would be a loud denial, since animals on this view have the right to be respected and not used as means to an end, including not being put at risk due to human profit. From a sentientist virtue ethics point of view, lastly, any kind of agriculture would have to exhibit an attitude of care, benevolence, generosity, and justice, etc., directly for the animals too. Consequently, such a theory would also require revisions of modern agriculture. Were we to extend the scope of direct moral considerability further, so as to take into consideration not only the sentient animals but also the environment, then these implications would most likely be further strengthened. For, we would then have to act directly for the sake of ecostability, biodiversity, and other aspects of the environment.

As this suggests, there is not only an overlapping consensus to be found between consequentialism, deontology, and virtue ethics, but also between anthropocentric and non-anthropocentric perspectives in the case of intensive agriculture and its risks of zoonoses. Here, mink farming might serve as an illustrative case. It has been criticized for impaired welfare for many years (Kornum *et al.,* 2017), and led to prohibition of farming in some countries (e.g. Austria, Belgium and Great Britain), whereas other countries, such as Denmark and the Netherlands, have announced a closing down after the spread of SARS-CoV-2. Here, the anthropocentric position contributed to the same goal as set from a sentientist position, much in parallel with stricter regulation on use of antibiotics, which has finally not been implemented for the sake of improving animal welfare, but rather to limit the effects on human health.

## Conclusion

In this paper, we have argued that the risks of zoonoses provides an additional reason – besides those related to climate change, biodiversity loss, antibiotic resistance, and animal welfare – to make changes to modern agriculture. Our main analysis adopted an anthropocentric perspective. We showed that the conclusion can be derived even with a focus only on human beings. However, we also showed that if a non-anthropocentric perspective is adopted instead, this conclusion would be further strengthened.

Given the difficulties to transform farming systems out of respect for the animals themselves, the current pandemic might provide a human-centred reason for such transformation. It might also increase respect for the interdependence of humans and animals, and evoke an interest in solid sustainability.

## References

Bajželj, B. *et al.* (2014). Importance of food-demand management for climate mitigation. Nature Clim Change 4, 924-929. https://doi.org/10.1038/nclimate2353

Bengtsson, B. & Greko, C. (2014). Antibiotic resistance-consequences for animal health, welfare, and food production. *Upsala journal of medical sciences*, vol. 119 (2), pp. 96-102 England: Informa Healthcare.

Brambell F.W.R. (1965). 'Report of the Technical Committee of Enquiry into the Welfare of Livestock Kept under Intensive Conditions'. HMSO; London, UK.

La Bonnardière, C. (2006). Milk yield, age at first calving, and the risk of BSE: An analysis at the farm level in France. *Preventive Veterinary Medicine*, 78, pp.67-78.

Ducrot, C. (2006). Milk production and BSE risk in France. Analysis at the farm level.|Production laitiere et risque ESB en France. *Epidémiologie et santé animale* (49), pp. 131-138 Maisons-Alfort: Association pour l'étude de l'épidémiologie des maladies animales.

Council Directive 98/58/EC of 20 July 1998 concerning the protection of animals kept for farming purposes.

Dhingra M.S. *et al.* (2018). Geographical and Historical Patterns in the Emergences of Novel Highly Pathogenic Avian Influenza (HPAI) H5 and H7 Viruses in Poultry. *Front. Vet. Sci.* 5:84.

Donaldson S. and Kymlicka W. (2013). Zoopolis: *A Political Theory of Animal Rights*. New York: Oxford University Press, 2013. 352 pp.

Farm Animal Welfare Council (1993), 'Second Report on Priorities for Research and Development in Farm Animal Welfare'. *DEFRA*; London, UK.

Harrison, R. (1964). *Animal machines: the new factory farming industry*. London: Stuart.

IPPC (2014) Climate Change 2014: Synthesis Report. Contribution of Working Groups I, II and III to the Fifth Assessment Report of the Intergovernmental Panel on Climate Change [IPCC, Geneva, Switzerland, 151 pp.

Kornum, A. L., *et al.* (2017). The concept of behavioural needs in contemporary fur science: do we know what American mink (Mustela vison) really need? *Animal Welfare.* 26(2). 151-164.

Oude Munninck B.B. *et al.* (2021). Transmission of SARS-CoV-2 on mink farms between humans and mink and back to humans. *Science.* Vol. 371, Issue 6525, pp. 172-177.

Regan, T and Singer, P. (1976), *Animal Rights and Human Obligations*. Englewood Cliffs, N.J. Prentice Hall, 1989 (second ed.).

Röös, E. *et al.* (2016). Limiting livestock production to pasture and by-products in a search for sustainable diets. Food policy, vol. 58, pp. 1-13 Elsevier Ltd.

Sibley, R. (2001). The effects of the BSE epidemic on cattle farms and on the veterinary profession in the UK from 1987 to 2001. Part 2. Changes for the cattle industry and the veterinary profession. *Large Animals Review* 8(2): 51-54.

Steinfeld, H. *et al.* (2006). *Livestock's long shadow*, FAO, Rome.

Torpman O. and Röcklinsberg H. (2021). Reinterpreting the SDGs: Taking Animals into Direct Consideration. *Sustainability* 2021, *13*, 843.

Visseren-Hamakers, I.J. (2020), 'The 18th Sustainable Development Goal', *Earth System Governance* 3 (2020).

Wallace, R. (2016) *Big Farms Make Big Flu: Dispatches on Infectious Disease, Agribusiness, and the Nature of Science quantity*. New York, NY Press.

# 65. Food insecurity, deepened poverty and ethics after Covid-19

*A. Inza-Bartolomé*[1*] *and L. Escajedo San-Epifanio*[2]
[1]*Department of Sociology and Social Work, Faculty of Labour Relations and Social Work, University of the Basque Country (UPV/EHU), Los Apraiz Street 2, 01006 Vitoria-Gasteiz, Spain;* [2]*Department of Constitutional Law and History of Political Thought, Faculty of Labour Relation and Social Work, University of the Basque Country (UPV/EHU), Postbox 644, 48080 Bilbao, Spain; amaia.inza@ehu.eus*

## Abstract

The status that charitable food aid resources have attained and the loss of responsibility by the Welfare State for the right to food undermines the latter doubly. In a first step, retrenchment and austerity reduce its operational space, which weakens it in its functions in the face of food insecurity. In a second step, the effect of charitable food aid resources undermines the idea about its redistributive function, which has been a basis for its legitimacy. The main hypothesis of this work is that this double undermining of the welfare state is constantly reinforced. It defends, through a logical-critical reflection based on specialized bibliography, that charitable food aid resources are a threat to the fundamental ethos of the welfare state, to its ability to respond to collective problems. The retrenchment and the austerity have strengthened the charitable food resources. Organizationally, the fact of not reaching the objective for which they arose, that is to say, ending food insecurity, paradoxically, reinforces them. Therefore, situations such as those triggered by Covid reinforce them even more. Closing the circle of dynamics, the effects of the work of these resources, where collective solidarity is channelled away from the traditional public redistributive mechanisms, represents a new motive of erosion. The main conclusion is that this phenomenon generates feelings of guilt or shame, and stigmatizes the recipients of food aid. It expels these people in vulnerable situations from the normalized consumption circuit. On the other hand, the institutionalization of food aid resources distracts from the origin of poverty, hides the true reasons for food need. Despite all this, it is important to bear in mind that it is the only resource available to many food-insecure people.

**Keywords:** food aid resources, successful failures, welfare state, neoliberalism

## Introduction

At present, the food aid provided by the third sector in Europe is the story of an emergency resource that has become an institutionalized remedy to feed people in poverty. There is a stark contrast if we consider that there is overproduction in advanced societies, yet surplus food is supplied to people in need through charitable resources. It serves to distract from the phenomenon of food waste, since surpluses are taken advantage of and the economic performance of the stocks of the companies dries up. People who have to resort to these resources, although they are subjects of full rights, come to have a stigmatizing relationship with whoever supplies them with food, causing them shame and guilt. This means that, for example, there are people in a precarious situation who do not even approach these resources. On the other hand, people are expelled from the consumer society, since they cannot choose the food they want to consume.

Several studies (Riches, 1997; Riches and Silvasti, 2014a) have argued that hunger in high-income countries could not be caused by the lack of provision of sufficient food and nutrients, but rather was a matter of distributive justice and human rights and the damaging impact of neoliberal economic ideology on the right to food of the most vulnerable people. Janet Poppendieck (1999) argues that the proliferation of charity contributes to the failure of society to deal with poverty, and creates a culture that normalizes destitution and legitimizes personal generosity as responses to the great social and economic


EurSafe 2021 – DOI 10.3920/978-90-8686-915-2_65, © Wageningen Academic Publishers 2021

dislocation. The institutionalization of charitable food banks makes it appear that the problem of hunger is being solved and the notion of hunger as a matter of charity, not of politics, is reinforced (Riches, 2011: 768). Meanwhile, hunger has been legitimized as a charity issue through speeches on community responsibility and participation, so that the lack of donation has become a problem and the solution is the responsibility of the community (Carson, 2013/14: 16).

In this text, it is first discussed what food security and the right to food of vulnerable groups entail, food aid resources as successful failures and food charity as a system to take advantage of food surpluses, and how it is constituted a network of shadow state institutions as part of a neoliberal project.

## Food security and the right to food for vulnerable groups in rich countries

According to FAO (1996), food security exists when 'all people, at all times, have physical and economic access to sufficient, safe and nutritious food that meets their dietary needs and food preferences for an active and healthy life'. Concretely, in high income countries, as Dowler and O'Connor (2012: 45) have stated:

> food security' implies that people have sufficient money to purchase the food they want to eat, to meet social as well as health and nutritional norms; that this money is not absorbed in other expenditure demands (rent, fuel, debt repayment, etc.); that people can reach shops or markets which stock appropriate food at affordable prices, or they can grow or otherwise obtain food in ways which are dignified and in keeping with social norms.

The right to food must be understood as a legal obligation (Riches and Silvasti, 2014b: 14). It is based on an a priori commitment to the value of human dignity and makes the individual an agent of change in a way that enables him or her to hold governments accountable and to seek redress for violations of his or her rights (Mechlem, 2004: 631). Paradoxically, it is expected that the need for food aid will be significant in the future, in countries that are food safe both through domestic production and imports, where hunger and food insecurity are increasing (Gentilini, 2013). The prevalence of household food insecurity is relatively high in some developed countries, ranging from 8 to 20% of the population, and a serious public health concern in rich countries with developed economies (Pollard and Booth, 2019).

Spending on food is the most flexible part of vulnerable people's budgets (Riches, 1997), and people living below the minimum income threshold, turn to food aid in kind so as to be able to devote their scant resources to non-food needs (Tarasuk and MacLean, 1990: 331), and not only targets the homeless or refugees from the richest countries, but increasingly people with jobs and housing, but who face sudden or continuous uncertainties (Dowler and O'Connor, 2012: 48). In 2019, the European Food Banks Federation assisted 9.5 million deprived people in 24 countries (FEBA, 2020a).

The pandemic caused by Covid-19 has had a great impact on existing inequalities and structural weaknesses, and has greatly affected those who are already poor or at high risk of being so (EAPN, 2020). To address the social impact of the coronavirus crisis, in April 2020, the Commission amended the FEAD (Fund for European Aid to the Most Deprived) in the context of the Coronavirus Plus Response Investment Initiative package (European Commission, 2020), introducing flexibility and liquidity for Member States. In July 2020, the demand for food had increased by 50% (FEBA, 2020b), and the trend of severely materially deprived people is expected to increase. In this way, it is evident that the right to food of an important part of the population remains in the hands of organizations that have to depend on volunteering.

## Food aid resources: successful failures and charity economy

Food aid is 'the phenomenon where non-governmental organizations provide free food to people who are living in poor social and economic situations' (Salonen *et al.*, 2018). Charity emergency food provision includes projects that help with an acute food crisis and includes food parcel and prepared food provision; this food provision is free, an outside the mainstream market (Lambie-Mumford and Silvasti, 2020: 222).

Food aid resources as Food Banks are 'a practical and simple solution to two important moral, social, and political problems: hunger and food waste' (Power, 2015). In general, charitable operations are easily understandable as practical and common sense responses, based on human compassion vis-a-vis the problem of hunger in homes (Silvasti and Riches, 2014: 204). The growing public discourse in favour of them is normalising, rendering natural their presence and social function (Perry *et al.,* 2016).

The third sector has had a great capacity both to establish relationships between different actors, responding at the same time to the diverse unmet needs, and has been even more prominent in a sector like food poverty, in which most welfare state regimes have almost completely returned their capacity and responsibility to societal actors (Baglioni *et al.*, 2016). Riches and Silvasti (2014b: 9), highlight the corporatization and establishment of Food Banks at a global level led by the European Federation of Food Banks and the Global Foodbanking Network in most of the rich countries of the OECD and the rapidly developing economies.

Ronson and Caraher (2016: 81) use Seibel's (1996) 'successful failures' concept to explain the strength these resources are acquiring. There are problems that are of special interest to voluntary organizations, eager to complete spaces where the State and the market may have failed. These organizations proliferate without anyone stressing that the results do not meet the needs of those they were intended to help, because it is believed that certain organizations do the best they can under difficult conditions, and their voluntary nature makes criticism socially unacceptable.

Food banks are one vital site taking on the work and filling the gaps in the emergent landscape of intervention that operates at ground-level to support people in negotiating the impacts of austerity (Strong, 2018). In a 'new charity economy' (Kessl *et al.*, 2020), people who are no longer or only partly able to participate in the given market economy are referred to a secondary supply system of surplus goods from the dominant capitalist market, and operates outside of the welfare state's framework. The charity economy has become an integral part of the modern welfare state arrangement and food banks and similar organizations can therefore be understood as a sign for the transformation of welfare state arrangements (Schoneville, 2018).

People who depend upon food aid lose part of their freedom of choice and inherent human dignity, because they have to accept donated food irrespective of their needs and preferences (Riches and Silvasti, 2014: 9), and the social stigma and the sensation of impotence associated with poverty is exacerbated by that dependence (Tarasuk and MacLean, 1990: 332). For some can lead to feelings of embarrassment, shame, fear and humiliation (Middleton *et al.*, 2018: 707).

## Food aid resources as shadow state institutions in a neoliberal project

As Starke has stated (2006: 104), most authors agree that socio-economic problems contribute to an atmosphere of permanent austerity which inspires cutbacks. Definitely, neo-liberal austerity marks a retreat from welfare state interventionism grounded in neo-Keynesian political economy (Taylor-Gooby *et al.*, 2017: 11). One marker of the success of a new neo-liberal settlement is that austerity has become the dominant global wisdom for addressing the problem of public debt in which public debt, public spending and public services are all viewed as problems to be overcome (Clarke and Newman, 2012: 300). And, rather than dealing with the structural causes of poverty and hunger, austerity involves a territorial reworking of the state through which forms of care, social reproduction and intervention are increasingly sourced from communities themselves (Strong, 2018). Notions of scarcity in a period of austerity have shaped materially insufficient and stigmatising welfare systems; under austerity, a particular moral economy of scarcity has become embedded at the level of common-sense, including in the common-sense of many of those distributing food aid (May *et al.*, 2019). While the illusion continues to exist that immediate food aid is available for people who cannot feed themselves, governments will be little disposed to evaluate the appropriateness of their welfare programmes, and in addition, there will be little pressure from the population for them to do so (Tarasuk *et al.,* 2014). The current focus distracts attention away from the ineffectiveness of government policies and responsibilities in addressing the social determinants of food insecurity, and depoliticises hunger (Pollard and Booth, 2019; Silvasti and Riches, 2014).

Following the theories of Mitchell (2010), food aid resources have become shadow state voluntary institutions, helping to consolidate the original economic policies of neoliberalism in a hegemonic process. These third sector organizations are playing an important role in the production and supply of public goods, discharge public responsibilities and transform the caring role of the State in society (Evans and Shields, 2000). On the other hand, as far as social support of basic needs is based on charity, it is legally not enforceable and it results in less social rights and reinforced social exclusion (Lorenz, 2014).

## Conclusions

The replacement of public redistributive mechanisms by community responses violates the legitimacy of the welfare state as the main protective resource. There are several dynamics that help this phenomenon to be at a point of no return. On the one hand, the charity economy, which makes it appropriate within the economic system as a way to take advantage of food waste. On the other, the role it plays in the face of the malfunction of the welfare state due to a context of permanent austerity. Its characteristics of successful failures, makes it an anchor element for the creation of the shadow state.

Therefore, starting from the need that food aid resources cover in austerity contexts, which dwarf the entity of welfare institutions, it can be seen that the activation of food aid mechanisms in the third sector results in a new attack to the foundations of the welfare state, since in part it invalidates the ability to give a public response to a collective problem. The function of these resources is uncritically accepted, when it has been shown that they are resources that do not meet adequate standards to care for the health of the recipients. They help overlook the right to food for all people, as well as a worthy way to get it.

However, these resources are necessary. There are many people in a precarious situation who need them since, due to the dysfunctions of the welfare state, they do not have other options in the face of food insecurity.

## References

Baglioni, S., De Pieri, B. and Tallrico, T. (2016). Surplus food recovery and food aid: the pivotal role of non-profit organisations. Insights from Italy and Germany. Voluntas 28: 2032-2053.

Carson, E.A. (2013/14). Canadian Food Banks and the depoliticization of food insecurity at the individual and community levels. Canadian Review of Social Policy 70: 7-21.

Clarke, J. and Newman, J. (2012). The alchemy of austerity. Critical Social Policy 32 (3): 299-319.

Dowler, E. A. and O'Connor, D. (2012). Rights-based approaches to addressing food poverty and food insecurity in Ireland and UK. Social Science&Medicine 74: 44-51.

EAPN (2020). The impact of COVID-19 on people experiencing poverty and vulnerability. Rebuilding Europe with a social heart. Available at: https://www.eapn.eu/wp-content/uploads/2020/07/EAPN-EAPN_REPORT_IMPACT_COVID19-4554.pdf.

European Commission (2020). Fund for European aid to the Most Deprived (FEAD). Available at: https://ec.europa.eu/social/main.jsp?catId=1089&langId=en.

Evans, B. M. and Shields, J. (2000). Neoliberal restructuring and the third sector: reshaping governance, civil society and local relations, Centre for Voluntary Sector Studies, Ryerson University Working Paper Series 13. Available at: http://www.ryerson.ca/~cvss/WP13.pdf.

FAO (1996). Rome Declaration on World Food Security. FAO: Rome. Available at: https://digitallibrary.un.org/record/195568?ln=es.

FEBA (2020a). European Food Banks Federation. Annual Report. Brussels: FEBA. Available at: https://www.eurofoodbank.org/en/annual-report.

FEBA (2020b). European Food Banks in a post COVID-19 Europe. Brussels: FEBA. Available at: https://lp.eurofoodbank.org/wp-content/uploads/2020/07/FEBA_Report_Survey_COVID_July2020.pdf.

Gentilini, U. (2013). Banking on food: The state of food banks in high-income countries. IDS Working Papers 415:1-18. Available at: http://www.ids.ac.uk/files/dmfile/Wp415.pdf.

Kessl, F., Oechler, M. and Schröder, T. (2020). Charity economy and social work. In: F. Kessl, F.; Lorenz, W.; Otto, H.-U. and White, S. (eds) European social work-A Compendium. Leverkussen/Farmington Hills: Bucrich Publishers, 361-378 pp.

Lambie-Mumford, H. and Silvasti, T. (2020). Conclusion: food charity in Europe. In: Lambie-Mumford, H. and Silvasti, T. (ed.). The rise of food charity in Europe. Briston: Policy Press, pp. 219-241.

Stephan L. (2012). Socio-ecological consequences of charitable food assistance in the affluent society: the German Tafel. International Journal of Sociology and Social Policy 32 (7/8), pp. 386-400.

May, J., Williams, A., Clocke, P. and Cherry, L. (2019). Food banks and the production of scarcity. Transactions of the Institute of British Geographers 45 (1): 208-222.

Mechlem, K. (2004). Food Security and the Right to Food in the Discourse of the United Nations. European Law Journal 10 (5): 631-648.

Middleton, G., Mehta, K., McNaughton, D. and Booth, S. (2018). The Experiences and Perceptions of Foodbank Amongst Users in High Income Countries: An international Scoping Review. Appetite 120: 698-708.

Mitchell, K. (2010). Transnationalism, neo-liberalism, and the rise of the shadow state. Economy and Society 30 (2): 165-189.

Perry, J., Williams, M., Seffton, T. & Haddad, M. (2014). Emergency Use Only. Understanding and reducing the use of Food Banks in the UK. London, CPAG – OXFAM. Available at: https://policy-practice.oxfam.org/resources/emergency-use-only-understanding-and-reducing-the-use-of-food-banks-in-the-uk-335731/.

Pollard, C.M. and Booth, S. (2019). Food insecurity and hunger in rich countries. It is time for action against inequality. International Journal of Environmental Research and Public Health 16, 1804.

Poppendieck, J. (1999). Sweet Charity: Emergency Food and the End of Entitlement. New York: Penguin Books. 354 pp.

Power, E.M. (2015). Food Banks. In: Albala, K. (ed.). The Sage encyclopedia of food issues. (vol.2). London: Sage Publications, pp. 552-7.

Riches, G. (1997). First world hunger. Food security and welfare politics. London: Macmillan Press. 200 pp.

Riches, G. (2011). Thinking and acting outside the charitable food box: Hunger and the right to food in rich societies. Development in Practice 21(4-5): 768-775.

Riches, G, and Silvasti, T. (eds). (2014a). First world hunger revisited. Food charity or the right to food?. Basingstoke, Palgrave Macmillan. 248 pp.

Riches G and Silvasti, T. (2014b). Hunger in the rich world: food aid and right to food perspectives. In: Riches, G, and Silvasti, T. (eds). First world hunger revisited. Food charity or the right to food?. Basingstoke: Palgrave Macmillan, pp. 1-14.

Ronson, D. and Caraher, M. (2015). Food Banks: Big society or Shunting Yards? Successful Failures. In: Caraher, M. and Coveney, J. (eds). Food Poverty and Insecurity: International Food Inequalities. Switzerland, Springer. https://doi.org/10.1007/978-3-319-23859-3_8

Salonen S A, Ohisalo M. and Laihiala T. (2018). Undeserving, Disadvantaged, Disregarded: Three Viewpoints of Charity Food Aid Recipients in Finland. International Journal of Environmental Research and Public Health 15, 2, 2896.

Schoneville, H. (2018). Poverty and the transformation of the welfare (state) arrangement. Food banks and the charity economy in Germany. Social Work and Society. International Online Journal 16, 2. Available at: https://ejournals.bib.uni-wuppertal.de/index.php/sws/article/view/570

Seibel, W. (1996). Successful failure. An alternative view on organizational coping. American Behavioural Scientist 39: 1011-1024.

Silvasti, T. and Riches, G. (2014). Hunger and food charity in rich societies: what hope for the right to food?. In: Riches, G. and Silvasti. T. (eds). First world hunger revisited. Food charity or the right to food?. Basingstoke: Palgrave Macmillan, pp.191-208

Starke, P. (2006). The politics of welfare state retrenchment: a literature review, Social Policy & Administration 40 (1): 104-120.

Strong, S. (2018). Food banks, actually existing austerity and the localisation of responsibility. Geoforum, Available at: https://www.sciencedirect.com/science/article/abs/pii/S0016718518302835.

Tarasuk, V. and MacLean, H. (1990). The institutionalization of food banks in Canada: a public health concern. Canadian Journal of Public Health 81: 331-332.

Tarasuk, V.S., Dachner, N. and Loopstra, R. (2014). Food Banks, welfare, and food insecurity in Canada. British Food Journal 116 (9): 1405-1417.

Taylor-Gooby, P., Leruth, B. and Chung, H. (2017). How European welfare states have responded to post-industrialism, ageing populations, and populist nationalism. In: Taylor-Gooby, P., Leruth, B. and Chung, H. (eds) After austerity. Welfare state transformation in Europe after the great recession. Oxford: Oxford University Press, pp. 1-26.

# 66. The eclipse of sharing within food ethics: unearthing blind spots during the Covid-19 pandemic

*R. Anthony*
*University of Alaska Anchorage, 3211 Providence Drive, Anchorage, AK, USA; rxanthony@alaska.edu*

## Abstract

In its most recent State of Food Security and Nutrition in the World, The United Nations warns of the meteoric rise of food and nutritional insecurity during the SARS-CoV-2 pandemic, potentially pushing approximately 130 million more people into chronic hunger by the end of 2020. Already in 2019, 10 million more people went undernourished than the previous year. Covid-19 has intensified the vulnerabilities and shed greater light on the inadequacies and social inequities of the global food system – food supply disruptions, lost livelihoods and the inability of people to send remittances home to their families. In a time of a global pandemic how shall we – individuals, governments and those dedicated to fighting poverty and undernourishment – shape a more other-regarding global food system so that it is more just and humane towards and inclusive of those who lie outside the conventional networks of production-consumption? My 'paper' explores the way that sharing is largely missing from the discourse in contemporary food ethics. In Section 1, I discuss the implicit bias in much of contemporary food ethics, which largely frames the ending of hunger and undernourishment in terms of an exercise of liberty. Preoccupation with what (should) we eat a la dietetic ethics and the rise of ethical consumerism/consumption as a form of moral redress to the ills of the global food complex/system have resulted in some significant blind spots, namely, a lack of representation, recognition and participatory parity, respectively. My contention in Section 2 is that we need an ethic of sharing that is prior to, rather than grounded in, consumption assessments to tackle the impact of Covid-19 and future pandemics (and the climate crisis for that matter) on some of the most vulnerable among us. Arguably, the tendency to align with notions of consumption has limited mainstream food ethics framings. I lay the contours of a response by considering how the collective wisdom of religion and indigenous traditions may provide mainstream philosophers of food with a blueprint for conscientiously cultivating and exercising private and public/political virtues and activities. My responses are framed in terms of the following moral categories: (1) attitude – shared vulnerability and solidarity; (2) approach – compassion, humility and gratitude; and (3) action – performing collective moral activities.

**Keywords:** food ethics, novel coronavirus pandemic, world religions, Indigenous subsistence cultures, food and nutritional security

## Introduction

In August 2020, World Health Organization (WHO) Director Dr Tedros Ghebreyesus, acknowledged that it was only natural for leaders to seek to protect 'their own people first,' but urged that the response to the pandemic 'has to be collective.' Cautioning against 'vaccine nationalism,' he argued that protecting the most vulnerable everywhere simultaneously was the best way to arrest the pandemic – 'No one is safe until everyone is safe.' (Tidey and Bauomy, 2020). Dr Ghebreyesus has since warned against 'a catastrophic moral failure' due to a 'me-first' approach, that would ultimately 'prolong the pandemic, the restrictions needed to contain it, and human and economic suffering' (WHO, 2021). His sentiments are echoed by Dag-Inge Ulstein, Norway's international development minister and co-chair of the global council tasked with accelerating access to Covid-19 vaccines for the world's poor (Wintour, 2021). Covid-19 is not just a health crisis. Stories of deep and pervasive hunger have surfaced (Swinnen and


EurSafe 2021 – DOI 10.3920/978-90-8686-915-2_66, © Wageningen Academic Publishers 2021

McDermott, 2020; UN, 2020). The pandemic places compounded stress on our communities and fragile food support systems, and our economic, social and political systems. How should we go about shaping a post-pandemic global food system that is more just and humane towards and inclusive of, especially, those who lie outside the conventional networks of production-consumption?

## Section 1: blind spots

The 'me-first' approach has also been increasingly evident in contemporary food ethics discourses and has stymied global food security initiatives. It has become evident that more 'collective solutions' are necessary to address hunger, undernourishment and poverty in the wake of this pandemic. The Covid-19 pandemic has reversed global food security and development gains made over the past decades and led to shortages in basic necessities (Swinnen and McDermott, 2020; UN, 2020). The World Food Program has indicated that, compared to half a million in Europe, 73 million people across 36 African countries are acutely food-insecure as a result of Covid-19 (WFP, 2020). Disruptions to consumption, food trade and the food supply chain, falling remittances, Covid-19 public health and labour related restrictions, and anaemic policy responses have exacerbated the plight of the hungry in low and middle income countries (Swinnen and McDermott, 2020). In Brazil, the Covid-19 health crisis and social distancing measures are 'dismantling' social protection policies like *Bolsa Familia*, and overwhelming assistance programs like the National School Food Program (PNAE), sending socially vulnerable households and populations, already faced by social inequality, especially among low-income people, i.e. the unemployed and those in the informal sector, into greater economic despair and food and nutritional shortages from which they can face mortality (De Cássia Ribeiro-Silva *et al.*, 2020). Examples like these are, unfortunately, prolific throughout the world. Also, individual government strategies (e.g. modest pay-outs, temporary increased 'emergency' food assistance) to minimize virus transmission have been insipid at best to address the impact on agriculture, food and nutritional security and livelihoods vis-a-vis social inequity. Many of the world's socially vulnerable populations (the unemployed and low-income earners), share tight living quarters that make adhering to personal hygiene recommendations during a pandemic near to impossible, and work in the informal economic sector, where they are already hard pressed to supplements their income. The most vulnerable members are in position to fare the worst (FAO and CELAC, 2020; UN, 2020). The pandemic has punctured gaping holes in self-interested policies, and individualistic ethical and nationalistic political frameworks. Addressing food insecurity will require (like on the health side) 'relational solidarity' that advances 'interconnections and a shared interest in survival, safety and security' (Kenny *et al.*, 2020: 11).

Contemporary food ethics, however, has been preoccupied with 'What (should) we eat and how we are to grow it' questions (Wolf, 2018). Dietetic ethics or ethical consumerism assume that all people are equally capable of exercising agency or are equally at liberty to care what's on their plates. The expression of 'my' liberty and self-identification may not truly benefit others, especially without adequate understanding of the impact of one's choices on the circumstances of those connected to them. Personalized ethical consumerism masks the need to build community relationships, assess impacts and exacerbating inequalities in food and nutritional security, advocate for structural changes and policy responses that are more equitable, and galvanize political will on behalf of the least among us. Also, there is enormous weight on consumers to set the moral tone for the food system instead of 'governments, corporations and other bigger actors' (Cook, 2006: 659). Producers, retailers and suppliers, seeking to win over consumers through assurance and certification schemes that tout quality foods bundled together with the ethical *préoccupations du jour* attributes, fail to register poverty alleviation as 'value added.'

Disaster related (pandemic) food ethics requires an approach that is itself relational rather than individualistic. The ethical social starting point for food ethics during a disaster must recognize the human community as foundational, and plan, from the perspective of our collective interests and inseparability from each other, on minimizing the impacts of the hazard (the virus) on food and nutritional security in addition to health. Doing so can mitigate the commodifying tendencies and alienating aspects of the global food system and industrial agriculture (Korthals, 2012) and advance participatory parity in discourse, decision, policy and action. Some individuals and groups lack voice or have limited agency (migrant and low wage workers) and some have disproportionate voice.[1] Absent participatory parity, some people are denied standing by institutionalized hierarchies or cultural value, and the 'opportunity to participate at all in the authorized contests over justice' (Fraser, 2012).

## Section 2: ethical food sharing – drawing inspiration from religious and indigenous traditions in a time of crisis

The ethics of sharing, is significant for the relatively simple reason that '[t]he food security and nutritional status of the most vulnerable population groups is likely to deteriorate further due to the health and socio-economic impacts of the Covid-19 pandemic' (FAO, 2020). An ethics of sharing seeks to minimize 'us' vs 'them' polarization. Rooted in vulnerability and compassion, it can shine a light on the perspectives of the invisible. Thus, we need an ethic that is prior to rather than grounded, in consumption assessments. Linking food ethics to relatively affluent consumption styles, identity politics, and corporate food interests is inherently reciprocal/transactional (Buller, 2010). The ethic of sharing is a necessary, if under emphasized, condition for the flourishing of the individual and our human community. It is an important supplement to the dominant way of thinking about food ethics today, namely, ethical consumerism. For a deeper sense of food sharing, contemporary food ethics can draw on the collective wisdom of the world's major religious traditions and indigenous subsistence cultures[2] for a guide-path towards the development of an ethic of sharing. Pope Francis recently invoked compassion for and identification with the most afflicted; 'We have realized that we are on the same boat, all of us fragile and disoriented, but at the same time important and needed, all of us called to row together, each of us in need of comforting the other' (Vatican News, 2020). While some individual religious persons and religious groups still struggle with 'othering' (Levinas, 1995), many of the world religions have strived to enhance ecumenicalism; confronting social polarization and fragmentation and providing an antidote to the identity politics and nationalism referenced above.

The religious and Indigenous subsistence traditions underscore an aspirational ethic that highlights an important non-consumptive framing that can serve to broaden our food ethics: we live in community with others and as this pandemic has made obvious, our existential condition is punctuated by uncertainty and shared vulnerability. There is a shared, general thrust of religions and their philosophies that have the community definition that puts sharing at the centre of responsibility to others. Religions stress compassion, social cohesion and justice, which implicitly includes address of unequal assets and need as values, which should lead to peace and mutual respect if not love. 'Religions are vehicles for self-transformation. Not only do they reassure us, they provide us with principles and precepts and practices that can change us or show us how to change ourselves' (Loy, 2003: 3). By looking at the intersection of food ethics-religious ethics, philosophers of food can nurture functions of religions (such as Buddhism, Hinduism, the Abrahamic religions and Indigenous spiritual perspectives) to cultivate private and public virtues and political action regarding more sharing. Three mutually reinforcing and interlocking impulses expressed by these traditions highlight how sharing is an implicit value:

[1] See, for example, the story which broke that 'Tyson Food managers bet on workers getting Covid-19,' https://www.bbc.com/news/world-us-canada-55009228.

[2] The descriptions discussed above are not meant to be universal or monolithic claims about all of the world's religion nor applicable to all religions. Rather, these broad strokes are illustrative of some significant and general threads from which deeper conversations may be struck.

1. Shared vulnerability and solidarity (attitude): The major religions of the world and indigenous spirituality stresses the oneness of humanity and sense of interconnectedness. They offer a blueprint for catalysing a sense of interconnectedness through equal vulnerability, some focusing on the worst off as their criterion for social action (e.g. Moore, 2020). The current pandemic challenges mainstream philosophers to inject food ethics discourse with clear logical links between shared vulnerability and solidarity; it is a question of feeling and seeing others as equal, as part of our human community, that should be inculcated. The pandemic highlights the need for a shared residue of universal peace, compassion, and justice as values and moral responses to help us collectively recognize values, interests and ethical viewpoints other than our own. The Golden Rule maxim (treat everyone with dignity or equitably), found in most of the world's religions (Hick, 1993), can be deployed privately or communally to the food system to help in the realization of the importance of sharing; catapulting us to take in the suffering of others and contributing more to meet the challenges posed by the pandemic and those that lay ahead post-Covid-19.
2. Internalizing compassion, gratitude and humility (approach): The ethics of sharing also requires envisioning the relationship between donor and recipient as non-reciprocal or non-transactional. World religions help adherents internalize key virtues or dispositions associated with this relationship through: 'Compassion', which is a moral sentiment associated with feeling for others and summons performance to improve the other's well-being (Singer and Klimecki, 2014). In Islam, one of the main pillars, Zakat or almsgiving, is a clear example of this (Ludwig, 2006); 'Gratitude', opens the psyche to be attentive to how things are in themselves without subjective inclinations. A grateful Buddhist self-embodies 'right view and right intention,' and comfortably sees how she belongs in the world and her connection to others (Baer, 2015; Berkwitz, 2003); and 'Humility', epistemic and moral, can be a relational disposition of being humble before one's divine creator and/or one's fellow human beings. There is a lot we still do not know about this virus, that has already resulted in over 98,820,000 confirmed Covid-19 cases worldwide and global deaths upwards of 2.1 million people.[3] The virus, oblivious to identity, has chipped away pretense of power and privilege and exposed the stark reality of our shared vulnerability and the need for heightened equity (Fuster and Turco, 2020). For Indigenous communities in the circumpolar north, for example, who practice subsistence based on an all-encompassing ecological philosophy, central community-oriented virtues associated with adaptation and humility include interdependence, sharing and cooperation, gratitude, balance, and mindfulness (Anthony and Varner, 2019). Here, Native elders in Alaska, have always recognized the need for revision in the face of fresh evidence and self-awareness of the limits to their understanding of natural phenomena (Barnhardt and Kawagley, 2005; Peter *et al.,* 2002).
3. Performing Collectively (Action): 'In ritual humans learn by doing... ritual both marks and assists in the process of change. It confirms and transforms people's worldviews, identities, and relationships with others' (Schirich, 2005: 2). This sense of ritual is important for thinking about framing and actualizing more inclusive forms of sharing. Devotional charity, sharing with those who are less well off or those who cannot reciprocate are plentiful in most of the world religions. In Hinduism, for example, the Bhagavad Gita encourages adherents to act selflessly for the welfare of others: 'Strive constantly to serve the welfare of the world; ... Do your work with the welfare of others in mind' (3:19-26) (see also Luke 10:25-39; Matthew 25:40 NRSV). The world's religions have a rich history and important narratives or commandments regarding food (Zwart, 2000), such as how food ought to be shared and what constitutes eating well. Religious-based rituals around food sharing has spilled over into the secular domain. Eating is a communal experience that reveals aspects of our identity and responsibility (Berry, 1990), where '[v]ulnerability is not a diminishment of the human experience, but is part of that experience' (Moore, 2020). This relational solidarity reflects deeply rooted communitarian leanings, i.e. the 'ethical ideals we strive for together,' the 'sacrifices we make for one another' (Sandel in Friedman, 2020) and sharing responsibilities as one humanity.

[3] https://coronavirus.jhu.edu/.

During this time of global hardship, the ethics of sharing requires making sure our governments know that everyone counts, they do not give up on anyone and no one should be left behind. This means political-social engagement that involves active concern for the well-being of others that manifests as direct involvement in civic and/or political life to bring about better conditions for them. This can translate into greater: 'accountability' by governments and other influencers to identify norms and policies for more equitable and inclusive solutions and to investigate exclusion or suppression of alternative food styles (Korthals, 2012); 'visibility' given to the realities of people living in poverty who rely on agriculture and natural resources to survive; connectivity, that governments 'enable us to help one another through this time,' and forge 'a vibrant civil society to empower people to care for one another' (Moore, 2020); 'transparency', making sure benefits and assistance programs are expanded to allow low-income and the unemployed to have access to more, culturally appropriate and better quality food; 'relationality', paying attention to how food aid may undercut local farmers' capacity to be self-reliant and food secure (Thompson, 2010) and minimizing negative impacts of food preferences in industrialized countries on commodity prices and farmers' capacity elsewhere to maintain food sovereignty and continue to farm as/what they have been (Korthals, 2012); and 'representation', by dismantling obstacles, such as institutionalized hierarchies and economic structures, that prevent equal voice in public deliberation and decision-making and access to affordable, culturally appropriate, healthy diets (FAO, 2020).

## Conclusion

Projections are that the world is 'not' on track to achieve Zero Hunger by 2030 and, despite some progress, most indicators are also not on track to meet global nutrition targets (FAO, 2020). Covid-19 has pushed millions further into poverty and food and nutritional insecurity. The decisions governments make about food assistance and policy as the world recovers will determine who lives and who dies. Current debates in food ethics largely instrumentalizes ethical concern by repackaging distribution, recognition, and representation in terms of private consumption choices or individual liberties apart from their social contexts; they also do not identify issues of poverty and starvation as in the pursuit of public good. The increasingly narrow emphasis on ethical eating in industrialized countries also effectively portrays those who lie outside the 'rhythms and circuits of the global agri-food system, as excludable' (Buller, 2010). The pandemic, however, has forced scores of unemployed and temporarily out of work citizens in more affluent countries to look beyond their circumstances or stand in the shoes of others. Religion can be a source of ethical guidance in advancing solidarity and action during a crisis and provide the impetus to action beyond charity. The blueprint above has acquired a particular potency today, which may help overcome self-interested nationalism and address our shared vulnerability; ensuring that all, especially the least well off, have access to adequate and healthy food.

## References

Anthony, R. and G. Varner (2019). 'Subsistence Hunting,' in B. Fischer (ed.) Routledge Animal Ethics Encyclopedia, New York: Routledge.

Baer, R. (2015). 'Ethics, Values, Virtues, and Character Strengths in Mindfulness-Based Interventions: a Psychological Science Perspective, *Mindfulness*, 6(4), 956-969.

Barnhardt, R. and A.O. Kawagley (2005). 'Indigenous Knowledge Systems and Alaska Native Ways of Knowing,' *Anthropology and Education Quarterly* 36(1): 8-23.

Berkwitz, S.C. (2003). History and Gratitude in Theravāda Buddhism, *Journal of the American Academy of Religion*, 71 (3): 579-604.

Berry, W. (1990). 'The Pleasures of Eating,' in What are People For? North Point Press. Available at: www.ecoliteracy.org/article/wendell-berry-pleasures-eating.

Bhagavadgita. (2007). Introduced and Translated by Eknath Easwaran-Second Editions. Tomales, CA: The Blue Mountain Center of Meditation.

Buller, H. (2010). 'Palatable Ethics,' *Environment and Planning A* 42: 1875-1880.

Cook, I. (2006). 'Geographies of Food: Following' *Progress in Human Geog.* 30(5): 655-66.

de Cássia Ribeiro-Silva, R., Pereira, M., Campello, T., Aragão, É., de Medeiros Guimarães, J.M., Ferreira, A., Barreto, M.L. Chaves dos Santos, S.M. (2020). Covid-19 Pandemic Implications for Food and Nutrition Security in Brazil, *Ciênc. Saúde Coletiva*, 25(9).

Food Agriculture Organization on the United Nations (FAO) (2020). The State of Food Security and Nutrition in the World: Transforming Food Systems for Affordable Healthy Diets. Rome. Available at: http://www.fao.org/3/ca9692en/CA9692EN.pdf.

FAO and CELAC (2020). Food security under the COVID-19 pandemic. Rome.

Fraser, N. (2012). Abnormal Justice in S. Dieleman, D. Rondel and C.J. Voparil (eds) Pragmatism and Justice. Oxford University Press: NY, 37-64.

Friedman, T.L (24 March 2020). 'Finding the 'Common Good' in a Pandemic,' *New York Times*.

Fuster, V. and Turco, J.V. (2020). COVID-19: A Lesson in Humility and an Opportunity for Sagacity and Hope, *J Am Coll Cardiol*, 75(20): 2625-2626.

Hick, J. (1993). Disputed Questions in Theology and the Philosophy of Religion, New Haven, Connecticut: Yale University Press, 93.

Kenny, N.P., Sherwin, S.B. and Baylis, F.E. (2020). Re-Envisioning Public Health Ethics: A Relational Perspective, in M. Schwartz (ed.) The Ethics of Pandemics. Broadview, 9-14.

Korthals, M. (2012). Two Evils in Food Country: Hunger and Lack of Representation, in D. M. Kaplan (ed.), The Philosophy of Food. Berkeley and Los Angeles, CA: University of California Press,103-121.

Levinas, E. (1995). Alterity and Transcendence, trans. Michael B. Smith, Columbia University Press, 1999 (in English).

Ludwig, T.M. (2006). The Sacred Paths: Understanding the Religions of the World, 4th Edition, Upper Saddle River, NJ: Pearson.

Loy, D.R. (2003). The Great Awakening: A Buddhist Social Theory, Wisdom Publications, MA.

Moore, R. (26 March 2020). 'God Does Not Want Us to Sacrifice the Old,' *New York Times*.

NRSV Bible. (1990). National Council of the Churches of Christ. Available at: https://bible.oremus.org/.

Peter, A., M. Ishulutak, J. Shaimaiyuk, N. Kisa, B. Kootoo, and S. Enuaraq (2002). 'The Seal: An Integral Part of Our Culture,' Études/Inuit/Studies 26(1): 167-174.

Schirch, L. (2005). Ritual and Symbol in Peacebuilding. Bloomfield, CT: Kumarian Press, Inc.

Singer, T. and Klimecki, O.M. (2014). Empathy and Compassion, *Current Biology*,24(18):r875-8

Swinnen, J. and McDermott, J. (eds) (2020). COVID-19 and Global Food Security. International Food Policy Research Institute (IFPRI).

Thompson, P.B. (2010). 'Food Aid and the Famine Relief Argument (Brief Return),' *Journal of Agricultural and Environmental* Ethics 23(3):209-227.

Tidey, A. and Bauomy, J. (2020). Coronavirus: 'No-one is Safe until Everyone is Safe,' says WHO Director. Available at: https://www.euronews.com/2020/08/18/coronavirus-19-european-countries-record-high-incidence-rates-as-surge-continues.

United Nations (UN) (2020). The Impact of COVID-19 on Food Security and Nutrition. Rome.

Vatican News. 2020. 'Pope at Urbi et orbi: Full text of his meditation,' 27 March 2020.

World Food Programme (WFP). (2020) Global Report on Food Crises. Available at: https://docs.wfp.org/api/documentsWFP-0000114546/download/?_ga=2.90198291.8236433.1611516333-1959533564.1611516333.

World Health Organization (WHO) (2021). WHO Director-General's Opening Remarks at 148th Session of the Executive Board, Available at: https://www.who.int/director-general/speeches/detail/who-director-general-s-opening-remarks-at-148th-session-of-the-executive-board.

Wintour, P. (2021). 'World's Poor Need Actions, Not Covid 'Vaccine Nationalism', say Experts' *The Guardian*, Available at: https://www.theguardian.com/world/2021/jan/22/worlds-poor-need-action-not-covid-vaccine-nationalism-say-experts.

Wolf, C. (2018). Sustainable Agriculture, Environmental Philosophy, and the Ethics of Food in A. Barnhill, M. Budolfson and T. Doggett (eds) The Oxford Handbook of Food Ethics, Oxford University Press, NY, 29-52.
Zwart, H. (2000). A Short History of Food Ethics, *Journal of Agricultural and Environmental Ethics* 12: 113-126.

# Author index